AF678365

Translational and Comparative Research on Innovative Anti-Cancer Therapies

Translational and Comparative Research on Innovative Anti-Cancer Therapies

Editors

Felisbina Luisa Queiroga
Bruno Cogliati

MDPI

Basel • Beijing • Wuhan • Barcelona • Belgrade • Novi Sad • Cluj • Manchester

Editors

Felisbina Luisa Queiroga
University of Trás-Os-Montes
and Alto Douro
Vila Real
Portugal

Bruno Cogliati
University of São Paulo (USP)
São Paulo
Brazil

Editorial Office
MDPI
St. Alban-Anlage 66
4052 Basel, Switzerland

This is a reprint of articles from the Special Issue published online in the open access journal *Cancers* (ISSN 2072-6694) (available at: https://www.mdpi.com/journal/cancers/special_issues/Innovative_Anti-Cancer_Therapies).

For citation purposes, cite each article independently as indicated on the article page online and as indicated below:

Lastname, A.A.; Lastname, B.B. Article Title. *Journal Name* **Year**, *Volume Number*, Page Range.

ISBN 978-3-03928-617-1 (Hbk)
ISBN 978-3-03928-618-8 (PDF)
doi.org/10.3390/books978-3-03928-618-8

© 2024 by the authors. Articles in this book are Open Access and distributed under the Creative Commons Attribution (CC BY) license. The book as a whole is distributed by MDPI under the terms and conditions of the Creative Commons Attribution-NonCommercial-NoDerivs (CC BY-NC-ND) license.

Contents

About the Editors

Felisbina Luisa Queiroga

Felisbina Luisa Queiroga has been an Associate Professor with Habilitation at the University of Trás-os-Montes and Alto Douro (UTAD), Portugal, in Veterinary Internal Medicine and Oncology since 2013. She was the Clinical Director of the Small Animal Service at Veterinary Teaching Hospital (2007–2010) and the Director of the Integrated Masters in Veterinary Medicine (2011–2013). She is an ESEVT Expert (since 2020) in the European Association of Establishments for Veterinary Education (EAEVE) for Basic and Clinical Sciences. She has worked in several scientific and clinical projects, involving researchers from foreign countries and the pharmaceutic industry. She is the Editor-in-Chief of the Oncology in Veterinary Medicine, a new Section of Frontiers in Veterinary Sciences. She is also an Associate Editor of several scientific indexed journals: *BMC Vet Research*; *Animals*—Companion Animals Section; *Ciencia Rural and Frontiers in Veterinary Science*—Comparative and Clinical Medicine Section. She held a position on the World Small Animal Veterinary Association (WSAVA) Executive Board (2019–2023), and is currently a member of the WSAVA Scientific Committee and a liaison for the WSAVA Oncology Working Group. She is also the Vice President of the General Assembly of the National Order of Veterinarians in Portugal.

Bruno Cogliati

Dr. Bruno Cogliati graduated in Veterinary Medicine at the University of Sao Paulo (USP) in 2005. He had a Doctorate degree in Comparative and Experimental Pathology from USP (Fellowship from Sao Paulo State Agency, FAPESP, 2010). Dr. Cogliati has done a Postdoc at the School of Medicine at USP on experimental nonalcoholic fatty liver disease (FAPESP Fellowship, 2011). Since 2011, Dr. Cogliati has been associate professor at the Department of Pathology at the School of Veterinary Medicine and Animal Science, USP, Sao Paulo, Brazil. He is the head of the Laboratory of Experimental and Comparative Liver Research (LIVER LAB). He was awarded with a prestigious Research Productivity Fellow from National Council for Scientific and Technological Development (CNPq). He is a member of the Vrije Universiteit Brussel-USP Alliance Research Group and FWO Scientific Research Network. He is a Visiting Researcher at the Department of Liver Diseases at the Icahn School of Medicine at Mount Sinai, NY, USA, under supervision of Prof. Dr. Scott L. Friedman.

cancers
MDPI

Editorial

Translational and Comparative Research on Innovative Anti-Cancer Therapies

Felisbina Queiroga [1,2,3,*] and Bruno Cogliati [4,*]

1 Department of Veterinary Sciences, University of Trás-os-Montes and Alto Douro, 5000-801 Vila Real, Portugal
2 Animal and Veterinary Research Centre (CECAV), University of Trás-os-Montes e Alto Douro, 5000-801 Vila Real, Portugal
3 Center for the Study of Animal Sciences, CECA-ICETA, University of Porto, 4200-465 Porto, Portugal
4 Department of Pathology, School of Veterinary Medicine and Animal Science, University of São Paulo, Av. Prof. Dr. Orlando Marques de Paiva 87, 05508-270 São Paulo, Brazil
* Correspondence: fqueirog@utad.pt (F.Q.); bcogliati@usp.br (B.C.)

Citation: Queiroga, F.; Cogliati, B. Translational and Comparative Research on Innovative Anti-Cancer Therapies. *Cancers* **2023**, *15*, 1335. https://doi.org/10.3390/cancers15041335

Received: 14 February 2023
Accepted: 16 February 2023
Published: 20 February 2023

Copyright: © 2023 by the authors. Licensee MDPI, Basel, Switzerland. This article is an open access article distributed under the terms and conditions of the Creative Commons Attribution (CC BY) license (https://creativecommons.org/licenses/by/4.0/).

Oncology research has received considerable attention in recent years due to the increasing prevalence of cancer in human and animal populations worldwide. The similarities between neoplastic diseases that affect both species have been identified in terms of etiopathogenesis, serum and immunohistochemical biomarkers, and gene expression, as well as their response to antineoplastic drugs, for example. Therefore, several translational and comparative studies are described in the literature, placing cancer in a one health perspective and simultaneously improving the anti-cancer therapies provided in human and veterinary oncology. In this Special Issue, eleven studies with relevant scientific information regarding innovative anti-cancer therapies were included in the context of comparative oncology.

This Special Issue starts with an original study in which the influence of folic acid was investigated in the genetic and epigenetic regulation of two colorectal cancer cell lines (HT-29 and SW480) [1]. Zsigrai et al. [1] showed that the short-term folic acid supplementation affected HT-29 cell proliferation, viability, and genomic stability, although similar findings were not found in the SW480 cell line. Moreover, gene downregulation and upregulation were identified in both cell lines, contributing to new information about the effect of this vitamin in different human colorectal cancer cell lines. Moving from a nutritional supplement to a cytotoxic drug, Ferrari et al. [2] first described the use of Bleosome, a new compound consisting of ultra-deformable liposomes with encapsulated bleomycin for topical administration, evaluating its ability to penetrate ex vivo skin explants from dogs and horses. The authors showed that Bleosome was able to penetrate the skin and release bleomycin into deeper epidermal layers, which could represent a more effective and safer anti-cancer therapy for non-melanoma skin cancer compared to the conventional systemic treatments.

In the context of comparative oncology, canine mammary gland neoplasia is commonly used as a translational study model for breast cancer in women due to its complex nature and biological behavior. In order to find potential prognostic biomarkers, a proteomic mass spectrometry imaging technique was performed by Cordeiro et al. [3] in mammary tumors surgically removed from dogs diagnosed with metastatic neoplastic disease. The authors associated the malignant tumor phenotype with alterations in the expression of five proteins: Fibronectin type III domain containing 1 protein (FNDC1), Alpha-1B-glycoprotein (A1BG), calnexin (CANX), heat-shock protein family A member 5 (HSPA5), and protein disulfide isomerase family A member 3 (PDIA3). Based on these findings, it was suggested that these five key proteins could be potential prognostic biomarkers for this type of cancer, but further investigation and validation are still required. In turn, Colombo et al. [4] showed the effectiveness of liquid biopsy as a diagnostic and prognostic tool for both women and female dogs diagnosed with breast cancer through the identification of several specific genetic mutations in the respective neoplastic fragments.

Head and neck squamous cell carcinoma (HNSCC) was also addressed in this Special Issue as Da Silva et al. [5] described a novel matriptase-dependent proteolytic pathway associated with the activation of the protease-activated receptor 3 (PAR-2) by the Kallikrein 5 (KLK5), which represents a modulatory mechanism of potential clinical interest in the carcinogenesis of this cell type. Moreover, as the authors identified an inhibitory action of the serine protease inhibitor lymphoepithelial Kazal-type-related inhibitor (LEKTI) on KLK5, it is expected that these results motivate the investigation of new and related targeted therapies in these tumors. Still focusing on HNSCC, Piotrowski et al. [6] added relevant diagnostic information, describing the potential clinical interest of microRNAs expression in differentiating between tumor and healthy tissues. In addition, the authors reported an association between this expression and the patient outcome, suggesting the potential use of these biomarkers as predictors of survival.

Regarding innovative anti-cancer therapies, Kung et al. [7] showed the anti-cancer effect of piperlongumine, an amide alkaloid, in human thyroid cancer mainly due to its ability to inhibit cell proliferation and to promote apoptosis by generating reactive oxygen species. As this study was performed in vitro and in vivo, its potential applicability as an effective and safe anti-cancer therapy in humans is very promising. In dogs, Voges et al. [8] reported the effectiveness of weekly intratumoral doses of polyinosinic-polycytidylic acid-poly-l-lysine carboxymethylcellulose (poly-ICLC) in a clinical study with advanced and unresectable canine tumors for the first time. This anticancer therapy improved the qualities of life of these dogs and was considered locally effective and well tolerated, with only mild adverse effects being described.

At the end, the readers will find three comprehensive review papers that contribute up-to-date evidence on different topics associated with translational and comparative oncology research. In the first one, Torres-Juárez et al. [9] addressed the neurobiology of cancer, the impact of the nervous system in carcinogenesis events, and the influence of this interaction on anti-cancer approaches. Secondly, Romualdo et al. [10] described the main in vivo and in vitro models of hepatocellular carcinoma, regarding their respective advantages, disadvantages, and applications. Finally, Wawruszak et al. [11] explored the impact of the single or combined use of valproic acid, a short-chain fatty acid, in the treatment of breast cancer, through the information available in the peer-reviewed literature.

In conclusion, we believe that this Special Issue of *Cancers* contributes with high-quality scientific evidence through novel data from preclinical and clinical studies performed in humans and/or animals. Thus, we hope that this set of articles will increase readers' knowledge within the scope of comparative and translational oncology and that it will encourage further research into these innovative anti-cancer therapies.

Conflicts of Interest: The authors declare no conflict of interest.

References

1. Zsigrai, S.; Kalmár, A.; Barták, B.K.; Nagy, Z.B.; Szigeti, K.A.; Valcz, G.; Kothalawala, W.; Dankó, T.; Sebestyén, A.; Barna, G.; et al. Folic Acid Treatment Directly Influences the Genetic and Epigenetic Regulation along with the Associated Cellular Maintenance Processes of HT-29 and SW480 Colorectal Cancer Cell Lines. *Cancers* **2022**, *14*, 1820. [CrossRef] [PubMed]
2. Ferrari, G.; Pang, L.Y.; De Moliner, F.; Vendrell, M.; Reardon, R.J.M.; Higgins, A.J.; Chopra, S.; Argyle, D.J. Effective Penetration of a Liposomal Formulation of Bleomycin through Ex-Vivo Skin Explants from Two Different Species. *Cancers* **2022**, *14*, 1083. [CrossRef] [PubMed]
3. Cordeiro, Y.G.; Mulder, L.M.; van Zeijl, R.J.M.; Paskoski, L.B.; van Veelen, P.; de Ru, A.; Strefezzi, R.F.; Heijs, B.; Fukumasu, H. Proteomic Analysis Identifies FNDC1, A1BG, and Antigen Processing Proteins Associated with Tumor Heterogeneity and Malignancy in a Canine Model of Breast Cancer. *Cancers* **2021**, *13*, 5901. [CrossRef] [PubMed]
4. Colombo, J.; Moschetta-Pinheiro, M.G.; Novais, A.A.; Stoppe, B.R.; Bonini, E.D.; Gonçalves, F.M.; Fukumasu, H.; Coutinho, L.L.; Chuffa, L.G.D.A.; Zuccari, D.A.P.D.C. Liquid Biopsy as a Diagnostic and Prognostic Tool for Women and Female Dogs with Breast Cancer. *Cancers* **2021**, *13*, 5233. [CrossRef] [PubMed]

5. da Silva, E.Z.M.; Fraga-Silva, T.F.d.C.; Yuan, Y.; Alves, M.G.; Publio, G.A.; da Fonseca, C.K.; Kodama, M.H.; Vieira, G.V.; Candido, M.F.; Innocentini, L.M.A.R.; et al. Kallikrein 5 Inhibition by the Lympho-Epithelial Kazal-Type Related Inhibitor Hinders Matriptase-Dependent Carcinogenesis. *Cancers* **2021**, *13*, 4395. [CrossRef] [PubMed]
6. Piotrowski, I.; Zhu, X.; Saccon, T.; Ashiqueali, S.; Schneider, A.; Nunes, A.d.C.; Noureddine, S.; Sobecka, A.; Barczak, W.; Szewczyk, M.; et al. miRNAs as Biomarkers for Diagnosing and Predicting Survival of Head and Neck Squamous Cell Carcinoma Patients. *Cancers* **2021**, *13*, 3980. [CrossRef]
7. Kung, F.-P.; Lim, Y.-P.; Chao, W.-Y.; Zhang, Y.-S.; Yu, H.-I.; Tai, T.-S.; Lu, C.-H.; Chen, S.-H.; Li, Y.-Z.; Zhao, P.-W.; et al. Piperlongumine, a Potent Anticancer Phytotherapeutic, Induces Cell Cycle Arrest and Apoptosis In Vitro and In Vivo through the ROS/Akt Pathway in Human Thyroid Cancer Cells. *Cancers* **2021**, *13*, 4266. [CrossRef]
8. Voges, A.R.N.; Ubukata, R.; Yazbek, K.Y.; Caballero, O.; Salazar, A.; Massoco, C.; Dagli, M. Intratumoral (Poly-ICLC) Therapy for Dogs with Advanced Cancers: First Report on Clinical Effectiveness, Quality of Life, and Adverse Events. *Cancers* **2021**, *13*, 2237. [CrossRef] [PubMed]
9. Torres-Juárez, K.V.; Queiroga, F.L.; Romero-Romero, L.P. The Nervous System as a Regulator of Cancer Hallmarks: Insights into Therapeutic Implications. *Cancers* **2022**, *14*, 4372. [CrossRef] [PubMed]
10. Romualdo, G.R.; Leroy, K.; Costa, C.J.S.; Prata, G.B.; Vanderborght, B.; da Silva, T.C.; Barbisan, L.F.; Andraus, W.; Devisscher, L.; Câmara, N.O.S.; et al. In Vivo and In Vitro Models of Hepatocellular Carcinoma: Current Strategies for Translational Modeling. *Cancers* **2021**, *13*, 5583. [CrossRef] [PubMed]
11. Wawruszak, A.; Halasa, M.; Okon, E.; Kukula-Koch, W.; Stepulak, A. Valproic Acid and Breast Cancer: State of the Art in 2021. *Cancers* **2021**, *13*, 3409. [CrossRef] [PubMed]

Disclaimer/Publisher's Note: The statements, opinions and data contained in all publications are solely those of the individual author(s) and contributor(s) and not of MDPI and/or the editor(s). MDPI and/or the editor(s) disclaim responsibility for any injury to people or property resulting from any ideas, methods, instructions or products referred to in the content.

cancers

MDPI

Article

Folic Acid Treatment Directly Influences the Genetic and Epigenetic Regulation along with the Associated Cellular Maintenance Processes of HT-29 and SW480 Colorectal Cancer Cell Lines

Sára Zsigrai [1,*], Alexandra Kalmár [1,2], Barbara K. Barták [1], Zsófia B. Nagy [1], Krisztina A. Szigeti [1], Gábor Valcz [1,2], William Kothalawala [1], Titanilla Dankó [3], Anna Sebestyén [3], Gábor Barna [3], Orsolya Pipek [4], István Csabai [4], Zsolt Tulassay [2,5], Péter Igaz [1,2,6], István Takács [1] and Béla Molnár [1,2]

1 Department of Internal Medicine and Oncology, Semmelweis University, 1083 Budapest, Hungary; kalmar.alexandra@med.semmelweis-univ.hu (A.K.); molnar.barbara_kinga@med.semmelweis-univ.hu (B.K.B.); nagy.zsofia@med.semmelweis-univ.hu (Z.B.N.); szigeti.krisztina_andrea@med.semmelweis-univ.hu (K.A.S.); valcz.gabor@med.semmelweis-univ.hu (G.V.); kothalawala.william@phd.semmelweis.hu (W.K.); igaz.peter@med.semmelweis-univ.hu (P.I.); takacs.istvan@med.semmelweis-univ.hu (I.T.); molnar.bela1@med.semmelweis-univ.hu (B.M.)

2 Molecular Medicine Research Group, Eötvös Loránd Research Network, 1083 Budapest, Hungary; tulassay.zsolt@med.semmelweis-univ.hu

3 1st Department of Pathology and Experimental Cancer Research, Semmelweis University, 1085 Budapest, Hungary; danko.titanilla@med.semmelweis-univ.hu (T.D.); sebestyen.anna@med.semmelweis-univ.hu (A.S.); barna.gabor@med.semmelweis-univ.hu (G.B.)

4 Department of Physics of Complex Systems, ELTE Eötvös Loránd University, 1117 Budapest, Hungary; pipeko@caesar.elte.hu (O.P.); csabai@phys-gs.elte.hu (I.C.)

5 Department of Internal Medicine and Hematology, Semmelweis University, 1088 Budapest, Hungary

6 Department of Endocrinology, Semmelweis University, 1083 Budapest, Hungary

* Correspondence: zsigrai.sara@med.semmelweis-univ.hu

Citation: Zsigrai, S.; Kalmár, A.; Barták, B.K.; Nagy, Z.B.; Szigeti, K.A.; Valcz, G.; Kothalawala, W.; Dankó, T.; Sebestyén, A.; Barna, G.; et al. Folic Acid Treatment Directly Influences the Genetic and Epigenetic Regulation along with the Associated Cellular Maintenance Processes of HT-29 and SW480 Colorectal Cancer Cell Lines. *Cancers* **2022**, *14*, 1820. https://doi.org/10.3390/cancers14071820

Academic Editors: Athanasios G. Papavassiliou and Éric Chastre

Received: 4 February 2022
Accepted: 1 April 2022
Published: 3 April 2022

Copyright: © 2022 by the authors. Licensee MDPI, Basel, Switzerland. This article is an open access article distributed under the terms and conditions of the Creative Commons Attribution (CC BY) license (https://creativecommons.org/licenses/by/4.0/).

Simple Summary: Folic acid (FA) participates in DNA synthesis and in DNA methylation; hence, it has a dual role in established neoplasms. We aimed to observe this phenomenon on FA-treated colorectal cancer cell lines (HT-29, SW480). Our results demonstrated that the maintenance processes, namely cell proliferation, cell viability, and DNA repair, were altered in HT-29 cells for short-term FA supplementation, while genetic and epigenetic regulations of SW480 cells were also affected. Despite the fact that FA is a precursor molecule in methyl donor formation, DNA methylation alterations were observed in both directions, primarily influencing the pathways of carcinogenesis. Moreover, behind the great number of differentially expressed genes, other FA-related effects than promoter methylation were suspected. All of our results point beyond the attributes related to FA so far. The different response of the two cell lines is worth considering in clinical practice to facilitate the effectiveness of therapy in the case of tumor heterogeneity.

Abstract: Folic acid (FA) is a synthetic form of vitamin B9, generally used as a nutritional supplement and an adjunctive medication in cancer therapy. FA is involved in genetic and epigenetic regulation; therefore, it has a dual modulatory role in established neoplasms. We aimed to investigate the effect of short-term (72 h) FA supplementation on colorectal cancer; hence, HT-29 and SW480 cells were exposed to different FA concentrations (0, 100, 10,000 ng/mL). HT-29 cell proliferation and viability levels elevated after 100 ng/mL but decreased for 10,000 ng/mL FA. Additionally, a significant ($p \leq 0.05$) improvement of genomic stability was detected in HT-29 cells with micronucleus scoring and comet assay. Conversely, the FA treatment did not alter these parameters in SW480 samples. RRBS results highlighted that DNA methylation changes were bidirectional in both cells, mainly affecting carcinogenesis-related pathways. Based on the microarray analysis, promoter methylation status was in accordance with FA-induced expression alterations of 27 genes. Our study demonstrates that the FA effect was highly dependent on the cell type, which can be attributed to the distinct molecular background and the different expression of proliferation- and DNA-repair-associated genes (*YWHAZ, HES1, STAT3, CCL2*). Moreover, new aspects of FA-regulated DNA methylation and consecutive gene expression were revealed.

Keywords: colorectal cancer; folic acid; genomic stability; DNA methylation; gene expression

1. Introduction

Folate, also known as vitamin B9, is mainly found in green leafy vegetables [1]. Its synthetic form, called folic acid (FA), is more stable chemically and has higher bioavailability than food folates [2]. FA is often used commercially in nutritional supplements and fortified products [3].

Folate mediates the transfer of one-carbon units for nucleotide synthesis and also for the formation of S-adenosylmethionine (SAM), the main methyl donor molecule responsible for most of the methylation reactions, including that of DNA [1]. Based on these facts, folate is involved in both genetic and epigenetic regulations, thereby being essential for normal cell growth and development [1]. Due to the dual modulatory effect, its role in carcinogenesis is still controversial [4], but it is becoming increasingly evident that the timing of folate supplementation is crucial [2]. Before neoplastic transformation, the FA treatment seems to be beneficial [4] because of its protective role against nucleotide imbalance, as well as the resulting DNA synthesis and repair defects [5]. In addition, FA supplementation in normal tissues can prevent global DNA hypomethylation, which is considered the hallmark of cancer by causing genomic instability [1,4]. On the other hand, in pre-existing neoplastic cells—where DNA replication is accelerated—folate has a conducive effect, since it serves as the precursor of nucleotides [2]; moreover, it may silence tumor suppressor genes by methylating their promoter region [1,6]. However, folate deficiency has the exact opposite impact to that mentioned before, both on healthy and on cancerous tissues [1]. Besides timing, the applied FA dose also has significance for tumor progression, as evidenced by numerous in vivo and in vitro experiments [7–12].

The colonic epithelium is considered to have one of the highest proliferation rates in the human body [13], and as such, it has a huge folate demand [14]. This tissue is greatly involved in tumor transformation caused by folate deficiency. As a result, the role of vitamin B9 in carcinogenesis has been intensively studied in colorectal cancer (CRC) [15]. Nevertheless, drugs interfering with folate metabolism, such as 5-fluorouracil (5-FU), are generally used in CRC treatment [16]. Chemotherapeutic regimens, such as FOLFIRI or FOLFOX, include a folate derivate, namely leucovorin, to increase the effect of 5-FU [17].

The main goal of our study was to observe the independent effect of FA on an already developed tumor by focusing on the processes in which FA has previously been reported to be involved. Hence, we investigated genetic and epigenetic regulation with transcriptome and methylome analyses, as well as the consequent maintenance mechanisms necessary for tumor growth and development, namely cell proliferation, cell viability, and DNA repair. To model the conditions of an FA-depleted diet and an adequate FA intake [8], we either exposed CRC cells to an FA-free environment (0 ng/mL) or treated them with 100 ng/mL FA. Additionally, we were interested in the beneficial or even adverse effects of extremely high FA doses on the cell lines; therefore, supraphysiological concentrations (10,000 ng/mL) were applied as well.

Since the majority of CRCs have arisen through the chromosomal instability (CIN) pathway [18], we used two CRC cell lines, namely HT-29 and SW480, representing this phenotype. Besides the similarity of CIN status, principal differences exist in respect of their molecular features, which made these cell lines suitable subjects for analyzing the dependency of the FA act on the genetic and epigenetic background. Among others, according to the consensus molecular subtype (CMS) and CpG island methylator phenotype (CIMP) classification systems, HT-29 is considered as a CMS3 (metabolic) and CIMP+, while SW480 is a CMS4 (mesenchymal) and CIMP- cell line [19–21]. Additionally, based on our previous whole-exome sequencing (WES) analysis [22], differences of driver mutations (HT-29: *BRAF*, SW480: *KRAS*), as well as altering mutation profiles of methylenetetrahydrofolate reductase (*MTHFR*), a key mediator gene involved in the folate cycle, were identified [22–24]. The latter can be accounted for approximately 50% and 70% reduction in the MTHFR activity in HT-29 and SW480 cells, respectively, thereby greatly influencing the bioavailability of the applied FA [25,26].

2. Materials and Methods

2.1. Cell Cultures

HT-29 (ATCC HTB-39) and SW480 (ATCC CCL-228) human colon adenocarcinoma cell lines were cultured in RPMI 1640 medium (LM-R1641, Biosera, Ringmer, UK) containing 10% fetal bovine serum (Biosera), 80 mg/2 mL gentamycin (Sandoz GmbH, Kundl, Austria), and 2 mM L-glutamine (Biosera). Cells were maintained at 37 °C in a 5% CO_2 humidified atmosphere.

Prior to FA treatment, 1.25×10^5 cells were seeded in triplicate in each well of a 6-well plate (Sarstedt, Nümbrecht, Germany) and incubated for 24 h in 2.5 mL growth media. For proliferation and viability assays, 96-well plates (Sarstedt) were used, and cells were seeded with a density of 3×10^3 cells/well in 100 µL media.

Following the 72 h long incubation period, washing with 1× phosphate-buffered saline (PBS) was performed 3 times, and the culture media were changed to FA-free RPMI 1640 (LM-R1642, Biosera). FA-depleted cells (HT-29_0, SW480_0) were kept in this type of media without any additional substances. Chronic supplementation can provide approximately 100 ng/mL FA concentration in human blood serum [8], and we determined the supraphysiological dose as 100 times this value; therefore, 100 ng/mL (HT-29_{100}, SW480_{100}) and 10,000 ng/mL (HT-$29_{10,000}$, SW$480_{10,000}$) FA concentrations were applied. According to the manufacturer's instruction, FA (Sigma-Aldrich, St. Louis, MO, USA) was dissolved in 1M NaOH before it was added to the medium. After 72 h, cells were harvested with TrypLE Express (Thermo Fisher Scientific, Carlsbad, CA, USA), then they were counted to monitor cell proliferation. Trypan blue dye exclusion technique was applied to monitor cell viability. Treatment with only 1 M NaOH was also carried out in the same amount used in the case of FA supplementation to detect its individual effect on the cells. The results of control samples were subtracted from those treated with FA in order to optimize the final values. The raw data of Figures 1, 2 and 3a,b can be found in Table S1.

2.2. Cell Viability and Proliferation Analysis with alamarBlue and Sulforhodamine B Assays

Cell viability was measured with alamarBlue. The obtained results refer to the activity of the electron transport chain; hence, it is a so-called indicator of cell health. Furthermore, Sulforhodamine B (SRB) assay was used for cell proliferation detection, as it can estimate the protein mass of cultured cells.

AlamarBlue (Thermo Fisher Scientific) was added to the wells 4 h prior to the 72 h long incubation period. Next, the fluorescence was measured in the 570–590 nm range for cell viability detection using a fluorimeter (Fluoroskan Ascent FL Microplate Fluorometer and Luminometer, Thermo Fisher Scientific).

Following the measurement, we carried out Sulforhodamine B assay on the same plate to detect cell proliferation changes. First, cells were fixed using trichloroacetic acid for 1 h at 4 °C, then washed with tap water. Dyeing was performed with 50 µL 0.4% *m/v* sulforhodamine B solution for 15 min at room temperature (RT), and the unbound dye was removed with 1% *v/v* acetic acid. After air drying, 150 µL of 10 mM, unbuffered tris base was added to the wells, and the samples were shaken thoroughly. Finally, absorbance was measured with a microplate reader at 570 nm wavelength using Transmit software (Multiskan MCC 355, Thermo Fisher Scientific). The percentage of cell viability and cell proliferation was compared to the samples kept in the RPMI 1640 growth media. One-way ANOVA, followed by Tukey's multiple comparisons test, was applied to assess statistical significances ($p \leq 0.05$) using Prism 8.0.2 software (GraphPad, San Diego, CA, USA).

2.3. Cell Cycle Analysis with Flow Cytometry

After harvesting around 1×10^6 HT-29 and SW480 cells, 1× PBS washing was performed, and cells were fixed by keeping them at −20 °C in 70% ethanol overnight. The next day, the samples were centrifuged, and the pellets were resuspended with 1× PBS. Following RNase (Thermo Fisher Scientific) treatment for 15 min, 2 µL propidium iodide (2 mg/mL) (Sigma-Aldrich) was added to the samples. FACSCalibur bench-top flow cytometer (Becton, Dickinson and Company, Franklin Lakes, NJ, USA) and CellQuest Pro software (Becton, Dickinson and Company) were used to measure the samples and analyze the results. Statistical significances ($p \leq 0.05$) were assessed by two-way ANOVA followed by Tukey's multiple comparisons tests using Prism 8.0.2 software (GraphPad).

2.4. Genomic Stability Detection with Micronucleus Scoring

Micronucleus scoring was applied to detect the level of genetic damage via the analysis of chromosomes or their fragments that lagged behind during cell division. HT-29 and SW480 cells were grown on glass coverslips, placed into 6-well plates, as described before by Valcz et al. [27], and, following a 24 h long incubation period, we performed the FA treatment. Cells were fixed with 10% buffered formalin for 10 min at 4 °C, and permeabilization was carried out with 0.2% Triton-X-100 for another 10 min at RT. Anti-γ-H2AX antibody (ab26350, Abcam, Cambridge, UK) (1 h; 1:150 dilution; RT; anti-mouse) labeled with Alexa 488 (Invitrogen, Carlsbad, CA, USA) (30 min; 1:200 dilution; RT) was used for γ-H2AX detection in the micronuclei. Cell nuclei and micronuclei were stained with DAPI staining (Thermo Fisher Scientific) (5 min; 1:1000 dilution; RT). Washing with 1× PBS was performed 3 times between each step of the immunostaining. Slides were digitized in 21 Z-axial confocal layers of 0.4 µm focus steps using Pannoramic Confocal scanner (3DHISTECH Ltd., Budapest, Hungary), and an area containing at least 4000 cells per sample was analyzed. CellQuant tool of CaseViewer software (3DHISTECH) was applied to count micronucleus (MN) numbers and to obtain the percentage of cells with MN. The range within the MN number was detected was in line with the data of previous studies [28,29]. Frequently detected chromatin-containing nuclear blebs, which were not evidently separated from the nuclei, were excluded from the analysis. Statistical significances ($p \leq 0.05$) were evaluated by one-way ANOVA followed by Tukey's multiple comparisons test using Prism 8.0.2 software (GraphPad).

2.5. Genomic Stability Detection with Comet Assay

Comet assay was used for the detection of DNA strand breaks, since free DNA segments have increased migration speed during electrophoresis. At the end of the 72 h long FA treatment period, we removed the cells from the bottom of the plate with a rubber policeman, then washed the samples with Dulbecco's PBS (Sigma-Aldrich). The Comet Assay Kit (Abcam) was applied according to the manufacturer's protocol with a minor modification: original slides were replaced with agarose-coated Superfrost Ultra Plus slides (Thermo Fisher Scientific). Single-cell electrophoresis was performed in alkaline electrophoresis solution at 150 mA for 45 min. Vista Green Dye (Abcam) was used to stain the DNA. Comets were captured by an AxioCam camera (Carl Zeiss AG, Oberkochen, Germany) attached to a fluorescence microscope (Carl Zeiss AG). At least 50 cells/sample were analyzed using Comet Score software, and tail DNA% was defined as the ratio of tail DNA and cell DNA length. Statistical analyses were carried out using Prism 8.0.2 software (GraphPad), and significances ($p \leq 0.05$) were assessed by one-way ANOVA test followed by Tukey's multiple comparisons test.

2.6. DNA Methylation Analysis with LINE-1 Pyrosequencing

DNA isolation was performed with High Pure PCR Template Preparation Kit (Roche, Mannheim, Germany). Qubit dsDNA HS Assay Kit (Thermo Fisher Scientific) was used on Qubit 1.0 fluorometer (Thermo Fisher Scientific) for measuring the concentration of the extracted DNA; then, samples were stored at −20 °C for later analyses. Bisulfite conversion of 100 ng DNA was performed with EZ DNA Methylation-Direct Kit (Zymo Research, Orange, FL, USA). PyroMark PCR Kit (Qiagen, Hilden, Germany) was used to amplify a 146-basepair-long region of long interspersed nuclear element 1 (LINE-1) retrotransposon, and the PCR product was visualized with gel electrophoresis using 2% agarose gel. According to the instructions of PyroMark Q24 CpG LINE-1 Handbook (Qiagen), samples were prepared for pyrosequencing on a PyroMark Q24 Vacuum Workstation (Qiagen). Pyrosequencing was performed by PyroMark Q24 System (Qiagen), and LINE-1 methylation level was quantified with PyroMark Q24 Software (Qiagen). The mean methylation level of three LINE-1 CpG (cytosines followed by guanine residues) sites was interpreted as the global DNA methylation level of the given sample. Two-way ANOVA followed by Tukey's multiple comparisons test was applied to determine statistical significances ($p \leq 0.05$) using Prism 8.0.2 software (GraphPad).

2.7. DNA Methylation Analysis with Reduced Representation Bisulfite Sequencing

The isolated DNA from HT-29 and SW80 cells kept in 0 and 10,000 ng/mL FA-containing media was used for genome-wide methylation profile analysis with Premium Reduced Representation Bisulfite Sequencing (RRBS) Kit (Diagenode Diagnostics, Seraing, Belgium). After enzymatic digestion, ends preparation, adaptor ligation, and size selection, the samples were quantified with qPCR. Based on their Ct value, 2 pools were created. Following the bisulfite conversion and amplification steps, DNA was quantified with Qubit 1.0 Fluorometer using Qubit dsDNA HS Assay Kit (Thermo Fisher Scientific). Then, quality control was performed with Agilent 2100 Bioanalyzer using Agilent High Sensitivity DNA Kit (Agilent Technologies, Santa Clara, USA). The samples were prepared according to the Next Seq System Denature and Dilute Libraries Guide (Illumina, San Diego, USA), then transferred to NextSeq 500/550 High Output Sequencing kit v2 (75 cycles) (Illumina). Finally, we used NextSeq 500 instrument (Illumina) for sequencing. Sequencing data were assembled into single FASTQ files by merging the results of different sequencing lanes. As a quality assessment step, fastQC [30] was run on all resulting FASTQ files. Adapter and quality trimming was performed with Trim Galore, with default settings optimized for Bisulfite-Seq files. Prior to alignment, the hg38 version of the human reference genome was bisulfite converted in silico and indexed using the Bismark software [31]. Alignment and methylation calling were performed by the same software with default parameter values. The lists of differentially methylated sites (DMSs), along with CpG content and coverage histograms in sample pairs, were obtained with the methylKit R package [32]. The default "SLIM" method was used for multiple testing correction. The list of potential DMSs was filtered out if either the methylation difference between the two compared samples was below 15%, or the q-value of the comparison was above 0.05. Annotation of DMSs was performed using the appropriate annotation databases downloaded from the UCSC Genome Browser website [33]. The Kyoto Encyclopedia of Genes and Genomes (KEGG) pathway enrichment analysis was performed using The Database for Annotation, Visualization and Integrated Discovery (DAVID) v6.8 tool [34]. The top 10 KEGG pathways significantly ($p \leq 0.05$) enriched for DMSs in both cell lines were determined and later illustrated with heatmaps.

2.8. Whole Genomic Expression Analysis with Microarray

HT-29 and SW480 cells kept in the FA-free medium or treated with 10,000 ng/mL folate were involved in this evaluation. Harvested cells were washed with 1× PBS; then, following a centrifugation step, pellets were resuspended with 350 µL RLT buffer containing β-mercaptoethanol. Total RNA was purified with RNeasy Mini Kit (Qiagen), and isolated RNA concentration was measured with Qubit 1.0 fluorometer using RNA HS Assay Kit (Thermo Fisher Scientific). Agilent 6000 Pico Assay Kit on Agilent 2100 Bioanalyzer system (Agilent Technologies) was used for determining the integrity of RNA. Samples with a higher RNA integrity number (RIN) than 8 were applied for further analysis. Using GeneChip WT PLUS Reagent Kit (Thermo Fisher Scientific), amplification, quantification, fragmentation, and terminal labeling were performed, followed by target hybridization to HTA 2.0 microarray (Human Transcriptome Array 2.0, Affymetrix, Santa Clara, CA, USA) according to the manufacturer's instructions. Washing, staining, and scanning steps were performed as previously described by Kalmar et al. [35].

The signal space transformation–robust multi-array average (SST-RMA) algorithm was applied for background subtraction, normalization, and signal summarization. Gene expression alterations were evaluated using Transcriptome Analysis Console 4.0 (TAC 4.0, Affymetrix) software. Transcript IDs showing significant alterations ($p \leq 0.05$) with a fold change (FC) equal or greater than 1.5 and equal or lower than −1.5 were annotated with Ensembl Genome Browser BioMart and used for further analyses. We obtained the normalized log2 values from the TAC 4.0 software, and with the use of the "prcomp" package of R, we performed principal component analysis (PCA). Protein–protein interaction (PPI) networks of differentially expressed genes were built with the StringApp of Cytoscape software. Color coding was applied for indicating the expression level of the genes (dark blue: FC ≤ -2, light blue: FC ≥ -2 and ≤ -1.5, light red: FC ≥ 1.5 and ≤ 2, dark red: FC ≥ 2). Top 10 genes showing significant ($p \leq 0.05$) up- and downregulation, as well as genes with a promoter methylation status alteration in accordance with their expression level ($p \leq 0.05$ and fold change $\geq |1.5|$), were visualized with volcano plots using TAC 4.0. Finally, Protein Analysis Through Evolutionary Relationships 16.0 classification system (PANTHER 16.0) was applied for determining the molecular pathways in which the affected genes were involved.

3. Results

3.1. *Effect of Folic Acid Treatment on Cell Proliferation, Cell Viability, and Cell Cycle*

As a first step, we examined the results of FA treatment on the proliferation of two CRC cell lines using SRB assay (Figure 1a). In HT-29 cells, the highest proliferation rate was detected following 100 ng/mL FA supplementation (HT-29_{100}: 128.43 ± 24.94%). Meanwhile, FA depletion (HT-29_{0}: 101.25 ± 13.53%) and 10,000 ng/mL FA concentration (HT-$29_{10,000}$: 86.06 ± 20.75%) caused significant ($p \leq 0.05$) reduction compared to this value. By contrast, remarkable alterations were not observed in SW480 cell line for different FA supplies ($SW480_{0}$: 90.96 ± 9.72%, $SW480_{100}$: 88.75 ± 2.69%, $SW480_{10,000}$: 84.15 ± 10.67%).

The tendency of cell viability was analogous to the results detected during cell proliferation analyses (Figure 1b). In HT-29 cells, the values were decreased in the case of FA depletion (HT-29_{0}: 91.57 ± 13.27%) and also for supraphysiological (HT-$29_{10,000}$: 64.06 ± 20.24%) treatment compared to 100 ng/mL FA level (HT-29_{100}: 115.81 ± 30.88%). However, in SW480 cells, the viability was around 90%, independent of the applied FA dose ($SW480_{0}$: 90.22 ± 9.55%, $SW480_{100}$: 90.05 ± 5.03%, $SW480_{10,000}$: 89.72 ± 11.22%).

In order to examine how FA treatment affects the cell cycle, fluorescence-activated cell sorting (FACS) measurement was performed. The proportion of cells in the different cell cycle phases was specific to the given cell type, with a G0/1 phase dominance in HT-29 and an S phase dominance in SW480 cells. Interestingly, it was not affected by any treatment conditions (an average of all HT-29 samples: 62.51 ± 4.76% in the G0/G1, 17.90 ± 4.32% in the G2/M, and 19.59 ± 2.12% in the S phase; an average of all SW480 samples: 41.31 ± 1.60% in the G0/G1, 0.81 ± 0.66% in the G2/M, and 57.88 ± 1.59% in the S phase).

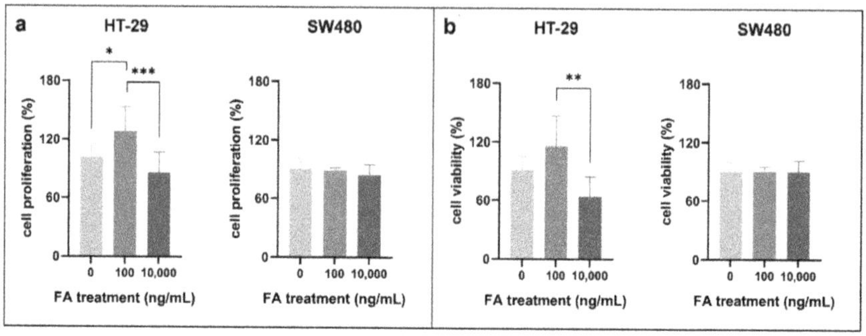

Figure 1. (**a**) Cell proliferation and (**b**) cell viability alterations of HT-29 and SW480 colorectal cancer cell lines following different folic acid (FA) supplies. Sulforhodamine B (SRB) was used for cell proliferation detection (* $p \leq 0.05$, *** $p \leq 0.001$), while cell viability data were obtained by alamarBlue assay (** $p \leq 0.01$). FA-depleted cells were kept in media containing 0 ng/mL FA, whereas treated cells were exposed to 100 and 10,000 ng/mL FA for 72 h. The percentages of cell proliferation and viability were given relative to samples kept in the normal growth media. FA: folic acid.

3.2. *Effect of Folic Acid Treatment on Genomic Stability*

We analyzed genomic stability alterations affected by FA treatment with MN scoring (Figure S1) and comet assay (Figure S2). According to immunocytochemistry (Figure 2a), 0.56 ± 0.05% of the HT-29 cells proved to have MN in the FA-depleted environment, while, as a result of FA treatment, this value significantly ($p \leq 0.01$) decreased (HT-29_{100}: 0.17 ± 0.05%, HT-$29_{10,000}$: 0.25 ± 0.09%). On the other hand, 0.79 ± 0.10% SW480 cells kept in FA-free medium had MN, and the ratio was barely affected by the FA treatment (SW480_{100}: 0.74 ± 0.07%, SW$480_{10,000}$: 0.77 ± 0.16%).

Besides DAPI staining, anti-γ-H2AX antibody was applied, and the rate of micronuclei showing γ-H2AX positivity compared to all micronuclei was determined. Approximately 79% of these particles were γ-H2AX positive, independent of the cell type or the treatment condition (HT-29: 76.68 ± 3.54%, SW480: 80.35 ± 6.46%).

In the case of comet assay, the percentage of DNA in the comet tail was the parameter used to describe DNA damage (Figure 2b). In FA-depleted HT-29 cells, tail DNA was 37.35 ± 3.45%, which was even higher in SW480 cells with a 58.79 ± 0.83% rate. Following FA supplementation, a reduction in comet tail length could be observed in HT-29 samples (HT-29_{100}: 31.23 ± 3.41%, HT-$29_{10,000}$: 20.07 ± 3.59%), while in FA-treated SW480 cells, prominent alterations were not detected (SW480_{100}: 59.84 ± 1.89%, SW$480_{10,000}$: 58.24 ± 2.56%).

3.3. *Effect of Folic Acid Treatment on DNA Methylation*

FA is involved in the DNA methylation process; thereby, the analysis of its alterations was needed. The methylation levels of three LINE-1 CpG sites were summarized (Figure 3a) and also visualized individually (Figure 3b) to assess global DNA methylation. Although HT-29 cells had an 11.59% higher DNA methylation level than SW480 samples, remarkable changes could not be detected for FA supplementation in any of the cell lines (HT-29_0: 59.30 ± 2.90%, HT-29_{100}: 61.68 ± 4.17%, HT-$29_{10,000}$: 60.42 ± 2.12%; SW480_0: 48.76 ± 4.93%, SW480_{100}: 49.15 ± 6.13%, SW$480_{10,000}$: 48.73 ± 7.27%).

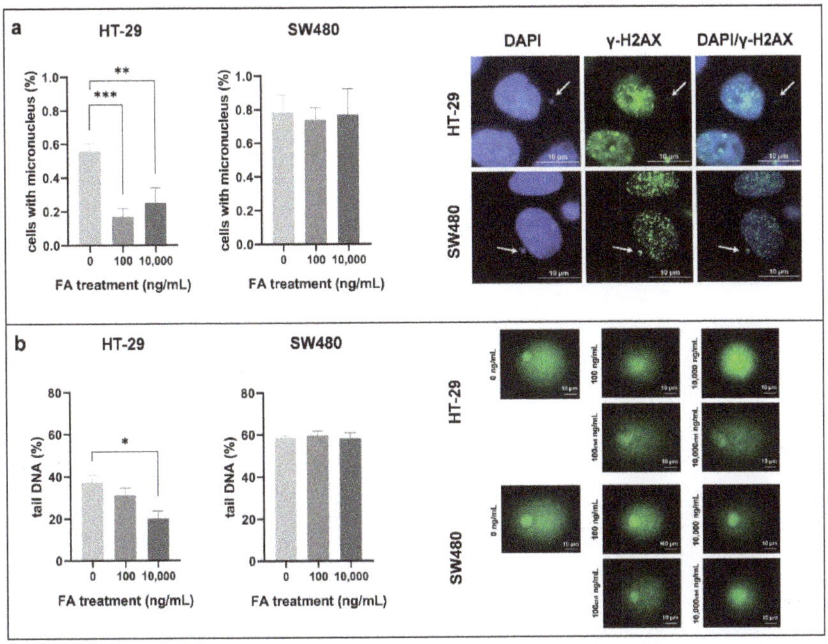

Figure 2. Genomic stability detection of HT-29 and SW480 cells exposed to different folic acid (FA) concentrations (0, 100, 10,000 ng/mL). (**a**) Micronucleus (MN) scoring was performed on DAPI- and anti-γ-H2AX-stained slides. Left: We obtained the results by proportioning the cells with MN with all cells counted (** $p \leq 0.01$, *** $p \leq 0.001$). Right: Representative γ-H2AX-positive micronuclei are indicated with arrows. (**b**) DNA integrity was evaluated with comet assay, additionally. Left: Graphs show the changes in genomic stability in consideration of comet tail DNA percentage (* $p \leq 0.05$). Right: Characteristic comets of both cell lines were captured following different treatments. FA: folic acid.

Next, we determined DNA methylation changes with RRBS in order to compare the genome-wide methylome map of FA-depleted and 10,000 ng/mL FA-treated samples. Following FA supplementation, the number of genes with methylated (HT-29$_{hyper}$: 1436, SW480$_{hyper}$: 1368) and unmethylated (HT-29$_{hypo}$: 1368, SW480$_{hypo}$: 1370) CpG sites was similar within a cell line and between the two cell types as well (Figure 3c). KEGG enrichment analysis revealed that genes that underwent methylation changes following the FA treatment were dominantly associated with carcinogenesis in both cell lines (Figure 3d). The gene numbers in different pathways were within the range of 40–117. Based on the localization of DMSs in distinct chromosomal states, the heterochromatin or low signal regions (HT-29: 27.57%, SW480: 31.61%) along with active (HT-29: 22.29%, SW480: 20.29%) and weak (HT-29: 18.57%, SW480: 17.81%) promoter sequences were mainly represented (Figure 3e).

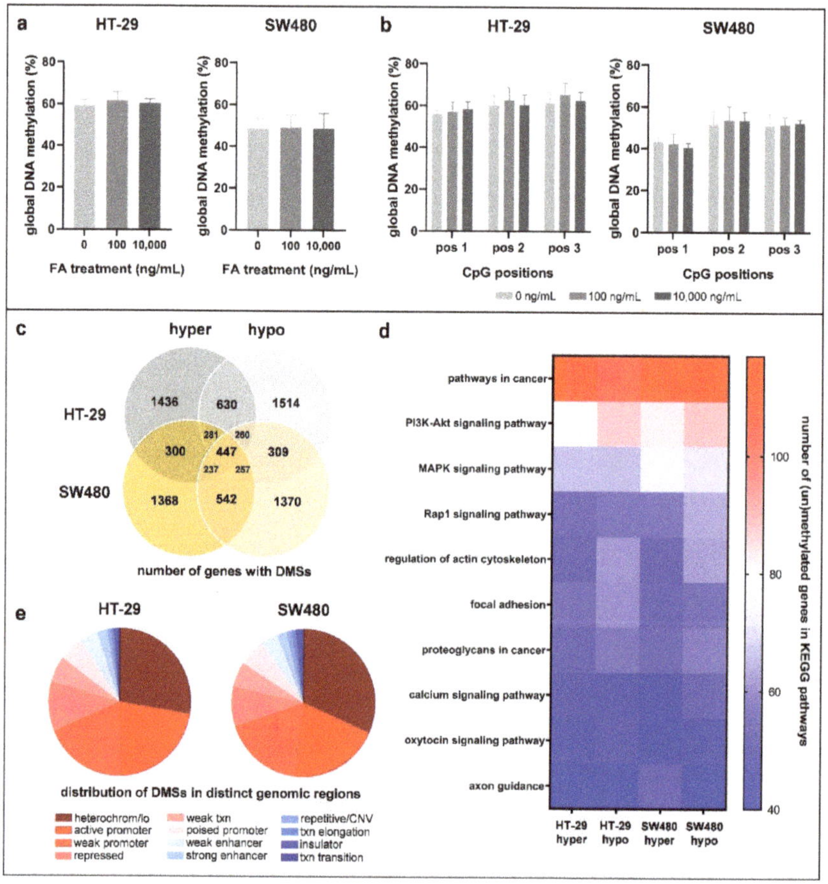

Figure 3. DNA methylation analysis of HT-29 and SW480 cell lines exposed to different folic acid (FA) concentrations. The methylation levels of long interspersed nuclear element 1 (LINE-1) CpG positions (pos 1, pos 2, pos 3) were (**a**) summarized and also (**b**) visualized individually to detect global DNA methylation changes. With the use of Reduced Representation Bisulfite Sequencing (RRBS) method, a genome-wide methylome profile of 10,000 ng/mL FA-treated cells was established in the comparison of cells kept in FA-free (0 ng/mL FA) media. (**c**) Firstly, the number of genes with altered methylation in the investigated CpG sites was assessed. "Hyper" and "hypo" sections indicate the number of genes with methylated and unmethylated CpG sites, respectively. The intersection of these two categories refers to the genes that possess both methylated and unmethylated CpG dinucleotides. (**d**) Heatmap shows the top 10 significantly ($p \leq 0.05$) enriched Kyoto Encyclopedia of Genes and Genomes (KEGG) pathways with the number of differentially methylated genes. (**e**) Pie charts represent the localization of differentially methylated sites (DMS) in distinct chromatin states. FA: folic acid; pos: CpG position; hyper: hypermethylation; hypo: hypomethylation; DMSs: differentially methylated sites; heterochrom/lo: heterochromatin or low signal region; txn: transcription; CNV: copy number variation; KEGG: Kyoto Encyclopedia of Genes and Genomes.

3.4. *Effect of Folic Acid Treatment on Gene Expression*

Folate can influence gene expression through its interplay in epigenetic modifications. Microarray analysis of non-treated and 10,000 ng/mL FA-treated cells was performed to investigate the above-mentioned phenomenon. We focused on transcripts owning gene symbols, according to Ensemble BioMart annotation, that showed significant alterations with a fold change (FC) equal or greater than 1.5 and equal or lower than −1.5. Following FA supplementation, genes with decreased expression predominated mainly in HT-29 cells (HT-29: 78.82% down- and 21.18% upregulation, SW480: 60.60% down- and 39.40% upregulation) (Figure 4a). However, the number of altering genes (HT-29: 458, SW480: 769) (Figure 4a) and also the extent of the alterations analyzed with PCA (Figure S3) were higher in the SW480 cell line. Moreover, according to PPI networks constructed by STRING, the predicted functional associations of proteins encoded by these genes were also more complex (HT-29: 113 nodes and 82 edges, SW480: 186 nodes and 221 edges). In HT-29 cells, YWHAZ (11 edges), while in SW480, TNF (27 edges) were the hub proteins based on the number of neighbors (Figure 4b).

In HT-29 cells, two transcripts (TC03001071.hg, TC03002732.hg) detecting the *HES1* gene showed the greatest increase in their expression level after FA treatment (FC: 4.98 and 3.27), followed by *FAM210B*, *TUBBP6*, *MAPKAP1*, and *TMEM185A* (FC: 2.85, 2.65, 2.49, 2.42, respectively). On the other hand, *FAM95B1* detected by two transcripts (TC01003063.hg, TC09001384.hg) was the most predominantly downregulated gene (FC: −3.15 and −3.15) in this cell line along with *LINC01783*, *LINC00905*, *TMOD2*, and *MIR450A2* (FC: −3.09, −2.34, −2.24, −2.17, respectively). In the SW480 cell line, *SLC7A11* had the highest expression change in response to FA supplementation with a 5.12 FC, and the following four genes (*RNU1-2*, *KRTAP2-3*, *RNY5*, and *RUNX1-IT1*) showed higher than 3 FC (4.12, 3.40, 3.18, 3.13, respectively). The greatest gene expression decrease was observed in the *CCL2* gene (FC: −5.90), then in *CIC*, *SNORA38B*, *H4C12*, and 5s ribosomal RNA genes or pseudogenes, such as *RNA5S1*, *RNA5SP150*, *RNA5SP226*, and *RNA5SP389* (FC: −4.22, −4.01, −3.76, −3.72, respectively) (Figure 4c; Table S2). In the common set of HT-29 and SW480 cells, seven genes, namely *LOC727896*, *LOC105371030*, *RN7SL677P*, *RPPH1*, *SFN*, *TNFSF10*, and *TXNIP* were included based on our criteria ($p \leq 0.05$ and FC $\geq |1.5|$).

According to the PANTHER classification system, the "platelet-derived growth factor (PDGF) signaling" pathway was the one wherein most of the analyzed genes ($p \leq 0.05$ and $\geq |1.5|$ FC) of HT-29 cells were represented (*SRGAP1*, *SRGAP2*, *SRGAP2B*, *SRGAP2C*, *STAT3*, *RAB11B*, and *FOSB*). In addition, all of these genes were overexpressed after FA treatment. Concerning SW480 cells, the "inflammation mediated by chemokine and cytokine signaling" pathway was mostly affected by FA treatment, as five genes (*CCL2*, *CCL7*, *CCL20*, *ITGA2*, and *PLCG2*) were involved. Except for *ITGA2*, all genes showed downregulation for FA supplementation.

Finally, RRBS and HTA data were linked as we evaluated the expression level of genes possessing DMSs within the promoter region. Promoter methylation caused downregulation of four (*NABP1*, *ATG16L1*, *RPPH1*, and *TCEAL1*) and five (*EIF4G2*, *RAB31*, *NEU3*, *SESN3*, *IFI6*) genes in HT-29 and SW480 cells, respectively. Meanwhile, hypomethylation of the promoter region corresponded with upregulation of 10 genes (*TUBB2B*, *BIVM*, *SLC39A8*, *SKIL*, *TMBIM6*, *DHRS3*, *ROR1*, *LGALS3*, *ERBB3*, and *MAPKAP1*) in HT-29 and 8 genes (*SNRNP70*, *FOSL1*, *WDR1*, *NBPF8*, *STC2*, *NR2F1*, *NTSR1*, and *SERPINB1*) in SW480 cells (Figure 5).

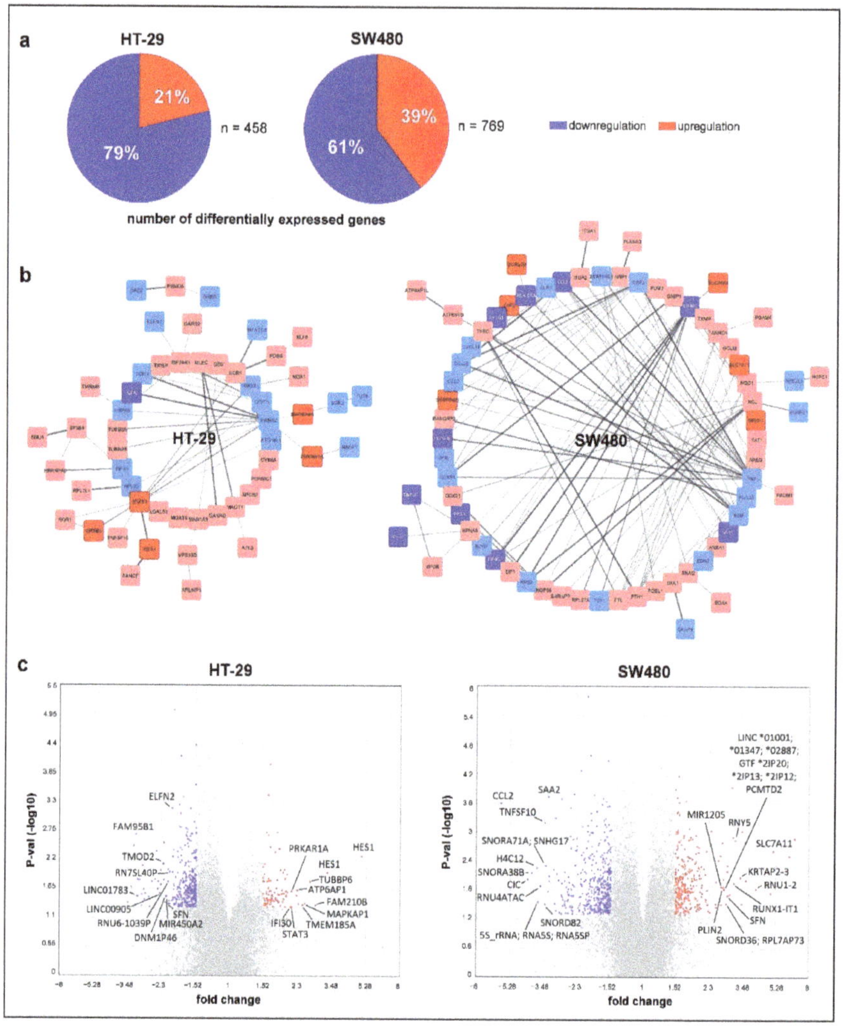

Figure 4. Genome-wide transcriptome alterations of HT-29 and SW480 cells following 10,000 ng/mL folic acid (FA) supplementation detected by Human Transcriptome Array 2.0 (HTA 2.0). (**a**) Pie charts represent the proportion of up- and downregulated genes. (**b**) Visual networks of protein–protein interactions were generated by the StringApp of Cytoscape software based on the list of genes with significant ($p \leq 0.05$) expression alterations and $\geq |1.5|$ fold change (FC). Colors refer to the expression level of protein-coding genes (dark blue: FC ≤ -2, light blue: FC ≥ -2 and ≤ -1.5, light red: FC ≥ 1.5 and ≤ 2, dark red: FC ≥ 2). (**c**) Top 10 genes showing significant ($p \leq 0.05$) up- and downregulation visualized with volcano plots. Gray points represent all the transcripts detected by HTA 2.0 microarray, while significantly ($p \leq 0.05$) altering genes with FC $\geq |1.5|$ value were marked with red and blue. P-val: p-value.

Associated gene expression and promoter methylation status in HT-29 cells

gene	met. status	met. diff. (%)	expr. status	fold change
NABP1	hyper	29.67	↓	−1.69
ATG16L1	hyper	100.00	↓	−1.67
RPPH1	hyper	22.59	↓	−1.59
TCEAL1	hyper	100.00	↓	−1.53
TUBB2B	hypo	−31.03	↑	1.52
BIVM	hypo	−100.00	↑	1.57
SLC39A8	hypo	−100.00	↑	1.58
SKIL	hypo	−44.58	↑	1.59
TMBIM6	hypo	−77.44	↑	1.62
DHRS3	hypo	−22.19	↑	1.67
ROR1	hypo	−16.85	↑	1.69
LGALS3	hypo	−100.00	↑	1.78
ERBB3	hypo	−37.42	↑	2.02
MAPKAP1	hypo	−100.00	↑	2.49

Associated gene expression and promoter methylation status in SW480 cells

gene	met. status	met. diff. (%)	expr. status	fold change
EIF4G2	hyper	22.73	↓	−2.55
RAB31	hyper	100.00	↓	−1.85
NEU3	hyper	17.86	↓	−1.62
SESN3	hyper	22.86	↓	−1.51
IFI6	hyper	36.84	↓	−1.50
SNRNP70	hypo	−100.00	↑	1.52
FOSL1	hypo	−80.56	↑	1.70
WDR1	hypo	−100.00	↑	1.73
NBPF8	hypo	−85.42	↑	1.81
STC2	hypo	−66.15	↑	1.82
NR2F1	hypo	−100.00	↑	2.02
NTSR1	hypo	−100.00	↑	2.09
SERPINB1	hypo	−23.40	↑	2.36

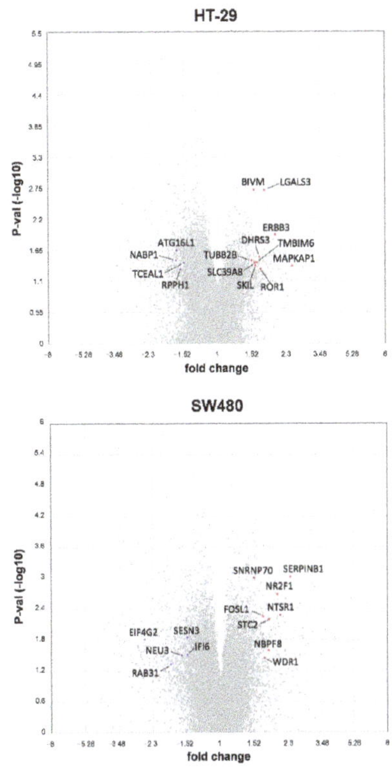

Figure 5. The intersection of genome-wide DNA methylation and gene expression data obtained by Reduced Representation Bisulfite Sequencing (RRBS) and Human Transcriptome Array (HTA) 2.0 analyses. Values represent the methylome and transcriptome pattern changes of 10,000 ng/mL folic acid (FA)-treated HT-29 and SW480 cells compared to non-treated samples (0 ng/mL FA). Only genes with promoter methylation status alteration in accordance with their expression level ($p \leq 0.05$ and fold change $\geq |1.5|$) were listed (left) and also visualized in volcano plots (right). Gray points represent all the transcripts detected by the microarray, while blue ones highlight down- and red ones show upregulating genes from the list. met. status: DNA methylation status; met. diff.: DNA methylation difference; expr. status: gene expression status; P-val: *p*-value.

4. Discussion

Folate deficiency has been shown to be associated with several disorders, such as megaloblastic anemia [36], cardiovascular diseases [37], obstetrical complications [38,39], neuropsychiatric conditions [40], and malignancies [41]. Its supplementation is considered to be beneficial in the prevention of all the diseases mentioned above. However, great care is required, as folate can promote the progression of established neoplasms, while the low intake has an exactly opposite effect [41,42]. FA is commonly used in CRC care as an adjunctive medication to promote the efficacy of chemotherapy medications [16,17]. However, its individual effect on FA-related cellular maintenance activities, along with the genetic and epigenetic regulation, has not been evaluated comprehensively. Therefore, we found it important to observe the short-term (72 h) impact of FA-free (0 ng/mL) and FA-supplemented (100 ng/mL and 10,000 ng/mL) environments on two CRC cell lines (HT-29 and SW480).

An adequate level of nucleotides is necessary for tumor progression, and FA can serve this need by participating in DNA synthesis [2]. This phenomenon can be in the background of the significantly ($p \leq 0.05$) elevated HT-29 cell proliferation for 100 ng/mL FA treatment, along with an increased cell viability compared to the FA-depleted environment. These observations were in line with the results of Pellis et al. [8]. Interestingly, supraphysiological FA concentration caused lower proliferation than in FA-depleted samples. Previous studies demonstrated that similarly high FA dose inhibited the proliferation of CRC cell lines (HT-29, COLO-205, LoVo), along with endothelial and nasopharyngeal cancer cells, supposedly via the activation of folate receptor α-mediated extracellular signal-regulated kinase (ERK) pathway [7,43,44]. Conversely to HT-29 samples, cell proliferation and viability changes could not be detected in the SW480 cell line. Farias et al. likewise verified that for various FA exposures, SW480 cells had a different proliferation rate than other CRC cell lines (HCT116, LS174T) [4].

It is already known that low folate levels can lead to genomic instability through the disturbances of DNA methylation and nucleotide imbalance [45–50]. Since FA intake might improve these conditions, we examined DNA integrity using MN scoring and comet assay methods. SW480 cells inherently had greater genomic instability than the HT-29 cell line, comparing the samples kept in FA-free environment. The micronucleus number significantly decreased for FA supplementation, and comets were shortened in parallel with the applied FA dose regarding the HT-29 cell line. Therefore, the treatment facilitated genomic stability, similarly to the results of Catala et al. using colonic epithelial cells and fibroblasts [45]. On the other hand, FA treatment did not cause any remarkable changes in the genomic stability of SW480 samples.

DNA methylation—one of the most intensely studied epigenetic modifications—is crucial to normal genome regulation [51–53]. After FA is converted into 5-methyltetrahydrofolate (5-methyl-THF) by the MTHFR enzyme, it participates in the synthesis of SAM, the main methyl donor molecule [42]. Hence, we investigated global DNA methylation alterations with LINE-1 bisulfite sequencing in response to FA treatment. HT-29 samples had an 11.59% higher baseline DNA methylation level than SW480 cells, in agreement with their CIMP statuses [19]. However, remarkable changes were not detected in any of the samples after FA supplementation. Stempak et al. came to similar results when analyzing folate-deficient and sufficient HCT116 and Caco-2 CRC cell lines [54].

Thereafter, DNA methylation was analyzed at a higher resolution on the single-nucleotide level with RRBS method. Our results highlighted that the methylation pattern did not differ much between the two cell lines following 10,000 ng/mL FA supplementation in regard to the affected gene numbers, pathways, and genomic regions. Additionally, FA equally caused DNA hyper- and hypomethylation, as the numbers of genes possessing methylated and unmethylated DMSs were relatively equivalent. This phenomenon may explain the non-significantly altering global DNA methylation results. In the background of the discrepancy between the facts that FA is a precursor in methyl donor formation, but it also modulates DNA methylation in both directions, we suspect the dual nature of SAM. Hence, this molecule is rather considered to be a methylome modulator than a pure hypermethylating agent, based on the hypothesis suggested by Wang et al. and our recent study results as well [22,55].

Enrichment analysis revealed that KEGG signaling pathways tightly related to carcinogenesis (such as "PI3K-Akt" [56], "MAPK" [57], "Rap1" [58], "regulation of actin cytoskeleton" [59], "focal adhesion" [60], and "proteoglycans in cancer" [61]) were the most dominantly affected ones by DNA methylation changes. The former two of the above-mentioned processes have been reported to be influenced by the folic acid status [62,63]. Considering the distribution of DMSs in distinct genomic regions, we observed an enrichment mainly in heterochromatin (or low signal) sequences, similarly to other RRBS studies [64,65]. Besides these regions, active and weak promoter sequences were also represented dominantly.

The methylation status of promoter-associated CpG islands is of particular importance because it can modulate gene expression [21,42]; therefore, our study was complemented with genome-wide transcriptome analysis of FA-depleted and 10,000 ng/mL FA-treated samples. We observed that genes showing a decreased expression after FA treatment outnumbered the upregulated ones. In the SW480 cell line, not only the number of genes with altering expression ($p \leq 0.05$ and FC $\geq |1.5|$) but also the extent of alterations and the predicted functional associations between the encoded proteins were more prominent than in HT-29 cells. *YWHAZ* and the cytokine-expressing *TNF* were considered as the coding genes of hub proteins in the HT-29 and SW480 cell lines, respectively, both possessing significant ($p \leq 0.05$) downregulation for FA supplementation. *YWHAZ* is a potential oncogene that can enhance proliferation, migration, and metastasis in many cancer types; therefore, its inhibition has strong clinical importance [66]. Moreover, ERK1/2 and PI3K/Akt/mTOR pathways mentioned in this study earlier have been reported to be the downstream targets of *YWHAZ* [66]. Not only the proliferation changes of HT-29 cells were supported by such gene expression alterations, but also the increased genomic stability. We examined the relevance of the top 10 up- and downregulated genes in DNA repair, and only *HES1* [67], *STAT3* [68], and *CCL2* [69] were reported to be involved in this process. The former two had an increased expression in HT-29 cells, while the latter was silenced in SW480 cells following the treatment, which could explain the different DNA repair abilities between the cell lines.

Finally, we combined the results of RRBS and microarray analyses to investigate whether FA actually affects gene expression by methylating or unmethylating the promoter regions. We observed that promoter methylation caused transcriptional silencing ($p \leq 0.05$ and FC $\geq |1.5|$) of four genes in HT-29 and five genes in SW480 cells following 10,000 ng/mL FA supplementation. On the other hand, promoter hypomethylation was accompanied by the upregulation ($p \leq 0.05$ and FC $\geq |1.5|$) of 10 and 8 genes in the HT-29 and SW480 cell lines, respectively. Due to the fact that the methylation status of the promoter region was not in concordance with the expression of numerous genes, we suppose that short-term FA treatment does not primarily exert its effect on gene regulation by altering promoter methylation. The involvement of other mechanisms, for example, histone methylation, can be implicated in this finding, as also hypothesized by Price et al. [70].

5. Conclusions

FA is a widely consumed vitamin known to be involved in an important epigenetic regulatory process, DNA methylation. However, our results suggest that in HT-29 and SW480 CRC cells, short-term FA supplementation can equally induce DNA hyper- and hypomethylation that mainly affects the pathways related to carcinogenesis. Promoter methylation changes could act on gene regulation, but we also suppose the involvement of other FA-associated effects behind the great number of differentially expressed genes. As opposed to SW480, FA treatment had a significant impact on the maintenance activities of HT-29 cells, including proliferation, viability, and DNA repair. Apart from the distinct molecular background between the two cell lines, expression alterations of genes contributing to the mentioned processes might explain this finding. Based on the results of this study, we can conclude that the short-term effect of FA on the observed cell lines goes far beyond the fundamental attributes connected to this vitamin.

Supplementary Materials: The following supporting information can be downloaded at: https://www.mdpi.com/article/10.3390/cancers14071820/s1, Figure S1: DAPI (blue) and anti-γ-H2AX (green) double-staining of HT-29 and SW480 cell lines; Figure S2: DNA integrity evaluation of HT-29 and SW480 cell lines with comet assay after exposing the samples to different folic acid (FA) concentrations (0, 100, 10,000 ng/mL); Figure S3: Principal component analysis (PCA) of Human Transcriptome Array 2.0 (HTA 2.0) gene expression data obtained from non-treated (0 ng/mL) and 10,000 ng/mL folic acid (FA)-supplemented HT-29 and SW480 cells; Table S1: The raw data of Figures 1–3; Table S2: The list of the top 10 differentially expressed genes.

Author Contributions: Conceptualization, S.Z., B.M.; methodology, S.Z., A.K., B.K.B., Z.B.N., K.A.S., G.V., T.D., A.S., G.B.; software, W.K., O.P., I.C.; writing—original draft preparation, S.Z.; writing—review and editing, A.K., B.K.B., Z.B.N., K.A.S., G.V., W.K., T.D., A.S., G.B., I.C., P.I., I.T., B.M.; supervision, Z.T., P.I., I.T., B.M.; funding acquisition, S.Z., I.C., P.I., B.M. All authors have read and agreed to the published version of the manuscript.

Funding: This research was funded by the National Research, Development and Innovation Office, Hungary (grant No. 2020-1.1.2-PIACI-KFI-2021-00298; grant No. FIEK_16-1-2016-0005). The study was also supported by the ÚNKP-21-4-I-SE-20 New National Excellence Program of the Ministry for Innovation and Technology from the Source of the National Research, Development and Innovation Fund, as well as the Higher Education Institutional Excellence Program of the Ministry of Human capacities in Hungary, within the framework of the molecular biology thematic program of Semmelweis University.

Institutional Review Board Statement: Not applicable.

Informed Consent Statement: Not applicable.

Data Availability Statement: The data presented in this study were uploaded to the Gene Expression Omnibus (GEO) repository, reference number [GSE186084].

Acknowledgments: The authors would like to acknowledge Kristina Gottschall for the language revision of the manuscript. We kindly thank Ildikó Felletár, Kornélia Kelemen, and Géza Antalffy for the excellent technical support.

Conflicts of Interest: The authors declare no conflict of interest.

References

1. Kim, Y.I. Current Status of Folic Acid Supplementation on Colorectal Cancer Prevention. *Curr. Pharmacol. Rep.* **2016**, *2*, 21–33. [CrossRef]
2. Kim, Y.I. Folate and colorectal cancer: An evidence-based critical review. *Mol. Nutr. Food Res.* **2007**, *51*, 267–292. [CrossRef] [PubMed]
3. Ulrich, C.M.; Potter, J.D. Folate supplementation: Too much of a good thing? *Cancer Epidemiol. Biomarkers Prev.* **2006**, *15*, 189–193. [CrossRef] [PubMed]
4. Farias, N.; Ho, N.; Butler, S.; Delaney, L.; Morrison, J.; Shahrzad, S.; Coomber, B.L. The effects of folic acid on global DNA methylation and colonosphere formation in colon cancer cell lines. *J. Nutr. Biochem.* **2015**, *26*, 818–826. [CrossRef] [PubMed]
5. Pufulete, M.; Emery, P.W.; Sanders, T.A. Folate, DNA methylation and colo-rectal cancer. *Proc. Nutr. Soc.* **2003**, *62*, 437–445. [CrossRef] [PubMed]
6. Price, R.J.; Lillycrop, K.A.; Burdge, G.C. Folic acid induces cell type-specific changes in the transcriptome of breast cancer cell lines: A proof-of-concept study. *J. Nutr. Sci.* **2016**, *5*, e17. [CrossRef]
7. Kuo, C.T.; Chang, C.; Lee, W.S. Folic acid inhibits COLO-205 colon cancer cell proliferation through activating the FRα/c-SRC/ERK1/2/NFκB/TP53 pathway: In vitro and in vivo studies. *Sci. Rep.* **2015**, *5*, 11187. [CrossRef] [PubMed]
8. Pellis, L.; Dommels, Y.; Venema, D.; Polanen, A.; Lips, E.; Baykus, H.; Kok, F.; Kampman, E.; Keijer, J. High folic acid increases cell turnover and lowers differentiation and iron content in human HT29 colon cancer cells. *Br. J. Nutr.* **2008**, *99*, 703–708. [CrossRef] [PubMed]
9. McCabe, J.; Chang, S.; Hajibandeh, J.; Tran, M.D.; Meeder, C.A.; Sharma, K.; Nguyen, D.H.; Moody, M.; Keiserman, M.A.; Bergman, C.J.; et al. Folate supplementation induces differential dose-dependent modulation of proliferative phenotypes among cancerous and noncancerous oral cell lines in vitro. *J. Diet. Suppl.* **2010**, *7*, 325–340. [CrossRef] [PubMed]
10. Farias, N. The Role of Folic Acid in Maintaining Colorectal Cancer Cell DNA Methylation Patterns and Cancer Stem Cell Phenotype In Vitro. Master's Thesis, University of Guelph, Guelph, ON, Canada, 2013.
11. Kim, Y.I. Role of folate in colon cancer development and progression. *J. Nutr.* **2003**, *133*, 3731S–3739S. [CrossRef] [PubMed]
12. Kim, Y.I. Folate and carcinogenesis: Evidence, mechanisms, and implications. *J. Nutr. Biochem.* **1999**, *10*, 66–88. [CrossRef]
13. Richardson, R.B.; Allan, D.S.; Le, Y. Greater organ involution in highly proliferative tissues associated with the early onset and acceleration of ageing in humans. *Exp. Gerontol.* **2014**, *55*, 80–91. [CrossRef] [PubMed]
14. Luebeck, E.G.; Moolgavkar, S.H.; Liu, A.Y.; Boynton, A.; Ulrich, C.M. Does folic acid supplementation prevent or promote colorectal cancer? Results from model-based predictions. *Cancer Epidemiol. Biomark. Prev.* **2008**, *17*, 1360–1367. [CrossRef] [PubMed]
15. Kim, Y.I. Folate: A magic bullet or a double edged sword for colorectal cancer prevention? *Gut* **2006**, *55*, 1387–1389. [CrossRef] [PubMed]
16. Vodenkova, S.; Buchler, T.; Cervena, K.; Veskrnova, V.; Vodicka, P.; Vymetalkova, V. 5-fluorouracil and other fluoropyrimidines in colorectal cancer: Past, present and future. *Pharmacol. Ther.* **2020**, *206*, 107447. [CrossRef] [PubMed]
17. Nielson, C.M.; Bylsma, L.C.; Fryzek, J.P.; Saad, H.A.; Crawford, J. Relative Dose Intensity of Chemotherapy and Survival in Patients with Advanced Stage Solid Tumor Cancer: A Systematic Review and Meta-Analysis. *Oncologist* **2021**, *26*, e1609–e1618. [CrossRef] [PubMed]
18. Pino, M.S.; Chung, D.C. The chromosomal instability pathway in colon cancer. *Gastroenterology* **2010**, *138*, 2059–2072. [CrossRef] [PubMed]

19. Berg, K.C.G.; Eide, P.W.; Eilertsen, I.A.; Johannessen, B.; Bruun, J.; Danielsen, S.A.; Bjørnslett, M.; Meza-Zepeda, L.A.; Eknæs, M.; Lind, G.E.; et al. Multi-omics of 34 colorectal cancer cell lines—A resource for biomedical studies. *Mol. Cancer* **2017**, *16*, 116. [CrossRef]
20. Menter, D.G.; Davis, J.S.; Broom, B.M.; Overman, M.J.; Morris, J.; Kopetz, S. Back to the Colorectal Cancer Consensus Molecular Subtype Future. *Curr. Gastroenterol. Rep.* **2019**, *21*, 5. [CrossRef]
21. Ogino, S.; Cantor, M.; Kawasaki, T.; Brahmandam, M.; Kirkner, G.J.; Weisenberger, D.J.; Campan, M.; Laird, P.W.; Loda, M.; Fuchs, C.S. CpG island methylator phenotype (CIMP) of colorectal cancer is best characterised by quantitative DNA methylation analysis and prospective cohort studies. *Gut* **2006**, *55*, 1000–1006. [CrossRef] [PubMed]
22. Zsigrai, S.; Kalmár, A.; Nagy, Z.B.; Barták, B.K.; Valcz, G.; Szigeti, K.A.; Galamb, O.; Dankó, T.; Sebestyén, A.; Barna, G.; et al. S-Adenosylmethionine Treatment of Colorectal Cancer Cell Lines Alters DNA Methylation, DNA Repair and Tumor Progression-Related Gene Expression. *Cells* **2020**, *9*, 1864. [CrossRef]
23. Pietrzik, K.; Bailey, L.; Shane, B. Folic acid and L-5-methyltetrahydrofolate: Comparison of clinical pharmacokinetics and pharmacodynamics. *Clin. Pharmacokinet.* **2010**, *49*, 535–548. [CrossRef] [PubMed]
24. Moll, S.; Varga, E.A. Homocysteine and MTHFR Mutations. *Circulation* **2015**, *132*, e6–e9. [CrossRef] [PubMed]
25. Zetterberg, H.; Regland, B.; Palmér, M.; Ricksten, A.; Palmqvist, L.; Rymo, L.; Arvanitis, D.A.; Spandidos, D.A.; Blennow, K. Increased frequency of combined methylenetetrahydrofolate reductase C677T and A1298C mutated alleles in spontaneously aborted embryos. *Eur. J. Hum. Genet.* **2002**, *10*, 113–118. [CrossRef]
26. Vidmar Golja, M.; Šmid, A.; Karas Kuželički, N.; Trontelj, J.; Geršak, K.; Mlinarič-Raščan, I. Folate Insufficiency Due to MTHFR Deficiency Is Bypassed by 5-Methyltetrahydrofolate. *J. Clin. Med.* **2020**, *9*, 2836. [CrossRef] [PubMed]
27. Valcz, G.; Buzás, E.I.; Kittel, Á.; Krenács, T.; Visnovitz, T.; Spisák, S.; Török, G.; Homolya, L.; Zsigrai, S.; Kiszler, G.; et al. En bloc release of MVB-like small extracellular vesicle clusters by colorectal carcinoma cells. *J. Extracell. Vesicles* **2019**, *8*, 1596668. [CrossRef] [PubMed]
28. Ribas, L.E.; Baravalle, M.E.; Gasser, F.B.; Renna, M.S.; Addona, S.; Ortega, H.H.; Savino, G.H.; Van de Velde, F.; Hein, G.J. Extraction of phenolic compounds from the shells of pecan nuts with cytotoxic activity through apoptosis against the colon cancer cell line HT-29. *J. Food Sci.* **2021**, *86*, 5409–5423. [CrossRef] [PubMed]
29. El Hosry, L.; Di Giorgio, C.; Birer, C.; Habib, J.; Tueni, M.; Bun, S.S.; Herbette, G.; De Meo, M.; Ollivier, E.; Elias, R. In vitro cytotoxic and anticlastogenic activities of saxifragifolin B and cyclamin isolated from Cyclamen persicum and Cyclamen libanoticum. *Pharm. Biol.* **2014**, *52*, 1134–1140. [CrossRef] [PubMed]
30. Wingett, S.W.; Andrews, S. FastQ Screen: A tool for multi-genome mapping and quality control. *F1000Research* **2018**, *7*, 1338. [CrossRef] [PubMed]
31. Krueger, F.; Andrews, S.R. Bismark: A flexible aligner and methylation caller for Bisulfite-Seq applications. *Bioinformatics* **2011**, *27*, 1571–1572. [CrossRef]
32. Akalin, A.; Kormaksson, M.; Li, S.; Garrett-Bakelman, F.E.; Figueroa, M.E.; Melnick, A.; Mason, C.E. MethylKit: A comprehensive R package for the analysis of genome-wide DNA methylation profiles. *Genome Biol.* **2012**, *13*, R87. [CrossRef]
33. Lee, B.T.; Barber, G.P.; Benet-Pagès, A.; Casper, J.; Clawson, H.; Diekhans, M.; Fischer, C.; Gonzalez, J.N.; Hinrichs, A.S.; Lee, C.M.; et al. The UCSC Genome Browser database: 2022 update. *Nucleic Acids Res.* **2022**, *50*, D1115–D1122. [CrossRef] [PubMed]
34. Nagy, Z.B.; Wichmann, B.; Kalmár, A.; Galamb, O.; Barták, B.K.; Spisák, S.; Tulassay, Z.; Molnár, B. Colorectal adenoma and carcinoma specific miRNA profiles in biopsy and their expression in plasma specimens. *Clin. Epigenet.* **2017**, *9*, 22. [CrossRef]
35. Kalmár, A.; Nagy, Z.B.; Galamb, O.; Csabai, I.; Bodor, A.; Wichmann, B.; Valcz, G.; Barták, B.K.; Tulassay, Z.; Igaz, P.; et al. Genome-wide expression profiling in colorectal cancer focusing on lncRNAs in the adenoma-carcinoma transition. *BMC Cancer* **2019**, *19*, 1059. [CrossRef]
36. Metz, J. Folates in megaloblastic anaemia. *Bull. World Health Organ.* **1963**, *28*, 517–529.
37. Lucock, M. Folic acid: Nutritional biochemistry, molecular biology, and role in disease processes. *Mol. Genet. Metab.* **2000**, *71*, 121–138. [CrossRef]
38. Molloy, A.M.; Kirke, P.N.; Brody, L.C.; Scott, J.M.; Mills, J.L. Effects of folate and vitamin B12 deficiencies during pregnancy on fetal, infant, and child development. *Food Nutr. Bull.* **2008**, *29*, S101–S115. [CrossRef]
39. Hibbard, B.M.; Hibbard, E.D.; Jeffcoate, T.N. Folic acid and reproduction. *Acta Obstet. Gynecol Scand.* **1965**, *44*, 375–400. [CrossRef]
40. Young, S.N. Folate and depression–a neglected problem. *J. Psychiatry Neurosci.* **2007**, *32*, 80–82. [PubMed]
41. Novakovic, P.; Stempak, J.M.; Sohn, K.J.; Kim, Y.I. Effects of folate deficiency on gene expression in the apoptosis and cancer pathways in colon cancer cells. *Carcinogenesis* **2006**, *27*, 916–924. [CrossRef] [PubMed]
42. Zsigrai, S.; Kalmár, A.; Valcz, G.; Szigeti, K.A.; Barták, B.K.; Nagy, Z.B.; Igaz, P.; Tulassay, Z.; Molnár, B. Physiological and pathophysiological significance of vitamin B9. Summary on the occasion of the 30-year introduction of folic acid as a die-tary supplement. *Orv. Hetil.* **2019**, *160*, 1087–1096. [CrossRef]
43. Lin, S.Y.; Lee, W.R.; Su, Y.F.; Hsu, S.P.; Lin, H.C.; Ho, P.Y.; Hou, T.C.; Chou, Y.P.; Kuo, C.T.; Lee, W.S. Folic acid inhibits endothelial cell proliferation through activating the cSrc/ERK 2/NF-κB/p53 pathway mediated by folic acid receptor. *Angiogenesis* **2012**, *15*, 671–683. [CrossRef] [PubMed]
44. Liu, Z.; Jin, X.; Pi, W.; Liu, S. Folic acid inhibits nasopharyngeal cancer cell proliferation and invasion via activation of FRα/ERK1/2/TSLC1 pathway. *Biosci. Rep.* **2017**, *37*, BSR20170772. [CrossRef] [PubMed]

45. Catala, G.N.; Bestwick, C.S.; Russell, W.R.; Tortora, K.; Giovannelli, L.; Moyer, M.P.; Lendoiro, E.; Duthie, S.J. Folate, genomic stability and colon cancer: The use of single cell gel electrophoresis in assessing the impact of folate in vitro, in vivo and in human biomonitoring. *Mutat. Res. Genet. Toxicol. Environ. Mutagen.* **2019**, *843*, 73–80. [CrossRef] [PubMed]
46. Meng, H.; Cao, Y.; Qin, J.; Song, X.; Zhang, Q.; Shi, Y.; Cao, L. DNA methylation, its mediators and genome integrity. *Int. J. Biol. Sci.* **2015**, *11*, 604–617. [CrossRef]
47. Szigeti, K.A.; Galamb, O.; Kalmár, A.; Barták, B.K.; Nagy, Z.B.; Márkus, E.; Igaz, P.; Tulassay, Z.; Molnár, B. Role and alterations of DNA methylation during the aging and cancer. *Orv. Hetil.* **2018**, *159*, 3–15. [CrossRef] [PubMed]
48. Hardy, L.W.; Finer-Moore, J.S.; Montfort, W.R.; Jones, M.O.; Santi, D.V.; Stroud, R.M. Atomic structure of thymidylate synthase: Target for rational drug design. *Science* **1987**, *235*, 448–455. [CrossRef] [PubMed]
49. Hazra, A.; Selhub, J.; Chao, W.H.; Ueland, P.M.; Hunter, D.J.; Baron, J.A. Uracil misincorporation into DNA and folic acid supplementation. *Am. J. Clin. Nutr.* **2010**, *91*, 160–165. [CrossRef] [PubMed]
50. Blount, B.C.; Mack, M.M.; Wehr, C.M.; MacGregor, J.T.; Hiatt, R.A.; Wang, G.; Wickramasinghe, S.N.; Everson, R.B.; Ames, B.N. Folate deficiency causes uracil misincorporation into human DNA and chromosome breakage: Implications for cancer and neuronal damage. *Proc. Natl. Acad. Sci. USA* **1997**, *94*, 3290–3295. [CrossRef]
51. Bailey, L.B.; Gregory, J.F. Folate metabolism and requirements. *J. Nutr.* **1999**, *129*, 779–782. [CrossRef]
52. Kulis, M.; Esteller, M. DNA methylation and cancer. *Adv. Genet.* **2010**, *70*, 27–56. [PubMed]
53. Crider, K.S.; Yang, T.P.; Berry, R.J.; Bailey, L.B. Folate and DNA methylation: A review of molecular mechanisms and the evidence for folate's role. *Adv. Nutr.* **2012**, *3*, 21–38. [CrossRef] [PubMed]
54. Stempak, J.M.; Sohn, K.J.; Chiang, E.P.; Shane, B.; Kim, Y.I. Cell and stage of transformation-specific effects of folate deficiency on methionine cycle intermediates and DNA methylation in an in vitro model. *Carcinogenesis* **2005**, *26*, 981–990. [CrossRef] [PubMed]
55. Wang, Y.; Sun, Z.; Szyf, M. S-adenosyl-methionine (SAM) alters the transcriptome and methylome and specifically blocks growth and invasiveness of liver cancer cells. *Oncotarget* **2017**, *8*, 111866–111881. [CrossRef]
56. Jiang, N.; Dai, Q.; Su, X.; Fu, J.; Feng, X.; Peng, J. Role of PI3K/AKT pathway in cancer: The framework of malignant behavior. *Mol. Biol. Rep.* **2020**, *47*, 4587–4629. [CrossRef] [PubMed]
57. Zhang, W.; Liu, H. MAPK signal pathways in the regulation of cell proliferation in mammalian cells. *Cell Res.* **2002**, *12*, 9–18. [CrossRef]
58. Zhang, Y.L.; Wang, R.C.; Cheng, K.; Ring, B.Z.; Su, L. Roles of Rap1 signaling in tumor cell migration and invasion. *Cancer Biol. Med.* **2017**, *14*, 90–99. [PubMed]
59. Yamaguchi, H.; Condeelis, J. Regulation of the actin cytoskeleton in cancer cell migration and invasion. *Biochim. Biophys. Acta* **2007**, *1773*, 642–652. [CrossRef] [PubMed]
60. Eke, I.; Cordes, N. Focal adhesion signaling and therapy resistance in cancer. *Semin. Cancer Biol.* **2015**, *31*, 65–75. [CrossRef] [PubMed]
61. Ahrens, T.D.; Bang-Christensen, S.R.; Jørgensen, A.M.; Løppke, C.; Spliid, C.B.; Sand, N.T.; Clausen, T.M.; Salanti, A.; Agerbæk, M.Ø. The Role of Proteoglycans in Cancer Metastasis and Circulating Tumor Cell Analysis. *Front. Cell. Dev. Biol.* **2020**, *8*, 749. [CrossRef]
62. Huang, X.; He, Z.; Jiang, X.; Hou, M.; Tang, Z.; Zhen, X.; Liang, Y.; Ma, J. Folic Acid Represses Hypoxia-Induced Inflammation in THP-1 Cells through Inhibition of the PI3K/Akt/HIF-1α Pathway. *PLoS ONE* **2016**, *11*, e0151553. [CrossRef]
63. Slattery, M.L.; Lundgreen, A.; Wolff, R.K. Dietary influence on MAPK-signaling pathways and risk of colon and rectal cancer. *Nutr. Cancer* **2013**, *65*, 729–738. [CrossRef]
64. Ollikainen, M.; Ismail, K.; Gervin, K.; Kyllönen, A.; Hakkarainen, A.; Lundbom, J.; Järvinen, E.A.; Harris, J.R.; Lundbom, N.; Rissanen, A.; et al. Genome-wide blood DNA methylation alterations at regulatory elements and heterochromatic regions in monozygotic twins discordant for obesity and liver fat. *Clin. Epigenetics* **2015**, *7*, 39. [CrossRef] [PubMed]
65. Chandra, S.; Baribault, C.; Lacey, M.R.; Ehrlich, M. Myogenic Differential Methylation: Diverse Associations with Chromatin Structure. *Biology* **2014**, *3*, 426–451. [CrossRef] [PubMed]
66. Gan, Y.; Ye, F.; He, X.X. The role of YWHAZ in cancer: A maze of opportunities and challenges. *J. Cancer* **2020**, *11*, 2252–2264. [CrossRef] [PubMed]
67. Zhang, H.; Jiang, H.; Chen, L.; Liu, J.; Hu, X.; Zhang, H. Inhibition of Notch1/Hes1 signaling pathway improves radiosensitivity of colorectal cancer cells. *Eur. J. Pharmacol.* **2018**, *818*, 364–370. [CrossRef] [PubMed]
68. Barry, S.P.; Townsend, P.A.; Knight, R.A.; Scarabelli, T.M.; Latchman, D.S.; Stephanou, A. STAT3 modulates the DNA damage response pathway. *Int. J. Exp. Pathol.* **2010**, *91*, 506–514. [CrossRef]
69. Redon, C.E.; Dickey, J.S.; Nakamura, A.J.; Kareva, I.G.; Naf, D.; Nowsheen, S.; Kryston, T.B.; Bonner, W.M.; Georgakilas, A.G.; Sedelnikova, O.A. Tumors induce complex DNA damage in distant proliferative tissues in vivo. *Proc. Natl. Acad. Sci. USA* **2010**, *107*, 17992–17997. [CrossRef]
70. Price, R.J.; Lillycrop, K.A.; Burdge, G.C. Folic acid supplementation in vitro induces cell type-specific changes in BRCA1 and BRCA 2 mRNA expression, but does not alter DNA methylation of their promoters or DNA repair. *Nutr. Res.* **2015**, *35*, 532–544. [CrossRef]

cancers

MDPI

Article

Effective Penetration of a Liposomal Formulation of Bleomycin through Ex-Vivo Skin Explants from Two Different Species

Giulia Ferrari [1,†], Lisa Y. Pang [1,*,†], Fabio De Moliner [2], Marc Vendrell [2], Richard J. M. Reardon [1], Andrew J. Higgins [3], Sunil Chopra [3] and David J. Argyle [1]

1 Roslin Institute, The Royal (Dick) School of Veterinary Studies, University of Edinburgh, Edinburgh EH25 9RG, UK; giuliaferrari1988@gmail.com (G.F.); richard.reardon@ed.ac.uk (R.J.M.R.); david.argyle@ed.ac.uk (D.J.A.)

2 Centre for Inflammation Research, The Queen's Medical Research Institute, University of Edinburgh, Edinburgh EH16 4TJ, UK; fdemoli@exseed.ed.ac.uk (F.D.M.); marc.vendrell@ed.ac.uk (M.V.)

3 The London Dermatology Centre, London W1G 8AS, UK; ajhvet@gmail.com (A.J.H.); sunil.chopra3@btinternet.com (S.C.)

* Correspondence: lisa.pang@ed.ac.uk; Tel.: +44-13-1651-9164

† These authors contributed equally to this work.

Simple Summary: Bleomycin, a chemotherapy drug, is currently injected into patients, but this can damage healthy tissues. Ideally, we would like to apply bleomycin directly onto a skin tumour but bleomycin is a big molecule and cannot pass through the skin or directly enter into cancer cells to kill them. Therefore, we need to find new ways of packaging the drug to get it inside cancer cells. Liposomes are small artificial bubbles made of from the same building blocks as our skin and cell membranes that can be filled with pharmaceutical drugs. In this study we propose that liposomes can assist with the delivery of bleomycin by improving penetration through the skin. We are using a new compound called Bleosome, which contains liposomes packed with bleomycin. We found that Bleosome penetrated more through the healthy skin of dogs and horses than bleomycin. These are promising results, indicating that Bleosome may be an effective treatment, with easy application and limited side-effects, to treat skin cancer.

Abstract: Bleomycin is a chemotherapy agent that, when administered systemically, can cause severe pulmonary toxicity. Bleosome is a novel formulation of bleomycin encapsulated in ultra-deformable (UD) liposomes that may be applicable as a topical chemotherapy for diseases such as non-melanoma skin cancer. To date, the ability of Bleosome to effectively penetrate through the skin has not been evaluated. In this study, we investigated the ability of Bleosome to penetrate through ex vivo skin explants from dogs and horses. We visualized the penetration of UD liposomes through the skin by transmission electron microscopy. However, to effectively image the drug itself we fluorescently labeled bleomycin prior to encapsulation within liposomes and utilized multiphoton microscopy. We showed that UD liposomes do not penetrate beyond the stratum corneum, whereas bleomycin is released from UD liposomes and can penetrate to the deeper layers of the epidermis. This is the first study to show that Bleosome can effectively penetrate through the skin. We speculate that UD liposomes are penetration enhancers in that UD liposomes carry bleomycin through the outer skin to the stratum corneum and then release the drug, allowing diffusion into the deeper layers. Our results are comparative in dogs and horses and warrant further studies on the efficacy of Bleosome as topical treatment.

Keywords: liposomes; Bleosome; Bodipy-FL; topical chemotherapy; skin penetration

Citation: Ferrari, G.; Pang, L.Y.; De Moliner, F.; Vendrell, M.; Reardon, R.J.M.; Higgins, A.J.; Chopra, S.; Argyle, D.J. Effective Penetration of a Liposomal Formulation of Bleomycin through Ex-Vivo Skin Explants from Two Different Species. *Cancers* **2022**, *14*, 1083. https://doi.org/10.3390/cancers14041083

Academic Editor: David J. Waxman

Received: 12 December 2021
Accepted: 14 February 2022
Published: 21 February 2022

Copyright: © 2022 by the authors. Licensee MDPI, Basel, Switzerland. This article is an open access article distributed under the terms and conditions of the Creative Commons Attribution (CC BY) license (https://creativecommons.org/licenses/by/4.0/).

1. Introduction

In both veterinary and human medicine there is an unmet clinical need for effective anti-cancer drugs that can be applied topically. Notably, the standard treatment of non-melanoma skin cancers (NMSC), including basal cell carcinoma and squamous cell carcinoma, are surgical modalities that can cause severe pain, inflammation, and scarring [1,2]. These side-effects can be significant given that the tumors may be spread over large areas of the body, therefore anti-cancer drugs that can be applied topically would be therapeutically and cosmetically valuable by reducing surgical costs and by avoiding undesirable scarring. However, a major challenge in developing these non-invasive modalities is the degree of penetration of the drug through the skin in sufficient quantities to target cancer cells [3].

Bleomycin belongs to a family of glycopeptide antibiotics and is used primarily as a cytotoxic chemotherapy drug. In human medicine it is an effective treatment against squamous cell carcinoma, malignant lymphomas, testicular cancers, germinal cell tumors, AIDS-associated Kaposi sarcoma and osteosarcoma [4], and is commonly used as part of combination chemotherapy regimens, as its toxicity does not overlap with that of other anti-cancer drugs [5]. The proposed mechanism of action of bleomycin is the inhibition of DNA synthesis by causing DNA single and double strand breaks [4]. At high concentrations, cellular RNA and protein synthesis are also suppressed [6,7]. Generally, bleomycin administration is via the parenteral route, as oral bioavailability is limited [8]. Bleomycin toxicity is dose-dependent and the most common adverse side-effect is pulmonary inflammation, which has been reported to develop into fatal lung fibrosis in 1% to 3% of patients [9]. In the case of NMSCs, topical application of bleomycin may avoid these systemic side-effects, but because bleomycin itself is a large hydrophilic molecule with a polar charge, it is unable to penetrate lipophilic skin barriers and efficiently diffuse across the plasma membrane to reach its site of action [4,10]. To overcome these issues, bleomycin has been administered by electrochemotherapy, which uses short electric impulses to transiently permeabilize the cell membrane allowing delivery of non-permeant drugs to the interior of the cell. This method has been successful in the management of skin tumors in both humans [11] and companion animals [12], and does reduce the major systemic side-effects of bleomycin but can also cause pain, muscle contractions, and discoloration of the area treated. Patients also generally undergo a general or local anesthetic, which, especially in veterinary patients, can be distressing for patients and owners.

In this study we have utilized a novel formulation of bleomycin called Bleosome, which negates a number of adverse side-effects of bleomycin administered systemically or by electrochemotherapy and does not require either local or general anesthetic. Bleosome consists of bleomycin encapsulated within ultra-deformable (UD) liposomes [13]. Liposomes have been heralded as 'magic bullets', as they are able to deliver a chemotherapy payload directly into cancer cells [14]. Liposomes have a strong tendency to form in aqueous solutions and are composed of spherical phospholipid bilayers that enclose an aqueous internal compartment which can be loaded with desired cargo [15]. The proposed benefits of carrier-mediated drugs include greater solubility, longer duration of exposure, selective delivery of the drug to the site of action, enhanced therapeutic index and the potential to overcome resistance associated with the free-form of the drug. Unfortunately, conventional liposomes cannot effectively penetrate through the skin barrier due to their rigidity [16]. However, UD liposomes, which include an edge activator, usually a single chain surfactant added to the phospholipid bilayer, can pass through the skin via pores that are much smaller in diameter than their own diameter by undergoing rapid changes in shape and fusing with lipids [17]. Skin is a complex organ consisting of a stratified cellular epidermis, an underlying dermis of connective tissue, and beneath the dermis there is a layer of subcutaneous fat separated from the rest of the body by a vestigial layer of striated muscle [18]. Beyond initial penetration of the skin, the mechanism by which UD liposomes improve drug delivery is yet to be fully elucidated. Two mechanisms have been proposed: (1) UD liposomes act as drug carrier systems, where intact UD liposomes enter the stratum corneum carrying the encapsulated drug deeper into the skin [19]; or (2) UD liposomes act as penetration enhancers, where liposome bilayers enter the stratum corneum and modify the intracellular lipid lamellae, releasing the encapsulated drug and enabling the penetration of the free drug deeper into the skin [20,21]. Here we show that

Bleosome can penetrate further through ex vivo skin explants than free bleomycin in both canine and equine species. To determine the mechanism by which UD liposomes improve drug delivery, we utilized a combination of transmission electron microscopy (TEM) and multiphoton microscopy (MP) to visualize the UD liposomes and fluorescently labeled drug, respectively. We found that UD liposomes act as penetration enhancers: liposomes are visible throughout stratum corneum but not beyond, whereas the fluorescently labeled drug is seen within the stratum corneum as dense spots as the drug is enclosed with the liposome, but in the deeper layers of the epidermis the drug is more diffuse, indicating that it has been released from the UD liposome. This is the first study to show that this novel formulation of bleomycin can effectively penetrate through the skin of dogs and horses highlighting its clinical potential.

2. Materials and Methods

2.1. Cell Culture

A canine melanoma cell line, CML10 (a kind gift from Dr Lauren Wolfe, Auburn University) was grown in Dulbecco's modified Eagle's medium (DMEM) low glucose supplemented with 10% foetal bovine serum (FBS) and 1% penicillin/streptomycin (ThermoFisher Scientific, Waltham, MA, USA). A canine mastocytoma cell line, C2 (a kind gift from Dr Richard Elders, University of Edinburgh) was grown is Eagle DMEM (ThermoFisher Scientific) supplemented with 5% FBS, 1% non-essential amino acids, 1% Glutamax and 25 mg of gentamicin. All cell cultures were maintained at 37 °C in a humidified 5% CO_2 incubator.

2.2. Stock Solutions of Bleosome and Bleomycin

Bleosome and bleomycin were provided at 0.2% in a cream. Both formulations were manufactured by Ascot Laboratories Ltd. (London, UK) and were a kind gift from SPS Animal Care Ltd. (London, UK). All compounds were stored at 4 °C.

2.3. Cell Viability

Cells were seeded in triplicate in opaque 96-well plates (Corning, Glendale, AZ, USA) at 500 cells/well. Cells were incubated 37 °C, 5% CO_2 for 24 h prior to being treated with the indicated dose of Bleosome or bleomycin. Cell viability was assayed 48 h after treatment using the CellTiterGlo® Luminescent Cell Viability Assay (Promega, Madison, WI, USA) according to the manufacturer's instructions. Luminescence was recorded using a Viktor3 luminometer (PerkinElmer, Waltham, MA, USA). Data were averaged and then normalized against the average signal of untreated/negative control samples.

2.4. Fluorescent Labelling of Bleomycin

We coupled bleomycin sulfate (A2–B2) (120 mg) to the green fluorophore boron-dipyrromethene (BODIPY-FL). This reaction was carried out twice: (1) at a ratio of 0.7 BODIPY-FL NHS (8.1 mg): 1 bleomycin sulfate (45 mg); and (2) at a ratio of 0.8 BODIPY-FL NHS (14.6 mg): 1 bleomycin sulfate (71 mg). Both compounds were dissolved in DMSO, then BODIPY-FL NHS solution (~5 mg/mL) was slowly added dropwise to the bleomycin solution (~20 mg/mL) over 30 min with a Pasteur pipette. The coupling reaction was monitored by high-performance liquid chromatography/mass spectrometry (HPLC/MS) at 30 min, 1.5 h and 3 h and allowed to continue overnight to achieve complete conversion. Two batches displayed a very similar reaction profile and an ~85% final purity of the crude. Upon reaction completion, the two batches were combined in a round bottom flask to give a 5 mL solution of crude labeled bleomycin in DMSO, which was diluted with water (25 mL), snap-frozen and then freeze-dried overnight to remove the solvents. The resulting red solid was then triturated in dichloromethane (DCM) in order to remove impurities, notably succinimide-OH, which is released by the BODIPY-FL NHS upon coupling, and additional fluorescent byproducts. Next, 5 mL of DCM was added and incubated until all undissolved solid material settled at the bottom of the flask (approximately after 15 min). The supernatant containing impurities and side-products dissolved in DCM was removed by means of a syringe. This process was repeated five times until the supernatant was colorless. This final product was re-dissolved in water (10 mL), transferred in a vial, snap-frozen and then freeze-dried overnight again, resulting in 121 mg of a fine, red powder that was characterized by HPLC/MS. The final product was sent at 4 °C, in the dark, to be encapsulated into UD liposomes and formulated into a 0.2% cream (F-Bleosome). A small amount was retained as an equivalent non-encapsulated fluorescently labeled bleomycin (F-bleomycin).

2.5. Skin Sample Preparation

Canine and equine skin samples were collected from cadavers donated to the Royal (Dick) School of Veterinary Studies (RDSVS) at The University of Edinburgh with consent to be used for teaching and/or research. Specimens were taken from the shaved flank and abdomen of cadavers, and stretched taut by attaching to a cork block; subcutaneous fat was removed using a scalpel. Each sample was then immediately wrapped in aluminum foil, sealed in a plastic bag, snap-frozen and stored at −20 °C. The Veterinary Ethical Review Committee of The University of Edinburgh deemed that this study raised no ethical concerns.

2.6. Transmission Electron Microscopy

Skin samples were thawed at RT in PBS and cut to 2 × 2 cm^2 sections. Each skin sample was treated with the indicated dose of drug for the indicated time. The appropriate sample was then cut into 1 cm thick slices that were fixed in 3% glutaraldehyde in 0.1 M sodium cacodylate buffer, pH 7.3 for 2 h then washed 3 × 10 min in 0.1 M sodium cacodylate buffer, pH 7.3. Samples were then post-fixed in 1% osmium tetroxide, 0.1 M sodium cacodylate for 45 min at RT, then washed 3 × 10 min in a 0.1 M sodium cacodylate buffer. Samples were then dehydrated in 50%, 70%, 90% and 100% ethanol for 3 × 15 min each followed by 2 × 10 min in propylene oxide and then embedded in TAAB812 resin. Ultrathin sections were cut using a Lecia Ultracut S ultramicrotome at a thickness of 60 nm, stained with 5% uranyl acetate and 2% lead citrate and imaged on a JEOL JEM-1400 TEM (JEOL, Ltd., Akishima, Japan) equipped with Gatan OneView camera (Gatan, Pleasanton, CA, USA) at King's Buildings at The University of Edinburgh.

2.7. Multiphoton Microscopy

Skin samples were thawed at RT in PBS and cut to 2 × 2 cm^2 sections. Each skin sample was treated with the indicated dose of drug for the indicated time. Each skin section was initially stained on both sides with Hoechst 33342 (Thermofisher Scientific) at 0.02 mg/mL in PBS by incubating for 30 min in the dark at RT on both sides. Skin samples were then washed in PBS and treated with the indicated dose of drug for the indicated time by spreading 0.1 mL cream evenly on the epidermal surface. Care was taken to avoid treating the sides of the section. After the indicated time point the residual cream was removed with a cotton swab and the sample was washed with PBS. All samples were positioned in 35 × 10 mm petri dishes (gridded Nunclon™ Delta, Thermofisher Scientific) with the epidermis side anchored to the bottom of the dish by 4% agarose with the subcutaneous side facing upward, toward the microscope lens. The dish with the sample was filled with PBS and positioned in the visualising plate of the MP. All samples were imaged with a Zeiss (Cambridge, UK) LSM7 MP intravital multiphoton microscope equipped with W Plan-Apochromat 20×/1.0 DIC D = 0.17 M27 75 mm objective lens and a Coherent chameleon ultraII pulsed laser with OPO (wavelength 690 to 1400 nm). Green (NDD R4 ET535/30) and blue (NDD R3 ET460/36) filters were used. Multidimensional acquisition was performed with the Z-stack tool, setting 150 slices as average with an interval of approximately 3 μm. 3D images were visualised, processed and interpreted using Imaris (Biplane) software (version 9.1.2, Oxford Instruments, Abingdon, UK).

2.8. Quantification

Imaris 9.1.2 (Bitplane) software (Oxford Instruments, Abingdon, UK) was used to analyze 3D images acquired by MP. Structures labelled with Hoechst 33342 staining that did not emit or emitted a low green fluorescent signal (including a majority of auto-fluorescent skin appendages such as connective tissue) were considered as "surfaces" ("surfaces" had NDD R3-blue source channel), while all the elements emitting high levels of the green signal, either labelled with Hoechst 33342 (such as cellular nuclei) or not (F-Bleosome and F-bleomycin), were labelled as "spots" ("spots" had NDD R4-green source channel). After this distinction, the program distinguished between the "spots far from surfaces" (green particles emitting high- intensity green fluorescence far from the particles emitting blue and low-intensity green fluorescence), from the "spots close to surfaces" (green particles emitting high-intensity green fluorescence, and blue fluorescence very close to auto-fluorescent skin and cellular structures, emitting low-intensity green fluorescence), using a specific threshold (0.5). The aim was to target the green particles that were far from the auto-fluorescent skin structures and that were not stained with Hoechst 33342 (cellular nuclei were always close to intracellular auto-fluorescent elements and thus recognised as "spots close to surfaces"). This system enabled us to identify F-bleomycin and F-Bleosome particles within the examined skin section, recognised as green emitting particles far enough from auto-fluorescent structures ("spots far from surfaces"). All images were visualised from the subcutaneous side of the skin section. Imaris automatically compensated for the inversion and depths were compared to fluorescent images. In the equine skin samples, there was a discrepancy between the computer compensated values and the images, and here we manually compensated for this inversion by making all particles relative to the deepest particle in each image.

2.9. Statistics

Data were analysed using Minitab® 17 Statistical Software (Minitab Ltd., Coventry, UK). p-values < 0.05 were considered statistically significant. When data followed a normal distribution, two-sample t-tests and Pearson's chi-squared tests were used to compare differences between two samples, or one-sample t-tests to determine whether the sample mean was statistically different from a known or hypothesised mean. A general linear model (2-way ANOVA- analysis of variance) was used to understand if in two groups there was an interaction between two independent variables on the dependent variable. A Fisher's post-hoc test was used to compare differences between time points post-treatment.

3. Results

3.1. Preparation of BODIPY-Labeled Bleomycin for Fluorescent Imaging of Bleosome through the Skin

In this study we were assessing if a novel formulation of bleomycin encapsulated in UD liposomes penetrates through the outer layers of the skin better than the free drug. Fluorescent labeling of a drug is a valuable tool to allow visualization within an enclosed biological system, and in this case to quantify the degree of penetration through the skin. When preparing fluorescent versions of drugs, labels must be introduced at appropriate amounts and positions so they do not interfere with activity. Here, we aimed to produce mono-labeled bleomycin as we considered that this would be less disruptive to the pharmacological activity of the drug than the incorporating of multiple labels (Figure 1A).

Preliminary experiments were carried out to determine the optimum fluorophore to bleomycin ratio to produce mono-labeled bleomycin. The experimental procedure is detailed in materials and methods and highlighted in Figure 1B. The coupling reaction was monitored by HPLC/MS (Figure 1C). The final compound contained 88% mono-labeled and 12% free-bleomycin (Figure 1C(vi)), and was subsequently encapsulated into UD liposomes to produce fluorescent Bleosome (F-Bleosome) and the equivalent un-encapsulated control (F-bleomycin).

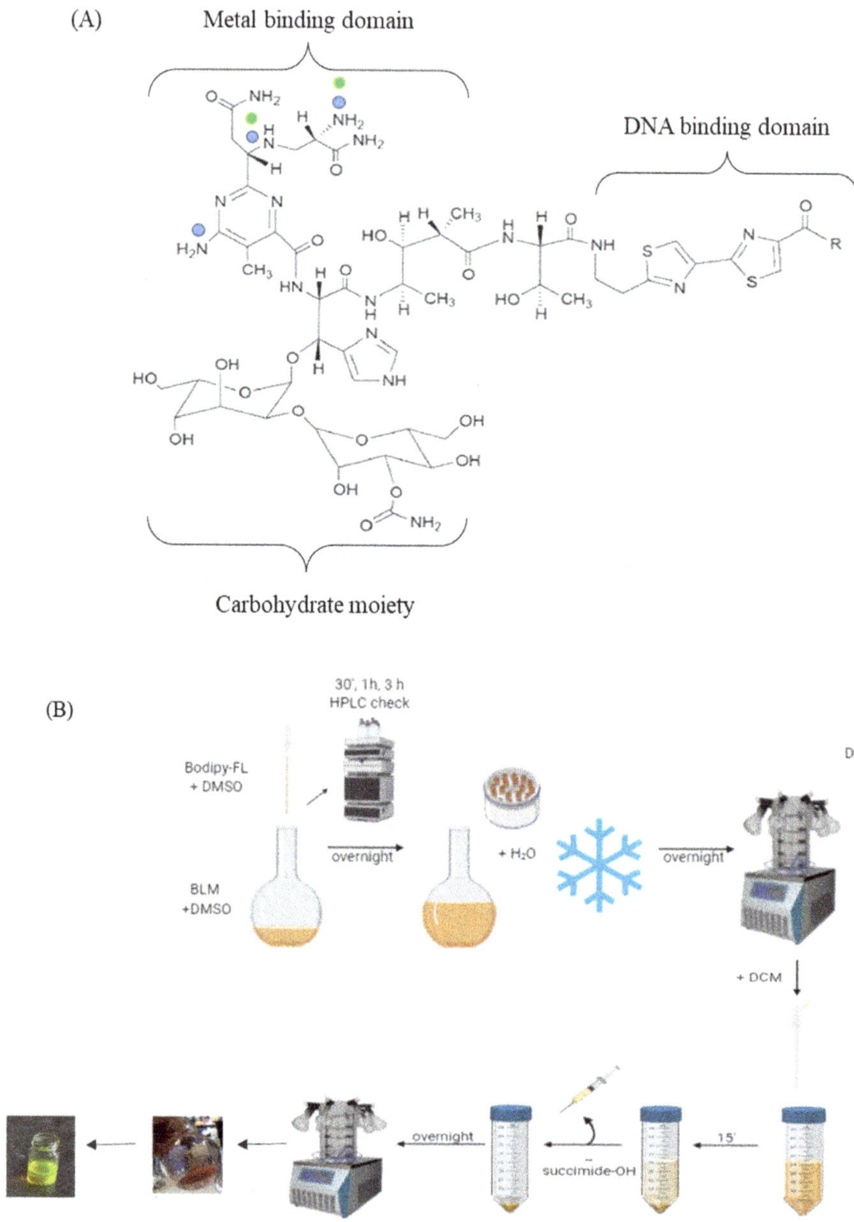

Figure 1. *Cont.*

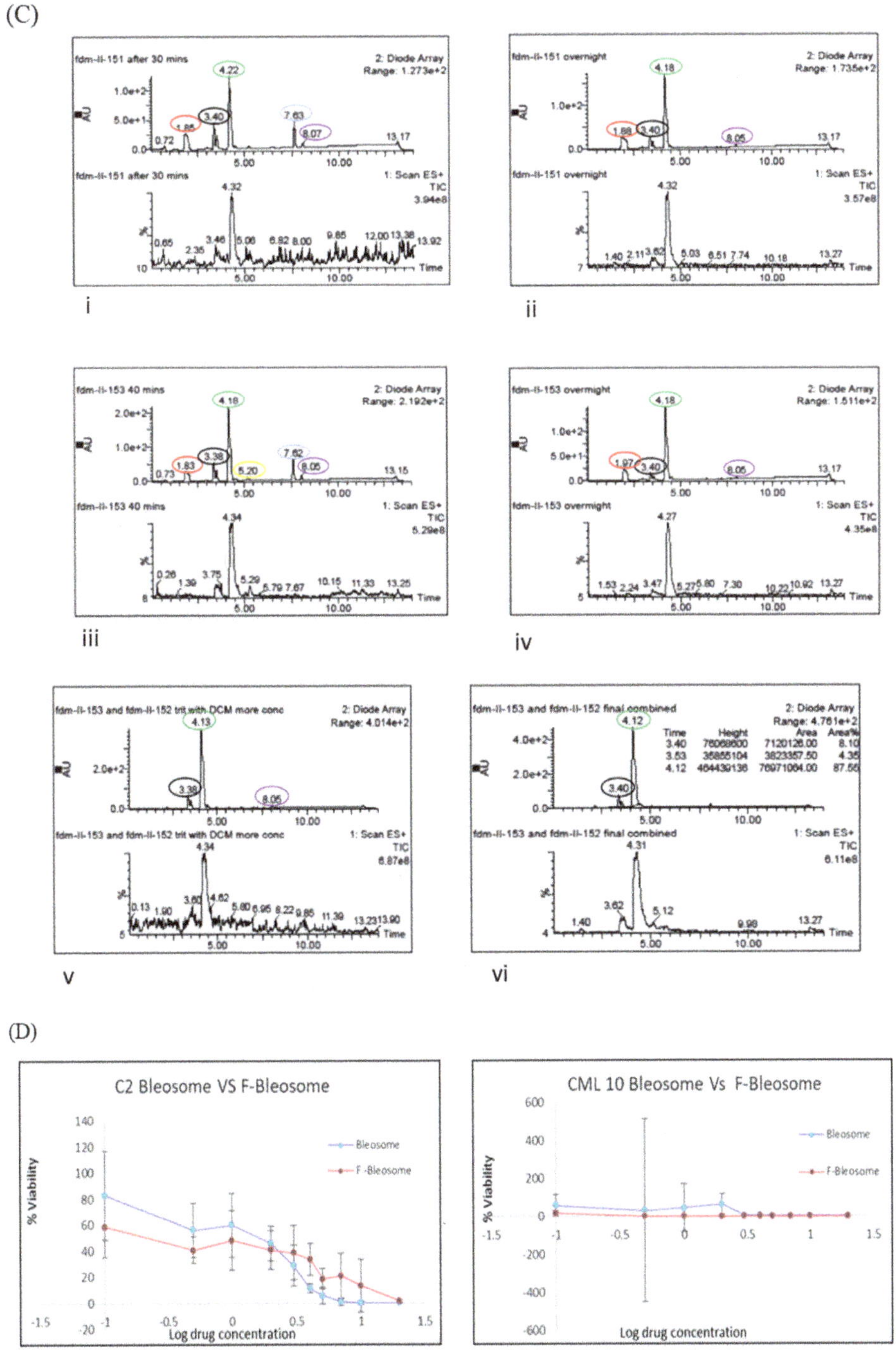

Figure 1. Fluorescent labeling of bleomycin with BODIPY-FL. (**A**) Molecular structure of bleomycin showing the metal binding domain with the potential amine groups available for BODIPY-FL binding highlighted (dots: blue dots indicate all potential binding sites; and green dots indicate preferred

binding sites); the DNA binding domain encompassing the bithiazole tail; and the carbohydrate moiety required for cellular uptake. (**B**) Schematic representation of the labeling process of bleomycin with BODIPY-FL. The reaction products of BODIPY-FL NHS added to bleomycin sulfate were analyzed by HPLC/MS and recovered from DMSO by lyophilization. Succinimide-OH and fluorescent impurities were removed by trituration with DCM. The compound was dried overnight and the final product obtained was a red powder (which turns fluorescent green in solution). (**C**) HPLC profiles of coupling of BODIPY-FL to bleomycin. Using 45 mg bleomycin after (**i**) 30 min and (**ii**) overnight. Using 71 mg of bleomycin after (**iii**) 30 min and (**iv**) overnight. Combined labelled bleomycin batches (**v**) before and (**vi**) after the final lyophilization step. According to their absorbance profile and mass spectra: peak 1.83 (red circle) corresponds to succinimide-OH; peak 3.38 (black circle) to free bleomycin; peak 4.22–4.12 (green circle) corresponds to the mono-labeled bleomycin; 5.20 (yellow circle) corresponds to double-labeled bleomycin; peak 7.63 (blue circle) corresponds to free (unbound) BODIPY-FL; 8.05 (violet circle) corresponds to a minor impurity of the fluorophore. (**vi**) the final compound consisting of 87.55% mono-labeled bleomycin and 12.45% free bleomycin. (**D**) F-Bleosome has comparable cytotoxic effects to unlabeled Bleosome. The cancer cell lines C2 (**i**) and CML10 (**ii**) cells were treated with the indicated concentrations of Bleosome and F-Bleosome, and cell viability was assayed after 48 h.

3.2. *F-Bleosome Retains the Same Efficacy as Unlabeled Bleosome*

To determine if the coupling of BODIPY-FL to bleomycin impaired the cytotoxicity of the drug, a canine mastocytoma cell line (C2) and a canine melanoma cell line (CML10) were treated with increasing doses of either F-Bleosome or Bleosome and cell viability was assayed 48 h after treatment (Figure 1D). There was no significant difference between the killing effect of F-Bleosome and Bleosome, indicating that the fluorophore coupled to bleomycin did not affect the cytotoxicity of the drug. These results were consistent in both cell lines tested.

3.3. *Liposomes Do Not Penetrate beyond the Stratum Corneum*

Canine skin explants were treated with 0.1 mL of Bleosome for 0 (vehicle control), 0.5, 2, 4, 6 and 8 h, fixed and visualized by TEM. All the epidermis was visible including the layers of the stratum corneum. Within the 0 h control, no liposomes were observed in any of the epidermal structures (Figure 2A), whereas liposomes were visible in all skin explants treated with Bleosome over the time course (Figure 2B–F). Liposomes do penetrate through the corneocytes of the stratum corneum but do not penetrate deeper over time and were not identified beyond the external keratinocyte layer (Figure 2B–F).

3.4. *Bleosome Penetrates Further through Skin Than Bleomycin*

Canine and equine skin sections were treated with 0.1 mL of either F-Bleosome or F-bleomycin for 0 h, 10 min, 4 h and 6 h and visualized by MP. The skin has several structures that are inherently auto-fluorescent such as collagen, hair follicles and cellular nuclei. All images were compared to the 0 h control to establish background fluorescence. Hoechst staining was used to stain nuclei blue, however the intensity of the blue was similar to the green of BODIPY-FL, to differentiate we set the 'blue' of Hoechst 33342 to 'magenta' during image acquisition. At the image analysis stage, in the merged image, when magenta and green overlap it appears as white. Hoechst 33342 does not stain liposomes or bleomycin, therefore objects appearing green are considered as the drug.

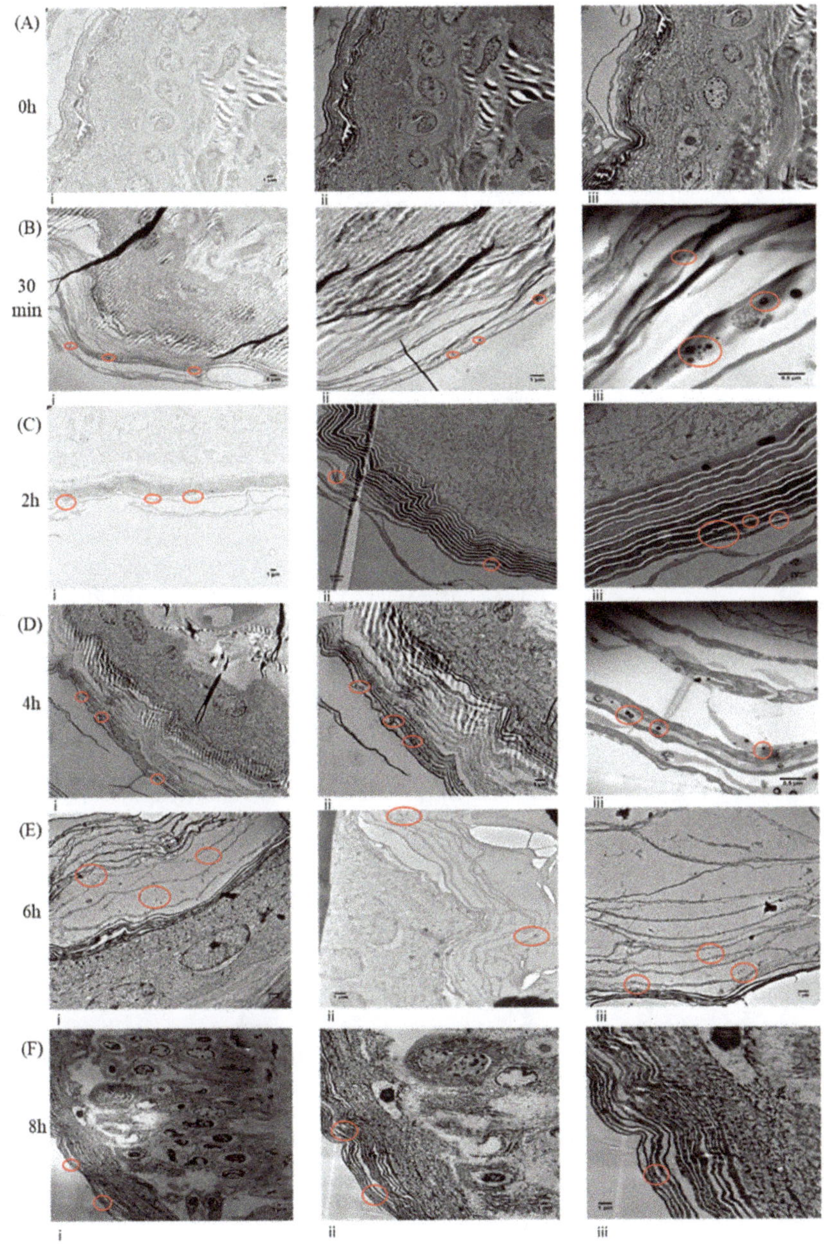

Figure 2. Transmission electron microscopy visualization of liposomes through canine skin at different time points after a single Bleosome administration. (**A**) Skin samples were treated for 0 h (control) with Bleosome, visualized at different magnification: (**i**) ×1200, (**ii**) ×1200, (**iii**) ×1200; (**B**) images of samples treated for 30 min with Bleosome: (**i**) ×1200, (**ii**) ×5000, (**iii**) ×10K; (**C**) images of samples treated for 2 h: (**i**) ×500, (**ii**) ×2500, (**iii**) ×5000; (**D**) images of samples treated for 4 h: (**i**) ×1200, (**ii**) ×2500 K, (**iii**) ×10 K; (**E**) images of samples treated for 6 h: (**i**) ×2500, (**ii**) ×2500, (**iii**) ×2500; (**F**) images of skin samples after 8 h of Bleosome treatment: (**i**) ×500, (**ii**) ×1200, (**iii**) ×2500. Representative liposomal nanoparticles are circled in red within the keratinocytes of the stratum corneum only.

Treated canine skin samples are shown from the side-view and the top-view in Figure 3A,B, respectively. The untreated control did not display any distinct solid green particles but there was a basal level of auto-fluorescence, mainly due to cutaneous structures, that were also stained by Hoechst and appeared white on the merged image (Figure 3A(i),B(i)). Distinct round green fluorescent dots, corresponding to fluorescent BODIPY-FL-labeled bleomycin encapsulated in UD liposomes were visible at a superficial level in the skin sample 10 min after treatment with F-Bleosome (Figure 3A(ii),B(ii)). Over time F-Bleosome particles penetrate deeper through the skin, at a higher overall number and increased in size with a less defined spherical shape (Figure 3A(iv,vi,vii),B(iv,vi,vii). In comparison, the fluorescently labelled free drug, F-bleomycin, is not readily detectable in the deeper layers of the skin and there is no notable change over treatment time of number, size or depth of F-bleomycin (Figure 3A(iii,v,vii),B(iii,v,vii). Imaris software was used to identify and quantify the number of fluorescent particles within each sample. Calibration of the software, as detailed in materials and methods, failed to identify the fluorescently labelled drug located close to auto-fluorescent skin structures, leading to a consistent underestimation of the overall number of fluorescent particles. Skin sections treated with F-Bleosome contained a significantly higher number of particles at all time points than those treated with F-bleomycin (Table 1, p-value 0.002). Ten minutes post-treatment F-Bleosome and F-bleomycin particles were found at a similar superficial level within the skin (Figure 3C(i)), but by 4 h and 6 h post-treatment F-Bleosome particles were identified significantly deeper (at an average depth of 307 μm and 452 μm, respectively) than F-bleomycin (at an average depth of 167 μm and 237 μm, respectively) (Figure 3C).

Table 1. Number of particles detected within canine skin treated with either F-Bleosome of F-bleomycin after 10 min, 4 h and 6 h.

Time Point	Treatment	Average Number of Particles (n = 3)
10 min	F-Bleosome	87
4 h	F-Bleosome	85
6 h	F-Bleosome	618
10 min	F-bleomycin	8
4 h	F-bleomycin	34
6 h	F-bleomycin	178

Equine skin showed similar results to canine skin whereby more F-Bleosome particles penetrated through the skin and to a greater depth than F-bleomycin at the indicated time points (Figure 4A,B). Using Imaris software we determined that the overall number of particles was significantly higher at each time point for those skin samples treated with F-Bleosome compared to F-bleomycin (Table 2, p-value 0.000) and significantly deeper at 4 h and 6 h post-treatment (at an average depth of 117 μm and 166 μm, respectively, for F-Bleosome compared to 67 μm and 65 μm, respectively, for F-bleomycin) (Figure 4C). Our results confirm that UD liposomes are effective drug carriers of bleomycin through the canine and equine skin.

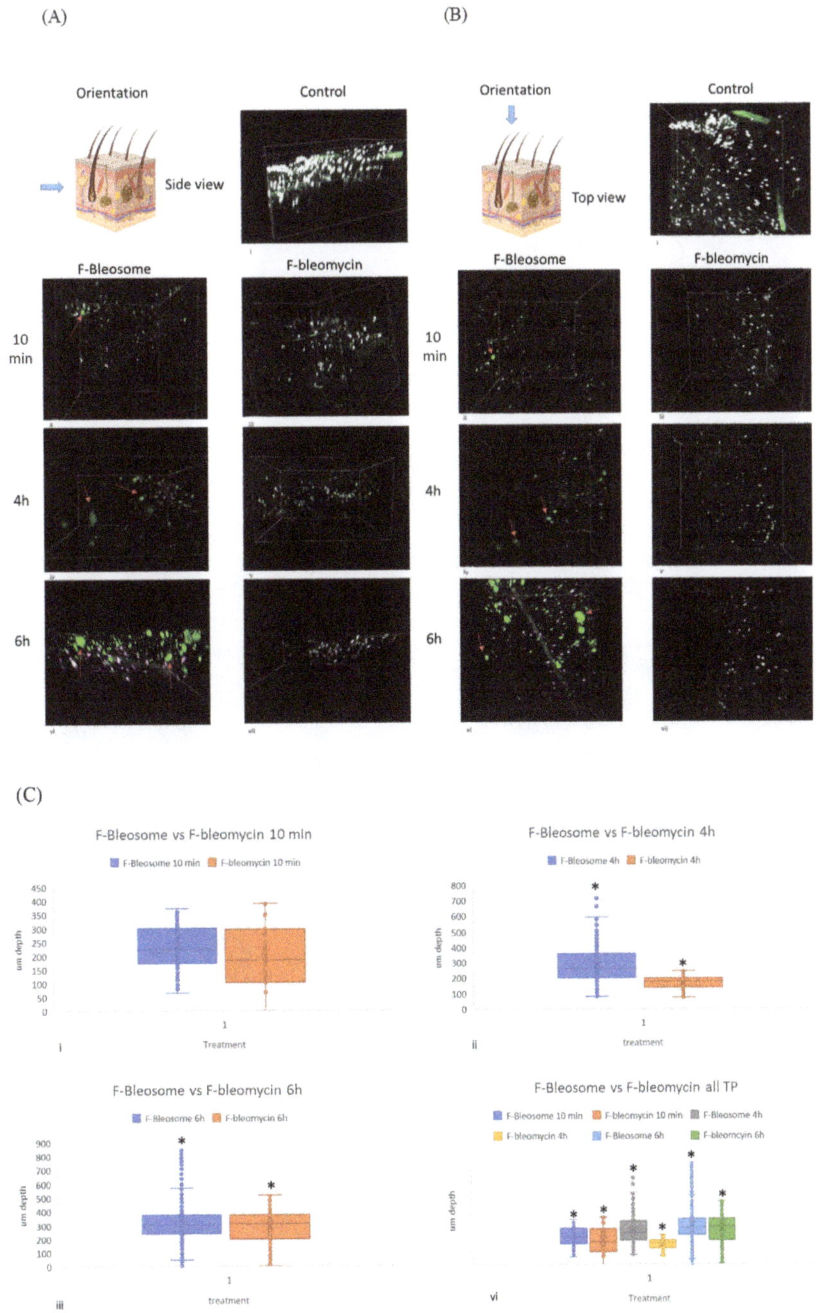

Figure 3. Bleosome penetrates deeper into canine skin than free-bleomycin. (**A**) Side view of canine skin sections treated with F-Bleosome and F-bleomycin at the indicated time points. (**i**) control (untreated skin section); (**ii**,**iv**,**vi**) correspond to the 3-D images of the skin sections treated with F-Bleosome after 10 min, 4 h and 6 h. Representative F-Bleosome molecules are indicated by red arrows. (**iii**,**v**,**vii**) correspond to the side view of the 3-D sections after 10 min, 4 h and 6 h of F-bleomycin treatment. (**B**) Top view of corresponding canine skin sections. (**i**) control (untreated skin

section); (**ii,iv,vi**) correspond to the 3-D images of the skin sections treated with F-Bleosome after 10 min, 4 h and 6 h. Representative F-Bleosome molecules are indicated by red arrows. (**iii,v,vii**) correspond to the side view of the 3-D sections after 10 min, 4 h and 6 h of F-bleomycin treatment. (**C**) Box plots comparing the depth of the drug molecules (μm) through the canine skin sections treated with F-Bleosome (blue plot) and F-bleomycin (orange plot) after (**i**) 10 min, (**ii**) 4 h and (**iii**) 6 h. (**iv**) Comparison of all three time points. * Indicates statistically significant difference (p-value < 0.05, two-way ANOVA).

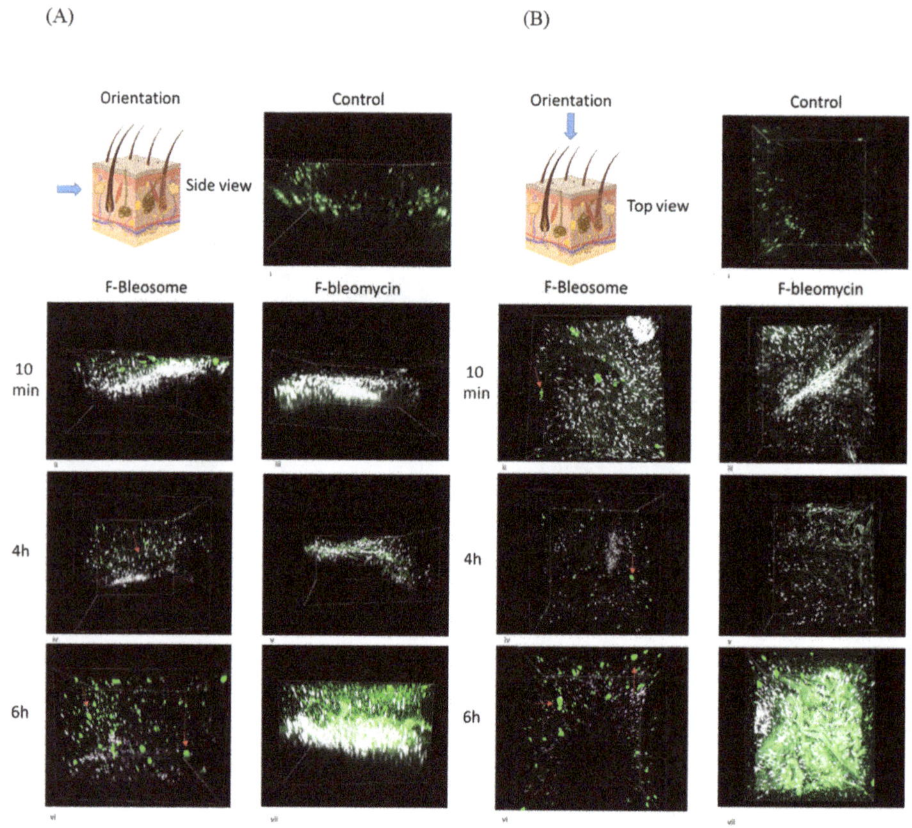

Figure 4. *Cont.*

(C)

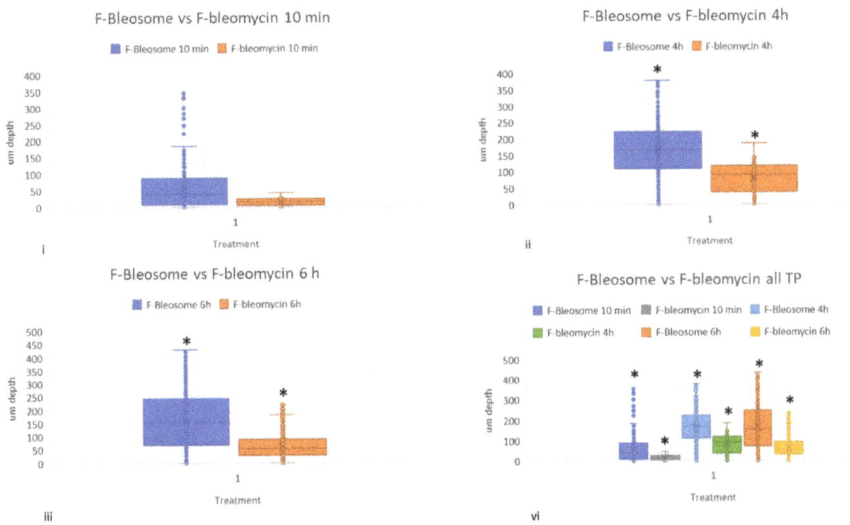

Figure 4. Bleosome penetrates deeper into equine skin than free-bleomycin. (**A**) Side view of equine skin sections treated with F-Bleosome and F-bleomycin at the indicated time points. (**i**) control (untreated skin section); (**ii**,**iv**,**vi**) correspond to the 3-D images of the skin sections treated with F-Bleosome after 10 min, 4 h and 6 h. Representative F-Bleosome molecules are indicated by red arrows. (**iii**,**v**,**vii**) correspond to the side view of the 3-D sections after 10 min, 4 h and 6 h of F-bleomycin treatment. (**B**) Top view of corresponding equine skin sections. (**i**) control (untreated skin section); (**ii**,**iv**,**vi**) correspond to the 3-D images of the skin sections treated with F-Bleosome after 10 min, 4 h and 6 h. Representative F-Bleosome molecules are indicated by red arrows. (**iii**,**v**,**vii**) correspond to the side view of the 3-D sections after 10 min, 4 h and 6 h of F-bleomycin treatment. (**C**) Box plots comparing the depth of the drug molecules (μm) through the equine skin sections treated with F-Bleosome (blue plot) and F-bleomycin (orange plot) after (**i**) 10 min, (**ii**) 4 h and (**iii**) 6 h. (**iv**) Comparison of all three time points. * Indicates statistically significant difference (p-value < 0.05, two-way ANOVA).

Table 2. Number of particles detected within equine skin treated with either F-Bleosome of F-bleomycin after 10 min, 4 h and 6 h.

Time Point	Treatment	Average Number of Particles (n = 3)
10 min	F-Bleosome	120
4 h	F-Bleosome	305
6 h	F-Bleosome	500
10 min	F-bleomycin	4
4 h	F-bleomycin	37
6 h	F-bleomycin	412

4. Discussion

Bleomycin is an anti-neoplastic antibiotic used in treatment regimens to treat several types of human cancer including cervical and uterine cancer, squamous cell carcinoma, testicular and penile cancer and certain types of lymphoma [22–24]. The clinical use of bleomycin as a stand-alone therapy is limited by its toxicity to the lung causing irreversible pulmonary damage and fibrosis in approximately one fifth of all patients [25]. However, bleomycin has potential as a topical treatment in both human and veterinary medicine for diseases such as NMSCs and equine sarcoids, due to its low toxicity against non-cancer cells and hematopoietic tissues [26,27], however its molecular structure impedes penetration through the skin [4]. The application of nanotechnology to overcome biological hurdles is overhauling biomedical research and clinical applications. Here, we utilized a novel formulation of bleomycin encapsulated in UD liposomes (Bleosome).

Bleosome has previously been shown to have a lethal effect on an immortalized human keratinocyte cell line and a cell line derived from a primary squamous cell carcinoma [13] and has potential as a clinically relevant modality for horses with occult and/or verrucose sarcoids [28]. However, to date there has been limited evidence that Bleosome can effectively penetrate through the skin compared to free-bleomycin and limited insight into the mechanism of the UD liposomes' enhanced skin penetration. In this study, we showed for the first time that Bleosome effectively penetrates deeper into the skin than free-bleomycin, and that these results are consistent in two model systems: canine and equine. We visualized UD liposomes approximately 150 nm in diameter using TEM. TEMs are powerful analytical tools that transmit a beam of electrons through a thin slice of a tissue to capture the very fine details of that specimen; they are of a magnification of up to thousands of times higher than that of a light microscope [29]. UD liposomes did not penetrate beyond the external keratinocyte layer of canine skin even after 8 h of treatment. Similar results were observed by Bouwstra et al. (2003) [30], whereby intact surfactant-based elastic vesicles were not found beyond the stratum corneum of human skin.

To complement the TEM, we fluorescently labelled bleomycin prior to encapsulation in UD liposomes to enable us to visualize the drug by MP (Figure 5A). BODIPY-FL was selected, as it is relatively non-polar and non-charged fluorophore that should have minimum effect on the functional properties of bleomycin after conjugation [31]. To our knowledge this is the first time that BODIPY-FL has been conjugated to bleomycin, and our main concern was that the reactive succinimidyl group of BODIPY-FL NHS could potentially bind to all available amine groups in the metal binding domain of bleomycin, which may impair the DNA binding and cleavage and affect the efficacy of the fluorescently labeled drug [4]. We effectively mono-labeled bleomycin with BODIPY-FL prior to encapsulation in UD liposomes (F-Bleosome) and confirmed the comparable cytotoxic effect on cell lines compared to unlabeled Bleosome. Using F-Bleosome we were able to visualize the drug within the UD liposomes as it penetrated through the skin. We found that F-Bleosome penetrated beyond the keratinocyte layer and penetrated deeper over time compared to F-bleomycin, which was mainly retained in the superficial layers of the skin. The epidermal thickness of the skin is estimated to be approximately 40 μm and 30 μm for canine [32] and equine [33] skin respectively, and in both models F-Bleosome penetrated beyond the epidermis and was identified within the dermal layers of the skin. In some instances we also identified F-Bleosome particles in close proximity to structures that are likely to be small blood vessels, indicating that F-Bleosome can penetrate beyond the epidermis, an avascular component of the skin. We noted that the shape of F-Bleosome particles changed over time: 10 min post-treatment they are well demarcated and spheroidal whereas at later time points there were less clearly defined and more diffuse. Given that UD liposomes, as visualized by TEM, were not observed beyond the stratum corneum, we propose that UD liposomes act as penetration enhancers. Our model is consistent with that proposed by Verma et al. (2003) [20], whereby within the superficial layers of the skin Bleosome consists of bleomycin encapsulated within UD liposomes; once Bleosome passes beyond the stratum corneum, we hypothesize that the UD liposomes are damaged by their interaction with hardy keratinocytes, which are highly keratinized and flat, causing disruption of the liposomal membrane. These damaged UD liposomes then become leaky, and the previously encapsulated bleomycin is released and able to freely penetrate deeper into the cellular and vascular dermal layer of the skin (Figure 5B). We therefore propose that UD liposomes are penetration enhancers that can transport encapsulated drugs across the epidermis and release them in the deeper layers of the skin.

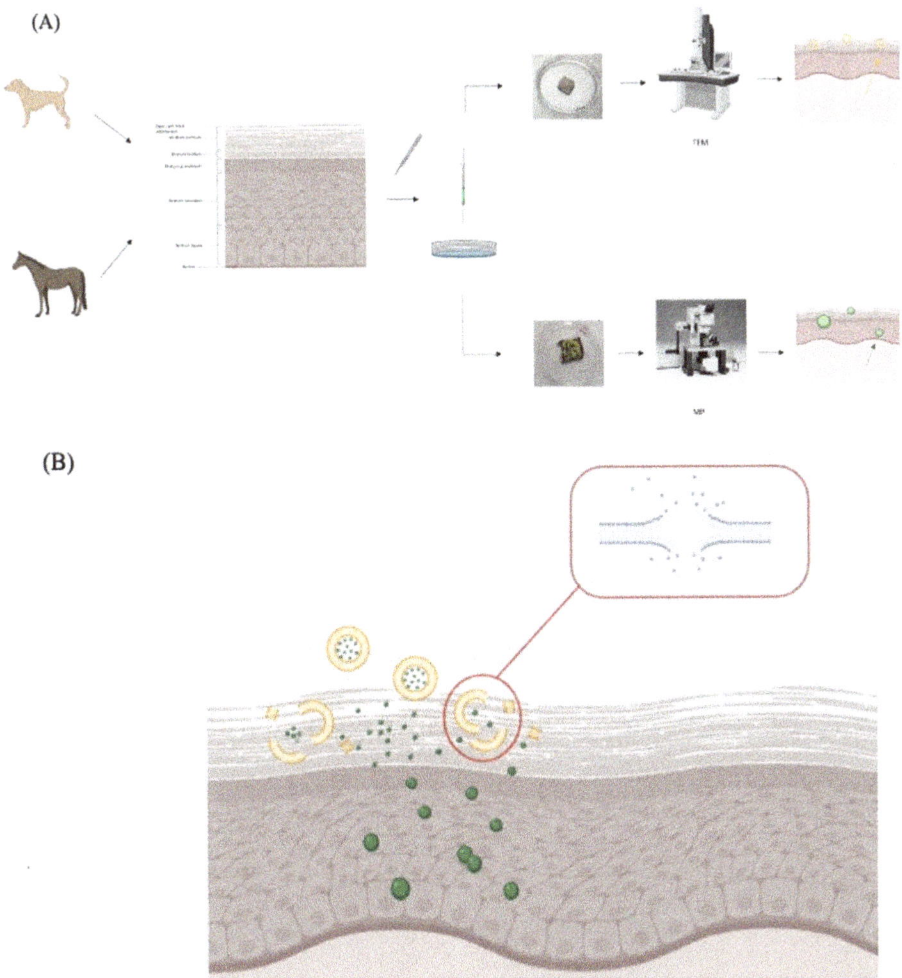

Figure 5. Proposed mechanism of penetration of Bleosome through the skin. (**A**) Schematic representation of the experimental techniques used to assess the penetration of Bleosome through the skin. Skin was collected from cadavers of dogs and horses. Subcutaneous fat was removed using a scalpel and hair (in the case of dogs and horses) was removed using electric clippers. Skin explants were cut into 2 × 2 cm^2 sections and treated with either Bleosome or free bleomycin and visualized by the TEM (top) for liposomes assessment (yellow arrow) or treated either with F-Bleosome or F-bleomycin and visualized by the MP (bottom) for drug particles assessment (green arrow). (**B**) UD liposomes act as penetration enhancers to increase the penetration of encapsulated bleomycin through animal and human skin. We propose that when liposomal vesicles interact with the outermost keratinocytes, they are disrupted, allowing the release of the entrapped bleomycin and consequently the enhanced penetration of the drug to the inner layers of the skin.

Our results support further studies to investigate the clinical application of Bleosome as a topical treatment for NMSCs. In both human and veterinary patients these types of tumors are usually not fatal, but their treatment represents a clinical challenge as they are often highly invasive and result in life-long disfiguration of the patient [1]. An effective topical treatment for these types of tumors would negate the need for patients to undergo disfiguring surgery and could be safely applied at home. Clinical use in human NMSCs has shown minimal side-effects and scarring (Chopra, S. personal communication). Within the context of veterinary medicine, topical treatments should be designed for rapid absorption to minimize companion animals from licking and swallowing any external treatments, which may be harmful if swallowed. Here we showed that Bleosome can more effectively penetrate through the healthy skin of dogs and horses, and can reach the deeper layers of skin better than the free drug. Future studies should focus on the rate of absorption, penetration through tumor samples, and efficacy against NMSCs. Two promising preliminary studies showed that Bleosome in combination with either 5-fluorouracil or tazarotene was more effective at treating horses with sarcoids than either treatment alone [28], and Bleosome alone has been used to treat humans with NMSC (Chopra, S. personal communication). The current study adds to a body of evidence that Bleosome is potentially an effective non-invasive treatment, with minimum side-effects, for NMSC in both human and veterinary medicine.

5. Conclusions

Innovation and novel formulations of established anti-neoplastic drugs have the potential to affect clinical practice and patient outcomes quicker than completely new drugs for which there is no prevailing pharmacological, pharmacokinetic or clinical data available. Bleosome is a novel formulation of bleomycin that has potential as a topical chemotherapy to treat NMSC and negate the adverse side-effects of pulmonary toxicity when administered systemically. In this study we show for the first time that Bleosome can effectively penetrate through the skin of dogs and horses better than the free drug which cannot pass beyond the stratum corneum. Using novel imaging techniques, we have also elucidated the mechanism of UD liposome enhanced skin penetration of bleomycin. From our results we conclude that Bleosome is able to penetrate through skin better than free-bleomycin as UD liposomes can penetrate through the protective, hardy, outer layers of skin which impair passage of the free-drug, and then once beyond the keratinocyte layer the UD liposomes break-up to release the drug, enabling it to diffuse deeper through the skin. This study provides a scientific underpinning to preliminary clinical observations that Bleosome may be an effective treatment of NMSCs.

6. Patents

S.C holds a patent (GB2398495) "A drug delivery preparation comprising at least one anti-tumor drug and a topical carrier for the drug" and has a patent pending for "Compounds, Compositions and Methods for the Treatment or Prevention of Hair Loss".

Author Contributions: Conceptualization, L.Y.P. and D.J.A.; methodology, G.F., L.Y.P., F.D.M. and M.V.; software, G.F.; validation, G.F.; formal analysis, G.F.; investigation, G.F.; resources, R.J.M.R. and D.J.A.; data curation, G.F.; writing—original draft preparation, L.Y.P.; writing—review and editing, all authors; visualization, G.F. and L.Y.P.; supervision, L.Y.P. and D.J.A. project administration, L.Y.P.; funding acquisition, D.J.A., A.J.H. and S.C. All authors have read and agreed to the published version of the manuscript.

Funding: D.J.A. acknowledges a charitable donation from SPS Animal Care Ltd. M.V. acknowledges funding from an ERC Consolidator Grant (DYNAFLUORS, 771443). TEM facility is supported by a Welcome Trust Multi User Equipment Grant (WT104915MA).

Institutional Review Board Statement: Not applicable.

Informed Consent Statement: Not applicable.

Data Availability Statement: The data presented in this study are available in article.

Acknowledgments: The authors would like to thank Rhona Muirhead, Stephen Mitchell and Bob Fleming for technical work and kind support throughout the research period.

Conflicts of Interest: S.C. and A.J.H. are directors of SPS Animal Care Ltd., which partially funded this project. Regular meetings were held to discuss the results. However, the funders had no role in design of the study; in the collection, analyses, or interpretation of data; in the writing of the manuscript, or in the decision to publish the results. The other authors declare no conflict of interest.

References

1. Caparrotti, F.; Troussier, I.; Ali, A.; Zilli, T. Localized non-melanoma skin cancer: Risk factors of post-surgical relapse and role of postoperative radiotherapy. *Curr. Treat. Options Oncol.* **2020**, *21*, 97. [CrossRef] [PubMed]
2. Lopez, R.F.; Lange, N.; Guy, R.; Bentley, M.V. Photodynamic therapy of skin cancer: Controlled drug delivery of 5-ala and its esters. *Adv. Drug Deliv. Rev.* **2004**, *56*, 77–94. [CrossRef] [PubMed]
3. Yang, M.; Gu, Y.; Tang, X.; Wang, T.; Liu, J. Advancement of lipid-based nanocarriers and combination application with physical penetration technique. *Curr. Drug Deliv.* **2019**, *16*, 312–324. [CrossRef] [PubMed]
4. Chen, J.; Stubbe, J. Bleomycins: Towards better therapeutics. *Nat. Rev. Cancer* **2005**, *5*, 102–112. [CrossRef] [PubMed]
5. Canellos, G.P.; Anderson, J.R.; Propert, K.J.; Nissen, N.; Cooper, M.R.; Henderson, E.S.; Green, M.R.; Gottlieb, A.; Peterson, B.A. Chemotherapy of advanced hodgkin's disease with mopp, abvd, or mopp alternating with abvd. *N. Engl. J. Med.* **1992**, *327*, 1478–1484. [CrossRef] [PubMed]
6. Hecht, S.M. Rna degradation by bleomycin, a naturally occurring bioconjugate. *Bioconjug. Chem.* **1994**, *5*, 513–526. [CrossRef] [PubMed]
7. Abraham, A.T.; Zhou, X.; Hecht, S.M. Metallobleomycin-mediated cleavage of DNA not involving a threading-intercalation mechanism. *J. Am. Chem. Soc.* **2001**, *123*, 5167–5175. [CrossRef] [PubMed]
8. Froudarakis, M.; Hatzimichael, E.; Kyriazopoulou, L.; Lagos, K.; Pappas, P.; Tzakos, A.G.; Karavasilis, V.; Daliani, D.; Papandreou, C.; Briasoulis, E. Revisiting bleomycin from pathophysiology to safe clinical use. *Crit. Rev. Oncol. Hematol.* **2013**, *87*, 90–100. [CrossRef] [PubMed]
9. Sleijfer, S. Bleomycin-induced pneumonitis. *Chest* **2001**, *120*, 617–624. [CrossRef]
10. Pron, G.; Mahrour, N.; Orlowski, S.; Tounekti, O.; Poddevin, B.; Belehradek, J., Jr.; Mir, L.M. Internalisation of the bleomycin molecules responsible for bleomycin toxicity: A receptor-mediated endocytosis mechanism. *Biochem. Pharmacol.* **1999**, *57*, 45–56. [CrossRef]
11. Zygogianni, A.; Kyrgias, G.; Scarlatos, J.; Koukourakis, M.; Souliotis, K.; Kouvaris, J.; Kelekis, N.; Kouloulias, V. Potential role of electrochemotherapy as anticancer treatment for cutaneous and subcutaneous lesions. *Asian Pac. J. Cancer Prev.* **2016**, *17*, 3753–3757.
12. Impellizeri, J.; Aurisicchio, L.; Forde, P.; Soden, D.M. Electroporation in veterinary oncology. *Vet. J.* **2016**, *217*, 18–25. [CrossRef]
13. Lau, K.G.; Hattori, Y.; Chopra, S.; O'Toole, E.A.; Storey, A.; Nagai, T.; Maitani, Y. Ultra-deformable liposomes containing bleomycin: In vitro stability and toxicity on human cutaneous keratinocyte cell lines. *Int. J. Pharm.* **2005**, *300*, 4–12. [CrossRef]
14. Utku, N. New approaches to treat cancer—What they can and cannot do. *Biotechnol. Healthc.* **2011**, *8*, 25–27.
15. Papahadjopoulos, D.; Cowden, M.; Kimelberg, H. Role of cholesterol in membranes. Effects on phospholipid-protein interactions, membrane permeability and enzymatic activity. *Biochim. Biophys. Acta* **1973**, *330*, 8–26. [CrossRef]
16. Bozzuto, G.; Molinari, A. Liposomes as nanomedical devices. *Int. J. Nanomed.* **2015**, *10*, 975–999. [CrossRef]
17. Cevc, G.; Blume, G. Lipid vesicles penetrate into intact skin owing to the transdermal osmotic gradients and hydration force. *Biochim. Biophys. Acta* **1992**, *1104*, 226–232. [CrossRef]
18. Bouwstra, J.A.; Ponec, M. The skin barrier in healthy and diseased state. *Biochim. Biophys. Acta* **2006**, *1758*, 2080–2095. [CrossRef]
19. Cevc, G.; Schatzlein, A.; Richardsen, H. Ultradeformable lipid vesicles can penetrate the skin and other semi-permeable barriers unfragmented. Evidence from double label clsm experiments and direct size measurements. *Biochim. Biophys. Acta* **2002**, *1564*, 21–30. [CrossRef]
20. Verma, D.D.; Verma, S.; Blume, G.; Fahr, A. Liposomes increase skin penetration of entrapped and non-entrapped hydrophilic substances into human skin: A skin penetration and confocal laser scanning microscopy study. *Eur. J. Pharm. Biopharm.* **2003**, *55*, 271–277. [CrossRef]
21. Elsayed, M.M.; Abdallah, O.Y.; Naggar, V.F.; Khalafallah, N.M. Deformable liposomes and ethosomes: Mechanism of enhanced skin delivery. *Int. J. Pharm.* **2006**, *322*, 60–66. [CrossRef]
22. Gordon, L.I.; Hong, F.; Fisher, R.I.; Bartlett, N.L.; Connors, J.M.; Gascoyne, R.D.; Wagner, H.; Stiff, P.J.; Cheson, B.D.; Gospodarowicz, M.; et al. Randomized phase iii trial of abvd versus stanford v with or without radiation therapy in locally extensive and advanced-stage hodgkin lymphoma: An intergroup study coordinated by the eastern cooperative oncology group (e2496). *J. Clin. Oncol.* **2013**, *31*, 684–691. [CrossRef]
23. Albers, P.; Siener, R.; Krege, S.; Schmelz, H.U.; Dieckmann, K.P.; Heidenreich, A.; Kwasny, P.; Pechoel, M.; Lehmann, J.; Kliesch, S.; et al. Randomized phase iii trial comparing retroperitoneal lymph node dissection with one course of bleomycin and etoposide plus cisplatin chemotherapy in the adjuvant treatment of clinical stage i nonseminomatous testicular germ cell tumors: Auo trial ah 01/94 by the german testicular cancer study group. *J. Clin. Oncol.* **2008**, *26*, 2966–2972.
24. Murad, A.M.; Triginelli, S.A.; Ribalta, J.C. Phase ii trial of bleomycin, ifosfamide, and carboplatin in metastatic cervical cancer. *J. Clin. Oncol.* **1994**, *12*, 55–59. [CrossRef]

25. Carver, J.R.; Shapiro, C.L.; Ng, A.; Jacobs, L.; Schwartz, C.; Virgo, K.S.; Hagerty, K.L.; Somerfield, M.R.; Vaughn, D.J.; Panel, A.C.S.E. American society of clinical oncology clinical evidence review on the ongoing care of adult cancer survivors: Cardiac and pulmonary late effects. *J. Clin. Oncol.* **2007**, *25*, 3991–4008. [CrossRef]
26. Tisman, G.; Herbert, V.; Go, L.T.; Brenner, L. Marked immunosuppression with minimal myelosuppression by bleomycin in vitro. *Blood* **1973**, *41*, 721–726. [CrossRef]
27. Dorr, R.T. Bleomycin pharmacology: Mechanism of action and resistance, and clinical pharmacokinetics. *Semin. Oncol.* **1992**, *19*, 3–8.
28. Knottenbelt, D.C.; Watson, A.H.; Hotchkiss, J.W.; Chopra, S.; Higgins, A.J. A pilot study on the use of ultra-deformable liposomes containing bleomycin in the treatment of equine sarcoid. *Equine Vet. Educ.* **2020**, *32*, 258–263. [CrossRef]
29. Jain, A.K.; Thareja, S. In vitro and in vivo characterization of pharmaceutical nanocarriers used for drug delivery. *Artif. Cells Nanomed. Biotechnol.* **2019**, *47*, 524–539. [CrossRef]
30. Bouwstra, J.A.; Honeywell-Nguyen, P.L.; Gooris, G.S.; Ponec, M. Structure of the skin barrier and its modulation by vesicular formulations. *Prog. Lipid Res.* **2003**, *42*, 1–36. [CrossRef]
31. Ulrich, G.; Ziessel, R.; Harriman, A. The chemistry of fluorescent bodipy dyes: Versatility unsurpassed. *Angew. Chem. Int. Ed. Engl.* **2008**, *47*, 1184–1201. [CrossRef] [PubMed]
32. Theerawatanasirikul, S.; Suriyaphol, G.; Thanawongnuwech, R.; Sailasuta, A. Histologic morphology and involucrin, filaggrin, and keratin expression in normal canine skin from dogs of different breeds and coat types. *J. Vet. Sci.* **2012**, *13*, 163–170. [CrossRef] [PubMed]
33. Jorgensen, E.; Lazzarini, G.; Pirone, A.; Jacobsen, S.; Miragliotta, V. Normal microscopic anatomy of equine body and limb skin: A morphological and immunohistochemical study. *Ann. Anat.* **2018**, *218*, 205–212. [CrossRef] [PubMed]

cancers

MDPI

Article

Proteomic Analysis Identifies FNDC1, A1BG, and Antigen Processing Proteins Associated with Tumor Heterogeneity and Malignancy in a Canine Model of Breast Cancer

Yonara G. Cordeiro [1], Leandra M. Mulder [2], René J. M. van Zeijl [2], Lindsay B. Paskoski [1], Peter van Veelen [2], Arnoud de Ru [2], Ricardo F. Strefezzi [1], Bram Heijs [2,†] and Heidge Fukumasu [1,*,†]

1 Laboratory of Comparative and Translational Oncology, Department of Veterinary Medicine, School of Animal Science and Food Engineering, University of São Paulo, Pirassununga 13635-900, Brazil; yo.gcordeiro@gmail.com (Y.G.C.); lindsaybiom@usp.br (L.B.P.); rstrefezzi@usp.br (R.F.S.)

2 Center of Proteomics and Metabolomics, Leiden University Medical Center, 2300 RC Leiden, The Netherlands; leandra.mulder@live.nl (L.M.M.); R.J.M.van_Zeijl@lumc.nl (R.J.M.v.Z.); P.A.van_Veelen@lumc.nl (P.v.V.); A.H.de_Ru@lumc.nl (A.d.R.); B.P.A.M.Heijs@lumc.nl (B.H.)

* Correspondence: fukumasu@usp.br; Tel.: +55-19-3565-6864

† Contributed equally.

Simple Summary: Comparative oncology is centered around the study of naturally occurring tumors in animals as a parallel and complementary model for human cancer research. Canine mammary tumors pose as excellent models since they share similarities in their spontaneous nature, histological subtypes, genetic background, and clinical course, which would be impossible to reproduce in murine models. Our study aimed to investigate cancer heterogeneity in primary tumors and metastasis, by applying bottom-up proteomics and mass spectrometry imaging to identify potential disease-state markers. We have demonstrated that the malignant phenotype may have arisen as a consequence of alterations in the expression of proteins involved in immune evasion facilitating metastatic events. To our knowledge, this is the first study to use mass spectrometry imaging in a dog model of breast cancer, that have demonstrated that poorly described proteins might play important roles in cancer spreading and should be further validated as potential early-stage tumor biomarkers.

Abstract: New insights into the underlying biological processes of breast cancer are needed for the development of improved markers and treatments. The complex nature of mammary cancer in dogs makes it a great model to study cancer biology since they present a high degree of tumor heterogeneity. In search of disease-state biomarkers candidates, we applied proteomic mass spectrometry imaging in order to simultaneously detect histopathological and molecular alterations whilst preserving morphological integrity, comparing peptide expression between intratumor populations in distinct levels of differentiation. Peptides assigned to FNDC1, A1BG, and double-matching keratins 18 and 19 presented a higher intensity in poorly differentiated regions. In contrast, we observed a lower intensity of peptides matching calnexin, PDIA3, and HSPA5 in poorly differentiated cells, which enriched for protein folding in the endoplasmic reticulum and antigen processing, assembly, and loading of class I MHC. Over-representation of collagen metabolism, coagulation cascade, extracellular matrix components, cadherin-binding and cell adhesion pathways also distinguished cell populations. Finally, an independent validation showed FNDC1, A1BG, PDIA3, HSPA5, and calnexin as significant prognostic markers for human breast cancer patients. Thus, through a spatially correlated characterization of spontaneous carcinomas, we described key proteins which can be further validated as potential prognostic biomarkers.

Keywords: breast cancer; canine model; comparative oncology; MALDI imaging; tumor progression

Citation: Cordeiro, Y.G.; Mulder, L.M.; van Zeijl, R.J.M.; Paskoski, L.B.; van Veelen, P.; de Ru, A.; Strefezzi, R.F.; Heijs, B.; Fukumasu, H. Proteomic Analysis Identifies FNDC1, A1BG, and Antigen Processing Proteins Associated with Tumor Heterogeneity and Malignancy in a Canine Model of Breast Cancer. *Cancers* **2021**, *13*, 5901. https://doi.org/10.3390/cancers13235901

Academic Editors: Bruno Cogliati and Felisbina Queiroga

Received: 13 October 2021
Accepted: 19 November 2021
Published: 24 November 2021

Publisher's Note: MDPI stays neutral with regard to jurisdictional claims in published maps and institutional affiliations.

Copyright: © 2021 by the authors. Licensee MDPI, Basel, Switzerland. This article is an open access article distributed under the terms and conditions of the Creative Commons Attribution (CC BY) license (https://creativecommons.org/licenses/by/4.0/).

1. Introduction

The estimated number of human breast cancer cases is still increasing worldwide, accounting for 11.6% of all new cases of cancer and 6.6% of deaths due to the disease until 2018 [1,2]. Even with a thorough histological examination and the use of well-recognized gene expression patterns for molecular subtyping [3], a reliable classification for clinical prognosis and therapy decision making might be challenging. In order to find new prognostic markers and develop more efficient and precise treatments, a deeper molecular understanding of the disease is necessary.

To gain new insights into tumor growth and revolutionary therapy research, a trustworthy animal model that mimics the fundamental aspects of human cancer is required. In this context, spontaneous canine mammary tumors are an excellent example of the intricate biology behind heterogeneous cancer progression. Pet dogs not only experience the same environment and risk factors as people, but they also share anatomical and histological similarities, comparable clinical features, as well as more homologous DNA and protein sequences compared to rodent models [4–7]. Moreover, the large population of dogs and improvements in the pet–owner relationship can simultaneously benefit both sides by helping researchers to improve the assessment of novel treatments for humans by treating pet animals with cancer [5].

It has been recognized that cancer is hardly a disease of tumor cells solely, but instead a disorder of imbalance in which tumor microenvironment plays crucial roles in development and malignancy [8,9]. Mass spectrometry imaging (MSI) enables the spatially correlated, label-free, and highly multiplexed molecular analysis directly from histological sections, and therefore is a useful tool for analyzing tumor heterogeneity in cancer tissues [10]. One advantage of MSI is the possibility of investigating cancer cells simultaneously with their immediately adjacent microenvironment, reducing the interference of other extra-tumor stroma components which could lead to misinterpretations. Despite limitations, such as the restricted analytical depth and the challenging high-dimensional datasets [11], the combination of bottom-up and top-down proteomics approaches, achieved by combining MSI and LC-MS/MS, represents a valuable tool for basic and translational oncology research [12].

Therefore, we aimed to explore tumor heterogeneity and molecular alterations in pathways involved in cancer progression, which could favor diagnosis, prognosis, and therapeutic strategies. To date, only a few studies have employed MSI in canine cancer, all of which comprised lipid profiling to identify molecules that might be used to discriminate between tumor types, or tumoral and non-tumoral areas [13–15]. To our knowledge, this is the first study to explore tumor heterogeneity using MSI in a dog model of breast cancer. Here, we have identified variations in peptide expression between cell populations in different degrees of differentiation, therefore malignancy, which may lead to the discovery and improvement of potential biomarkers.

2. Materials and Methods

2.1. Tissue Samples

All procedures were approved by FZEA/USP Ethics Committee (CEUA n° 8632101017). Mammary tumor samples from five unspayed dogs submitted to elective mastectomies were obtained in a collaboration with the Veterinary Hospital of the Faculty of Animal Science and Food Engineering (FZEA/USP) in Pirassununga, Brazil, and the Veterinary Hospital Dr. Vicente Borelli, from the Octávio Bastos Teaching Foundation (UNIFEOB) in São João da Boa Vista, Brazil. Patients had no previous history of either surgical or chemotherapy treatments.

2.2. TMA Construction

All tumor samples were divided into smaller fragments of $\cong 1\ cm^3$ after surgical excision. Fragments were fixed in a 10% buffered formalin solution and processed according to standard procedures. The diagnosis of mammary cancer was confirmed by histological evaluation following the classification proposed by Goldschmidt et al. [16].

For tissue microarray (TMA) arrangement, formalin-fixed, paraffin-embedded (FFPE) blocks were sectioned into 4 μm thick pieces, hematoxylin and eosin (HE) stained, and regions were marked for a good depiction of intratumor heterogeneity across the solid mass. Areas containing epithelial and myoepithelial cells, as well as stromal regions including vessels, inflammatory cells, and connective tissues were chosen. From each tumor, up to 5 punches/cm^3 were taken, and the 2 mm cores were assembled with at least 1 mm spacing between samples. The number of cores varied between 40 and 56 cores per TMA. In addition, metastatic cells in lymph nodes or other secondary sites (i.e., lungs) as well as non-neoplastic mammary glands from different animals were also placed asymmetrically at one end of each block for control purposes. After assembly, TMAs were sectioned and stained with HE to assess the integrity of the samples.

2.3. Mass Spectrometry Imaging

For matrix-assisted laser desorption/ionization mass spectrometry imaging (MALDI-MSI), FFPE TMAs were sectioned into 4 μm-thick sections and mounted on indium-tin-oxide (ITO)-coated glass slides (Bruker Daltonics, Bremen, Germany) additionally coated with 0.05% poly-L-lysine (Sigma-Aldrich, St. Louis, MO, USA) and 0.1% Nonidet P-40 (Sigma-Aldrich, St. Louis, MO, USA). Sections were dewaxed in xylene, rehydrated, and washed with deionized water after an hour on a hot plate at 60 °C. Heat-induced antigen retrieval was performed in citrate buffer (10 mM at pH 6) at 121 °C for 20 min, then cooled to room temperature for 120 min. Trypsin (0.02 μg/mL in deionized water, Trypsin Gold, Promega Corporation, Wisconsin, WI, USA) was applied for tissue digestion using the SunCollect (SunChrom, Friedrichsdorf, Germany) automated pneumatic spray platform (15 layers at 10 μL/min). Slides were then incubated for 18 h using the SunDigest (SunChrom) incubator at 37 °C and 95% of relative humidity. Following digestion, MALDI-matrix (25 mg/mL of 2,5-dihydroxybenzoic acid in 50% aqueous acetonitrile and 0.1% trifluoroacetic acid) was also applied using the SunCollect (7 layers at (1) 10 μL/min, (2) 20 μL/min, (3) 30 μL/min, (4+) 40 μL/min).

MALDI-FT-ICR-MSI data acquisition was performed in positive ion mode on a 9.4 T solariX equipped with CombiSource™ and dynamically harmonized ParaCell™ (Bruker Daltonics). Data were acquired over an *m/z* range between 600–3500 Da, 200 laser shots per pixel, and the small laser focus (~50 μm diameter) with a pixel size of 100 × 100 μm^2. Instrument calibration was performed using Peptide Calibration Standard (PepMix, Bruker Daltonics).

2.4. Histopathological Staining and Annotation

Excess MALDI-matrix was removed with 70% ethanol after MALDI-MSI. TMA slides were stained with HE and later scanned (Philips Slide Scanner, Andover, MA, USA) on 5, 10, 20, and 40× magnification. Cores that were eventually lost or damaged during tissue processing and/or staining were not included in the analyses. Using flexImaging (v5.0, Bruker Daltonics), scanned images were co-registered to the MSI data allowing for virtual microdissection. The annotation of regions of interest (ROIs), mainly comprised cancer epithelial cell populations. Annotation step avoided major vessels and ducts along with necrotic and desmoplastic areas, in order to minimize the interference caused by the influence of MALDI-matrix signals accumulating in empty spaces. Tumor ROIs were categorized into two different groups based on morphological criteria of the malignancy of canine mammary cancer adapted from the well-established grading system published by Goldschmidt et al. [16], as follows: (i) well-differentiated tumor populations (WD), presenting uniform cell morphology to a moderate degree of cell and nuclei pleomorphism, none to occasional hyperchromatic nuclei and none to occasional nucleoli. Differentiated tubules in >25% of region area and rare to occasional mitoses. (ii) poorly differentiated tumor populations (PD), showing a high degree of pleomorphism, hyperchromatic nuclei, presence of prominent and/or multiple nucleoli. Differentiated tubules in <25% of the region assuming areas of solid growth, where frequent mitoses could be observed.

2.5. MALDI-MSI Data Processing

Raw data were exported as an imzML file for preprocessing and peak picking using the SCiLS Lab Software (v2016b, Bruker Daltonics, Bremen, Germany). In the R environment (www.r-project.org accessed 15 July 2021), data processing was conducted using the rMSIproc package [17]. Spectra were aligned using a maximum shift of 50 ppm, alignment iterations = 3, and alignment oversampling = 2. Data were root-mean-squared (RMS)-normalized, and peak picking was conducted setting a signal to noise ratio (S/N) = 7 and a binning tolerance of 12 scans. Peaks detected in less than 5% of the total number of spectra were removed. After statistical analysis (described in the following paragraph), the list with significant peaks was checked for the presence of isotope peaks using mMass 5.5.0 [18]. Deisotoping was performed using a max charge of 1+ and isotope mass tolerance of 0.02 *m/z* for monoisotopic mass selection.

2.6. Liquid Chromatography-Tandem Mass Spectrometry (LC-MS/MS)

For LC-MS/MS, tissue tryptic digestion was performed as previously described in the MALDI-MSI methods. A solution of 50% aqueous acetonitrile and 0.1% trifluoroacetic acid was used for peptides extraction from the TMAs. Then, extracted peptides were desalted using C18 ZipTips (Millipore, Billerica, MA, USA) according to the standard protocol, and later dried. Lyophilized peptides were dissolved in 95/3/0.1 (% *v/v/v*) water/acetonitrile/formic acid and analyzed by online C18 nano HPLC MS/MS with a system consisting of an Easy-nLC 1200 gradient HPLC system (Thermo Scientific, Bremen, Germany), and an Orbitrap Fusion Lumos mass spectrometer (Thermo Scientific). Next, the sample was injected onto a precolumn (100 µm × 15 mm, C18 Reprosil-Pur C18-AQ, 3 µm, 120 A, and eluted via a homemade analytical nano-HPLC column (50 cm × 75 µm; Reprosil-Pur C18-AQ 1.9 µm, 120 A (Dr. Maisch, Ammerbuch, Germany). The gradient was run from 2 to 36% solvent B (20/80/0.1 water/acetonitrile/formic acid (% *v/v/v*) in 120 min. The nano-HPLC column was drawn to a tip of ~10 µm acting as the electrospray needle of the MS source. The mass spectrometer was operated for a cycle time of 3 s in data-dependent MS/MS mode, with an HCD collision energy at 32 V and recording of the MS2 spectrum in the orbitrap, with a quadrupole isolation width of 1.2 Da. The resolution was set to 120,000, the scan range 400–1500, at an AGC target of 4,000,000 at a maximum fill time of 50 ms in the master scan (MS1). Precursors were dynamically excluded after $n = 1$ with an exclusion duration of 60 s, and with a precursor range of 20 ppm. The charge states 2–4 were included. For MS2, the first mass was set to 110 Da, and the MS2 scan resolution was 30,000 at an AGC target of 40,000 at a maximum fill time of 60 ms.

As post-analysis, raw data were converted to peak lists using Proteome Discoverer version 2.2 (Thermo Electron), and protein identification was carried out by comparing the lists to the Canis2019 database (24,669 entries) on Mascot v. 2.2.04 (www.matrixscience.com, accessed on 15 July 2021). Mascot searches were performed using 10 ppm and 0.02 Da tolerance for precursor and fragment mass, respectively, and trypsin was defined as the protease. Methionine oxidation and the main formaldehyde reaction product on Lysine (+12.000 *m/z*) were set as a variable modification, while Carbamidomethyl (C) was a fixed modification. Peptides with an FDR $\leq$ 1% in combination with a mascot ion score $\geq$25 were accepted. For peptide identity assignment (IDs), observed MALDI-MSI signals were matched to LC-MS/MS data considering a maximum mass error of ±10 ppm. Single peptides assigned to different proteins from the same family were kept since they may be helpful for a full comprehension of the results.

2.7. Statistical Analysis

2.7.1. Discriminative *m/z* Signals

Groups were pre-adjusted so that the number of spectra obtained from each animal was randomly—and equally—sampled before statistical analysis. Discriminative peaks between annotated WD and PD tumor regions were examined using a receiver operating characteristic (ROC) analysis and then tested for significance using the Mann–Whitney Wilcoxon Test followed by Benjamini–Hochberg (BH) post-test correction. For that, the mean intensity of each *m/z* feature in every ROI was calculated and tested. Then, the same number of spectra from each condition ($n = 3000$) were used for the ROC curve, and signals with AUC ≥ 0.70 in at least three out of five ROC analyses were tested. Only *m/z* signals presenting a p-value ≤ 0.05 and FDR ≤ 0.05 were considered significant.

2.7.2. Functional Enrichment Analysis

For a more comprehensive knowledge base about protein function, the human genome-coding database was employed as a reference set. Thus, gene ontology (GO) and pathway enrichment analysis were determined through the overrepresentation test on the PANTHER classification system (www.pantherdb.org, accessed 15 November 2021), using the REACTOME (version 65) database [19]. Protein symbols retrieved from peptides with unique matches were used as input and statistical analysis was performed using Fisher's exact test followed by FDR correction. Enriched GO terms and pathways were considered significant if FDR ≤ 0.05.

2.7.3. Independent Comparative Validation

To further evaluate the significance of relevant proteins found in canine samples as prognostic factors in human breast cancer, an in silico validation was performed using an online survival analysis tool [20] (www.kmplot.com, accessed 15 July 2021). The datasets include gene expression (Affymetrix microarrays) and survival data from Gene Expression Omnibus (GEO and The Cancer Genome Atlas (TCGA). Overall survival (OS) and distant metastasis-free survival (DMFS) were analyzed using protein IDs as input, and the optimal probe [21] was selected for the assessment of the expression levels. All possible cutoff values between the lower and upper quartile were computed, and the best performing threshold was used to divide patients into two different cohorts. No restrictions were made regarding tumor subtypes such as ER, PR, and HER2 status, grade, or stage. No restrictions were made regarding patient treatment. Survival curve, number-at-risk, hazard ratio (and 95% confidence intervals), and log-rank p were displayed on the Kaplan–Meier plot. A protein was considered a significant marker when p-value ≤ 0.05.

3. Results

3.1. Patients, Tumor Samples and Tissue Annotation

Canine mammary gland FFPE tissue samples were examined using MSI. The study included 17 mammary tumors surgically removed from five female dogs diagnosed with metastatic disease, confirmed by positive anti-cytokeratin immunostaining in lymph nodes and other metastatic sites (patient information and histological classification of samples are detailed in Supplementary Table S1). Epithelial and myoepithelial populations, stromal areas consisting of connective tissue and peritumoral inflammatory cell infiltration, as well as metastatic regions and non-neoplastic glands were assembled in ten TMA blocks. A total of 453 cores, from which 320 were viable, were submitted to pathological classification and annotation. From annotated regions, a total of 282 tumor ROIs were selected and further categorized as WD and PD (Figure 1). We obtained 70 WD and 212 PD ROIs, along with 21 metastasis regions and seven ROIs of non-neoplastic mammary glands. In addition, 108 stroma-only regions were annotated, but not included in any statistical analysis.

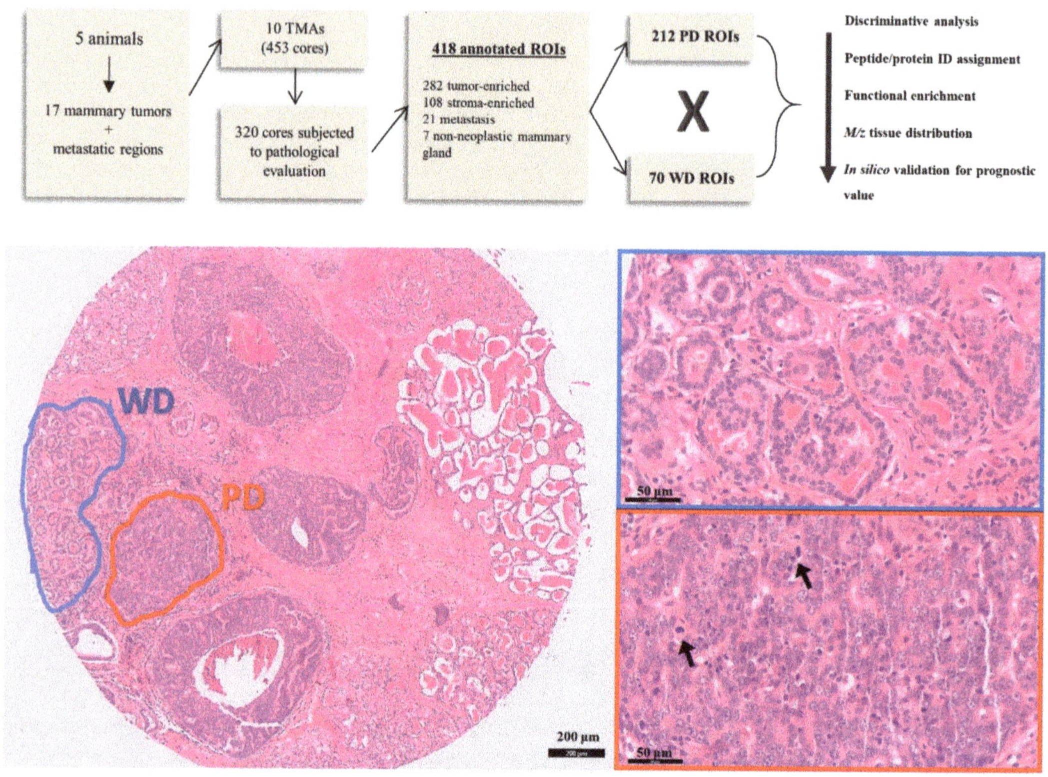

Figure 1. Peptide expression was compared between neoplastic cell populations in two distinct levels of differentiation (well-differentiated, WD, and poorly differentiated, PD), from 17 tumor samples assembled in FFPE tissue microarrays. Protein identity assignment of significant features was performed based on mass matching to an LC-MS/MS derived peptide database. Protein IDs were then submitted to functional enrichment analysis and promising prognostic marker candidates were validated in silico using a human gene expression database. HE stained TMA core shows the annotation of a WD (blue) and a PD (red) region of interest (ROI), detailed in the images on the right. ROIs were classified based on morphological criteria such as tubule differentiation, cell, nuclei, and nucleoli pleomorphism. Arrows point to atypical mitoses inside a PD tumor region.

3.2. *Protein Identification and Discriminative m/z Signals between Low and High Grade Intratumor Regions*

After data processing and peak picking, a list containing 1780 *m/z* features was obtained from the examined tissues. Next, molecular dissimilarities between WD and PD subpopulations that could be contributing to tumor malignancy were determined. After deisotoping, 168 differentially expressed *m/z* signals were found (Supplementary Table S2), where the majority of features (162) were more intensely expressed in WD tumor regions compared to PD (Figure 2A).

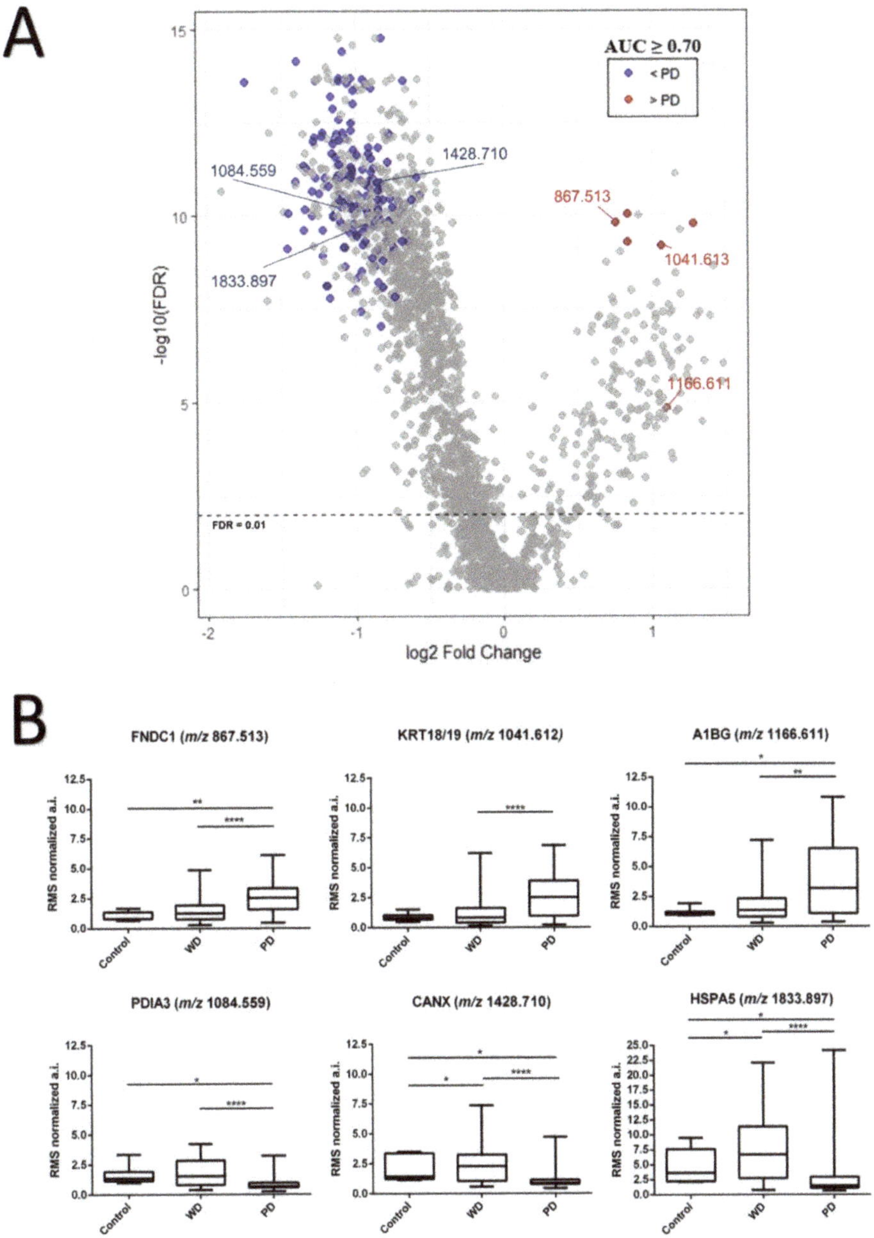

Figure 2. (**A**) Volcano plot shows the discriminative peaks (AUC $\geq$ 0.70 and FDR $\leq$ 0.05) in WD vs. PD group. Higher intensity levels in PD tumors are indicated in red, lower intensity levels are indicated in blue (**B**) Boxplot analysis of representative peptide expression of six discriminative *m/z* signals in canine mammary carcinoma tissues. Normalized absolute intensity data and error bars are shown for WD and PD tumor areas, as well as non-neoplastic mammary gland used as control. Ions at *m/z* 867.513, 1041.613, and 1166.611 were more expressed in PD ROIs, while signals at *m/z* 1084.559, 1428.710, and 1833.897 presented a lower expression in the same regions. (*) $p < 0.05$ (**) $p < 0.01$ (****) $p < 0.0001$.

Protein identity assignment of the 168 discriminating features based on mass matching to the LC-MS/MS derived peptide database resulted in 38 proteins IDs with unique matches (from 47 corresponding peptides), and 10 extra IDs from four signals which were assigned to two or more proteins belonging to the same family/type (keratin 18 and 19, KRT18/KRT19 at *m/z* 1041.613 with a mass error of 7.37 ppm; heat shock protein 90 alpha family class B member 1 and class A member 1, HSP90AB1/HSP90AA1 at *m/z* 1311.563 with a mass error of 5.11 ppm; tropomyosin 1 and 2, TPM1/TPM2 at *m/z* 1332.648 with a mass error of 7.42 ppm; and beta tubulins, TUBB4B/TUBB/TUBB2A/TUBB1 at *m/z* 1636.829 with a mass error of 0.42 ppm). A full list of identity assignments can be found in Supplementary Table S3. Peptides whose identities were assigned to Fibronectin type III domain containing 1 protein (FNDC1, at *m/z* 867.513, mass error = 9.61 ppm), Alpha-1B-glycoprotein (A1BG, at *m/z* 1166.611, mass error = 2.25 ppm), and double-matching keratins 18 and 19 (KRT18/KR19 at *m/z* 1041.613, mass error = 7.37 ppm) were among *m/z* signals showing higher intensity in PD regions (Figure 2B). Besides those peptides, ions at *m/z* 679.547, *m/z* 742.305, and *m/z* 1158.647 were found to be also overexpressed in PD regions, but the identities could not be retrieved from LC-MS/MS data.

3.3. Functional Enrichment Analysis of Differentially Expressed Peptides between Well and Poorly Differentiated Tumor Populations

To increase our understanding of biological and molecular pathways distinguishing WD and PD populations, gene ontology (GO) was performed using protein IDs retrieved from LC-MS/MS data of discriminative *m/z* signals between conditions. GO of unique IDs significantly enriched for terms such as protein folding in endoplasmic reticulum, extracellular matrix structural constituent, molecule activity, cadherin binding, and cell adhesion (Table 1).

Table 1. Biological processes and molecular function terms enriched in GO analysis using 38 proteins IDs retrieved from LC/MS-MS data.

Biological Process	Fold Enrichment	IDs	FDR
Protein folding in endoplasmic reticulum	>100	CANX, HSPA5, PDIA3	2.7×10^{-2}
Extracellular matrix structural constituent conferring tensile strength	52.16	COL1A1, COL1A2, COL6A3, COL12A1	1.73×10^{-3}
Extracellular matrix structural constituent	18.98	COL1A1, COL1A2, COL6A3, COL12A1, VCAN, EMILIN2	1.3×10^{-3}
Structural molecule activity	6.81	COL1A1, COL1A2, COL6A3, COL12A1, VCAN, EMILIN2, LMNA, EPB41L2, CTNNA1	4.95×10^{-3}
Cadherin binding Cell adhesion molecule binding	17.75 11.01	LIMA1, EEF1D, CALD1, TNKS1BP1, HSPA5, FLNA, TAGLN2, SERBP1, DDX3X, CTNNA1, BAG3	1.07×10^{-7} 9.50×10^{-6}

Besides a general overview of GO processes, we also performed a functional pathway enrichment of differentially expressed peptides using the open-access and peer-reviewed REACTOME database, which describes possible reactions if all annotated proteins were present and active simultaneously in a cell. Pathways directly involved in collagen metabolism, coagulation cascade, extracellular matrix proteoglycans, and signaling by receptor tyrosine kinase showed as significant in enrichment analysis (Table 2).

A1BG, which absolute intensity was higher in the malignant phenotype, was significant for three different pathways involved with platelet mechanisms, such as platelet activation, aggregation, and degranulation. In addition, calnexin (CANX), heat-shock protein family A member 5 (HSPA5), and protein disulfide isomerase family A member 3 (PDIA3), which in GO enriched for protein processing in endoplasmic reticulum, also played an important role for antigen presentation associated to the class I MHC, presenting the second-highest fold enrichment value among all significant pathway classes (Table 2 and Figure 2B). All three proteins were found to be less intense in PD tumor regions (Figure 3A).

Table 2. Significant REACTOME enriched pathways. Results are sorted by the hierarchical relations between over-represented functional classes, which may be interpreted as a group rather than individually.

Pathway Name	Fold Enrichment	IDs	FDR
GP1b-IX-V activation signaling	>100	COL1A1, COL1A2, FLNA	2.47×10^{-3}
Platelet activation signaling and aggregation	12.39	COL1A1, COL1A2, FLNA, A1BG, HSPA5, TAGLN2	4.13×10^{-3}
Antigen presentation: folding, assembly and peptide loading of class I MHC	61.69	CANX, HSPA5, PDIA3	4.85×10^{-3}
Collagen chain trimerization	48.60	COL1A1, COL1A2, COL6A3, COL12A1	4.34×10^{-3}
Collagen biosynthesis and modifying enzymes	31.92	COL1A1, COL1A2, COL6A3, COL12A1	3.51×10^{-3}
Collagen formation	24.03	COL1A1, COL1A2, COL6A3, COL12A1	5.58×10^{-3}
Assembly of collagen fibrils and other multimeric structures	35.64	COL1A1, COL1A2, COL6A3, COL12A1	4.63×10^{-3}
Collagen degradation	33.42	COL1A1, COL1A2, COL6A3, COL12A1	4.43×10^{-3}
Degradation of the extracellular matrix	15.28	COL1A1, COL1A2, COL6A3, COL12A1	2.41×10^{-2}
ECM proteoglycans	28.14	COL1A1, COL1A2, COL6A3, VCAN	4.83×10^{-3}
Platelet degranulation	16.84	A1BG, HSPA5, FLNA, TAGLN2	1.95×10^{-2}
Response to elevated platelet cutosolic Ca2+	16.20	A1BG, HSPA5, FLNA, TAGLN2	2.08×10^{-2}
Signaling by receptor tyrosine kinases	8.19	COL1A1, COL1A2, COL6A3, STMN1, HNRNPM, CSN2, CTNNA1	5.38×10^{-3}

3.4. Tissue Distribution of FNDC1 and A1BG Peptides

Next, we further explored the spatial distribution of m/z signals with a higher intensity in PD regions and whose identities could be assigned from the LC-MS/MS data. First, we compared ion intensities between tumor regions presenting an aggressive morphological phenotype with non-neoplastic mammary gland tissues. Both peptides at *m/z* 867.513 and 1166.611, respectively assigned to FNDC1 and A1BG proteins, were significantly more intense in PD tumor over control regions ($p = 0.0006$ for *m/z* 867.513 and $p = 0.0395$ for *m/z* 1166.611). Then, we examined tissue distributions of both peaks across the entire sample, including tumor microenvironment, lymph nodes, and metastatic sites. Besides tumor regions, the FNDC1 peptide was also of high intensity across metastatic areas in lymph nodes and other secondary tumor sites, as lungs and abdominal masses (Figure 3B). Similarly, the A1BG peptide was detected across primary tumor tissue in both WD and PD regions, as well as in immune cell clusters, as demonstrated in Figure 3C. Ion abundance varied between intratumor infiltrating cells and tissue-resident leukocytes in the lymph nodes. In the latter, the peptide distributions seemed anatomically homogeneous, since no distinction between cortex and medulla could be observed.

3.5. Prognostic Value of FNDC1, A1BG, CANX, HSPA5 and PDIA3 in Human Breast Cancer Patients

As demonstrated by molecular tissue distributions and functional enrichment analyses, FNDC1, A1BG, CANX, HSPA5, and PDIA3 were related to the immune system response against tumor cells and/or cancer spreading to distant sites in canine cancer, and thus being promising candidates for further validation as comparative prognostic biomarkers. Therefore, we intended to study the relationship between gene expression of such proteins and clinical outcomes of human breast cancer patients using the Kaplan–Meier plotter online database. In silico validation explored the overall survival (OS, $n = 943$ patients) and distant metastasis-free survival (DMFS, $n = 958$ patients) prognostic values that were obtained according to the low and high expression of each gene. We observed that the high expression of FNDC1 (probe 226930_at, $p = 8.3 \times 10^{-6}$), as well as the low expression of A1BG (probe 229819_at, $p = 3.4 \times 10^{-6}$) and HSPA5 (probe 230031_at, $p = 0.0017$) were associated with worse DMFS. In addition, high expression of FNDC1 (probe 226930_at, $p = 0.019$) and the low expression of PDIA3 (probe 227033_at, $p = 0.0037$), HSPA5 (probe

230031_at, $p = 0.0024$) and CANX (probe 238034_at, $p = 0.0013$) were also associated with a worse overall survival in breast cancer patients (Figure 4).

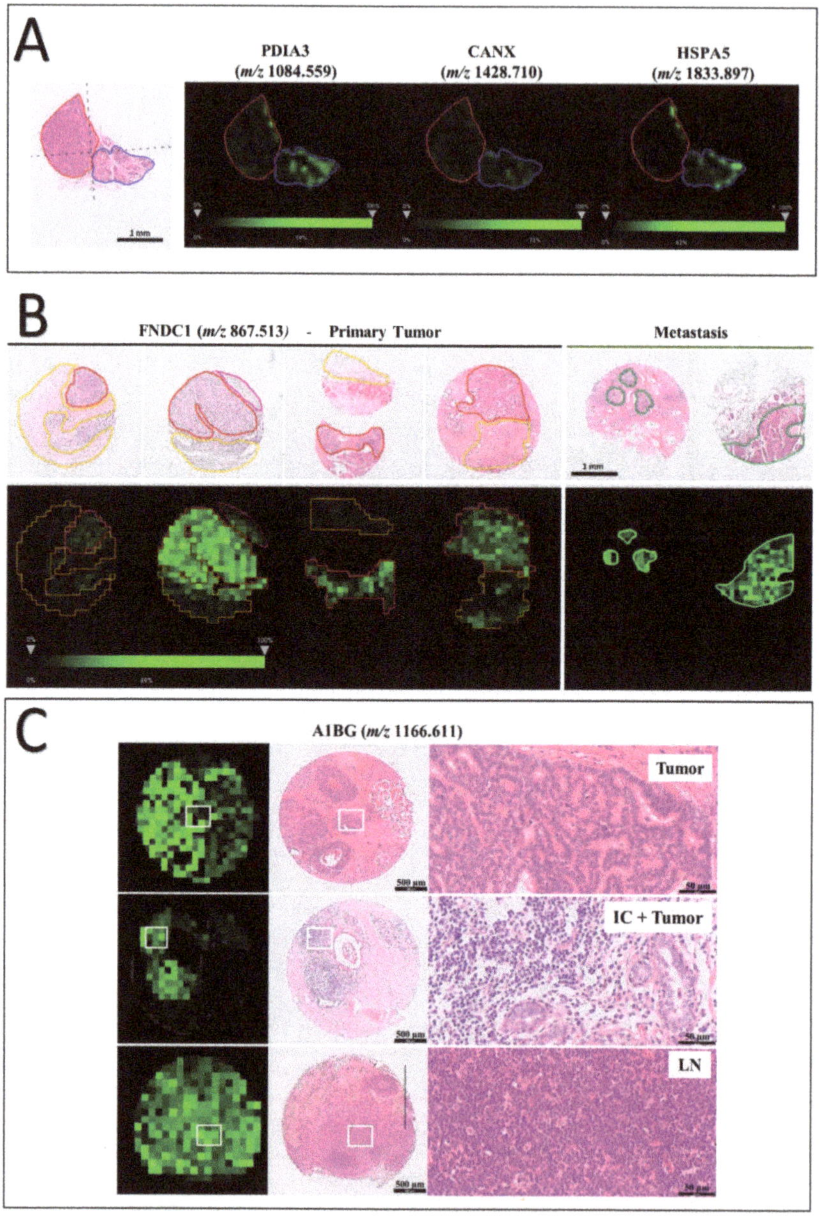

Figure 3. MSI data of canine samples. (**A**) WD (blue line) and PD (red line) tissue distribution of *m/z* signals assigned to proteins enriched in protein folding in endoplasmic reticulum, and folding, assembly, and loading of MHC class I pathway: protein disulfide isomerase family A member 3 (*m/z* 1084.559), calnexin (*m/z* 1215.622) and heat-shock protein family A member 5 (*m/z* 2042.051). (**B**) Signal of FNDC1 peptide at *m/z* 867.513 is strong across primary tumor and metastatic areas. (**C**) A1BG peptide at *m/z* 1166.611 was found to be expressed not only across primary tumor areas but also by inflammatory cells (IC) in canine samples. Corresponding HE staining shows signal distribution in tumor cells, tumor areas with infiltrating inflammatory cells, and across the lymph node (LN).

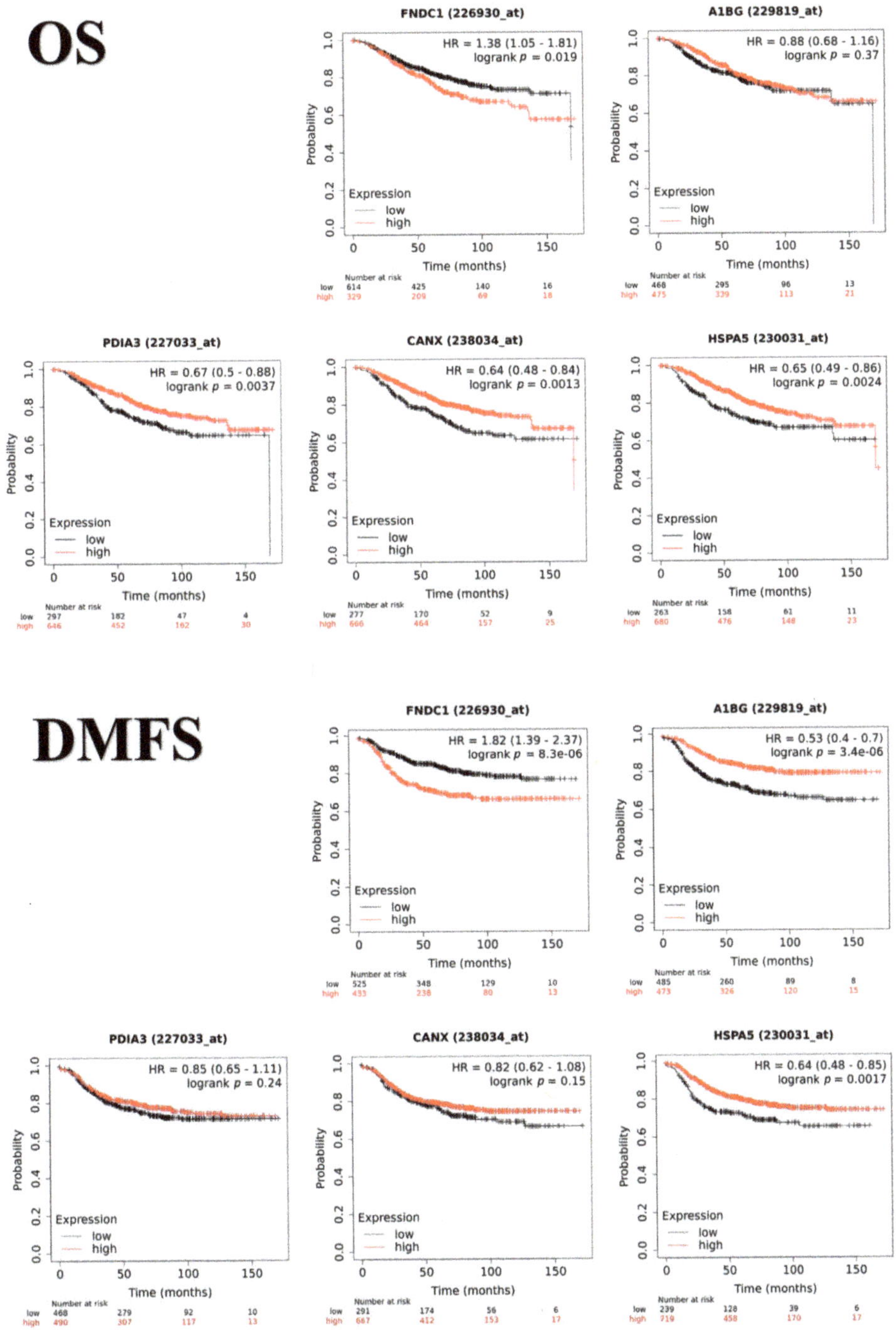

Figure 4. Prognostic values of FNDC1, A1BG, PDIA3, CANX and HSPA5 for OS and DMFS in human breast cancer patients. Gene expression and clinical outcome data were analyzed using Kaplan–Meier plotter. Patients with expression above the threshold are indicated in red line, and patients with expression below the threshold in black line. HR = hazard ratio.

4. Discussion

In the present study, we performed MSI to assess and compare peptide expression in 282 cancer regions annotated within 17 tumor samples from five different patients, minimizing the effect of genetic background and macroenvironment influence. By analyzing heterogeneous intratumor subpopulations and metastatic sites, we demonstrated that FNDC1, A1BG, CANX, HSPA5, and PDIA3 may be key factors to tumor malignancy in a canine model of breast cancer. A strong indication would be that peptides assigned to these proteins were not only significant to discriminate between phenotypes, but they also presented a significant prognostic value, as demonstrated in the survival analysis based on gene expression data of human breast cancer.

The degree of differentiation generates relevant information regarding clinical behavior, and it still has a direct impact to determine patient diagnosis and prognosis [22–25]. Here, the histological grading system was used to divide intratumor populations into two distinct groups based on their level of differentiation: well- and poorly differentiated tumor regions (WD and PD). A total of 168 *m/z* signals were able to discriminate between WD and PD groups, in which the majority of ions presented a lower intensity in tumor regions with a more malignant morphological phenotype. As it may be, a small part of this effect might have been due to a greater amount of stroma elements in WD populations, since well-differentiated breast carcinomas do not show as many solid grow patterns as in PD regions. However, the importance of such molecules in the context of tumor development and progression should not be diminished. It is well known that the tumor microenvironment, which includes extracellular matrix (ECM) molecules, inflammatory cells, cancer-associated fibroblast, and blood vessels, is closely associated with carcinogenesis, cell proliferation, survival, invasion, and metastasis [8,26,27]. Collagen, for example, is one of the most abundant ECM proteins and withstands consecutive stages of degradation, redeposition, and remodeling, where both the increase and decrease in its deposition may be associated with the increase in tumor malignancy [28–30]. Here, peptides assigned to collagens types I, VI, and XII were recurrent in several significant terms in GO and pathway enrichment analysis, representing an important link between tumor microenvironment and cancer cells.

Peptides with a lower intensity in PD populations were assigned to proteins involved in molecular functions such as the coagulation cascade, ECM structure, and cell adhesion. Interestingly, we also observed a decreased expression in PD regions of *m/z* signals assigned to proteins such as calnexin, HSPA5, and PDIA3, which enriched for protein processing in the endoplasmic reticulum (ER) and peptide folding, assembly, and loading of class I MHC. Cells undergoing malignant progression have increased proliferation and the ability to adapt under adverse environments, frequently leading to ER stress resulting in protein misfolding, reduced ER processing, and the unfolded protein response (UPR) activation [31,32]. In normal physiology, misfolded and mutated proteins are cleaved into peptides which will be transported from the cytosol to the ER and later loaded into the MHC complex for antigen presentation triggering an immune response through CD8+ T cells [33]. Hence, losses or defects in transport and peptide loading in MHC may disrupt and impede antigen processing and presentation, thus having a profound effect on patient therapy and tumor cell survival against the immune system [34]. Accordingly, in our study, the low gene expression of calnexin, HSPA5, and PDIA3 had a significant prognostic value for OS and DMFS in human breast cancer patients.

PDIA3 interacts with lectin chaperones calreticulin and calnexin to modulate folding of newly synthesized glycoproteins [35], and the low expression of PDIA3 has been associated with poor overall survival for non-small lung cancer and gastric cancer, due to the formation of a complex with MHC class I [35–37], corroborating with our results. HSPA5 (also known as BiP) is a master regulator of the UPR [38]. Fae and collaborators demonstrated PDIA3 and HSPA5 to be recognized by heart infiltrating and peripheral T cells in chronic rheumatic heart disease patients, suggesting that these proteins may be involved in an autoimmune-mediated tissue response as autoantigen targets of antibodies [39]. In cancer, future studies could be performed to determine whether the same mechanism also participates in tumor progression and metastasis due to evasion from infiltrating immune cells.

Moreover, during protein folding, HSPA5 associates with calnexin, a highly abundant transmembrane chaperone that folds synthesized glycoproteins in the ER lumen, playing a central role in glycoprotein quality control [40–44]. The upregulation of calnexin has been associated with a worse prognosis in several types of cancer, but results are still controversial. Although its increased expression was shown to negatively regulate tumor MHC I surface expression and promote STAT3 oncogene activation leading to tumor malignancy [45,46], calnexin expression levels were significantly higher only in Stage I lung cancer patients when compared to healthy controls, while in colorectal cancer (CRC), no associations were found between higher expressions of calnexin in CRC Stages II or III compared to normal tissue samples [47,48]. Additionally, low or defective expression of calnexin in primary breast cancer was reported to be associated with a higher risk of brain metastases, due to defects in T-cell-based immunosurveillance [49]. These results corroborate our theory that downregulation of PDIA3, HSPA5, and calnexin, at a certain point of tumor progression, may contribute to malignancy by promoting tumor-immune system evasion through a dysregulation in peptide processing and antigen presentation, having a direct impact on patient outcome and therapy efficiency.

Among the six peptides with a higher intensity in PD regions, tissue distribution of *m/z* signal at 867.513, assigned to the Fibronectin type III domain-containing 1 protein (FNDC1, also known as ASG8), revealed that this ion was not only more intense in tumors over control regions, but also highly expressed across metastatic sites. In addition, its high expression was associated with a shorter DMFS in human patients. The fibronectin III domain-containing proteins are found in tandem arrays in ECM proteins such as fibronectin and tenascin [50], sharing an identical protein fold similarly to the immunoglobulin domain, but with a limited amino acid sequence identity [51]. Although fibronectin III domains role in cell adhesion and migration has been described [52,53], specifically FNDC1 expression and function still remain poorly understood. In normal tissues, FNDC1 seems to be involved in angiogenic events, such as hypoxia-induced apoptosis of cardiomyocytes and vascular epidermal growth factor-mediated signal processing during angiogenesis [54,55]. In cancer, FDNC1 expression was shown to be altered by deregulated epigenetic CpG island methylation in salivary gland adenoid cystic carcinomas [56]. In addition, FNDC1 knockdown inhibited gastric and prostate tumor cell proliferation, invasion and also downregulated the expression of important proteins involved in epithelial-to-mesenchymal transition (EMT), but the mechanisms have still not been elucidated [57,58]. These findings, combined with the results observed in our work, strongly suggest that FNDC1 may be a potential cancer biomarker candidate. To our knowledge, this is the first report to suggest that FNDC1 may also play an important role in breast cancer metastatic disease.

Besides FNDC1, LC-MS/MS analysis of peptides with a higher intensity in PD also identified a signal corresponding to the Alpha-1B-glycoprotein (A1BG). Significantly enriching for pathways of the coagulation cascade, tissue distribution of the *m/z* signal at 1166.611 showed by MSI analysis demonstrated that A1BG peptide was mainly expressed by inflammatory cells infiltrating tumor areas and also in the lymph nodes. Additionally, the low expression of A1BG showed to predict a worse DMFS prognosis in human breast cancer patients. A1BG, with still unknown functions, shows homology to the immunoglobulin supergene family and has been implicated in immune response inflammation processes [59–61]. Mainly found in the serum, plasma, and other bio-fluids such as pancreatic juice and urine, A1BG expression has been pointed out as a relevant biomarker for several types of cancer [62–66]. A1BG together with Complement C3 was shown to be overexpressed in serum samples from patients with squamous cell carcinoma of the cervix and grade III cervical intraepithelial cancer, compared to healthy control women [67,68]. Moreover, sialylation changes of A1BG and Complement C3 were demonstrated in blood samples of control and breast cancer patients, suggesting that such modification in the protein glycosylation pattern of these proteins may also serve as a potential biomarker [69].

The capacity of tumor cells to produce and release A1BG is still uncertain. To date, only a few studies have demonstrated its expression in cancer cells [64,67], and therefore, further experiments are needed to validate A1BG expression and prognostic significance in tumor tissue samples.

5. Conclusions

Through the comparison of the proteomic profile of intratumor cell populations in different degrees of differentiation, we demonstrated that the malignant phenotype may have arisen as a consequence of alterations in the expression of key proteins such as FNDC1, A1BG, CANX, HSPA5, and PDIA3 and that most of these variations may be involved in tumor evasion against inflammatory cells facilitating cancer spreading. A strong indication would be that all five proteins were presented as significant prognostic values for overall survival and metastasis-free survival in human patients, as demonstrated in an independent validation. Now, further investigations are necessary to deeply characterize how such variations evolve with malignancy. Therefore, dog samples are useful proof of concept experiments, and novel comparative studies can be applied for many other types of cancer, and other disease conditions.

Supplementary Materials: The following are available online at https://www.mdpi.com/article/10.3390/cancers13235901/s1. Table S1: Patient Information, Table S2: Discriminative analysis, Table S3: Identity Assignment.

Author Contributions: Y.G.C. and H.F. conceived the project; Y.G.C., P.v.V., B.H. and H.F. conceived, planned, and supervised the experiments; Y.G.C., L.B.P., R.J.M.v.Z. and A.d.R. contributed to sample preparation and carried out the experiments; Y.G.C., L.M.M., A.d.R. and B.H. planned, carried out and supervised data analyses; Y.G.C., R.F.S., B.H. and H.F. contributed to the interpretation of the results; Y.G.C. wrote the manuscript in consultation with B.H., R.F.S. and H.F.; B.H. and H.F. supervised the project. All authors have read and agreed to the published version of the manuscript.

Funding: This research was funded by the São Paulo Research Foundation (FAPESP grant 2014/02493-7), Fundação de Estudos Agrários Luiz de Queiroz (FEALQ project 103743) and the Brazilian Coordination for Improvement of Higher Education Personnel (CAPES scholarship 88882.376316/2007-1).

Institutional Review Board Statement: All procedures were conducted following the guidelines approved by FZEA/USP Ethics Committee (CEUA no 8632101017).

Informed Consent Statement: Not applicable.

Data Availability Statement: The data presented in this study are available on request from the corresponding author.

Acknowledgments: We would like to thank our colleagues from the veterinary hospitals (UDCH/FZEA/USP and UNIFEOB), our colleagues from the Laboratory of Comparative and Translational Oncology/USP, and our colleagues from the Center of Proteomics and Metabolomics and Pathology Department at LUMC for the contribution to this study.

Conflicts of Interest: The authors declare no conflict of interest. The funders had no role in the design of the study, in the collection, analyses, or interpretation of data; in the writing of the manuscript, or in the decision to publish the results.

References

1. Ferlay, J.; Laversanne, M.; Ervik, M.; Lam, F.; Colombet, M.; Mery, L.; Piñeros, M.; Znaor, A.; Soerjomataram, I.; Bray, F. *Global Cancer Observatory: Cancer Tomorrow*; International Agency for Research on Cancer: Lyon, France, 2020; Available online: https://gco.iarc.fr/tomorrow (accessed on 18 November 2021).
2. Bray, F.; Ferlay, J.; Soerjomataram, I.; Siegel, R.L.; Torre, L.A.; Jemal, A. Global cancer statistics 2018: GLOBOCAN estimates of incidence and mortality worldwide for 36 cancers in 185 countries. *CA. Cancer J. Clin.* **2018**, *68*, 394–424. [CrossRef] [PubMed]
3. Parker, J.S.; Mullins, M.; Cheang, M.C.U.; Leung, S.; Voduc, D.; Vickery, T.; Davies, S.; Fauron, C.; He, X.; Hu, Z.; et al. Supervised Risk Predictor of Breast Cancer Based on Intrinsic Subtypes. *J. Clin. Oncol.* **2009**, *27*, 1160–1167. [CrossRef]

4. Lindblad-Toh, K.; Wade, C.M.; Mikkelsen, T.S.; Karlsson, E.K.; Jaffe, D.B.; Kamal, M.; Clamp, M.; Chang, J.; Kulbokas, E.J.; Zody, M.; et al. Genome sequence, comparative analysis and haplotype structure of the domestic dog. *Nature* **2005**, *438*, 803–819. [CrossRef]
5. Paoloni, M.; Khanna, C. Translation of new cancer treatments from pet dogs to humans. *Nat. Rev. Cancer* **2008**, *8*, 147–156. [CrossRef] [PubMed]
6. Rowell, J.L.; McCarthy, D.O.; Alvarez, C.E. Dog models of naturally occurring cancer. *Trends Mol. Med.* **2011**, *17*, 380–388. [CrossRef] [PubMed]
7. Liu, D.; Xiong, H.; Ellis, A.E.; Northrup, N.C.; Rodriguez, C.O.; O'Regan, R.M.; Dalton, S.; Zhao, S. Molecular Homology and Difference between Spontaneous Canine Mammary Cancer and Human Breast Cancer. *Cancer Res.* **2014**, *74*, 5045–5056. [CrossRef] [PubMed]
8. Mao, Y.; Keller, E.T.; Garfield, D.H.; Shen, K.; Wang, J. Stromal cells in tumor microenvironment and breast cancer. *Cancer Metastasis Rev.* **2013**, *32*, 303–315. [CrossRef]
9. Swartz, M.A.; Iida, N.; Roberts, E.W.; Sangaletti, S.; Wong, M.H.; Yull, F.E.; Coussens, L.M.; DeClerck, Y.A. Tumor Microenvironment Complexity: Emerging Roles in Cancer Therapy. *Cancer Res.* **2012**, *72*, 2473–2480. [CrossRef] [PubMed]
10. Balluff, B.; Hanselmann, M.; Heeren, R.M.A. *Mass Spectrometry Imaging for the Investigation of Intratumor Heterogeneity*, 1st ed.; Elsevier Inc.: Amsterdam, The Netherlands, 2017; Volume 134, ISBN 9780128052495.
11. Vaysse, P.M.; Heeren, R.M.A.; Porta, T.; Balluff, B. Mass spectrometry imaging for clinical research—Latest developments, applications, and current limitations. *Analyst* **2017**, *142*, 2690–2712. [CrossRef] [PubMed]
12. Maier, S.K.; Hahne, H.; Gholami, A.M.; Balluff, B.; Meding, S.; Schoene, C.; Walch, A.K.; Kuster, B. Comprehensive identification of proteins from MALDI imaging. *Mol. Cell Proteom.* **2013**, *12*, 2901–2910. [CrossRef]
13. Dill, A.L.; Ifa, D.R.; Manicke, N.E.; Costa, A.B.; Ramos-Vara, J.A.; Knapp, D.W.; Cooks, R.G. Lipid profiles of canine invasive transitional cell carcinoma of the urinary bladder and adjacent normal tissue by desorption electrospray ionization imaging mass spectrometry. *Anal. Chem.* **2009**, *81*, 8758–8764. [CrossRef]
14. Jarmusch, A.K.; Kerian, K.S.; Pirro, V.; Peat, T.; Thompson, C.A.; Ramos-Vara, J.A.; Childress, M.O.; Cooks, R.G. Characteristic lipid profiles of canine non-Hodgkin's lymphoma from surgical biopsy tissue sections and fine needle aspirate smears by desorption electrospray ionization-mass spectrometry. *Analyst* **2015**, *140*, 6321–6329. [CrossRef]
15. D'Hue, C.A.; Dhawan, D.; Peat, T.; Ramos-Vara, J.; Jarmusch, A.; Knapp, D.W.; Cooks, R.G. Fatty Acid Patterns Detected by Ambient Ionization Mass Spectrometry in Canine Invasive Urothelial Carcinoma from Dogs of Different Breeds. *Bl. Cancer* **2018**, *4*, 283–291. [CrossRef] [PubMed]
16. Goldschmidt, M.; Peña, L.; Rasotto, R.; Zappulli, V. Classification and grading of canine mammary tumors. *Vet. Pathol.* **2011**, *48*, 117–131. [CrossRef] [PubMed]
17. Ràfols, P.; Heijs, B.; Del Castillo, E.; Yanes, O.; McDonnell, L.A.; Brezmes, J.; Pérez-Taboada, I.; Vallejo, M.; García-Altares, M.; Correig, X. rMSIproc: An R package for mass spectrometry imaging data processing. *Bioinformatics* **2020**, *36*, 3618–3619. [CrossRef] [PubMed]
18. Strohalm, M.; Hassman, M.; Košata, B.; Kodíček, M. mMass data miner: An open source alternative for mass spectrometric data analysis. *Rapid Commun. Mass Spectrom.* **2008**, *22*, 905–908. [CrossRef] [PubMed]
19. Fabregat, A.; Jupe, S.; Matthews, L.; Sidiropoulos, K.; Gillespie, M.; Garapati, P.; Haw, R.; Jassal, B.; Korninger, F.; May, B.; et al. The Reactome Pathway Knowledgebase. *Nucleic Acids Res.* **2018**, *46*, D649–D655. [CrossRef] [PubMed]
20. Györffy, B.; Lanczky, A.; Eklund, A.C.; Denkert, C.; Budczies, J.; Li, Q.; Szallasi, Z. An online survival analysis tool to rapidly assess the effect of 22,277 genes on breast cancer prognosis using microarray data of 1809 patients. *Breast Cancer Res. Treat.* **2010**, *123*, 725–731. [CrossRef] [PubMed]
21. Li, Q.; Birkbak, N.J.; Gyorffy, B.; Szallasi, Z.; Eklund, A.C. Jetset: Selecting the optimal microarray probe set to represent a gene. *BMC Bioinform.* **2011**, *12*, 474. [CrossRef] [PubMed]
22. Elston, C.W.; Ellis, I.O. Pathological prognostic factors in breast cancer. I. The value of histological grade in breast cancer: Experience from a large study with long-term follow-up. *Histopathology* **1991**, *19*, 403–410. [CrossRef] [PubMed]
23. Rakha, E.A.; Reis-Filho, J.S.; Baehner, F.; Dabbs, D.J.; Decker, T.; Eusebi, V.; Fox, S.B.; Ichihara, S.; Jacquemier, J.; Lakhani, S.R.; et al. Breast cancer prognostic classification in the molecular era: The role of histological grade. *Breast Cancer Res.* **2010**, *12*, 207. [CrossRef] [PubMed]
24. Canadas, A.; França, M.; Pereira, C.; Vilaça, R.; Vilhena, H.; Tinoco, F.; Silva, M.J.; Ribeiro, J.; Medeiros, R.; Oliveira, P.; et al. Canine Mammary Tumors: Comparison of Classification and Grading Methods in a Survival Study. *Vet. Pathol.* **2018**, *56*, 208–219. [CrossRef] [PubMed]
25. Kuppusamy, K.; Rajan, A.; Warrier, A.; Nadhan, R.; Patra, D.; Srinivas, P. Cytological Grading of Breast Tumors—The Human and Canine Perspective. *Front. Vet. Sci.* **2019**, *6*, 283. [CrossRef]
26. Tao, L.; Huang, G.; Song, H.; Chen, Y.; Chen, L. Cancer associated fibroblasts: An essential role in the tumor microenvironment (Review). *Oncol Lett.* **2017**, *14*, 2611–2620. [CrossRef] [PubMed]
27. Allinen, M.; Beroukhim, R.; Cai, L.; Brennan, C.; Lahti-Domenici, J.; Huang, H.; Porter, D.; Hu, M.; Chin, L.; Richardson, A.; et al. Molecular characterization of the tumor microenvironment in breast cancer. *Cancer Cell* **2004**, *6*, 17–32. [CrossRef]
28. Levental, K.R.; Yu, H.; Kass, L.; Lakins, J.N.; Egeblad, M.; Erler, J.T.; Fong, S.F.T.; Csiszar, K.; Giaccia, A.; Weninger, W.; et al. Matrix Crosslinking Forces Tumor Progression by Enhancing Integrin Signaling. *Cell* **2009**, *139*, 891–906. [CrossRef]

29. Fang, M.; Yuan, J.; Peng, C.; Li, Y. Collagen as a double-edged sword in tumor progression. *Tumour Biol. J. Int. Soc. Oncodevelopmental. Biol. Med.* **2014**, *35*, 2871–2882. [CrossRef] [PubMed]
30. Provenzano, P.P.; Eliceiri, K.W.; Campbell, J.M.; Inman, D.R.; White, J.G.; Keely, P.J. Collagen reorganization at the tumor-stromal interface facilitates local invasion. *BMC Med.* **2006**, *4*, 38. [CrossRef] [PubMed]
31. Arensdorf, A.M.; Diedrichs, D.; Rutkowski, D.T. Regulation of the transcriptome by ER stress: Non-canonical mechanisms and physiological consequences. *Front. Genet.* **2013**, *4*, 256. [CrossRef] [PubMed]
32. Clarke, H.J.; Chambers, J.E.; Liniker, E.; Marciniak, S.J. Endoplasmic Reticulum Stress in Malignancy. *Cancer Cell* **2014**, *25*, 563–573. [CrossRef] [PubMed]
33. Leone, P.; Shin, E.-C.; Perosa, F.; Vacca, A.; Dammacco, F.; Racanelli, V. MHC Class I Antigen Processing and Presenting Machinery: Organization, Function, and Defects in Tumor Cells. *JNCI J. Natl. Cancer Inst.* **2013**, *105*, 1172–1187. [CrossRef] [PubMed]
34. De Charette, M.; Marabelle, A.; Houot, R. Turning tumour cells into antigen presenting cells: The next step to improve cancer immunotherapy? *Eur. J. Cancer* **2016**, *68*, 134–147. [CrossRef]
35. Zhang, Y.; Baig, E.; Williams, D.B. Functions of ERp57 in the Folding and Assembly of Major Histocompatibility Complex Class I Molecules. *J. Biol. Chem.* **2006**, *281*, 14622–14631. [CrossRef] [PubMed]
36. Wang, K.; Li, H.; Chen, R.; Zhang, Y.; Sun, X.-X.; Huang, W.; Bian, H.; Chen, Z.-N. Combination of CALR and PDIA3 is a potential prognostic biomarker for non-small cell lung cancer. *Oncotarget* **2017**, *8*, 57. [CrossRef] [PubMed]
37. Shimoda, T.; Wada, R.; Kure, S.; Ishino, K.; Kudo, M.; Ohashi, R.; Fujita, I.; Uchida, E.; Yoshida, H.; Naito, Z. Expression of protein disulfide isomerase A3 and its clinicopathological association in gastric cancer. *Oncol. Rep.* **2019**, *41*, 2265–2272. [CrossRef]
38. Malhotra, J.D.; Kaufman, R.J. The endoplasmic reticulum and the unfolded protein response. Semin. *Cell Dev. Biol.* **2007**, *18*, 716–731. [CrossRef]
39. Faé, K.C.; da Silva Diefenbach, D.; Bilate, A.M.B.; Tanaka, A.C.; Pomerantzeff, P.M.A.; Kiss, M.H.; Silva, C.A.A.; Cunha-Neto, E.; Kalil, J.; Guilherme, L. PDIA3, HSPA5 and vimentin, proteins identified by 2-DE in the valvular tissue, are the target antigens of peripheral and heart infiltrating T cells from chronic rheumatic heart disease patients. *J. Autoimmun.* **2008**, *31*, 136–141. [CrossRef]
40. Aebi, M.; Bernasconi, R.; Clerc, S.; Molinari, M. N-glycan structures: Recognition and processing in the ER. *Trends Biochem. Sci.* **2010**, *35*, 74–82. [CrossRef] [PubMed]
41. Shibata, Y.; Shemesh, T.; Prinz, W.A.; Palazzo, A.F.; Kozlov, M.M.; Rapoport, T.A. Mechanisms determining the morphology of the peripheral ER. *Cell* **2010**, *143*, 774–788. [CrossRef] [PubMed]
42. Molinari, M.; Helenius, A. Chaperone Selection During Glycoprotein Translocation into the Endoplasmic Reticulum. *Science* **2000**, *288*, 331–333. [CrossRef]
43. Molinari, M.; Galli, C.; Vanoni, O.; Arnold, S.M.; Kaufman, R.J. Persistent Glycoprotein Misfolding Activates the Glucosidase II/UGT1-Driven Calnexin Cycle to Delay Aggregation and Loss of Folding Competence. *Mol. Cell* **2005**, *20*, 503–512. [CrossRef]
44. Hammond, C.; Helenius, A. Folding of VSV G protein: Sequential interaction with BiP and calnexin. *Science* **1994**, *266*, 456–458. [CrossRef] [PubMed]
45. Lakkaraju, A.K.K.; van der Goot, F.G. Calnexin Controls the STAT3-Mediated Transcriptional Response to EGF. *Mol. Cell* **2013**, *51*, 386–396. [CrossRef] [PubMed]
46. Alam, A.; Taye, N.; Patel, S.; Thube, M.; Mullick, J.; Shah, V.K.; Pant, R.; Roychowdhury, T.; Banerjee, N.; Chatterjee, S.; et al. SMAR1 favors immunosurveillance of cancer cells by modulating calnexin and MHC I expression. *Neoplasia* **2019**, *21*, 945–962. [CrossRef]
47. Ryan, D.; Carberry, S.; Murphy, Á.C.; Lindner, A.U.; Fay, J.; Hector, S.; McCawley, N.; Bacon, O.; Concannon, C.G.; Kay, E.W.; et al. Calnexin, an ER stress-induced protein, is a prognostic marker and potential therapeutic target in colorectal cancer. *J. Transl. Med.* **2016**, *14*, 196. [CrossRef] [PubMed]
48. Kobayashi, M.; Nagashio, R.; Jiang, S.-X.; Saito, K.; Tsuchiya, B.; Ryuge, S.; Katono, K.; Nakashima, H.; Fukuda, E.; Goshima, N.; et al. Calnexin is a novel sero-diagnostic marker for lung cancer. *Lung Cancer* **2015**, *90*, 342–345. [CrossRef] [PubMed]
49. Liu, Y.; Komohara, Y.; Domenick, N.; Ohno, M.; Ikeura, M.; Hamilton, R.L.; Horbinski, C.; Wang, X.; Ferrone, S.; Okada, H. Expression of antigen processing and presenting molecules in brain metastasis of breast cancer. *Cancer Immunol. Immunother.* **2012**, *61*, 789–801. [CrossRef] [PubMed]
50. Bork, P.; Doolittle, R.F. Proposed acquisition of an animal protein domain by bacteria. *Proc. Natl. Acad. Sci. USA* **1992**, *89*, 8990–8994. [CrossRef] [PubMed]
51. Erickson, H.P. Reversible unfolding of fibronectin type III and immunoglobulin domains provides the structural basis for stretch and elasticity of titin and fibronectin. *Proc. Natl. Acad. Sci. USA* **1994**, *91*, 10114–10118. [CrossRef] [PubMed]
52. Bian, T.; Zheng, L.; Jiang, D.; Liu, J.; Zhang, J.; Feng, J.; Zhang, Q.; Qian, L.; Qiu, H.; Liu, Y.; et al. Overexpression of fibronectin type III domain containing 3B is correlated with epithelial-mesenchymal transition and predicts poor prognosis in lung adenocarcinoma. *Exp. Med.* **2019**, *17*, 3317–3326. [CrossRef] [PubMed]
53. Obara, M.; Sakuma, T.; Fujikawa, K. The third type III module of human fibronectin mediates cell adhesion and migration. *J. Biochem.* **2010**, *147*, 327–335. [CrossRef] [PubMed]
54. Sato, M.; Jiao, Q.; Honda, T.; Kurotani, R.; Toyota, E.; Okumura, S.; Takeya, T.; Minamisawa, S.; Lanier, S.M.; Ishikawa, Y. Activator of G Protein Signaling 8 (AGS8) Is Required for Hypoxia-induced Apoptosis of Cardiomyocytes: ROLE OF $G\beta\gamma$ AND CONNEXIN 43 (CX43). *J. Biol. Chem.* **2009**, *284*, 31431–31440. [CrossRef] [PubMed]

55. Hayashi, H.; Al Mamun, A.; Sakima, M.; Sato, M. Activator of G-protein signaling 8 is involved in VEGF-mediated signal processing during angiogenesis. *J. Cell Sci.* **2016**, *129*, 1210–1222. [CrossRef] [PubMed]
56. Bell, A.; Bell, D.; Weber, R.S.; El-Naggar, A.K. CpG island methylation profiling in human salivary gland adenoid cystic carcinoma. *Cancer* **2011**, *117*, 2898–2909. [CrossRef]
57. Liu, Y.-P.; Chen, W.-D.; Li, W.-N.; Zhang, M. Overexpression of FNDC1 Relates to Poor Prognosis and Its Knockdown Impairs Cell Invasion and Migration in Gastric Cancer. Technol. *Cancer Res. Treat.* **2019**, *18*, 1533033819869928. [CrossRef]
58. Das, D.K.; Naidoo, M.; Ilboudo, A.; Park, J.Y.; Ali, T.; Krampis, K.; Robinson, B.D.; Osborne, J.R.; Ogunwobi, O.O. miR-1207-3p regulates the androgen receptor in prostate cancer via FNDC1/fibronectin. *Exp. Cell Res.* **2016**, *348*, 190–200. [CrossRef] [PubMed]
59. Cubedo, J.; Padró, T.; Badimon, L. Coordinated proteomic signature changes in immune response and complement proteins in acute myocardial infarction: The implication of serum amyloid P-component. *Int. J. Cardiol.* **2013**, *168*, 5196–5204. [CrossRef] [PubMed]
60. Ishioka, N.; Takahashi, N.; Putnam, F.W. Amino acid sequence of human plasma alpha 1B-glycoprotein: Homology to the immunoglobulin supergene family. *Proc. Natl. Acad. Sci. USA* **1986**, *83*, 2363–2367. [CrossRef]
61. Cortelazzo, A.; de Felice, C.; Leoncini, S.; Signorini, C.; Guerranti, R.; Leoncini, R.; Armini, A.; Bini, L.; Ciccoli, L.; Hayek, J. Inflammatory protein response in CDKL5-Rett syndrome: Evidence of a subclinical smouldering inflammation. *Inflamm. Res.* **2017**, *66*, 269–280. [CrossRef] [PubMed]
62. Zhang, H.; Cao, J.; Li, L.; Liu, Y.; Zhao, H.; Li, N.; Li, B.; Zhang, A.; Huang, H.; Chen, S.; et al. Identification of urine protein biomarkers with the potential for early detection of lung cancer. *Sci. Rep.* **2015**, *5*, 11805. [CrossRef] [PubMed]
63. Tian, M.; Cui, Y.-Z.; Song, G.-H.; Zong, M.-J.; Zhou, X.-Y.; Chen, Y.; Han, J.-X. Proteomic analysis identifies MMP-9, DJ-1 and A1BG as overexpressed proteins in pancreatic juice from pancreatic ductal adenocarcinoma patients. *BMC Cancer* **2008**, *8*, 241. [CrossRef]
64. Kreunin, P.; Zhao, J.; Rosser, C.; Urquidi, V.; Lubman, D.M.; Goodison, S. Bladder Cancer Associated Glycoprotein Signatures Revealed by Urinary Proteomic Profiling. *J. Proteome Res.* **2007**, *6*, 2631–2639. [CrossRef] [PubMed]
65. Kopylov, A.T.; Stepanov, A.A.; Malsagova, K.A.; Soni, D.; Kushlinsky, N.E.; Enikeev, D.V.; Potoldykova, N.V.; Lisitsa, A.V.; Kaysheva, A.L. Revelation of Proteomic Indicators for Colorectal Cancer in Initial Stages of Development. *Molecules* **2020**, *25*, 619. [CrossRef] [PubMed]
66. Liu, Y.; Luo, X.; Hu, H.; Wang, R.; Sun, Y.; Zeng, R.; Chen, H. Integrative Proteomics and Tissue Microarray Profiling Indicate the Association between Overexpressed Serum Proteins and Non-Small Cell Lung Cancer. *PLoS ONE* **2012**, *7*, e51748.
67. Canales Angélica Galicia, N.; Marina Madrid, V.; Castro Salmerón, J.; Jiménez Antúnez, A.; Mendoza-Hernández, G.; McCarron Langley, E.; Roman Bahena, M.; Castro-Romero Ivone, J. A1BG and C3 are overexpressed in patients with cervical intraepithelial neoplasia III. *Oncol. Lett.* **2014**, *8*, 939–947. [CrossRef] [PubMed]
68. Jeong, D.H.; Kim, H.K.; Prince, A.-E.B.; Lee, D.S.; Kim, Y.N.; Han, J.; Kim, K.T. Plasma proteomic analysis of patients with squamous cell carcinoma of the uterine cervix. *J. Gynecol. Oncol.* **2008**, *19*, 173–180. [CrossRef] [PubMed]
69. Zeng, Z.; Hincapie, M.; Haab, B.B.; Hanash, S.; Pitteri, S.J.; Kluck, S.; Hogan, J.M.; Kennedy, J.; Hancock, W.S. The development of an integrated platform to identify breast cancer glycoproteome changes in human serum. *J. Chromatogr. A* **2010**, *1217*, 3307–3315. [CrossRef] [PubMed]

cancers

MDPI

Article

Liquid Biopsy as a Diagnostic and Prognostic Tool for Women and Female Dogs with Breast Cancer

Jucimara Colombo [1], Marina Gobbe Moschetta-Pinheiro [1], Adriana Alonso Novais [1], Bruna Ribeiro Stoppe [1], Enrico Dumbra Bonini [1], Francine Moraes Gonçalves [1], Heidge Fukumasu [2], Luiz Lehmann Coutinho [3], Luiz Gustavo de Almeida Chuffa [4] and Debora Aparecida Pires de Campos Zuccari [1,*]

1 Laboratory of Molecular Investigation in Cancer (LIMC), Department of Molecular Biology, Faculdade de Medicina de São José, São José do Rio Preto 15090-000, Brazil; jucimara.colombo@famerp.br (J.C.); marina.moschetta@docente.unip.br (M.G.M.-P.); adriana.novais@ufmt.br (A.A.N.); bruna.stoppe@edu.famerp.br (B.R.S.); enrico.bonini@edu.famerp.br (E.D.B.); francine.goncalves@unifipa.com.br (F.M.G.)
2 Laboratory of Comparative and Translational Oncology (LOCT), Department of Veterinary Medicine, Faculty of Animal Science and Food Engineering, University of Sao Paulo, Pirassununga 13635-900, Brazil; fukumasu@usp.br
3 Luiz de Queiroz College of Agriculture (ESALQ), University of São Paulo, Piracicaba 13418-900, Brazil; llcoutinho@usp.br
4 Department of Structural and Functional Biology, Institute of Biosciences of Botucatu, Universidade Estadual Paulista, Botucatu 18618-689, Brazil; luiz-gustavo.chuffa@unesp.br
* Correspondence: debora.zuccari@famerp.br; Fax: +55-17-3201-5928

Citation: Colombo, J.; Moschetta-Pinheiro, M.G.; Novais, A.A.; Stoppe, B.R.; Bonini, E.D.; Gonçalves, F.M.; Fukumasu, H.; Coutinho, L.L.; Chuffa, L.G.d.A.; Zuccari, D.A.P.d.C. Liquid Biopsy as a Diagnostic and Prognostic Tool for Women and Female Dogs with Breast Cancer. *Cancers* **2021**, *13*, 5233. https://doi.org/10.3390/cancers13205233

Academic Editors: Felisbina Queiroga and Bruno Cogliati

Received: 28 June 2021
Accepted: 17 September 2021
Published: 19 October 2021

Publisher's Note: MDPI stays neutral with regard to jurisdictional claims in published maps and institutional affiliations.

Copyright: © 2021 by the authors. Licensee MDPI, Basel, Switzerland. This article is an open access article distributed under the terms and conditions of the Creative Commons Attribution (CC BY) license (https://creativecommons.org/licenses/by/4.0/).

Simple Summary: Breast cancer (BC) has common characteristics in women and female dogs, such as high recurrence, metastasis, and mortality rate. In both species, BC prognosis is limited due to its heterogeneous molecular aspects. Although conventional biopsy remains the gold standard for BC diagnosis, liquid biopsy is a very promising tool, especially for patient follow-up. We investigated the effectiveness of liquid biopsy in the diagnosis and follow-up of women and female dogs with BC, using both core biopsy and plasma samples processed by next generation sequencing (NGS) assay. We noted that NGS is a sophisticated technique generating multiple and complex results, which must be validated. Notably, the number of genetic variants increased as the disease progressed. We conclude that liquid biopsy can be considered more effective when performed from the onset of the disease and continues to be applied for monitoring the follow up of BC patients, helping to drive the clinician's decision for medical intervention.

Abstract: Introduction: Breast cancer (BC) is the malignant neoplasm with the highest mortality rate in women and female dogs are good models to study BC. Objective: We investigated the efficacy of liquid biopsy to detect gene mutations in the diagnosis and follow-up of women and female dogs with BC. Materials and Methods: In this study, 57 and 37 BC samples were collected from women and female dogs, respectively. After core biopsy and plasma samples were collected, the DNA and ctDNA of the tumor fragments and plasma were processed for next generation sequencing (NGS) assay. After preprocessing of the data, they were submitted to the Genome Analysis ToolKit (GATK). Results: In women, 1788 variants were identified in tumor fragments and 221 variants in plasma; 66 variants were simultaneously detected in tumors and plasma. Conversely, in female dogs, 1430 variants were found in plasma and 695 variants in tumor fragments; 59 variants were simultaneously identified in tumors and plasma. The most frequently mutated genes in the tumor fragments of women were *USH2A*, *ATM*, and *IGF2R*; in female dogs, they were *USH2A*, *BRCA2*, and *RRM2*. Plasma of women showed the most frequent genetic variations in the MAP3K1, BRCA1, and GRB7 genes, whereas plasma from female dogs had variations in the *NF1*, *ERBB2*, and *KRT17* genes. Mutations in the *AKT1*, *PIK3CA*, and *BRIP* genes were associated with tumor recurrence, with a highly pathogenic variant in *PIK3CA* being particularly prominent. We also detected a gain-of-function mutation in the *GRB7*, *MAP3K1*, and *MLH1* genes. Conclusion: Liquid biopsy is useful to identify specific genetic variations at the beginning of BC manifestation and may be accompanied over the entire follow-up period, thereby supporting the clinicians in refining interventions.

Keywords: breast cancer; circulating tumor DNA (ctDNA); liquid biopsy; next generation sequencing (NGS); gene variant

1. Introduction

Breast cancer (BC) in women and dogs shares similarities, such as high rate of recurrence and metastasis and a high mortality rate [1,2]. BC represents the second leading cause of cancer death in women worldwide [3]. In both species, the prognosis of the disease is often limited since it is a very heterogeneous type of neoplasia, characterized by the coexistence of different cell clones in the tumor. Such heterogeneity is considered the essential driving force for tumor clonal evolution, which can be induced by chemotherapy as a result of the expansion of resistant cell clones; these cell clones are the main cause of tumor recurrence and repeated failures in cancer treatment [4,5].

BC is a complex disease encompassing several distinct entities at the molecular and clinical presentation. Hereditary BC is represented mainly by mutations in the *BRCA1* and *BRCA2* genes [6,7] and accounts for only a small percentage of cases; otherwise, most BCs are sporadic and result from the accumulation of somatic changes [8]. Recent advances in sequencing-based technologies have provided understanding of genetic changes and deregulation of oncogenic signaling pathways, involving growth signaling, stress-related response, metabolism, and cell–cell communication, which affect the growth and progression of cancer. Together with the host's response to cancer, these somatic changes determine the clinical course of the disease [9,10].

Although conventional biopsy remains the gold standard for the diagnosis of BC [11], liquid biopsy is a very promising tool, especially for patient monitoring. Genetic diversity and dynamic changes in genomic profiles can be determined and accompanied through liquid biopsy, which allows for a better precision of prognosis and treatment [12].

Therefore, using circulating tumor DNA (ctDNA), it is possible to detect genetic alterations in the tumor, such as specific mutations that arise from the disease and during therapy, both in human and canine BC. Somatic mutations in *AKT1*, *PIK3CA*, *PTEN* and *TP53* genes are found at high frequency in human BC, representing 26.4% in *PIK3CA*, 24.7% in *TP53*, 3.8% in *PTEN*, and 2.8% in *AKT1*, according to the Somatic Mutations in Cancer Catalog (COSMIC) [13]. These changes are indicative of progression or may be responsible for the chemoresistance and consequent recurrence and spread of tumor, resulting from the process of clonal evolution [14,15].

There is a growing need to monitor the genomic profiles of tumor cells at the beginning of the process and in the patient follow-up, especially to detect disease recurrence and metastasis. However, repeat tissue biopsy is not practicable, whereas circulating tumor cells detached from a primary tumor are present in the bloodstream and can be easily obtained [16,17]. Liquid biopsy using ctDNA is a non-invasive and replicable method, being useful for tumor cell counting, pathological characterization and molecular assay. In addition, liquid biopsy with ctDNA may replace tissue biopsy to predict drug sensitivity and resistance, monitor drug responsiveness, and to examine metastasis [18–20].

In this context, one of the techniques of special interest is next-generation sequencing (NGS), which has been incorporated into clinical practice to identify mutations in cancer patients while targeting treatment with specific drugs [21,22]. The advent of NGS panels in clinical practice favors novel therapeutic choices, particularly for patients with limited therapy options [23].

Therefore, the objective of this work was to verify the effectiveness of liquid biopsy in detecting variants of interest using plasma and tumor fragments obtained during the diagnosis and also during the monitoring of women and female dogs with BC using NGS technique.

2. Materials and Methods

2.1. Ethics

The study was approved by the Research Ethics Committee (CEP) (Protocol number CAAE 83446118.5.00005415) and by the Ethics Committee on the Use of Animals (CEUA) (Protocol number 001-003244/2013) of the Faculdade de Medicina de São José do Rio Preto, (FAMERP).

2.2. Sample Collection

Women attended and were diagnosed with BC at the Gynecology and Obstetrics Section of the Medical School of São José do Rio Preto, (FAMERP). A total of 57 samples were analyzed: plasma (n = 16) and core biopsy (n = 16) from women newly diagnosed with BC and plasma (n = 4) from women, comprising four relapses and 11 remissions; 10 control women without previous disease were also evaluated. The inclusion criterion was the presence of a malignant breast tumor, confirmed by histopathological examination. The exclusion criterion for the control group was the presence or history of cancer. All information regarding the prognosis (response to therapy, development of metastases, recurrences or deaths), and results of anatomopathological and immunohistochemical tests of patients were obtained through consultation in the electronic medical record of the patients.

For canine BC, 37 samples were examined: 11 from tumor fragments and plasma of female dogs with newly diagnosed BC. We also studied eight follow-up dogs (mastectomized) and seven dogs without neoplasms obtained from the Veterinary Clinics in São José do Rio Preto (SP), Jaboticabal (SP), Catanduva (SP), São Paulo (SP), and Sinop (MT), Brazil. The inclusion criteria for the canine species were: presence of breast tumor(s) with or without metastasis and absence of comorbidities. For the control group, the exclusion criteria were: presence or history of neoplasia. Blood and fragment collections for the newly diagnosed group were performed at the time of surgical excision and blood collection from the control group and the follow-up group was carried out in routine and return visits, respectively. In the canine species, information regarding the prognosis (response to therapy, development of metastases, relapses or deaths) were obtained from clinical form.

2.3. DNA and ctDNA Extraction

The DNA and ctDNA of tumor fragments and plasma from women and female dogs were extracted using the AllPrep® DNA/RNA/Protein Mini Kit (Qiagen, Hilden, Germany) and QIAmp Circulating Nucleic Acid Kit (Qiagen®), respectively. The integrity and quality of the extracted DNAs were verified by the Qubit Fluorometric Quantitation equipment (Thermo Fisher Scientific, Santa Clara, CA, USA).

2.4. Preparation of DNA and ctDNA Sequencing Libraries for Illumina System

The genomic libraries were built with the SureSelectXT Low Input Target Enrichment System for Illumina Paired-End Multiplexed Sequencing Library Kit (Agilent Technologies, Santa Clara, CA, USA), following the manufacturer's recommendations. After DNA and ctDNA were fragmented, and end repair and 3'ends adenylation of fragments was performed by adding a nucleotide (a-tailing Mix) to the 3´end in order to prevent them from binding to each other during the ligation of adapters. Bar-coded adapters were ligated to the DNA fragments and a PCR reaction was performed to produce the sequencing libraries. The quality of libraries and quantification were performed using Agilent 2100 Bioanalyser and qPCR with KAPA Library Quantification kit (KAPA Biosystems, Foster City, CA, USA).

2.5. Next-Generation Sequencing

A 20 pM sample pool was loaded into specific cartridges for sequencing on the Illumina MiSeq® with the Illumina® MiSeq Reagent Kits v2 (300 cycles) (Illumina, San Diego, CA, USA). The kit generates around 4.5–5 Gb and running time was approximately 24 h. The sequencing data were extracted in fastQC format, making it possible to check the quality metrics. The samples were distributed in five cartridges. A panel of 168 genes involved in breast carcinogenesis were analyzed (Table S1), being 89 from humans and 79 from dogs. The Illumina Experiment Manager® program (Illumina, San Diego, CA, USA) was oriented to associate each identified read. Vertical and horizontal sequencing coverage was 200 times for DNA samples extracted from tumor fragments and 2000 times for samples extracted from free circulating plasma DNA. This criterion was determined since tumor samples theoretically have higher quality and quantity than free DNA samples circulating in plasma, allowing the number of base readings to be lower.

2.6. Bioinformatics Analysis

The quality values of the sequences were obtained using FastQC. After pre-processing the results, in BAM format, data were subjected to workflow, according to the good practices of GATK (Genome Analysis ToolKit, from the Broad Institute, USA) (Figure 1). The hg38 version of human genome and the CanFam3.1 version of canine genome were used as references for all data processing of women and dogs, respectively. In both species, a pool of normal breast samples (PoN) was used to filter out non-tumor variants. using the tools Mutect2, GenomicsDBImport, and CreateSomaticPanelOfNormals from the GTAK package. For the women's samples, the vcf file of the genomAD project (germline variants) was used to reduce the contamination of germline variants. For the dog's samples, a normal panel of germline variants (https://data.broadinstitute.org/vgb/dog/dog/canFam3/variation/broad.canine.pon.germline.snps.vcf.gz; accessed on 20 May 2021) was also used to reduce contamination of these variants. Only variants which passed through the GATK filters, that is, the somatic variants, were included. The annotation of women and dogs was performed using the Annovar and VEP programs, respectively.

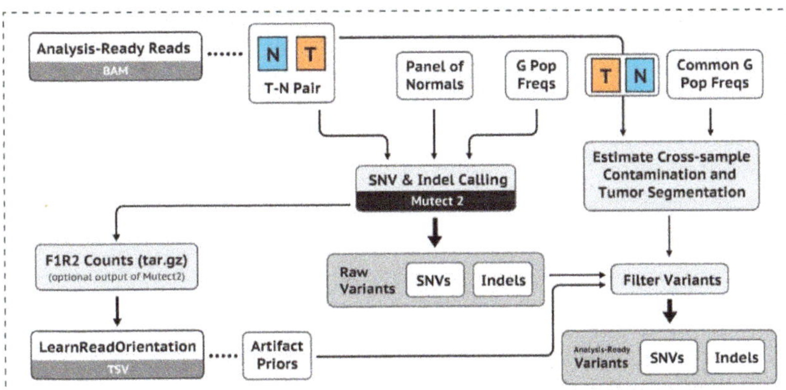

Figure 1. Presentation of the Workflow GAKT (https://gatk.broadinstitute.org/hc/en-us/articles/360035894731-Somatic-short-variant-discovery-SN Vs-Indels-; accessed on 20 May 2021).

The output from the previous step (filtered vcf file) was used for annotation, using the Annovar program, in the case of women, and the VEP program (from Ensembl) in the case of female dogs.

2.7. Statistical Analysis

The results were analyzed by R test (version 3). The association between the clinical outcome of women and the occurrence of mutations was evaluated using Fisher's exact test. Statistical significance was set at $p < 0.05$. Graphics were generated from GATK (Genome Analysis Toolkit, Broad Institute, Cambridge, MA, EUA).

3. Results

3.1. Characterizing Human and Canine Population with BC

Considering the newly diagnosed group, the mean age was 58.18 years, with the majority of women displaying invasive ductal carcinoma (26.66% grade 1 and 73.33% grade 2). The molecular subtype was based on immunohistochemical tests (12.50% luminal subtype A, 31.25% luminal B, 12.50% hybrid luminal, and 37.50% triple negative. The sample P12/C12 was diagnosed as fibroadenoma; however, it was a borderline tumor. The patient had high Ki-67 and underwent radiotherapy. Therefore, despite being diagnosed as fibroadenoma, the tumor had an aggressive behavior and showed important genetic variants in the NGS analysis (Table 1). The mean age of the follow-up group (disease recurrence) was 52.25 years (50% had the luminal molecular subtype B, 25% had HER2 positive and 25% had triple negative) (Table 2). Regarding the patients in remission stage, the mean age was 60.90, and 1 patient (9.09%) was in remission for 4 months, two patients (18.18%) for 3 years, and eight patients (72.72%) for more than five years. Among the treatments used in the remission group, three patients (27.27%) underwent chemotherapy combined with radiotherapy, three (27.27%) received only chemotherapy, 1 (9.09%) received only radiotherapy, one (9.09%) received hormone therapy, one (9.09%) underwent only mastectomy, one (9.09%) underwent mastectomy followed by chemotherapy and radiotherapy, and one (9.09%) was treated with radiotherapy combined with hormone therapy (Table 3). The mean age of the control group was 59.50 years (Table 4).

Table 1. Histopathological features of BC samples in newly diagnosed patients.

Plasma	Core Biopsy	Age	Pathology	TNM	Molecular Subtype	Immunohistochemistry
P01	C01	60	Invasive ductal carcinoma	T2N0M0	Luminal B	C.D.I Grade 2 (ER+/PR+/HER2−/Ki-67+ (20–25%)/E-cadherin+/Cytokeratin 5/6)−
P02	C02	57	Invasive ductal carcinoma	TNM1	Luminal Hybrid	C.D.I. Grade 2 (ER+/PR−/HER2+/Ki-67+ (40%)/E-cadherin+/Cytokeratin 5/6−)
P03	C03	42	Invasive ductal carcinoma	T2N0M0	Luminal Hybrid	C.D.I. Grade 2 (ER+/PR+/HER2+/K-67+ (25%)/E-cadherin+/Cytokeratin 5/6−)
P04	C04	44	Invasive ductal carcinoma	T2N0M0	Luminal B	C.D.I. Grade 2 (ER+/PR+/HER−/KI-67+(90%)/E-cadherin+/Cytokeratin 5/6−)
P05	C05	43	Invasive ductal carcinoma	T4bN3M0	Triple Negative	C.D.I. Grade 2 (ER−/PR−/Her2−/Ki-67+(95%)/E-cadherin+/Cytokeratin 5/6+)
P06	C06	61	Invasive ductal carcinoma	T2N0M0	Luminal B	C.D.I. Grade 2 (ER+/PR+/HER2+/KI-67+/E-cadherin+/Cytokeratin 5/6−)
P07	C07	61	Invasive ductal carcinoma	T3N3M0	Triple Negative	C.D.I. Grade 2 (ER−/PR−/Her2−/Ki-67+(70%)/E-cadherin+/Cytokeratin 5/6+)
P08	C08	88	Invasive ductal carcinoma	T4N1MX	Triple Negative	C.D.I. Grade 2 (ER-/PR-/HER2-/KI-67+(40%)/E-cadherin+/Cytokeratin 5/6+)
P09	C09	53	Invasive ductal carcinoma	T4N1M 0-IIIB	Triple Negative	C.D.I. Grade 2 (ER−/PR−/HER2−/Ki-67+(50%)/ E-cadherin +/Cytokeratin5/6−)

Table 1. *Cont.*

Plasma	Core Biopsy	Age	Pathology	TNM	Molecular Subtype	Immunohistochemistry
P10	C10	42	Invasive ductal carcinoma	T2N0M0	Luminal B	C.D.I. Grade 2 (ER+/PR+/HER+/Ki-67+(50%)/ E-cadherin+/Cytokeratin 5/6+)
P11	C11	62	Invasive ductal carcinoma	T4N3 M0-IIIC	Luminal B	C.D.I. Grade 2 (ER+/PR+ /HER2−Ki-67+(40%)/E-cadherin+/Cytokeratin 5/6−)
P12	C12	46	Fibroadenoma	Not applicable	Not applicable	C.DI. Grade 2 ER+/Ki67+ (50% stroma and 50% epithelium)
P13	C13	68	Invasive ductal carcinoma	T2N0M0	Luminal A	C.D.I. Grade 1 (ER+/PR+/Her2−/Ki-67+(<10%)/E-cadherin+/Cytokeratin 5/6−)
P14	C14	90	Invasive ductal carcinoma	T3N1M0	Luminal A	C.D.I. Grade 1 (ER+/PR+/HER2−/Ki-67+(10%)/ E-cadherin+/Cytokeratin 5/6−)
P15	C15	51	Invasive ductal carcinoma	T3N1M0	Triple Negative	C.D.I. Grade 1 (ER−/PR−/HER2−/Ki-67+(25%)/E-cadherin+/Cytokeratin 5/6−)
P16	C16	63	Invasive ductal carcinoma	T4BN1M0	Triple Negative	C.D.I. Grade 1 (ER−/PR−/HER2−/Ki-67+(>30%)/E-cadherin+/Cytokeratin 5/6−)

TNM classification of malignant tumors: T—primary tumor, N—regional lymph nodes, M—distant metastasis.

Table 2. Histopathological and clinical data of women with BC followed-up with disease recurrence.

Plasma	Age	Pathology	TNM	Molecular Subtype	Immunohistochemistry
Rec1	60	Invasive Breast carcinoma	T2N2MX	Luminal B	C.D.I. Grade 2 (ER+/PR+/HER2−/Ki-67+(40-50%)/E-cadherin+/Cytokeratin 5/6−)
Rec2	47	Invasive ductal carcinoma	T4N0M 0	Luminal B	C.D.I. Grade 2 (ER+/PR−/HER2−/ Ki-67+(70%)/E-cadherin+/Cytokeratin 5/6−)
Rec3	51	Invasive ductal carcinoma	T3N1M0	Triple Negative	C.D.I. Grade 1 (ER−/PR−/HER2−/Ki-67+(25%)/E-cadherin+/Cytokeratin 5/6−)
Rec4	51	Metastatic ductal breast cancer	T2N1Mx	HER2 positive	C.D.I. Grade 2 (ER+/PR+/HER2+/Ki-67+(40-50%)/E-cadherin+/Cytokeratin 5/6−)

TNM classification of malignant tumors: T—primary tumor, N—regional lymph nodes, M—distant metastasis.

Table 3. Data from women with BC in remission.

Plasma	Age	Remission	Treatment
Rem1	83	6 years +	Chemotherapy + Radiotherapy
Rem2	58	5 years +	Chemotherapy + Radiotherapy
Rem3	55	5 years +	Chemotherapy
Rem4	46	5 years +	Mastectomy
Rem5	47	5 years +	Chemotherapy
Rem6	47	5 years +	Hormone therapy
Rem7	65	3 years +	Radiotherapy + Hormone therapy
Rem8	63	5 years +	Radiotherapy
Rem9	53	4 months +	Chemotherapy + Radiotherapy
Rem10	74	5 years +	Chemotherapy + Radiotherapy + Mastectomy
Rem11	79	3 years +	Chemotherapy

Table 4. Data of women used as a control group.

Plasma	Age	Information
Control1	63	Healthy women without a family history of breast cancer and without any other disease
Control2	78	
Control3	41	
Control4	70	
Control5	56	
Control6	73	
Control7	70	
Control8	62	
Control9	32	
Control10	50	

The age of newly diagnosed female dogs ranged from 7 to 16 years, with an average age of 11.72 years. Due to the miscegenation of the population, most animals were SRD, i.e., without a defined race. Based on the histopathological information, three patients were classified as T1, 5 as T2, and 3 as T3; 10 as N0 and 1 as N1; 1 as M1. The histopathological features of breast tumors from 11 female dogs were multiple: represented by three mixed carcinomas, grade I; two complex carcinomas, grade II; one tubular carcinoma, grade I; one breast osteoma, one breast osteosarcoma, one ductal apocrine carcinoma, one complex adenoma, and one mast cell tumor (Table 5). It is noteworthy that ductal apocrine and mast cell carcinomas were initially diagnosed as breast cancer, but later, they were characterized as skin cancers. They were maintained in the study due to its high frequency of mutations. The age female dogs in follow-up ranged from 8 to 13 years, with an average age of 10.62 years. Based on the histopathological information, four patients were classified as T2, 2 as T3; 2 as N0, 3 as N1 and 1 as N2 as N1; 4 as M1. The histopathological features of breast tumors from 8 female dogs were: one papillary carcinoma, grade II; one tubular carcinoma, grade III; one carcinosarcoma, one mixed carcinoma, grade II; one osteosarcoma; and one complex carcinoma, grade I (Table 6).

Table 5. Histopathological and clinical data of female dogs with newly diagnosed BC.

	Plasma	Age (years)	TNM	Pathology	Castrated	Breed
F1	P1c	10	T2N0M0	Mixed Carcinoma, grade I	Yes	MBD
F2	P2c	15	T2N0M1	Osteosarcoma	Yes	Pekinese
F3	P3c	7	T2N0M0	Complex carcinoma, grade II	No	MBD
F4	P4c	10	T1N0M0	Complex adenoma	No	MBD
F5	P5c	10	T1N0M0	Osteoma	No	Poodle
F6	P6c	7	T3N0M0	Mast cell tumor	Yes	Beagle
F7	P7c	12	T3N0M0	Ductal apocrine carcinoma	No	Dachshund
F8	P8c	16	T2N0M0	Mixed carcinoma, grade I	Yes	MBD
F9	P9c	13	T1N1M0	Tubular carcinoma, grade I	Yes	Yorkshire
F10	P10c	16	T3N0M0	Mixed carcinoma, grade I	Yes	MBD
F11	P11c	13	T2N0M0	Complex carcinoma, grade II	Yes	Belgian Shepherd

TNM classification of malignant tumors: T—primary tumor, N—regional lymph nodes, M—distant metastasis. MBD = mixed-breed dogs.

Table 6. Histopathological data of follow-up female dogs with BC.

Plasma	Age (years)	TNM	Pathology	Castrated	Breed
Fu1	13	*	*	Yes	Poodle
Fu2	12	*	*	No	Maltese
Fu3	10	T2N0M1	Papillary carcinoma, grade II	Yes	Labrador
Fu4	10	T2N1M1	Tubular carcinoma, grade III	Yes	SRD
Fu5	8	T2N1M0	Carcinosarcoma	Yes	Beagle
Fu6	9	T3N1M0	Mixed carcinoma, grade II	Yes	Poodle
Fu7	13	T3N0M1	Osteosarcoma	Yes	Cocker sp
Fu8	10	T2N2M1	Complex Carcinoma, grade I	No	MDB

* Information not available. TNM classification of malignant tumors: T—primary tumor, N—regional lymph nodes, M—distant metastasis. MBD = mixed-breed dogs.

3.2. Detection of Individual Relevant Genetic Variants in Tumor Fragments and in Plasma of Women with BC

To examine the efficacy and functionality of liquid biopsy, high quality samples were extracted from plasma and tumor fragments to generate genomic library construction. In women, a total of 1788 and 221 gene variants were detected in tumor fragments and plasma, respectively; 66 gene variants were simultaneously identified in patients' tumors and plasma. Among these variants, 24 were located in the exon regions (exonic variants) (Table S2).

The most commonly mutated variant found in women fragments was the *missense* variant. The main single nucleotide variant in tumor fragments of women was C > A (57.49%) and C > T (29.73%) (Figure 2A,B). The most frequent top mutated genes in tumor fragments of women were *USH2A, ATM, IGF2R, MKI67,* and *MAP3K1* (median variants per sample = 20; missense, nonsense, and splice site mutations) (Figure 2C,D).

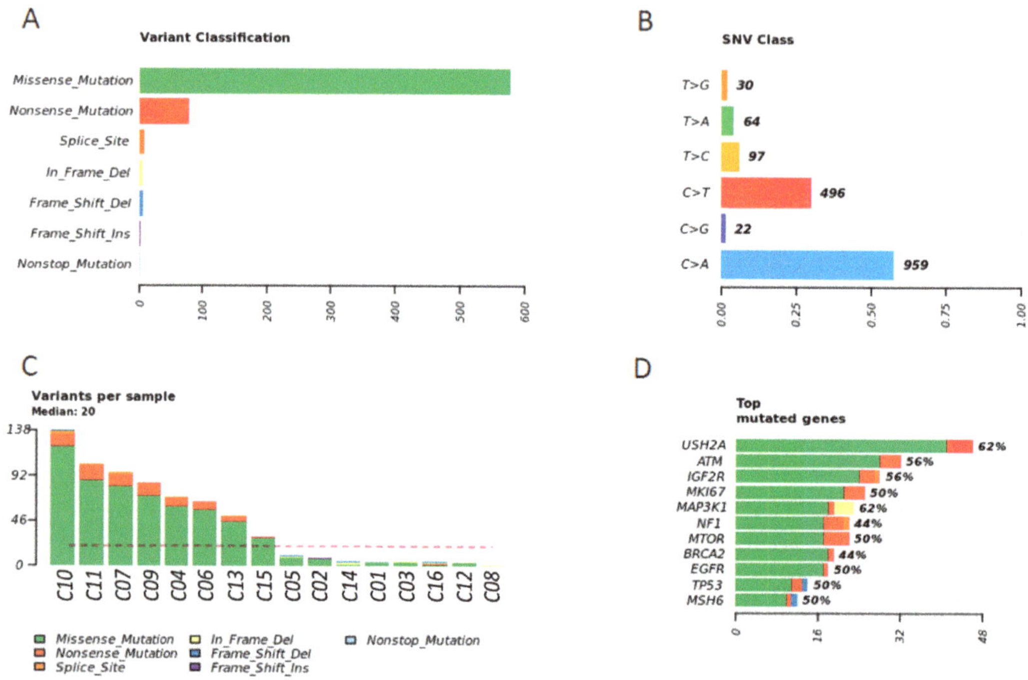

Figure 2. Genetic variation in tumor fragments of women with BC. (**A**) Representative variant classification. (**B**) Single-nucleotide variant (SNV) classification. (**C**) Representative variants per fragment sample. (**D**) Top mutated genes showing multiple variant mutations. Graphics were generated in the R language, using the Maftools package.

After examining the plasma of BC patients, we observed that missense variation was the most common or abundant mutated variant in women. The main single-nucleotide variant in plasma of women was C > T (53.65%) and C > A (17.07%) (Figure 3A,B). The most frequent top mutated gene in the plasma of women were *MAP3K1, BRCA1, GRB7, USH2A*, and *TP53* (median variants per sample = 2; missense, inframe deletion, and frameshift mutations) (Figure 3C,D).

3.3. Detection of Individual Relevant Genetic Variants in Tumor Fragments and in Plasma of Dogs with BC

In female dogs, a total of 695 and 1430 gene variants were detected in tumor fragments and plasma, respectively; 59 gene variants were simultaneously identified in female dogs' tumors and plasma (Table S3).

The most commonly mutated variant found in female dog fragments was the *missense* variant. The predominant nucleotide variants in female dogs was T > C (23.89%) and C > T (20.68%) (Figure 4A,B). The most frequent top mutated genes in female dogs, *USH2A, BRCA2* and *RM2, CEP55*, and *GATA3* genes were the most representative mutated genes (median variants per sample = 31; missense, nonsense, splice site, and frameshift mutations) (Figure 4C,D).

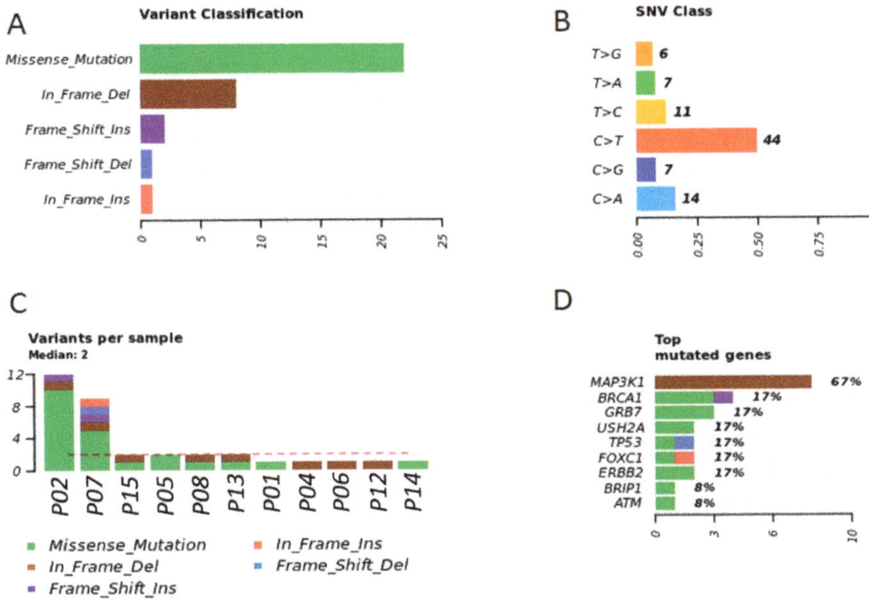

Figure 3. Genetic variation in plasma samples of women with BC. (**A**) Representative variant classification. (**B**) Single-nucleotide variant (SNV) classification. (**C**) Representative variants per plasma sample. (**D**) Top mutated genes showing multiple variant mutations. Graphics were generated in the R language, using the Maftools package.

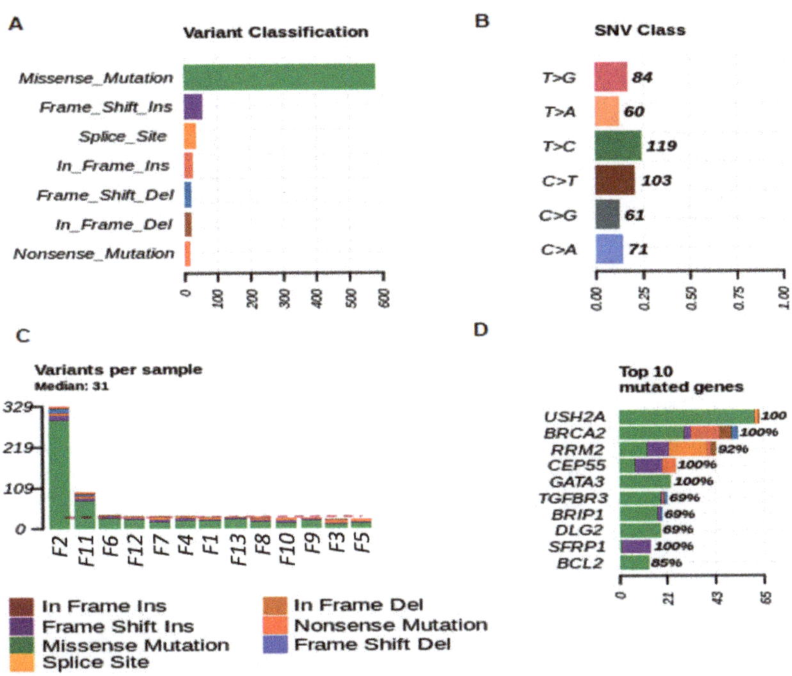

Figure 4. Genetic variation in tumor fragments of dogs with BC. (**A**) Representative variant classification. (**B**) Single-nucleotide variant (SNV) classification. (**C**) Representative variants per fragment sample. (**D**) Top mutated genes showing multiple variant mutations. Graphics were generated in the R language, using the Maftools package.

After examining the plasma of BC patients, we observed that *missense* variation was the most common or abundant mutated variant in canine samples. The main single nucleotide variant in plasma of female dogs was C > T (29.27%) and T > C (26.42%) (Figure 5A,B). The most frequent top mutated genes in the plasma of women were *MAP3K1, BRCA1, GRB7, USH2A*, and *TP53* (median variants per sample = 2; missense, inframe deletion, and frameshift mutations). In female dogs, *NF1, ERBB2, KRT17, PMS2*, and *MAP3K1* genes were the most representative mutated genes (median variants per sample = 56; missense and frameshift mutations) (Figure 5C,D).

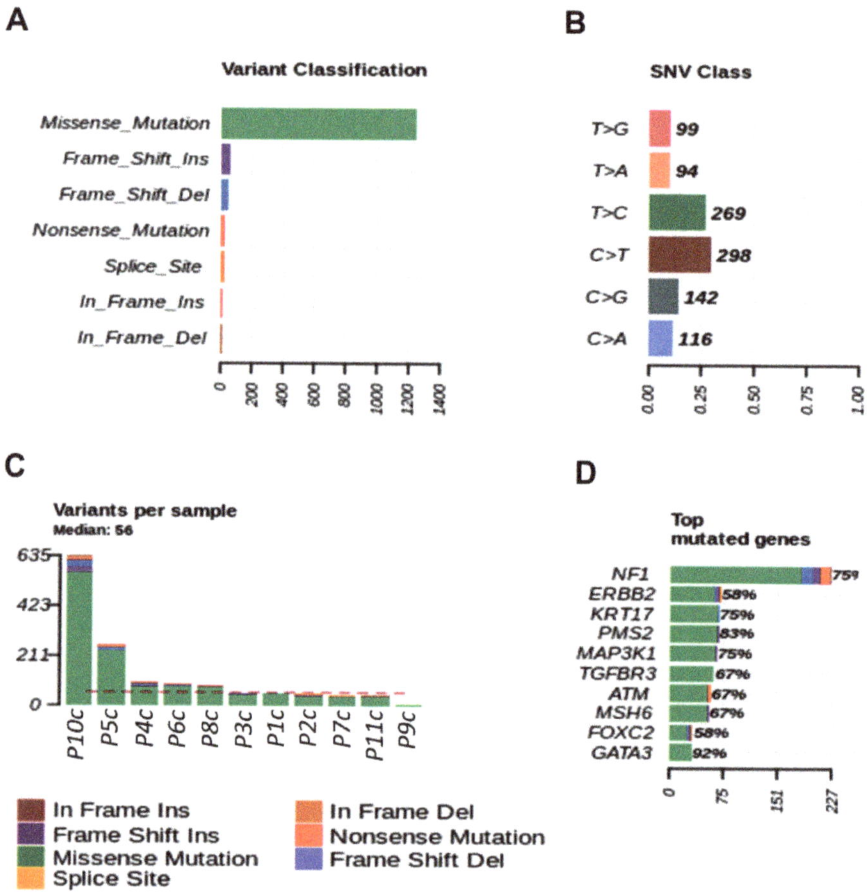

Figure 5. Genetic variation in plasma samples of dogs with BC. (**A**) Representative variant classification. (**B**) Single-nucleotide variant (SNV) classification. (**C**) Representative variants per plasma sample. (**D**) Top mutated genes showing multiple variant mutations. Graphics were generated in the R language, using the Maftools package.

3.4. Breast Cancer Driver Genes in Women Share Similar Variant Mutations in Tumor Fragments and Plasma

We sequenced 89 BC-related genes in 15 newly diagnosed breast cancer patients. Among the sequenced gene mutations, 40 genes exhibited at least one driver somatic mutation that included missense (64.18%), nonsense (8.89%), inframe (0.44%), frameshift (0.66%), synonymous (25.58%), startloss (0.11%) and stoploss (0.11%) mutations. The five most affected genes were *MAP3K1* (altered in 10 samples, 62%), followed by *USH2A* (10 cases, 62%), *ATM* (9 cases, 56%), *IGF2R* (9 cases, 56%), and *EGFR* (8 cases, 50%) (Figure 6A). Moreover, 89 genes were sequenced in plasma samples of 16 newly diagnosed breast cancer patients, and

18 genes presented missense (53.19%), inframe (19.14%), frameshift (6.38%) and synonymous (21.27%) mutations. The most affected genes were *MAP3K1* (altered in eight samples, 67%), *BRCA1, ERBB2, FOXC1,* and *GRB7* (2 cases, 17%) (Figure 6B).

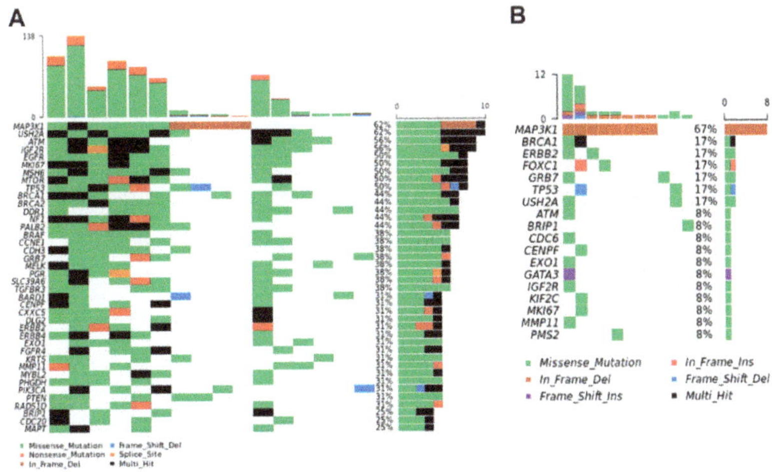

Figure 6. Oncoplots depicting the distribution of most representative variants in BC-associated genes regarding the individual tumor fragments (**A**) and plasma samples (**B**) of women. The upper plot shows the frequency of mutation for each tumor sample and lower left plots exhibit the mutations in each tumor sample (most deleterious mutation types are shown). The lower right plots indicate the frequency of samples mutated in fragments and plasma of patients. Graphics were generated from the R language, using the Maftools package.

Considering the common variants shared by tumor fragments and plasma of women, we identified important gene variations in *ACTR3B* (C > T mutation), *BRCA1* (T > A), *CDC6* (C > T), *CENPF* (T > A), *CHEK2* (T > C), *EXO1, GATA3,* and *GRB7* (G > A), *IGF2R* (G > A and T > C), *KIF2C* (G > A), *KRT5* (G > T and A > TG), *MAP3K1* (C > T and CAA > -), *MKI67* (T > C), *MMP11* (T > C), *MYBL2* (C > T), *PMS2* (C > T), *TP53* (T > C), *TYMS* (T > C), *USH2A* (G > A, G > T, and T > G) (Figure 7A). Notably, while the somatic mutation rate of *BRCA1* was 43.7%, the mutation rate of *TP53* was 50% (Figure 7B).

We also performed the analysis of genetic variants in the plasma of women with BC who are in remission or with recurrence. Thus, in women in remission, we observed missense (25.14%), nonsense (0.28%), inframe (0.57%), frameshift (0.57%) and synonymous (73.40%) mutations. On the other hand, women with recurrent disease presented missense (85.71%) and inframe (14.28%) mutations. We further identified 10 variants coexisting between patients in remission and after relapsed disease which were detected in *AKT1, NF1, AURKA, MSH6, MAP3K1, ALOX12-A, MTOR, TYMS, CDH1,* and *DLG2* genes (Figure 7C). Moreover, in women, there was also an association between mutations in *AKT1* (CA > TG, $p = 0.01$), *PIK3CA* (A > G, $p = 0.02$) and *BRIP* (T > C, $p = 0.02$) genes with tumor recurrence, highlighting a highly pathogenic variant in *PIK3CA* gene, according to CLINVAR. We observed that women in remission had more variants than women with recurrence. This probably occurred because some patients classified within the remission group did not show, at the time of sample collection, clinical evolution of disease; however, the disease progressed during the course of the study.

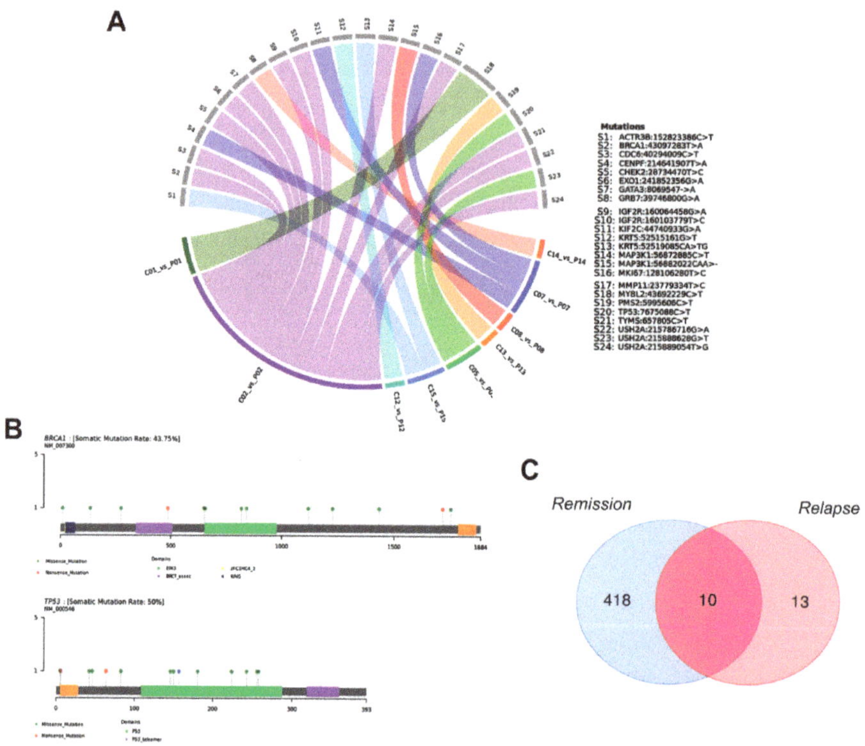

Figure 7. Gene variants shared by tumor fragments and plasma samples of women with BC. (**A**) Circos plot depicting the correlation between fragment and plasma shared samples and gene variants in patients. (**B**) Gene map of the somatic mutation in *BRCA* and *TP53* genes. (**C**) Total variants detected during remission and relapse of the disease. Graphics were generated from the R language, using the VennDiagram, circlize, dplyr and reshape2 packages.

3.5. Breast Cancer Driver Genes in Dogs Share Similar Variant Mutations in Tumor Fragments and Plasma

Regarding the female dogs, we sequenced 79 BC-related genes, and those with high number of variants shared by tumor fragments and plasma were *GATA3* and *mTOR* (missense), *SFRP1* (frameshift insertion), *BRCA2* (multi hit), *FOXC2, ATM, TGFBR3,* and *BRIP1* (missense and multi hit mutations). A total of 56 genes exhibited at least one somatic mutation in tumor fragments including missense (76.83%), nonsense (1.87%), inframe (5.88%), frameshift (10.92%) and splice (4.46%) mutations. The most affected genes were *BRCA2* (altered in 13 samples, inframe and multi hit mutations), *CEP55* (13 cases, frameshift, missense, and multi hit mutations), *GATA3* (13 cases, missense mutation), *SFRP1* (13 cases, frameshift and multi hit mutations), and *USH2A* (13 cases, missense and multi hit mutations) (Figure 8A). Additionally, from 79 genes sequenced in plasma samples of dogs with BC, we identified 61 genes harboring missense (87.27%), nonsense (1.88%), inframe (1.87%), frameshift (7.47%), and splice (1.46%) mutations. The most affected genes were *GATA3* (altered in 11 samples, missense mutation), *PMS2* (10 cases, missense and multi hit mutations), *KRT17* (9 cases, missense and multi hit mutations), *MAP3K1* (9 cases, missense, splice, and multi hit mutations), and *NF1* (9 cases, missense and multi hit mutations) (Figure 8B).

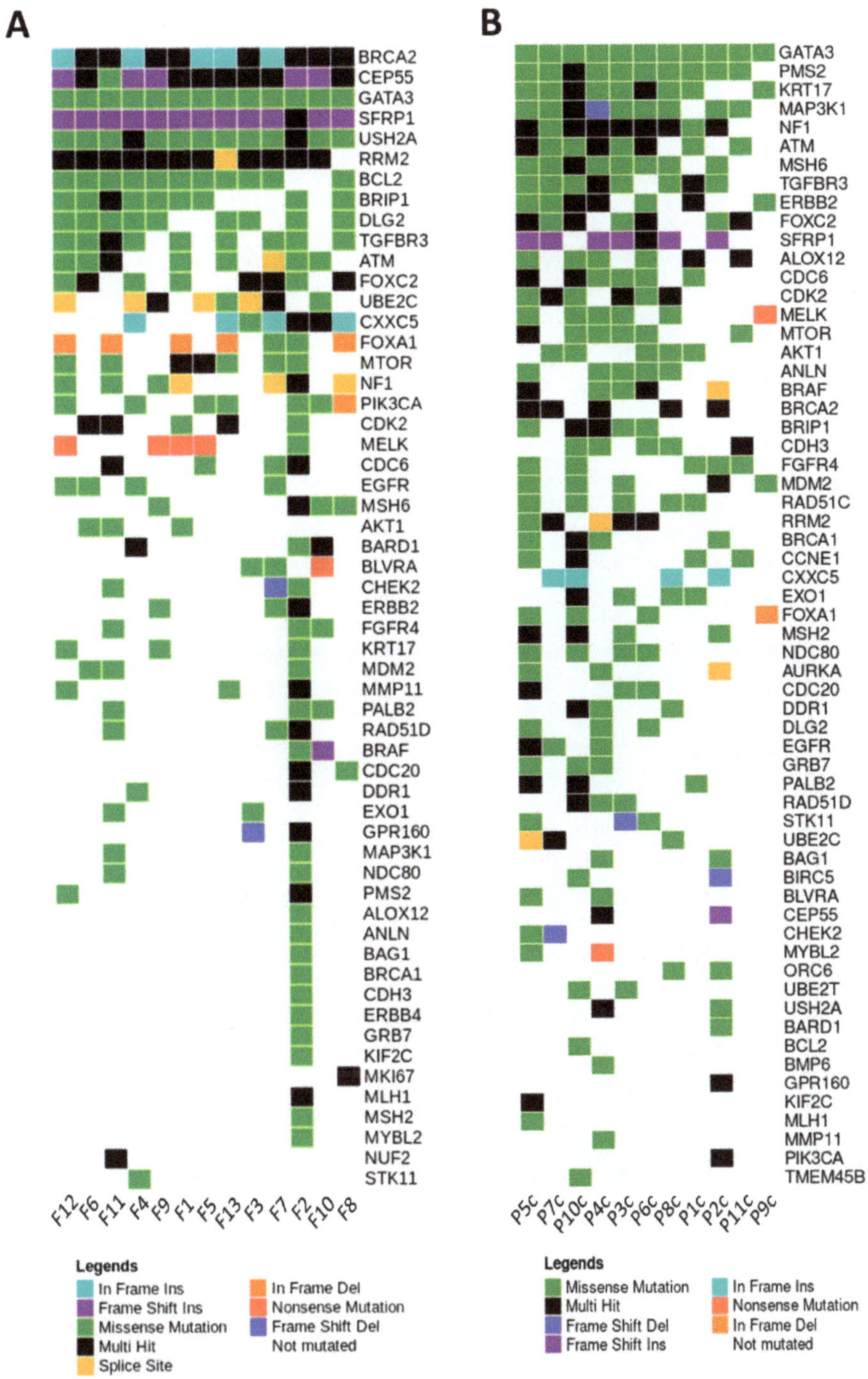

Figure 8. Oncoplots depicting the distribution of most representative variants in BC-associated mutated genes regarding the individual tumor fragments (**A**) and plasma samples (**B**) of dogs. The plots exhibit the mutations in each tumor sample (most deleterious mutation types are shown). Graphics were generated from the R language, using the Maftools package.

Considering the most representative variants shared by tumor fragments and plasma of dogs, we identified important gene variations in *ATM* (C > A mutation), *CCNE1* (G > C), *FGFR4* (C > T), and *GATA3* (G > A and C > CT) (Figure 9). The genetic variants in the plasma of dogs with BC were evaluated during the follow-up. The most common mutations were missense (75.37%), inframe (12.67%), frameshift (7.46%) and splice (4.47%). In dogs, the G > C variant (missense mutation) of *CCNE1* gene was detected in all paired samples of fragments and plasma (100.00%). In addition, the G > A and C > CT variants (missense mutation) of *GATA3* gene were commonly found in nine (81.8%) and eight (72.7%) paired fragment and plasma samples, respectively (Figure 9A). Figure 9B highlights the correlations between gene variants and female dogs' samples. We further explored blood samples of dogs to identify gene variants during the patient follow-up. Of note, missense mutation in *AKT1, ATM, CDK2, GATA3,* and *ERBB2* genes, and inframe deletion in *mTOR* gene were commonly detected in both disease remission and metastasized disease (Figure 10), thereby revealing potential targets to be screened during the course of disease.

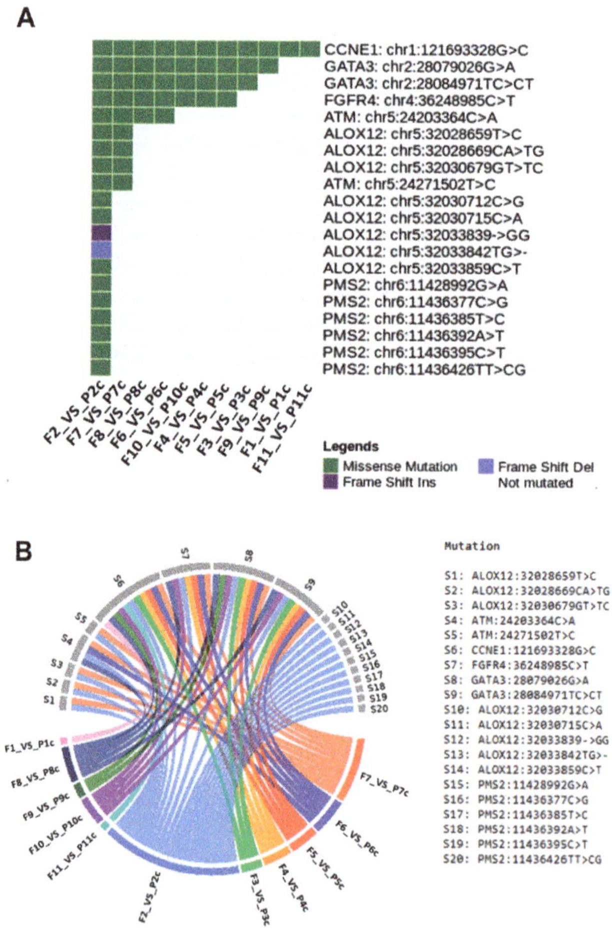

Figure 9. Genetic mutations shared by tumor fragments and plasma samples of dogs. (**A**) Common genetic variants in 11 corresponding samples. (**B**) Circos plot depicting the correlation between samples and gene variants in female dogs. Graphics were generated from the R language, using the Maftools, circlize, dplyr and reshape2 packages.

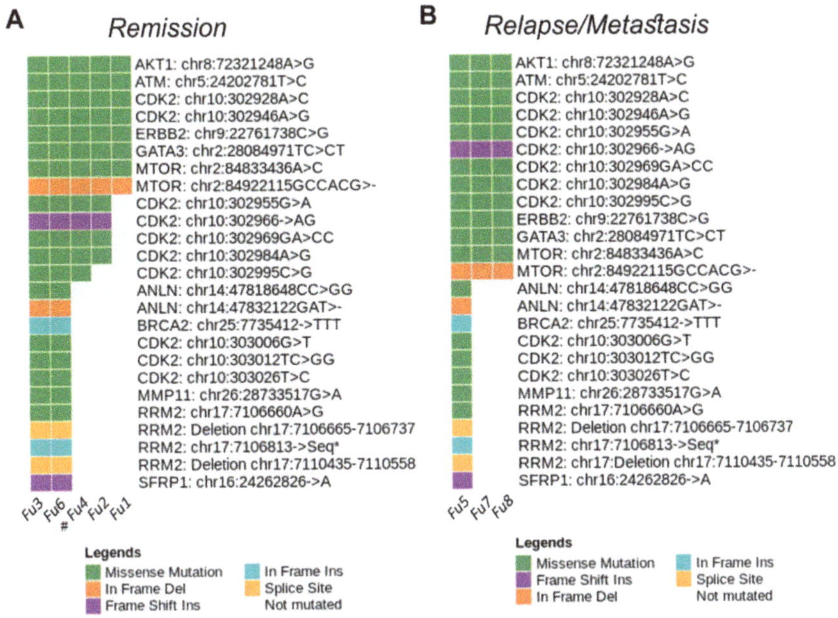

Figure 10. Genetic variant mutations identified in the plasma samples of dogs obtained during remission (**A**) and after disease relapse (**B**). Graphics were generated in the R language, using the Maftools package.

3.6. Main Variants Associated with Disease Progression Shared by Both Species

We found that the number of variants increased with disease progression. Women classified in the remission group (Rem5, Rem6, Rem7, Rem8 and Rem9) did not show clinical development of the disease at the time of sample collection. However, liquid biopsy analysis revealed a high number of mutations in these patients. This observation was consistent with the subsequent clinical evolution, as in these patients the disease progressed with the appearance of metastases.

After comparing the data between women and female dogs, both in the fragment and plasma samples, we observed important genes which were commonly altered. In tumor fragments, the *BRCA2* and *USH2A* genes showed many variants in both species. Otherwise, in the plasma, we detected several variants in the *ATM*, *ERBB2* and *MAP3K1* genes shared between women and female dogs.

4. Discussion

We performed genomic analysis in 57 BC samples in women, using *core biopsy* of fragments and plasma. By recognizing the similarities of BC molecular background in human and canine species, we further evaluated 37 BC samples (biopsy of fragments and plasma) in dogs. The canine mammary tumor (CMT) has been proposed as a promising model resembling woman BC, since they share common components regarding the physiopathology of the disease [2,24]. We extracted DNA from these samples to construct a genomic library for NGS, ensuring the high quality of the procedures with respect to the efficacy and functionality of liquid biopsy. While women had more mutations in the fragments vs. plasma, the dogs showed more mutations in the plasma vs. tumor fragments; this fact could be explained by individual tumor heterogeneity. The mammary tumors of female dogs are generally bulkier and more numerous than those of women. Therefore, fragments removed from female dogs will not represent all existing tumor clones. On the contrary, since BC is smaller in women, the fragment achieves a greater representation of the tumor as a whole [2,25].

Women with BC classified into the remission group had a high number of mutations. This fact was consistent with their later clinical evolution, as these patients presented disease progression with the occurrence of metastasis. This supports the importance of liquid biopsy in detecting mutations that can predict the clinical course of disease. The investigation of tumor variants in ctDNA, isolated from plasma, may provide substantial information regarding the tumor, such as the development of chemoresistance and presence of residual and recurrent disease [19,26].

These events occur because the genomic characteristics of a metastatic tumor are different from those found in a primary tumor, due to the interval between occurrences. Furthermore, genomic differences are intensified after treatment, including chemotherapy [27]. In order to monitor the mutational dynamics during a treatment, the acquisition of tissue biopsies would be inadequate and difficult to perform, since they are invasive procedures with potential clinical complications. Thus, the assessment of ctDNA is indisputably an ideal method, allowing the monitoring of response to treatment by evaluating the dynamics of genetic changes that occur in the tumor during exposure, for example, to a chemotherapeutic agent, being excellent for anticipating the recognition of relapse. In addition, conventional biopsy has shown limitations in terms of tumor representation, since neoplastic cells are heterogeneous [12,16].

Our data showed that most BC analyzed was classified as luminal type and missense mutations were the most frequent. In fact, this is in agreement with previous studies which documented that primary BC had more missense mutations in luminal/ER+ and epidermal growth receptor 2 positive (HER2+) subtypes while in triple negative breast cancers (TNBCs), nonsense, frameshift, and complex mutations were more common variations [8].

The *USH2A* gene was one of the most frequently mutated genes observed in tumor fragments from women and dogs, as well as in the plasma of women. This gene encodes the Usherin protein, which contains laminin EGF motifs, a pentraxin domain and several type III fibronectin motifs [28]. The mutation spectrum is very heterogeneous and includes over 1500 mutations with more than 690 variants presumed to be pathogenic, which span the whole USH2A gene, consisting of nonsense, missense, deletions, duplications, splicing variants and pseudo-exon inclusion variants [29].

Li et al. [30] detected *USH2A* variants in families that did not carry germline mutations of the *BRCA1* and *BRCA2* genes. They further concluded that these variants did not increase the risk of BC. In turn, Natraja et al. [31] verified the expression of chimeric transcripts in a chromosomal region comprising the *USH2A* gene in samples of micropapillary breast carcinomas. Furthermore, Santarpia et al. [8] also found variants in the USH2 gene in women with triple negative BC, which has a more reserved prognosis. In a recent work by Kim et al. [32], pathogenic germline variants of this gene were found in patients with BC. This supports the need for further studies that correlate variants of *USH2A* gene with BC risks.

In women, the *ACTR3B* gene showed concordance and a high frequency of somatic mutation in most tumor/plasma pairs. This gene is related to cytoskeleton and cell motility [33]. However, to date, there are no studies correlating *ACTR3B* changes with BC. The *AURKA* gene also showed concordance and high frequency of somatic mutation in most tumor/plasma pairs in women being further associated with tumor recurrence. This proto-oncogene encodes a protein kinase that plays a central role in mitosis and its overexpression or amplification was observed in several types of tumors, including BC. *AURKA* overexpression has been associated with a more aggressive phenotype and worse prognosis in BC patients [34].

The *CCNE1* and *GATA3* genes harbored somatic variants in most of the fragment/plasm a pairs in dogs. The *CCNE1* gene encodes cyclin E, a key kinase complex for cell cycle regulation from G1 to S phase. *CCNE* gene amplification is highly associated with the development of BC, especially TNBC [35]. According to Huang et al. [7], *CCNE1* amplification represents an independent risk factor in non-*BRCA* carriers with TNBC. Moreover,

Zhao et al. [35] reported that *CCNE1* amplification may confer resistance to chemotherapy and is associated with poor overall survival in patients with TNBC.

GATA3 has emerged as a prominent transcription factor required for maintaining mammary gland homeostasis, and its loss is associated with aggressive BC development. Yu et al. [36] reported that *GATA3* and *UTX* expressions are positively correlated and demonstrated that GATA3/UTX complex synergistically regulates a cohort of genes including *DICER* and *UTX*, which are critically involved in the epithelial-to-mesenchymal transition (EMT). The *GATA3* gene encodes a transcription factor and is mutated in about 10% of BC. Recent genomic analysis of human BCs revealed high-frequency mutation in *GATA3* in luminal tumors, suggesting an important driver function [37]. Notably, the alterations found in tumor and plasma of women with BC were also observed in the work of D'Amico et al. [38]. According to the authors, the apparent lack of agreement between some specific alterations is potentially determined by intratumoral heterogeneity and can be overcome by plasma-based analysis of variants.

In the present study, we detected a pathogenic variant in the *PIK3CA* gene correlated with tumor recurrence. According to the Catalog of Somatic Mutations in Cancer (COSMIC), somatic mutations in the *PIK3CA* gene occur in about 26.4% of BCs in women [13]. The *PIK3CA* gene is a proto-oncogene highly mutated in many tumor types. The PIK3CA protein is activated by growth factors via direct interaction with receptors in the presence of adapter proteins such as the IRS proteins. These interactions recruit PI3K to its substrate, phosphatidyl-inositol 4,5-bisphosphate (PIP2), allowing the generation of the second lipid messenger phosphatidyl-inositol 3,4,5-triphosphate. Mutations in *PIK3CA* gene are associated with increased risk of advanced BC, resistance to hormonal treatment, increased risk of metastasis, and worse prognosis. The same mutations are correlated with being hormone receptor positive and HER2 negative in about 40% of BC patients and are associated with tumor growth, resistance to endocrine treatment, and poor overall prognosis. More recent studies have shown that the use of *PIK3CA* inhibitors has protective effects in women with advanced BC [39–41].

A pathogenic variant in *PIK3CA* gene was found in canine plasma following recurrent CMT and in newly diagnosed female dogs. Lee et al. [42] reported that *PIK3CA* was the most frequently mutated gene in CMT (45% of cases). Furthermore, canine *PIK3CA* A3140G (H1047R), which is known as the mutational hotspot of human BC, was also a hotspot in CMT. Targeted sequencing confirmed that 29% of CMTs had the same *PIK3CA* A3140G mutation. Integration of the transcriptome suggested that *PIK3CA* (H1047R) induces cell metabolism and cell cycle via an increase in *PCK2* and a decrease in *CDKN1B* without affecting apoptosis. The authors also identified other significantly mutated genes in the dogs, including *SCRN1* and *CLHC1*, which were not reported in the human BC. Nevertheless, we couldn't identify *SCRN1* and *CLHC1* variants in the present study [42].

According to Sobhani et al. [43], the presence of a *PIK3CA* mutation represents an independent negative prognostic factor in BC in women. *PIK3CA* mutations in canine tumors may alter downstream molecules of PI3K/Akt/mTOR signaling pathway [44]. Alsaihati et al. [45] studied 182 samples from dogs and 886 samples from human BC, and described that CMT harbors frequent PI3K pathway alteration and *PIK3CA* H1047R mutation. They reinforced PI3K signaling as the most frequently altered pathway in both human BC and CMT.

The *BRCA2* gene also showed a large number of variants in CMT. The canine *BRCA2* is a tumor suppressor gene which encodes the BRCA2 protein, involved in DNA repair through interaction with RAD51 recombinase. This process is mediated by eight BRC repeats that are encoded by *BRCA2* exon 11. Maués et al. [46] investigated the frequency of variants in *BRCA2* exon 11 using 48 blood and tissue DNA samples from CMT. Seven single nucleotide polymorphisms (SNPs) were identified, three of which were evaluated as possibly or probably deleterious variants. Importantly, a total of 97.9% of dogs had one to three polymorphisms considering the seven SNPs identified in this study, suggesting a possible correlation between the canine *BRCA2* exon 11 polymorphisms and mammary carcinogenesis. According to a recent work by Oliveira et al. (2021) [47], SNPs are common in the *BRCA2* gene of female dogs with mammary tumors. In this work, the group studied, through liquid biopsy, germline genetic variants in 20 plasma samples from dogs with mammary cancer. Thus, eleven single nucleotide polymorphisms (SNPs) were detected, most of them in the exon 11, and two indels (deletion/insertion) in the *BRCA2* gene.

Previously, Yoshikawa et al. [48,49] studied the expression level of canine *BRCA2* gene and confirmed a reduced level in mammary tumor samples compared with healthy mammary gland, thereby associating this occurrence with canine mammary tumorigenesis. The open reading frame contained four missense variations, one insertion variation, and one silent variation, some of which were located in functional domains. Huskey et al. [50] performed whole genome sequencing on 14 purebred dogs diagnosed with mammary tumors from four breed-specific pedigrees (Golden Retriever, Siberian Husky, Dalmatian, and Standard Schnauzer) and highlighted variants in orthologs of human BC susceptibility genes. They identified variants in *BRCA2* and *STK11* genes that appear to be associated with CMT risk and demonstrated that these variants were identified in both fragment and plasma of dogs, being not frequent in women.

On the basis of liquid biopsy precision, our premise is to create clinical strategy options in oncology, with the objective of providing a more accurate and effective treatment to each patient, based on the individual genetic profile of the malignancy. Genetic diversity and dynamic change in genomic profiles of patients may be determined and accompanied by liquid biopsy, which allows for better treatment efficacy, structuring individualized therapeutic strategies [51,52].

Caveats and limitations of the study are related to the number of control samples. Because some control samples were not in sufficient quality for NGS analysis, we prioritize the evaluation of tumor and plasma samples obtained from BC patients. However, we emphasize that the number of control samples (n = 10) satisfies all statistical criteria by providing a reliable comparative data analysis. Moreover, among the tumor samples, there is a benign tumor, fibroadenoma. This sample remained in the study because it showed an aggressive behavior, with high expression of Ki-67 and important genetic alterations, which demonstrates the need for a more detailed study of this type of tumor.

5. Conclusions

We showed that liquid biopsy is useful for characterizing genetic variants and can help physicians choose a more appropriate medical intervention. In addition, liquid biopsy proved to be efficient in identifying the similarity of mutations in specific genes in both human and canine mammary tumors. Finally, liquid biopsy is an excellent method to detect new genetic mutations in the early stages and follow-up of breast cancer patients.

Supplementary Materials: The following are available online at https://zenodo.org/record/5033846#.YUlJD7gzZPZ, Table S1: Panel of 168 genes involved in breast carcinogenesis in women and female dogs. Table S2: Genetic variants detected in fragments, in plasma and in both, in women with breast cancer. Table S3: Genetic variants detected in fragments, in plasma and in both, in female dogs with breast cancer.

Author Contributions: D.A.P.d.C.Z. conceived the study, designed the experiments and drafted the manuscript. J.C., M.G.M.-P., B.R.S., E.D.B. and F.M.G. carried out the experiments. L.L.C. helped in NGS. J.C., A.A.N., H.F. and L.G.d.A.C. assisted in the analysis of data and with the production of the manuscript. All authors have read and agreed to the published version of the manuscript.

Funding: The present study was supported by FAPESP (Fundação de Amparo à Pesquisa do Estado de São Paulo, grant number 2017/15006-5).

Institutional Review Board Statement: The study was conducted according to the guidelines of the Declaration of Hensinki, and approved by the Research Ethics Committee (CEP) (Protocol number CAAE 83446118.5.00005415) and by the Ethics Committee on the Use of Animals (CEUA) (Protocol number 001-003244/2013) of the College of Medicine of São José do Rio Preto—FAMERP.

Informed Consent Statement: Informed consent was obtained from all subjects involved in the study.

Data Availability Statement: NGS data will be available by request from the corresponding author.

Acknowledgments: Tiago Henrique for scientific technical support.

Conflicts of Interest: The authors declare no conflict of interest.

References

1. Gelaleti, G.B.; Borin, T.F.; Maschio-Signorini, L.B.; Moschetta, M.G.; Hellmén, E.; Viloria-Petit, A.M.; Zuccari, D.A.P.C. Melatonin and IL-25 modulate apoptosis and angiogenesis mediators in metastatic (CF-41) and non-metastatic (CMT-U229) canine mammary tumour cells. *Vet. Comp. Oncol.* **2017**, *15*, 1572–1584. [CrossRef]
2. Amirkhani Namagerdi, A.; d'Angelo, D.; Ciani, F.; Iannuzzi, C.A.; Napolitano, F.; Avallone, L.; De Laurentiis, M.; Giordano, A. Triple-Negative Breast Cancer Comparison With Canine Mammary Tumors From Light Microscopy to Molecular Pathology. *Front. Oncol.* **2020**, *10*, 2499. [CrossRef]
3. Harbeck, N.; Penault-Llorca, F.; Cortes, J.; Gnant, M.; Houssami, N.; Poortmans, P.; Ruddy, K.; Tsang, J.; Cardoso, F. Breast cancer. *Nat. Rev. Dis. Prim.* **2019**, *5*, 66. [CrossRef] [PubMed]
4. Jamal-Hanjani, M.; Quezada, S.A.; Larkin, J.; Swanton, C. Translational implications of tumor heterogeneity. *Clin. Cancer Res.* **2015**, *21*, 1258–1266. [CrossRef] [PubMed]
5. Cherdyntseva, N.; Litviakov, N.; Denisov, E.; Gervas, P.; Cherdyntsev, E. Circulating tumor cells in breast cancer: Functional heterogeneity, pathogenetic and clinical aspects. *Exp. Oncol.* **2017**, *39*, 2–11. [CrossRef]
6. Palmero, E.I.; Carraro, D.M.; Alemar, B.; Moreira, M.A.M.; Ribeiro-Dos-Santos, Â.; Abe-Sandes, K.; Galvão, H.C.R.; Reis, R.M.; De Pádua Souza, C.; Campacci, N.; et al. The germline mutational landscape of BRCA1 and BRCA2 in Brazil. *Sci. Rep.* **2018**, *8*, 9188. [CrossRef]
7. Huang, X.; Shao, D.; Wu, H.; Zhu, C.; Guo, D.; Zhou, Y.; Chen, C.; Lin, Y.; Lu, T.; Zhao, B.; et al. Genomic Profiling Comparison of Germline BRCA and Non-BRCA Carriers Reveals CCNE1 Amplification as a Risk Factor for Non-BRCA Carriers in Patients With Triple-Negative Breast Cancer. *Front. Oncol.* **2020**, *10*, 2168. [CrossRef] [PubMed]
8. Santarpia, L.; Bottai, G.; Kelly, C.M.; Győrffy, B.; Székely, B.; Pusztai, L. Deciphering and Targeting Oncogenic Mutations and Pathways in Breast Cancer. *Oncologist* **2016**, *21*, 1063–1078. [CrossRef] [PubMed]
9. Redig, A.J.; Jänne, P.A. Basket trials and the evolution of clinical trial design in an era of genomic medicine. *J. Clin. Oncol.* **2015**, *33*, 975–977. [CrossRef]
10. O'Leary, E.; Iacoboni, D.; Holle, J.; Michalski, S.T.; Esplin, E.D.; Yang, S.; Ouyang, K. Expanded Gene Panel Use for Women With Breast Cancer: Identification and Intervention Beyond Breast Cancer Risk. *Ann. Surg. Oncol.* **2017**, *24*, 3060–3066. [CrossRef]
11. Vogelstein, B.; Papadopoulos, N.; Velculescu, V.E.; Zhou, S.; Diaz, L.A.; Kinzler, K.W. Cancer genome landscapes. *Science* **2013**, *340*, 1546–1558. [CrossRef]
12. de Figueiredo Barros, B.D.; Kupper, B.E.; Aguiar Junior, S.; de Mello, C.A.; Begnami, M.D.; Chojniak, R.; de Souza, S.J.; Torrezan, G.T.; Carraro, D.M. Mutation detection in tumor-derived cell free DNA anticipates progression in a patient with metastatic colorectal cancer. *Front. Oncol.* **2018**, *8*, 306. [CrossRef]
13. Forbes, S.A.; Beare, D.; Gunasekaran, P.; Leung, K.; Bindal, N.; Boutselakis, H.; Ding, M.; Bamford, S.; Cole, C.; Ward, S.; et al. COSMIC: Exploring the world's knowledge of somatic mutations in human cancer. *Nucleic Acids Res.* **2015**, *43*, D805–D811. [CrossRef] [PubMed]
14. Das, M.; Law, S. Role of tumor microenvironment in cancer stem cell chemoresistance and recurrence. *Int. J. Biochem. Cell Biol.* **2018**, *103*, 115–124. [CrossRef]
15. Atashzar, M.R.; Baharlou, R.; Karami, J.; Abdollahi, H.; Rezaei, R.; Pourramezan, F.; Zoljalali Moghaddam, S.H. Cancer stem cells: A review from origin to therapeutic implications. *J. Cell. Physiol.* **2020**, *235*, 790–803. [CrossRef] [PubMed]
16. Olsson, E.; Winter, C.; George, A.; Chen, Y.; Howlin, J.; Tang, M.E.; Dahlgren, M.; Schulz, R.; Grabau, D.; Westen, D.; et al. Serial monitoring of circulating tumor DNA in patients with primary breast cancer for detection of occult metastatic disease. *EMBO Mol. Med.* **2015**, *7*, 1034–1047. [CrossRef] [PubMed]
17. Franken, A.; Honisch, E.; Reinhardt, F.; Meier-Stiegen, F.; Yang, L.; Jaschinski, S.; Esposito, I.; Alberter, B.; Polzer, B.; Huebner, H.; et al. Detection of ESR1 Mutations in Single Circulating Tumor Cells on Estrogen Deprivation Therapy but Not in Primary Tumors from Metastatic Luminal Breast Cancer Patients. *J. Mol. Diagnost.* **2020**, *22*, 111–121. [CrossRef] [PubMed]
18. Maltoni, R.; Casadio, V.; Ravaioli, S.; Foca, F.; Tumedei, M.M.; Salvi, S.; Martignano, F.; Calistri, D.; Rocca, A.; Schirone, A.; et al. Cell-free DNA detected by "liquid biopsy" as a potential prognostic biomarker in early breast cancer. *Oncotarget* **2017**, *8*, 16642–16649. [CrossRef] [PubMed]
19. Otandault, A.; Anker, P.; Al Amir Dache, Z.; Guillaumon, V.; Meddeb, R.; Pastor, B.; Pisareva, E.; Sanchez, C.; Tanos, R.; Tousch, G.; et al. Recent advances in circulating nucleic acids in oncology. *Ann. Oncol.* **2019**, *30*, 374–384. [CrossRef] [PubMed]
20. Yoo, T.K. Liquid Biopsy in Breast Cancer: Circulating Tumor Cells and Circulating Tumor DNA. *Adv. Exp. Med. Bio.* **2021**, *1187*, 337–361. [CrossRef]
21. Housman, G.; Byler, S.; Heerboth, S.; Lapinska, K.; Longacre, M.; Snyder, N.; Sarkar, S. Drug resistance in cancer: An overview. *Cancers* **2014**, *6*, 1769–1792. [CrossRef] [PubMed]

22. Yuan, Y.; Yost, S.E.; Yuan, Y.C.; Solomon, N.M.; Mambetsariev, I.; Pal, S.; Frankel, P.; Salgia, R.; Neuhausen, S.L.; Mortimer, J. Genomic mutation-driven metastatic breast cancer therapy: A single center experience. *Oncotarget* **2017**, *8*, 26414–26423. [CrossRef]
23. Rolfo, C.; Manca, P.; Salgado, R.; Van Dam, P.; Dendooven, A.; MacHado Coelho, A.; Ferri Gandia, J.; Rutten, A.; Lybaert, W.; Vermeij, J.; et al. Multidisciplinary molecular tumour board: A tool to improve clinical practice and selection accrual for clinical trials in patients with cancer. *ESMO Open* **2018**, 3, e000398. [CrossRef] [PubMed]
24. Kaszak, I.; Ruszczak, A.; Kanafa, S.; Kacprzak, K.; Król, M.; Jurka, P. Current biomarkers of canine mammary tumors. *Acta Vet. Scand.* **2018**, *60*, 66. [CrossRef] [PubMed]
25. Sorenmo, K.U.; Rasotto, R.; Zappulli, V.; Goldschmidt, M.H. Development, anatomy, histology, lymphatic drainage, clinical features, and cell differentiation markers of canine mammary gland neoplasms. *Vet. Pathol.* **2011**, *48*, 85–97. [CrossRef] [PubMed]
26. Chen, M.; Zhao, H. Next-generation sequencing in liquid biopsy: Cancer screening and early detection. *Hum. Genom.* **2019**, *13*, 34. [CrossRef]
27. Lee, C.H.; Lee, S.J.; Choi, S.H.; Ahn, S.H.; Son, B.H.; Lee, J.W.; Yu, J.H.; Kwon, N.J.; Lee, W.C.; Yang, K.S.; et al. Cancer panel analysis of circulating tumor cells in patients with breast cancer. *Oncol. Lett.* **2018**, *16*, 612–618. [CrossRef] [PubMed]
28. Yu, D.; Zou, J.; Chen, Q.; Zhu, T.; Sui, R.; Yang, J. Structural modeling, mutation analysis, and in vitro expression of usherin, a major protein in inherited retinal degeneration and hearing loss. *Comput. Struct. Biotechnol. J.* **2020**, *18*, 1363–1382. [CrossRef]
29. Toualbi, L.; Toms, M.; Moosajee, M. USH2A-retinopathy: From genetics to therapeutics. *Exp. Eye Res.* **2020**, *201*, 108330. [CrossRef]
30. Li, J.; Li, H.; Makunin, I.; Thompson, B.A.; Tao, K.; Young, E.L.; Lopez, J.; Camp, N.J.; Tavtigian, S.V.; John, E.M.; et al. Panel sequencing of 264 candidate susceptibility genes and segregation analysis in a cohort of non-BRCA1, non-BRCA2 breast cancer families. *Breast Cancer Res. Treat.* **2017**, *166*, 937–949. [CrossRef]
31. Natrajan, R.; Wilkerson, P.M.; Marchiò, C.; Piscuoglio, S.; Ng, C.K.Y.; Wai, P.; Lambros, M.B.; Samartzis, E.P.; Dedes, K.J.; Frankum, J.; et al. Characterization of the genomic features and expressed fusion genes in micropapillary carcinomas of the breast. *J. Pathol.* **2014**, *232*, 553–565. [CrossRef]
32. Kim, S.H.; Ahn, S.; Suh, K.J.; Kim, Y.J.; Park, S.Y.; Kang, E.; Kim, E.K.; Kim, I.A.; Chae, S.; Choi, M.; et al. Identifying germline APOBEC3B deletion and immune phenotype in Korean patients with operable breast cancer. *Breast Cancer Res. Treat.* **2020**, *183*, 697–704. [CrossRef] [PubMed]
33. Yu, C.; Zhang, F. LncRNA AC009022.1 enhances colorectal cancer cells proliferation, migration, and invasion by promoting ACTR3B expression via suppressing miR-497-5p. *J. Cell. Biochem.* **2020**, *121*, 1934–1944. [CrossRef]
34. Korobeynikov, V.; Borakove, M.; Feng, Y.; Wuest, W.M.; Koval, A.B.; Nikonova, A.S.; Serebriiskii, I.; Chernoff, J.; Borges, V.F.; Golemis, E.A.; et al. Combined inhibition of Aurora A and p21-activated kinase 1 as a new treatment strategy in breast cancer. *Breast Cancer Res. Treat.* **2019**, *177*, 369–382. [CrossRef]
35. Zhao, Z.M.; Yost, S.E.; Hutchinson, K.E.; Li, S.M.; Yuan, Y.C.; Noorbakhsh, J.; Liu, Z.; Warden, C.; Johnson, R.M.; Wu, X.; et al. CCNE1 amplification is associated with poor prognosis in patients with triple negative breast cancer. *BMC Cancer* **2019**, *19*, 96. [CrossRef] [PubMed]
36. Yu, W.; Huang, W.; Yang, Y.; Qiu, R.; Zeng, Y.; Hou, Y.; Sun, G.; Shi, H.; Leng, S.; Feng, D.; et al. GATA3 recruits UTX for gene transcriptional activation to suppress metastasis of breast cancer. *Cell Death Dis.* **2019**, *10*, 832. [CrossRef] [PubMed]
37. Takaku, M.; Grimm, S.A.; Wade, P.A. GATA3 in breast cancer: Tumor suppressor or oncogene? *Gene Expr.* **2015**, *16*, 163–168. [CrossRef] [PubMed]
38. D'Amico, P.; Corvaja, C.; Gerratana, L.; Reduzzi, C.; Curigliano, G.; Cristofanilli, M. The use of liquid biopsy in early breast cancer: Clinical evidence and future perspectives. *J. Cancer Metastasis Treat.* **2021**, *7*, 1–16. [CrossRef]
39. Ellis, H.; Ma, C.X. PI3K Inhibitors in Breast Cancer Therapy. *Curr. Oncol. Rep.* **2019**, *21*, 110. [CrossRef]
40. Verret, B.; Cortes, J.; Bachelot, T.; Andre, F.; Arnedos, M. Efficacy of PI3K inhibitors in advanced breast cancer. *Ann. Oncol. Off. J. Eur. Soc. Med. Oncol.* **2019**, *30*, x12–x20. [CrossRef]
41. André, F.; Ciruelos, E.M.; Juric, D.; Loibl, S.; Campone, M.; Mayer, I.A.; Rubovszky, G.; Yamashita, T.; Kaufman, B.; Lu, Y.S.; et al. Alpelisib plus fulvestrant for PIK3CA-mutated, hormone receptor-positive, human epidermal growth factor receptor-2–negative advanced breast cancer: Final overall survival results from SOLAR-1. *Ann. Oncol.* **2021**, *32*, 208–217. [CrossRef] [PubMed]
42. Lee, K.H.; Hwang, H.J.; Noh, H.J.; Shin, T.J.; Cho, J.Y. Somatic mutation of PIK3ca (H1047R) is a common driver mutation hotspot in canine mammary tumors as well as human breast cancers. *Cancers* **2019**, *11*, 2006. [CrossRef]
43. Sobhani, N.; Roviello, G.; Corona, S.P.; Scaltriti, M.; Ianza, A.; Bortul, M.; Zanconati, F.; Generali, D. The prognostic value of PI3K mutational status in breast cancer: A meta-analysis. *J. Cell. Biochem.* **2018**, *119*, 4287–4292. [CrossRef]
44. Kim, J.H. PIK3CA mutations matter for cancer in dogs. *Res. Vet. Sci.* **2020**, *133*, 39–41. [CrossRef] [PubMed]
45. Alsaihati, B.A.; Ho, K.-L.; Watson, J.; Feng, Y.; Wang, T.; Zhao, S. Canine tumor mutation rate is positively correlated with TP53 mutation across cancer types and breeds. *Biorxiv* **2020**, *7*, 1–23. [CrossRef]
46. Maués, T.; El-Jaick, K.B.; Costa, F.B.; Araujo, G.E.F.; Soares, M.V.G.; Moreira, A.S.; Ferreira, M.L.G.; Ferreira, A.M.R. Common germline haplotypes and genotypes identified in BRCA2 exon 11 of dogs with mammary tumours and histopathological analyses. *Vet. Comp. Oncol.* **2018**, *16*, 379–384. [CrossRef] [PubMed]
47. Oliveira, J.R.; Colombo, J.; Gonçalves, F.M.; Carvalho, L.A.L.; Costa, D.S.; Henrique, T.; Novais, A.A.; Moscheta-Pinheiro, M.G.; Chuffa, L.G.A.; Coutinho, L.L.; et al. Liquid biopsy can detect BRCA2 gene variants in female dogs with mammary neoplasia short title: Liquid biopsy detects malignant gene variants. *Vet. Comp. Oncol.* **2021**, in press. [CrossRef] [PubMed]

48. Yoshikawa, Y.; Morimatsu, M.; Ochiai, K.; Nagano, M.; Tomioka, Y.; Sasaki, N.; Hashizume, K.; Iwanaga, T. Novel variations and loss of heterozygosity of BRCA2 identified in a dog with mammary tumors. *Am. J. Vet. Res.* **2008**, *69*, 1323–1328. [CrossRef]
49. Yoshikawa, Y.; Morimatsu, M.; Ochiai, K.; Ishiguro-Oonuma, T.; Wada, S.; Orino, K.; Watanabe, K. Reduced canine BRCA2 expression levels in mammary gland tumors. *BMC Vet. Res.* **2015**, *11*, 159. [CrossRef]
50. Huskey, A.L.W.; Goebel, K.; Lloveras-Fuentes, C.; McNeely, I.; Merner, N.D. Whole genome sequencing for the investigation of canine mammary tumor inheritance—An initial assessment of high-risk breast cancer genes reveal BRCA2 and STK11 variants potentially associated with risk in purebred dogs. *Canine Med. Genet.* **2020**, *7*, 8. [CrossRef]
51. Amoroso, V.; Generali, D.; Buchholz, T.; Cristofanilli, M.; Pedersini, R.; Curigliano, G.; Daidone, M.G.; Di Cosimo, S.; Dowsett, M.; Fox, S.; et al. International expert consensus on primary systemic therapy in the management of early breast cancer: Highlights of the Fifth Symposium on Primary Systemic Therapy in the Management of Operable Breast Cancer, Cremona, Italy (2013). *J. Natl. Cancer Inst. -Monogr.* **2015**, *2015*, 90–96. [CrossRef] [PubMed]
52. Poulet, G.; Massias, J.; Taly, V. Liquid Biopsy: General Concepts. *Acta Cytol.* **2019**, *63*, 449–455. [CrossRef] [PubMed]

cancers

MDPI

Article

Kallikrein 5 Inhibition by the Lympho-Epithelial Kazal-Type Related Inhibitor Hinders Matriptase-Dependent Carcinogenesis

Elaine Zayas Marcelino da Silva [1], Thais Fernanda de Campos Fraga-Silva [2], Yao Yuan [3], Márcia Gaião Alves [1], Gabriel Azevedo Publio [4], Carol Kobori da Fonseca [1], Márcio Hideki Kodama [1], Gabriel Viliod Vieira [1], Marina Ferreira Candido [1], Lara Maria Alencar Ramos Innocentini [5], Mateus Gonçalves Miranda [1], Alfredo Ribeiro da Silva [6,†], Jose Carlos Alves-Filho [4], Vania Luiza Deperon Bonato [2], Ramiro Iglesias-Bartolome [3] and Katiuchia Uzzun Sales [1,*]

1 Department of Cell and Molecular Biology and Pathogenic Bioagents, Ribeirao Preto Medical School, University of São Paulo, Ribeirao Preto 14049-900, SP, Brazil; ezmdasilva@alumni.usp.br (E.Z.M.d.S.); alvesg.marcia@gmail.com (M.G.A.); kobori@usp.br (C.K.d.F.); marciohkodama@gmail.com (M.H.K.); gabrielviliod@gmail.com (G.V.V.); mahfcandido@gmail.com (M.F.C.); mateusgm@usp.br (M.G.M.)

2 Basic and Applied Immunology Program, Department of Biochemistry and Immunology, Ribeirao Preto Medical School, University of São Paulo, Ribeirao Preto 14049-900, SP, Brazil; thaisfragasilva@usp.br (T.F.d.C.F.-S.); vlbonato@fmrp.usp.br (V.L.D.B.)

3 Laboratory of Cellular and Molecular Biology, Center for Cancer Research, National Cancer Institute, National Institutes of Health, Bethesda, MD 20892, USA; yao.yuan2@nih.gov (Y.Y.); ramiro.iglesias-bartolome@nih.gov (R.I.-B.)

4 Departament of Pharmacology, Ribeirao Preto Medical School, University of São Paulo, Ribeirao Preto 14049-900, SP, Brazil; gabriel.publio@usp.br (G.A.P.); jcafilho@usp.br (J.C.A.-F.)

5 Dentistry and Stomatology Division, Ophthalmology, Otolaryngology, and Head and Neck Surgery Department, Clinical Hospital of Ribeirao Preto Medical School, University of São Paulo, Ribeirao Preto 14049-900, SP, Brazil; lara.m.alencar@hotmail.com

6 Department of Pathology and Legal Medicine, Ribeirao Preto Medical School, University of São Paulo, Ribeirao Preto 14049-900, SP, Brazil; arsilva@fmrp.usp.br

* Correspondence: salesk@fmrp.usp.br; Tel.: +55-16-3315-9113

† Deceased 28 July 2021.

Citation: da Silva, E.Z.M.; Fraga-Silva, T.F.d.C.; Yuan, Y.; Alves, M.G.; Publio, G.A.; da Fonseca, C.K.; Kodama, M.H.; Vieira, G.V.; Candido, M.F.; Innocentini, L.M.A.R.; et al. Kallikrein 5 Inhibition by the Lympho-Epithelial Kazal-Type Related Inhibitor Hinders Matriptase-Dependent Carcinogenesis. *Cancers* **2021**, *13*, 4395. https://doi.org/10.3390/cancers13174395

Academic Editor: David Wong

Received: 5 August 2021
Accepted: 25 August 2021
Published: 31 August 2021

Publisher's Note: MDPI stays neutral with regard to jurisdictional claims in published maps and institutional affiliations.

Copyright: © 2021 by the authors. Licensee MDPI, Basel, Switzerland. This article is an open access article distributed under the terms and conditions of the Creative Commons Attribution (CC BY) license (https://creativecommons.org/licenses/by/4.0/).

Simple Summary: Head and neck squamous cell carcinomas (HNSCC) are among the most common cancers worldwide. In contrast to the advances in prevention and treatment of other types of cancer, the five-year survival rate for HNSCC is only about 50% and it has not changed for the past 50 years. This poor prognosis is mainly due to a shortage of suitable markers for early detection, delayed diagnosis and/or referral, and ineffectiveness of chemotherapy. The aim of this study was to explore the inhibitory role of LEKTI in matriptase-dependent squamous cell carcinogenesis and to investigate additional players operating in this pathway. We found that Kallikrein-5 is necessary for PAR-2-mediated IL-8 release, YAP1-TAZ/TEAD activation, and matriptase-mediated oral squamous cell carcinoma migration. This knowledge can contribute for the development of future targeted therapy in HNSCC.

Abstract: Head and neck squamous cell carcinoma remains challenging to treat with no improvement in survival rates over the past 50 years. Thus, there is an urgent need to discover more reliable therapeutic targets and biomarkers for HNSCC. Matriptase, a type-II transmembrane serine protease, induces malignant transformation in epithelial stem cells through proteolytic activation of pro-HGF and PAR-2, triggering PI3K-AKT-mTOR and NFKB signaling. The serine protease inhibitor lympho-epithelial Kazal-type-related inhibitor (LEKTI) inhibits the matriptase-driven proteolytic pathway, directly blocking kallikreins in epithelial differentiation. Hence, we hypothesized LEKTI could inhibit matriptase-dependent squamous cell carcinogenesis, thus implicating kallikreins in this process. Double-transgenic mice with simultaneous expression of matriptase and LEKTI under the keratin-5 promoter showed a prominent rescue of K5-Matriptase$^{+/0}$ premalignant phenotype. Notably, in DMBA-induced SCC, heterotopic co-expression of LEKTI and matriptase delayed matriptase-driven tumor incidence and progression. Co-expression of LEKTI reverted altered Kallikrein-5 expression observed in the skin of K5-Matriptase$^{+/0}$ mice, indicating that matriptase-dependent proteolytic pathway inhibition by LEKTI occurs through kallikreins. Moreover, we showed that Kallikrein-5 is necessary for PAR-2-mediated IL-8 release, YAP1-TAZ/TEAD activation, and matriptase-mediated oral squamous cell carcinoma migration. Collectively, our data identify a third signaling pathway for matriptase-dependent carcinogenesis in vivo. These findings are critical for the identification of more reliable biomarkers and effective therapeutic targets in Head and Neck cancer.

Keywords: LEKTI; SPINK5; KLK5; OSCC; matriptase

1. Introduction

Head and neck squamous cell carcinomas (HNSCC) are among the most common cancers worldwide, with over 800,000 new cases diagnosed yearly [1]. In contrast to the advances in prevention and treatment of other types of cancer, the five-year survival rate for HNSCC is only about 50% and is unchanged for 50 years [2,3]. Generally, this poor prognosis is due to a shortage of suitable markers for early detection, delayed diagnosis and/or referral, and ineffectiveness of chemotherapy [3–5]. Accordingly, there is an urgent need to identify more reliable therapeutic targets and biomarkers of prognosis in HNSCC.

Several studies demonstrated that dysregulated expression of Type II Transmembrane Serine Proteases (TTSPIIs) is associated with different types of cancers [6]. Matriptase, a TTSPII, is indeed dysregulated in several cancers [7]. List et al. (2005) showed that Matriptase induces malignant transformation when expressed in epithelial stem cells [8]. It has been shown that PI3K-Akt-mTor signaling, elicited by proteolytic activation of pro-hepatocyte growth factor (pro-HGF), is a molecular pathway by which matriptase promotes malignant transformation [9]. Another essential component of matriptase-mediated oncogenesis is the upregulation of NFkB-induced inflammatory cytokines dependent on Protease Activated Receptor 2 (PAR-2) proteolytic cleavage by matriptase [10].

Matriptase was also implicated in Netherton Syndrome (NS), an epidermal disorder caused by mutations in the SPINK5 gene, which encodes for the Lympho-Epithelial Kazal-Type-related Inhibitor (LEKTI). LEKTI inhibits matriptase-dependent skin desquamation in NS by the direct inhibition of Kallikreins 5 (KLK5) and 7 (KLK7), which are known for their role in the degradation of corneodesmosomes at the outermost layers of the stratum corneum [11]. Kallikrein-related peptidases (KLKs) comprise a large family of secreted serine proteases expressed in several tissues [12,13]. Kallikreins are secreted as inactive zymogens and generally depend on proteolytic cleavage at the carboxy-terminal end of either arginine or lysine for activation [14] and are often dysregulated in inflammatory skin disorders and many human cancers [12,15]. Abnormal functional levels of the KLKs in diseased states are a result of unbalanced activation and inhibition events. These peptidases function cooperatively in signaling cascades or complex regulatory networks, often bridging multiple protease families and classes [16] Accordingly, matriptase is able to activate pro-KLK5 and pro-KLK7 [11].

In this study, we found a decrease in LEKTI expression in poorly differentiated OSCCs. This led us to the hypothesis that LEKTI could be a tumor suppressor by regulating KLK activation by matriptase. Indeed, double-transgenic mice with simultaneous expression of matriptase and LEKTI under the keratin-5 promoter led to the inhibition of matriptase-dependent carcinogenesis through downregulation of KLK5 expression. Significantly, KLK5 promoted the release of IL-8 and activated the Yap1-TAZ/TEAD transcription network through PAR-2 activation. In conclusion, our study identifies a novel proteolytic pathway that contributes to the matriptase-driven malignant transformation.

2. Material and Methods

2.1. Human Tissue Samples

In this study, 127 cases of human Oral Squamous Cell Carcinomas (OSCCs) diagnosed between 2005 and 2016 were selected from Pathology Service (SERPAT) of Ribeirao Preto Medical School, University of São Paulo, to compose a tissue microarray (TMA). Formalin-fixed and paraffin-embedded TMAs were composed of well-differentiated OSCCs, W.D.C., $n = 37$; moderately-differentiated OSCCs, M.D.C., $n = 59$; and poorly-differentiated OSCCs, P.D.C., $n = 31$. Only cases whose diagnoses were made in the following regions of the oral cavity were considered: jaw, floor of mouth, retromolar, mouth, oral cavity, tongue, cheek mucosa, gingiva, gum, larynx/pharynx, hard palate, and lips.

2.2. Tissue Microarray Construction

An experienced pathologist (ARS) reviewed the hematoxylin and eosin slides to delineate the most significant area of each tumor to prepare the TMA. From the preselected donor paraffin blocks, 2 mm diameter cylinders were removed and sorted into a paraffin block recipient using the TMA Builder Kit (Histopathology Ltd., Pécs, Hungary).

2.3. Ethics Statement

The study was performed in accordance with the Declaration of Helsinki [17] and approved by the Ethics Committee on Human Research of Ribeirao Preto Clinical Hospital and Ribeirao Preto Medical School, University of São Paulo (protocol: 50533515.6.0000.5440, approved on 21 June 2016). Histopathological analysis was performed on paraffin-embedded non-identified samples comprising incisional biopsies. In this situation, patient consent is waived according to Brazilian laws.

The K5-LEKTI mice were generated in accordance with protocols approved by the National Institute of Dental and Craniofacial Research Animal Care and Use Committee. All mice were kept and bred in the animal facility of the Department of Cell and Molecular Biology and Pathogenic Bioagents, Ribeirao Preto Medical School, University of São Paulo, Brazil. All experiments involving mice were approved by the Ethics Committee on Animal Research of Ribeirao Preto Medical School, University of São Paulo (protocol: 003/2015-1, approved on 29 June 2015) and were performed in accordance with the Guidelines of the Brazilian College of Animal Experimentation.

2.4. Mice

The generation of Keratin5-matriptase transgenic mice (K5-Matriptase$^{+/0}$) has been described [8]. Keratin5-LEKTI transgenic mice (K5-LEKTI$^{+/0}$) were generated by cloning the full-length 3-kb mouse LEKTI cDNA (NM_001081180.1) into the pBK5-vector containing the 5.2-kb bovine keratin-5 regulatory sequences, beta-globin intron-2, and 3′-polyadenylation sequences [18]. The linearized vectors were microinjected into the male pronucleus of FVB zygotes, which were implanted into pseudopregnant mice. Founder animals carrying the transgene were identified by PCR analysis of genomic DNA extracted from tail biopsies. WT and K5-Mariptase$^{+/0}$/K5-LEKTI$^{+/0}$ were generated by interbreeding of K5-Mariptase$^{+/0}$ with K5-LEKTI$^{+/0}$ mice. The study was strictly littermate controlled to avoid genetic background differences from confounding data interpretation. The transgenic mice were genotyped by PCR of genomic DNA extracted from tail biopsies as described previously [19].

2.5. Cells

The HNSCC cell line Cal 27 was a kind gift from Dr. J. Silvio Gutkind (University California San Diego, San Diego, CA, USA). Cal 27 are derived from a human tongue SCC [20]. Generation of Cal 27 KLK5 knockout cells was previously reported [21]. Both WT and KLK5 knockout Cal 27 cells were authenticated by STR profiling (DATAPEP-FMUSP, Sao Paulo, SP, Brazil) with the following results: TH01 6, 9.3; D5S818 11, 12; D13S317 10, 11; D7S820 10; D16S539 11, 12; CSF1PO 10, 12; vWA 14, 17; TPOX 8; Amelogenin X. HEK293T cells are from ATCC and were not further authenticated (Catalog# CRL-3216, Manassas, VA, USA). Cells were cultured as previously described [21].

2.6. Antibodies and Recombinant Proteins

The following primary antibodies were used for IHC: anti-SPINK5 (3 μg/mL; clone HPA009067, Merck, Darmstadt, Germany), anti-Matriptase-ST14 (1:400 dilution, Catalog#

AF3946, R&D Systems, Minneapolis, MN, USA) and anti-mouse Kallikrein 5 (1:800 dilution; Catalog# MAB7236, R&D Systems). The following biotinylated secondary antibodies were used: goat anti-mouse IgG, goat anti-rat IgG, rabbit anti-sheep IgG (1:400; Vector Laboratories, Burlingame, CA, USA). The following conjugated primary antibodies were used for flow cytometry: anti-CD45-PeCy7 (clone 30-F11), anti-Ly6G-APC (clone IA8), anti-CD11c-FITC (clone HL3), anti-MHCII-BB700 (clone M5/114.15.2) from BD Bioscience (Franklin Lakes, NJ, USA), and anti-CD64-APC-Cy7 (clone X54-5/7.1) and anti-CD16/CD32 monoclonal antibody (clone 93) from eBioscience (San Diego, CA, USA). The following human recombinant proteases were used: Recombinant Human Matriptase/ST14 Catalytic Domain (Catalog# 3946-SE) and Recombinant Human Kallikrein 5 Protein (Catalog# 1108-SE), from R&D Systems.

2.7. SPINK5 mRNA Expression Analysis in K5-LEKTI$^{+/0}$ Mouse Epidermis

Newborn WT and K5-LEKTI$^{+/0}$ FVB/NJ mice were euthanized by decapitation. Dissected skin was incubated in PBS containing 10 mM EDTA for 5 min at 56 °C to separate the dermis from the epidermis. The epidermis was immediately covered with RNALater (MilliporeSigma, Burlington, MA, USA), and incubated overnight at 4 °C. For the extraction of total RNA, we used TRIzol (Thermo Fisher Scientific, Waltham, MA, USA). Epidermis was added to 1 mL of TRIzol buffer and 100 mg of Precellys® zirconium oxide beads (Bertin Technologies, Montigny-le-Bretonneux, France). The samples were homogenized using a Precellys-24 tissue homogenizer (Bertin Technologies) for three pulses at 6800 rpm for lysis and homogenization. The samples were then transferred to a new RNase/DNase-free tube for chloroform extraction. The upper phase was transferred to a new tube for isopropanol precipitation. Samples were vortexed and centrifuged at 12,000× *g* for 30 min. The pellet was then washed with 70% ethanol in DEPC water, and the material was dried. RNA was resuspended in DEPC water. RNA samples were quantified, and cDNA synthesis was performed from 1 μg total RNA using the Kit ™cDNA Synthesis SuperScript® VILO (Catalog# 11754, Thermo Fisher Scientific). The qRT-PCR reaction was performed in the 7500 Real-Time PCR System according to the protocol of TaqMan Master Mix (Applied Biosystems, Thermo Fisher Scientific) manufacturer. Gapdh and Hprt1 were used as internal controls. LEKTI gene expression was evaluated using mouse SPINK5 TaqMan probes (Mm00486343_cn). All reactions were repeated three times, and the experiments were validated with the use of negative controls.

2.8. Histological and Immunohistochemical Analysis

Adult mice were euthanized by CO_2 inhalation, and newborn mice were euthanized by decapitation. Mouse tissues were fixed overnight in 4% paraformaldehyde and embedded in paraffin. Both mice and TMA paraffin blocks were cut into sections 6 μm thick and mounted on glass slides. Tissue sections were processed for histology and stained either with hematoxylin and eosin (H&E), toluidine blue, or immunostained for LEKTI, Matriptase, or KLK5. For IHC staining, antigen retrieval was performed by boiling samples in 0.01 M sodium citrate buffer, pH 6, for 12 min. After non-specific antigen blocking was performed, the sections incubated overnight at 4 °C with the primary antibodies followed by incubation with appropriate biotin-conjugated secondary antibodies and the Vectastain-ABC Elite kit (Vector Laboratories, Burlingame, CA, USA). Staining was developed by incubation with 3,3′-diaminobenzidine (25 mg/mL, Merck) and H_2O_2 and the sections were counterstained with hematoxylin. Non-immune anti-rabbit IgG was used as negative control (3 μg/mL, Catalog# 011-000-003, Jackson ImmunoResearch Laboratories Inc., West Grove, PA, USA) for LEKTI, Matriptase, and KLK5. Images were obtained on an Olympus VS120 slide scanner (Olympus Corporation, Tokyo, Japan). Quantification of the total epithelial area, the stained epithelial area of each sample and counts of metachromatically stained mast cells were performed using the Image J software [22]. Briefly, for TMA slides stained for LEKTI or Matriptase, ratios between the total epithelial and immunolabeled epithelial area (μm^2) were calculated and plotted as a percentage (%); zero values were included where samples lacked LEKTI staining. For mice skin samples, the height of the epidermis (epithelial thickness) and mast cell count per 10^3 μm^2 were quantified.

2.9. Mouse Skin Dissociation

Mouse dorsal skin was extracted from euthanized healthy WT, K5-LEKTI$^{+/0}$, K5-Matriptase$^{+/0}$, and K5-LEKTI$^{+/0}$/K5-Matriptase$^{+/0}$ FVB/NJ mice. The dorsal skin was shaved and a depilatory cream was applied to remove remaining fur. The depilated skin was cleaned with PBS before extraction. An area of 15 mm × 35 mm of back skin was excised and cut into three pieces before being put in a 50 mL tube with 20 mL of HBSS and quickly washed twice by vortexing. The tissue was drained and placed into a 50 mL conical tube containing 10 mL of pre-dissociation buffer (RT HBSS with phenol red, without calcium or magnesium, 5 mM EDTA, and 10 mM HEPES), incubated for 30 min at 37 °C with agitation, and vortexed for 10 s before being strained through a 70 μm strainer. Tissue pieces were transferred to a Petri dish and finely cut into small fragments (2.2 mm × 2.2 mm) that were transferred to a new tube containing fresh pre-dissociation buffer and incubated again for further 30 min, vortexed vigorously for 10 s, strained and washed with 50 mL HBSS to remove excess EDTA. Collagenase digestion of skin was performed by incubating the strained tissue pieces in a new 2 mL Eppendorf tube containing 1.5 mL fresh digestion buffer (HBSS supplemented with 1 mg/mL collagenase D (Roche, Basel, Switzerland), 1 mg/mL collagenase type IV (Worthington Biochemical Co., Lakewood, NJ, USA), 100 μg/mL DNAse, 1 mM $CaCl_2$) and 100 mg of Precellys® Zirconium-Oxide Beads of 1.4 mm (Bertin Technologies). The tissue was incubated for 1–2 h at 37 °C with agitation and was briefly vortexed every 15 min. After dissociation, the contents were filtered through a 70 μm strainer into a new 50 mL conical tube, and flow-through containing the digested tissue was washed with 50 mL of DMEM with 10% FBS, centrifuged for 5 min at 350× *g*, and the cell pellet was resuspended in PBS for flow cytometry analysis.

2.10. Flow Cytometry

Skin isolated total cells were counted using an Automated Cell Counter (Countess I, Invitrogen, Thermo Fisher Scientific). The cell suspension (2.1–13 × 10^5 cells/sample) was incubated for 15 min at RT with fixable dead cell stain (Thermo Fisher Scientific) and washed with AutoMACS Rinsing Solution containing 0.5% BSA (Mylteni Biotec, Bergisch Gladbach, Germany). The cell suspensions were then incubated for 10 min at 4 °C with anti-CD16/CD32 monoclonal antibody, followed by incubation for 30 min at 4 °C with anti-CD45, anti-Ly6G, anti-CD11c, anti-MHCII, and anti-CD64. The samples were acquired using a BD FACSCanto II cytometer and CellQuest software (BD Biosciences). Total events per sample were collected and analyzed according to size and granularity, single events, live cells, and fluorescence intensity using FlowJo software (BD Biosciences). The following gating strategy was used: leukocytes (CD45$^+$) in total live and single cells; Ly6G$^+$ (neutrophils) and Ly6G$^-$ in CD45$^+$ cells; dendritic cells (DC–CD11c$^+$MHCII^{High}) and myeloid cells (MY–CD11c$^+$MHCII$^+$) in total Ly6G$^-$ cells, and macrophages (CD64$^+$) in total myeloid cells.

2.11. Chemical Carcinogenesis

The shaved dorsal area of mice was treated with five applications of 25 μg of 7,12-dimethylbenzanthracene (DMBA) diluted in 100 μL of acetone. Applications started at five weeks of age and were performed every three weeks. The tumor incidence and size in the carcinogen-treated mice were monitored every three weeks. Mice with ulcerating tumors or tumors reaching a diameter of >2 cm were euthanized before the termination of the study at 48 weeks of age. Tumors and organs were collected and fixed and processed for histology.

2.12. PAR-2 Activation Assay

The PAR-2 activation assay was performed as previously described [10]. Briefly, HEK293T cells were cultured in DMEM and co-transfected with pCDNA3.1-PAR-2 (50 ng), pRL-Renilla luciferase (20 ng), and SRE-Firefly luciferase (50 ng) plasmids. At 16 h post-transfection, cells were treated for 6 h with 100 nM hrKLK5, or 15 nM hrMatriptase, which was used as a positive control. PAR-2 activation was measured using the Dual-Glo Luciferase Assay System according to the manufacturer's instructions (Promega, Madison, WI, USA). Luminescence was measured using a Victor X3 plate reader (Perkin Elmer, Waltham, MA, USA), and SRE activation was determined as the ratio of firefly to Renilla luciferase counts.

2.13. TEAD Activation Assay

To measure TEAD activity, HEK293T cells in 12-well plates were co-transfected overnight with 8 × TEAD-Luc (0.25 μg/cm^2) and pCDNA3.1-PAR-2 (50 ng). The next day, cells were serum-starved overnight and stimulated with 10 nM hrKLK5 for 6 h; luciferase activity was measured using a Dual-Glo Luciferase Assay Kit (Promega) and a Microtiter plate luminometer (SpectrMax iD3, Molecular Devices LLC, San Jose, CA, USA). Luciferase normalization was performed in every case by co-transfecting a Renilla Luciferase Vector (0.025 μg/cm^2) (Promega).

2.14. Cytokine Release

WT or KLK5 KO Cal 27 cells were plated at 2 × 10^5 cells per well in 24-well plates and cultivated for 16 h in DMEM containing 10% FBS. Cells were then starved for 2 h before stimulation with hrMatripase for 24 h. Supernatants from cell cultures were collected and measured by ELISA using the DuoSet ELISA kits (R&D Systems) according to manufacturer's protocols for human IL-8/CXCL8 (Catalog# PD8000C), TNF-alpha (Catalog# DY210), and IL1-beta (Catalog# DBL50).

2.15. Scratch Assay

The Scratch assay was performed as previously described[21]. Briefly, WT or KLK5 KO Cal 27 cells (2 × 10^5 cells) were cultured on glass coverslips, serum-starved for 16 h and treated or not with hrMatriptase. At time "0" a 10 μL plastic tip was used to scratch the cells in the middle of each coverslip and wound closure was evaluated for up to 48 h. The coverslips were imaged at 0, 24, and 48 h after scratching using an DMI4000 Leica epifluorescence microscope. The FIJI image processing package [22] was used to measure the area between the edges of the scratch in all time points. The following formula was used to calculate the wound closure area: Wound Closure Area = Area [T0] − Area [Tn].

2.16. Statistical Analysis

GraphPad Prism 8 (GraphPad Software, Inc., San Diego, CA, USA) was used for data analysis and preparation of graphs. Normality tests were performed. Data from experiments with two groups were analyzed using a *t*-test. Data from experiments with three or more groups were analyzed using one-way analysis of variance test (ANOVA) and post-test as indicated in the figure legends.

3. Results

3.1. LEKTI but Not Matriptase Is Differentially Expressed in Human OSCCs

Matriptase expression and activity are dysregulated in several human cancers [7]. In particular, matriptase was previously found to be ubiquitously expressed in HNSCCs of different anatomical locations and different presumed etiology [9]. Conversely, a recent study from our group showed that LEKTI was downregulated in OSCCs [21]. Because LEKTI has also been shown to inhibit a matriptase-driven proteolytic pathway in terminal epithelial differentiation [11], we hypothesized that it could play a role in matriptase mediated squamous cell carcinogenesis. To investigate this, we performed an immunohistochemical evaluation of TMAs comprised of human OSCCs of different anatomical locations of the oral cavity and of the following histopathological grades: well-differentiated carcinomas (W.D.C., n = 37), moderately differentiated carcinomas (M.D.C., n = 59), and poorly differentiated carcinomas (P.D.C., n = 31). While LEKTI expression is remarkably reduced in M.D.C.s and P.D.C.s, where the majority of the samples were negative (Figure 1A, top panel), matriptase expression was found in 70% of the samples in all three groups (Figure 1A, bottom panel). Quantification of the immunostained area in all samples confirmed decreased expression of LEKTI in less differentiated samples (Figure 1B), while matriptase expression remained unaltered among the different groups (Figure 1C). Matriptase expression was found throughout in both W.D.C. and P.D.C. samples (Figure 1D–G), and LEKTI expression was limited to well-differentiated cells in W.D.C.s (Figure 1I,J). These results are consistent with our previous work where increased protease/inhibitor ratio was associated with increased pathological grades and poor prognosis [21].

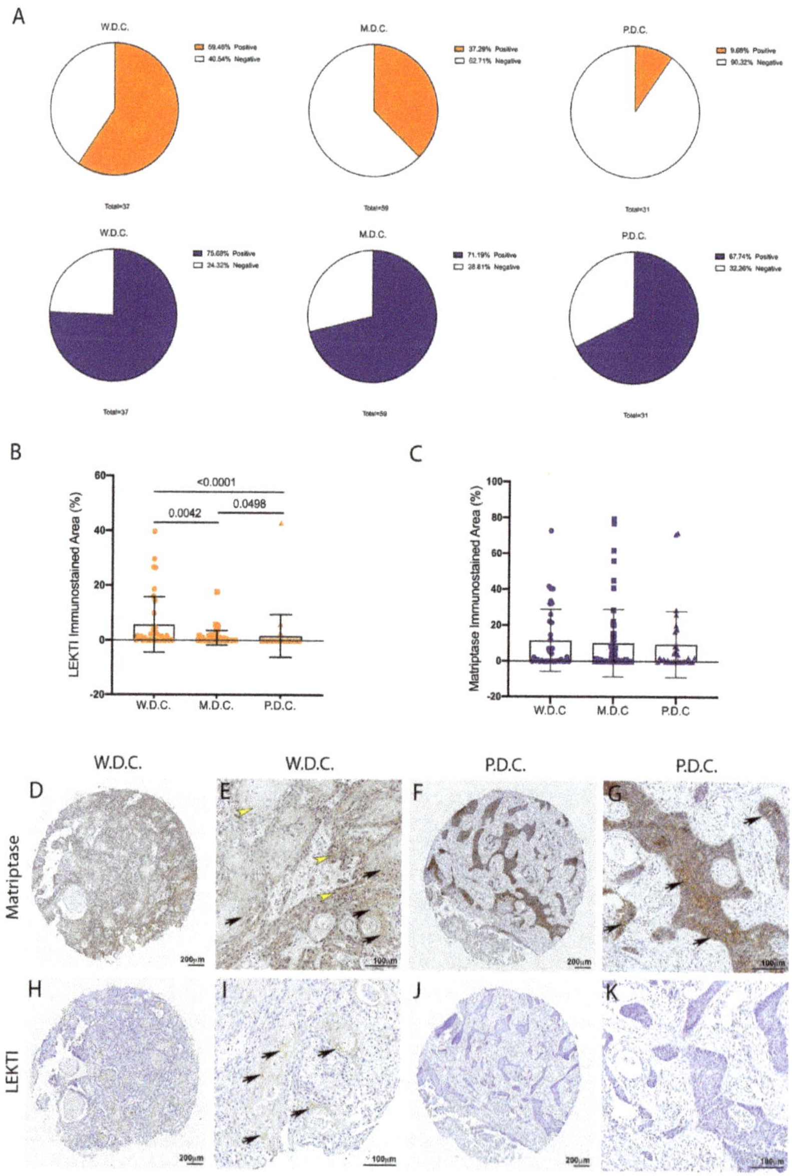

Figure 1. Unlike LEKTI, matriptase is not modulated in poorly differentiated carcinomas. (**A**) IHC staining for LEKTI

(orange) and matriptase (purple) in human OSCCs TMAs (W.D.C. n = 37, M.D.C. n = 59, and P.D.C. n = 31) showed that the number of positive samples for LEKTI (top panel) prominently decreases in the less differentiate samples, while for matriptase (bottom panel) this number remains similar. (**B**) Quantification of stained area confirmed that LEKTI is significantly decreased in M.D.Cs. and P.D.Cs.; p-values (One-Way ANOVA) are displayed in the graph. (**C**) Quantification of stained area shows that matriptase expression does not vary among W.D.Cs., M.D.Cs., and P.D.Cs.; (**B**,**C**) Data are expressed in mean ± SD. (**D–G**) Representative images of matriptase IHC staining in well-differentiated and poorly differentiated carcinomas. (**H–K**) Representative images of LEKTI IHC staining in well-differentiated and poorly differentiated carcinomas. Black arrows show deeper staining, while yellow arrowheads show diffuse staining. Lower magnifications (**D**,**F**,**H**,**J**)-bar = 100 µm; Higher magnifications (**E**,**G**,**I**,**K**) bar = 200 µm; Counterstaining with hematoxylin to visualize tissue architecture.

3.2. Generation of K5-LEKTI Mice

To further investigate the role of LEKTI in matriptase-dependent squamous cell carcinogenesis, we generated transgenic mice that express LEKTI in basal keratinocytes. Full-length murine LEKTI cDNA was cloned under the control of bovine keratin-5 promoter in the pBK5 vector (Figure 2A) [18]. Pronuclear injection of the transgene into multiple two-cell embryos generated five founders (Table 1). The keratin-5-LEKTI transgene was detected in two mice by PCR amplification of genomic tail DNA (Figure 2B). The transgenic founders were fertile and transmitted the transgene to the next generation. The mouse line FVB-K5-LEKTI-B1 was selected to establish the colony (K5-LEKTI$^{+/0}$, Table 1). K5-LEKTI$^{+/0}$ mice do not display any particular phenotype as evidenced by representative images of 3 and 11-month-old WT and K5-LEKTI$^{+/0}$ mice (Figure 2C–F). qPCR analysis of RNA extracted from the epidermis of newborn WT and K5-LEKTI$^{+/0}$ (Figure 2G) mice showed an average five-fold increase of LEKTI mRNA expression in the transgenic mice.

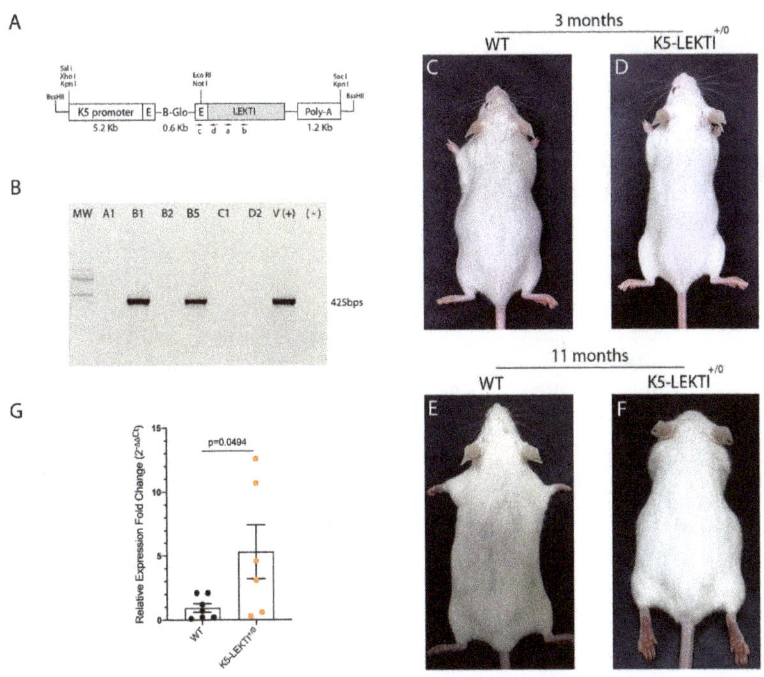

Figure 2. Generation of Keratin5-LEKTI transgenic mice. (**A**) Schematic structure of the K5-LEKTI transgene comprised of a bovine keratin-5 promoter (K5), rabbit-globin exons, a rabbit-globin intron, the mouse LEKTI cDNA and a rabbit-globin

polyadenylation signal (PolyA). (a,b) position of primers used for qPCR and (c,d) genotyping. The linearized transgene vector was microinjected into the male pronucleus of FVB/NJ zygotes, which then were implanted into pseudopregnant mice. (**B**) LEKTI transgenic founders were genotyped by PCR using genomic DNA from tail biopsies with the primer pair indicated on the vector (c,d). Genotyping gel showing positive 425 bps amplified bands from founder mice. pBK5-LEKTI vector was used as template for positive control. (**C–F**) Images of 3 (**C**,**D**) and 11 months (**E**,**F**) old WT (**C**,**E**) and K5-LEKTI$^{+/0}$ (**D**,**F**) mice showing no differences on the outward phenotype. (**G**) qPCR analysis of the epidermis of newborn WT (black dots) and K5-LEKTI$^{+/0}$ (orange dots) mice show a five-fold increase of LEKTI mRNA expression in transgenic compared to WT mice. WT $n = 7$ and K5-LEKTI$^{+/0}$ $n = 6$; values are expressed in mean ± SD. p-values (two-tailed unpaired parametric t-test) are displayed in the graph.

Table 1. Generation of Keratin5-LEKTI transgenic mice.

Transgenic Founder	Gender [a]	Skin Phenotype	LEKTI Expression
FVB-K5-LEKTI-A1	F	No	No
FVB-K5-LEKTI-B1	F	No	Yes [b]
FVB-K5-LEKTI-B2	F	No	No
FVB-K5-LEKTI-B5	F	No	Yes
FVB-K5-LEKTI-C1	M	No	No
FVB-K5-LEKTI-D2	F	No	No

[a]—F: female; M: male. [b]—Transgenic K5-LEKTI$^{+/0}$ mice with higher expression of LEKTI.

3.3. Co-Expression of LEKTI in Basal Keratinocytes Attenuates Matriptase-Dependent Premalignant Phenotype

Matriptase induces malignant transformation when expressed in epithelial stem cells, and this process is preceded by a premalignant phenotype characterized by hyperplasia, dysplasia, and dermal inflammation [8]. To investigate whether LEKTI could play an inhibitory role in this context, we induced concomitant expression of both Matriptase and LEKTI in basal keratinocytes of transgenic mice. For that, transgenic mice expressing matriptase cDNA under the control of Keratin 5 promoter (K5-Matriptase$^{+/0}$) were interbred with the K5-LEKTI$^{+/0}$ mice (Figure 3A). Double-transgenic mice (K5-LEKTI$^{+/0}$/K5-Matriptase$^{+/0}$) showed a prominent rescue of matriptase-dependent premalignant phenotype. The animals were analyzed at 3 and 11 months of age, and co-expression of LEKTI and matriptase partially rescued matriptase-driven alopecia and ichthyosis (3-month-old mice, Figure 3A, and 11-month-old mice, Figure S1A). In addition, epidermal hyperplasia (Figure 3B and Figure S1B, top panels, yellow dashed line delimitates epidermal thickness) and dermal mast cell recruitment (Figure 3B and Figure S1B, bottom panels, black arrows) were also diminished, as verified by H&E histological evaluation and toluidine blue metachromatic staining, respectively.

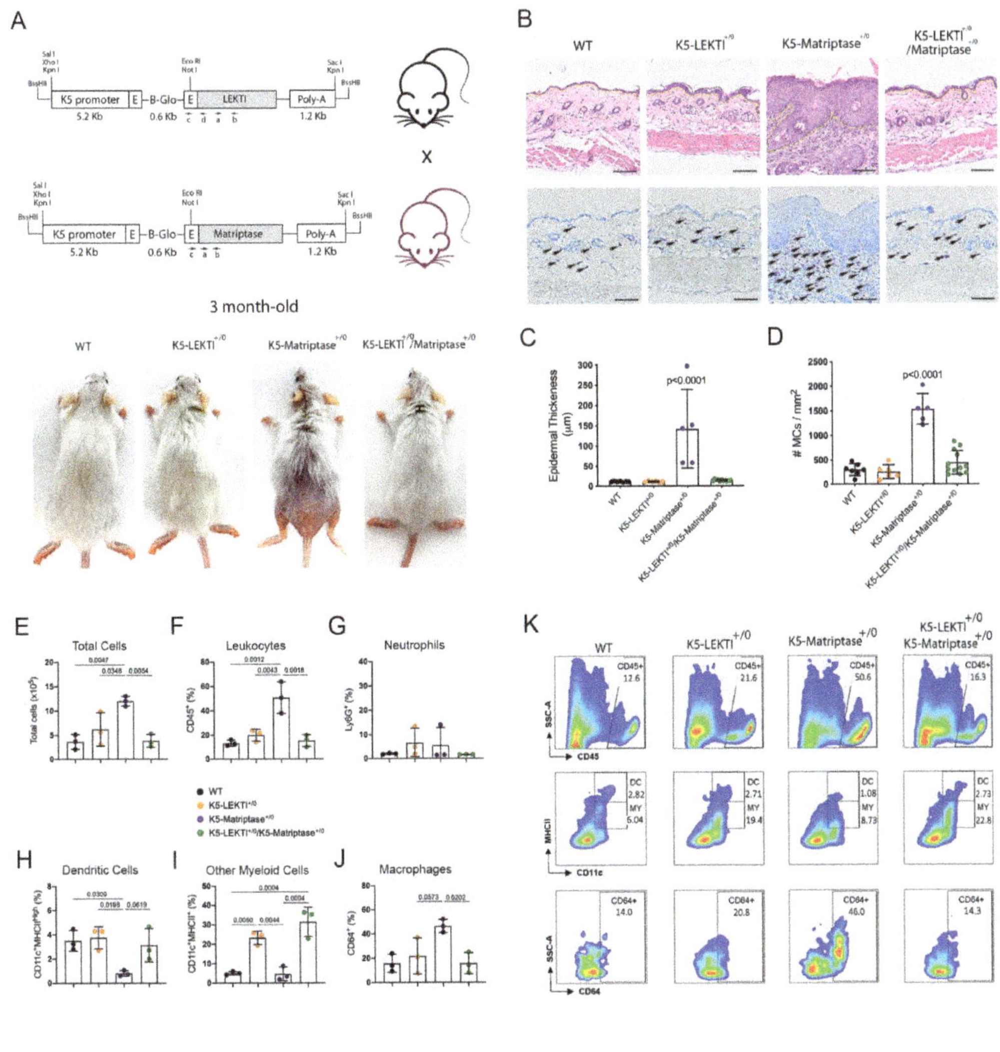

Figure 3. Co-expression of LEKTI attenuates matriptase-mediated premalignant skin phenotype. (**A**) The scheme shows the breeding of K5-LEKTI$^{+/0}$ with K5-Matriptase$^{+/0}$ mice and the resulting litter of WT, K5-LEKTI$^{+/0}$, K5-Matriptase$^{+/0}$, and K5-Matriptase$^{+/0}$/K5-LEKTI$^{+/0}$ mice. Images show the outward appearance of these mice at 3 months of age. Matriptase-induced alopecia and ichthyosis are considerably attenuated by co-expression of LEKTI in Matriptase$^{+/0}$/K5-LEKTI$^{+/0}$. LEKTI$^{+/0}$ in 3-month-old mice. (**B**) Representative histological appearance of dorsal skin of littermate WT (first column), K5-LEKTI$^{+/0}$ (second column), K5-Matriptase$^{+/0}$ (third column), and K5-Matriptase$^{+/0}$/K5-LEKTI$^{+/0}$ mice (forth column) stained by H&E (top panels) and Toluidine Blue (bottom panels) at 3 months of age. Bars = 100 µm. Yellow dashed lines show the limits between epidermis and dermis. Black arrows show metachromatically stained dermal mast cells. (**C**) Quantification of epidermal thickness in littermate WT (n = 7, black dots), K5-LEKTI$^{+/0}$ (n = 6, orange dots), K5-Matriptase$^{+/0}$ (n = 5, purple dots), and K5-Matriptase$^{+/0}$/K5-LEKTI$^{+/0}$ (n = 11, green dots) at 3 months of age. Data are

expressed in mean ± SD. (**D**) Quantification of the dermal mast cell accumulation in the skin of littermate WT (black dots), K5-LEKTI$^{+/0}$ (orange dots), K5-Matriptase$^{+/0}$ (purple dots), and K5-Matriptase$^{+/0}$/K5-LEKTI$^{+/0}$ (green dots) at 3 months of age. Data are expressed in mean ± SD. (**E–J**) Myeloid cellular infiltration was evaluated in skin samples by flow cytometry. (**E**) Total cells ($\times 10^5$) were assessed by automated cell counter using trypan blue. The percentage of (**F**) leukocytes (CD45$^+$), (**G**) neutrophils (Ly6G$^+$ gated on CD45$^+$), (**H**) dendritic cells (DC–CD11c$^+$MHCII^{High} gated on CD45$^+$Ly6G), (**I**) other myeloid cells (MY–CD11c$^+$MHCII$^+$ gated on CD45$^+$Ly6G$^-$), and (**J**) macrophages (CD64$^+$ gated on myeloid cells) in the skin. WT (black dots), K5-LEKTI$^{+/0}$ (orange dots), K5-Matriptase$^{+/0}$ (purple dots), and K5-Matriptase$^{+/0}$/K5-LEKTI$^{+/0}$ (green dots). (**K**) Representative flow cytometry of skin samples by groups. Data are representative of one experiment (n = 3/group) and are expressed as means ± SD. p-values (One-Way ANOVA with Tukey's post-hoc test) displayed in the graphs.

Further quantification of epidermal thickness and dermal mast cell recruitment showed that matriptase-dependent premalignant phenotype is indeed ameliorated by concomitant LEKTI expression (Figure 3C,D and Figure S1C,D). To further investigate the role of LEKTI in the attenuation of matriptase induced dermal inflammation and myeloid cells infiltration into the skin, we performed flow cytometry analysis of cells isolated from digested dorsal skin samples of 3-month-old mice. In accordance with histopathological data, our results demonstrated that K5-Matriptase$^{+/0}$ mice present a significant increase in the total number of inflammatory cells in the skin (Figure 3E). The inflammatory infiltration was confirmed by an elevated percentage of leukocytes (CD45$^+$ cells) in K5-Matriptase$^{+/0}$ mice, which decreased in K5-LEKTI$^{+/0}$/K5-Matriptase$^{+/0}$ mice (Figure 3F). Concerning myeloid cells, we analyzed neutrophils, dendritic cells, and macrophages. Although there was no difference in neutrophils (Ly6G$^+$) among groups (Figure 3G), K5-Matriptase$^{+/0}$ mice showed a significant reduction in the percentage of dendritic cells (D11c$^+$MHCII^{High}), not observed in the double-transgenic group (Figure 3H). Notably, a significant increase in the percentage of macrophages (CD64$^+$) was observed in K5-Matriptase$^{+/0}$ mice and rescued in double-transgenic mice (Figure 3J). Considering other myeloid cells (CD11c$^+$MHCII$^+$), LEKTI expression in both K5-LEKTI$^{+/0}$ and K5-LEKTI$^{+/0}$/K5-Matriptase$^{+/0}$ mice induced a significant increase of this population in comparison to WT and K5-Matriptase$^{+/0}$ mice (Figure 3I). Representative flow cytometry differences among groups are displayed in Figure 3K.

Overall, our data confirms that LEKTI can act as a tumor suppressor downstream of matriptase activation, as evidenced by its ability to modulate matriptase induced hyperplasia and inflammatory cell recruitment.

3.4. Co-Expression of LEKTI with Matriptase in Basal Keratinocytes Delays the Onset and Progression of Chemically Induced Carcinogenesis and Decreases KLK5 Expression

Ras-dependent SCC is potentiated by the topical application of the genotoxic agent DMBA in K5-Matriptase$^{+/0}$ mice [8]. To explore the inhibitory role of LEKTI in the initiation and progression of matriptase driven carcinogenesis, we used a one-stage carcinogenesis model where DMBA was applied to the dorsal skin of WT, K5-LEKTI$^{+/0}$, K5-Matriptase$^{+/0}$, and K5-Matriptase$^{+/0}$/K5-LEKTI$^{+/0}$ mice every three weeks for a total of five applications (Figure 4A). While WT and K5-LEKTI$^{+/0}$ mice did not develop any lesions, all K5-Matriptase$^{+/0}$ mice developed lesions at around 14–17 weeks of age. More importantly, there was an average 3-week delay in the emergence of lesions in K5-Matriptase$^{+/0}$/K5-LEKTI$^{+/0}$ mice compared to K5-Matriptase$^{+/0}$ mice (Figure 4B). Tumor progression was also hampered by co-expression of LEKTI, as evidenced by a smaller number of lesions and the reduced average size compared to lesions of K5-Matriptase$^{+/0}$ mice (Figure 4C,D). At 26 weeks of age, Matriptase$^{+/0}$ mice display an increase in tumor size compared to the double-transgenic mice (Figure 4E). The observed delay in onset and progression of DMBA-induced carcinogenesis indicates that LEKTI could be working as a suppressor of squamous cell carcinogenesis in the context of dysregulated matriptase.

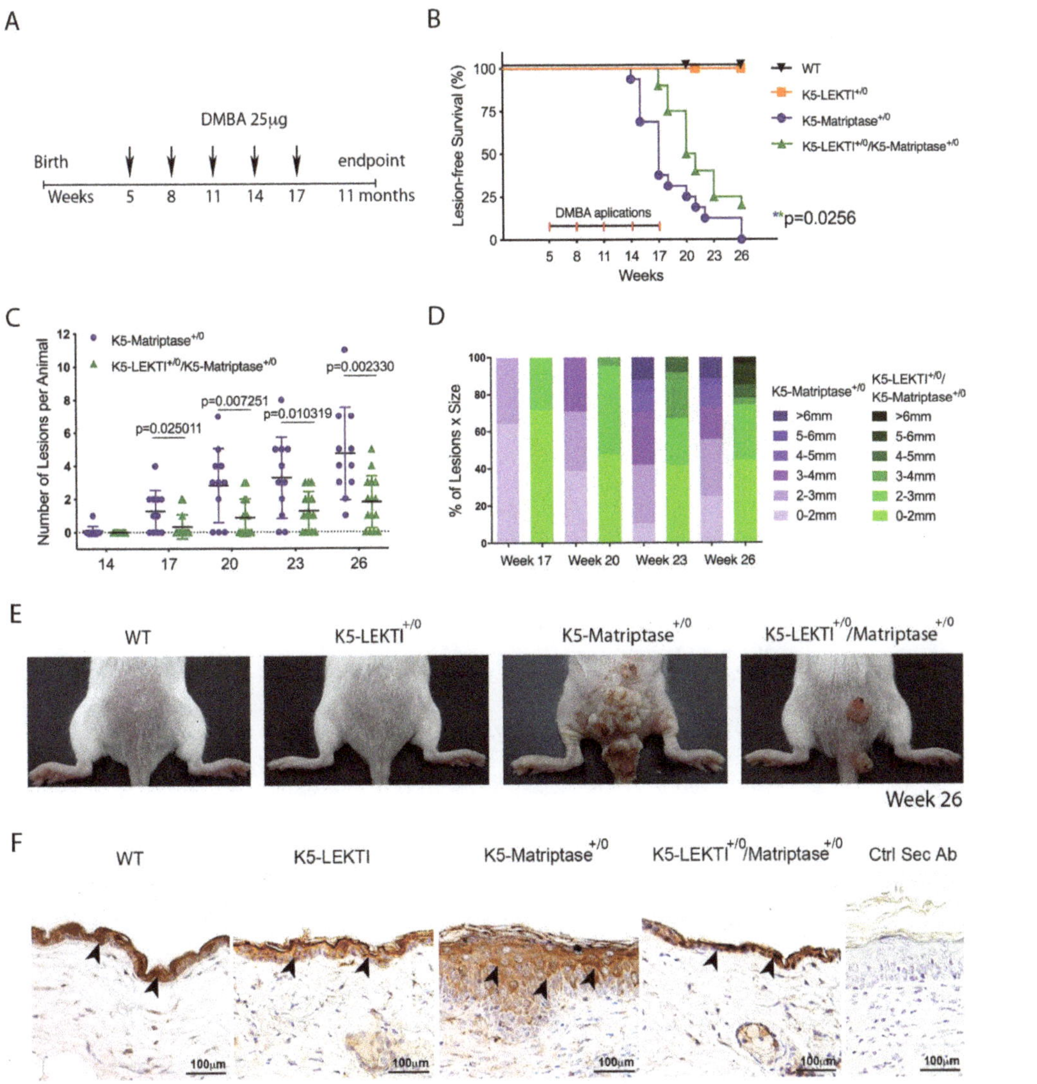

Figure 4. Co-expression of LEKTI with matriptase in basal keratinocytes delays the onset and progression of chemically induced carcinogenesis. (**A**) One-stage chemical carcinogenesis scheme: dorsal skin of mice was exposed 5 times to 25 μg of DMBA, starting at week 5 of age, every 3 weeks, and were followed for up to 48 weeks of age. WT (n = 20), K5-LEKTI$^{+/0}$ (n = 17), K5-Matriptase$^{+/0}$ (n = 11), and K5-Matriptase$^{+/0}$/K5-LEKTI$^{+/0}$ (n = 16). (**B**) Kaplan–Meier analysis of tumor-free survival. WT (black upside-down triangle), K5-LEKTI$^{+/0}$ (orange squares), K5-Matriptase$^{+/0}$ (purple dots), and K5-Matriptase$^{+/0}$/K5-LEKTI$^{+/0}$ (green triangles). (**C**,**D**) Matriptase induced tumor progression in K5-Matriptase$^{+/0}$, and K5-Matriptase$^{+/0}$/K5-LEKTI$^{+/0}$ mice. Data are expressed in mean ± SD. (**C**) number of lesions and (**D**) percentage of lesion of each size in littermate K5-Matriptase$^{+/0}$ (purple dots) and K5-Matriptase$^{+/0}$/K5-LEKTI$^{+/0}$ (green triangles) mice from 14 to 26 weeks of age. p-values (multiple t-tests) are displayed in the graphs. (E) Representative images of the outward appearance of littermate WT, K5-LEKTI$^{+/0}$, K5-Matriptase$^{+/0}$, and K5-Matriptase$^{+/0}$/K5-LEKTI$^{+/0}$ mice at 26 weeks of age. (**F**) Klk5 IHC staining of the skin of 3-month-old WT, K5-LEKTI$^{+/0}$, K5-Matriptase$^{+/0}$, and K5-Matriptase$^{+/0}$/LEKTI$^{+/0}$ mice. Black arrowheads indicate stained areas; Negative secondary antibody control; Bar = 100 μm.

Since LEKTI is unable to inhibit matriptase directly [11], we hypothesized that there is another matriptase protease substrate in squamous cell carcinogenesis. In this respect, LEKTI inhibition of the matriptase-dependent proteolytic pathway could occur through the inhibition of epithelial kallikreins. In fact, immunohistochemical analysis of KLK5 expression in skin biopsies of 3-month-old mice showed that this protease is increased in K5-Matriptase$^{+/0}$ mice. Notably, concomitant expression of LEKTI and matriptase in basal layer keratinocytes is sufficient to revert the aberrant KLK5 expression found in K5-Matriptase$^{+/0}$ mice (Figure 4F).

3.5. *KLK5 Activates YAP1-TAZ/TEAD Transcription via PAR-2 and Induces Matriptase-Mediated Release of IL-8 and Cell Migration in OSCC Cells*

Because PAR-2 was previously shown to be essential for matriptase-driven premalignant progression and squamous cell carcinogenesis, we next investigated whether KLK5 activates PAR-2. For this purpose, we used a reconstituted cell-based assay in which HEK293 cells were transfected with a PAR-2 expression vector and a serum response element (SRE)-luciferase reporter plasmid. The transfected cells were then exposed to hrKLK5 to determine PAR-2 activation. As expected, KLK5 treatment was able to activate PAR-2 (Figure 5A, matriptase-dependent activation of PAR-2 was used as positive control). Importantly, KLK5-dependent PAR-2 activation was also responsible for an increase in Hippo-YAP1/TEAD transcriptional activity, as measured by luciferase assay with a reporter containing tandem TEAD-binding sites in HEK293T cells transfected with PAR-2 and treated with hrKLK5 (Figure 5B). Because PAR-2 activation also leads to NFkB activation, we sought to investigate the extent to which KLK5 contributes to the release of proinflammatory cytokines. To this aim, we used WT and KLK5 KO OSCC cell lines [21] treated or not with hrMatriptase for 24 h, and proinflammatory cytokine release was evaluated by ELISA. Matriptase-dependent release of IL-8 occurred only in cells expressing KLK5 (Figure 5C). TNF-α release was stimulated by matriptase in both WT and KLK5 KO cells (Figure 5D). Because both NFkB and Hippo signaling pathways may lead to increased cell migration in cancer [23,24], a scratch wound healing assay was used. WT and KLK5 KO OSCC Cal 27 monolayers were treated or not with hrMatriptase, scratched, and wound closure was evaluated for up to 48 h. Interestingly, matriptase-dependent wound closure after 24 h took place solely in cells expressing KLK5. After 48 h, however, hrMatriptase treatment induced wound closure in both WT and KLK5 KO cells (Figure 5E, third and fourth columns) and this effect was partially inhibited in KLK5 KO cells (Figure 5E, fourth column) compared to WT cells (Figure 5E, third column). Wound closure was also delayed in untreated KLK5 KO cells (Figure 5E, second column) compared to WT cells (Figure 5E, first column). Quantification of the wound closure confirmed a significant delay in both matriptase-treated (striped columns) and untreated (unstriped columns) KLK5 KO cells (red columns) when compared to WT cells (white columns) after 24 and 48 h (Figure 5F).

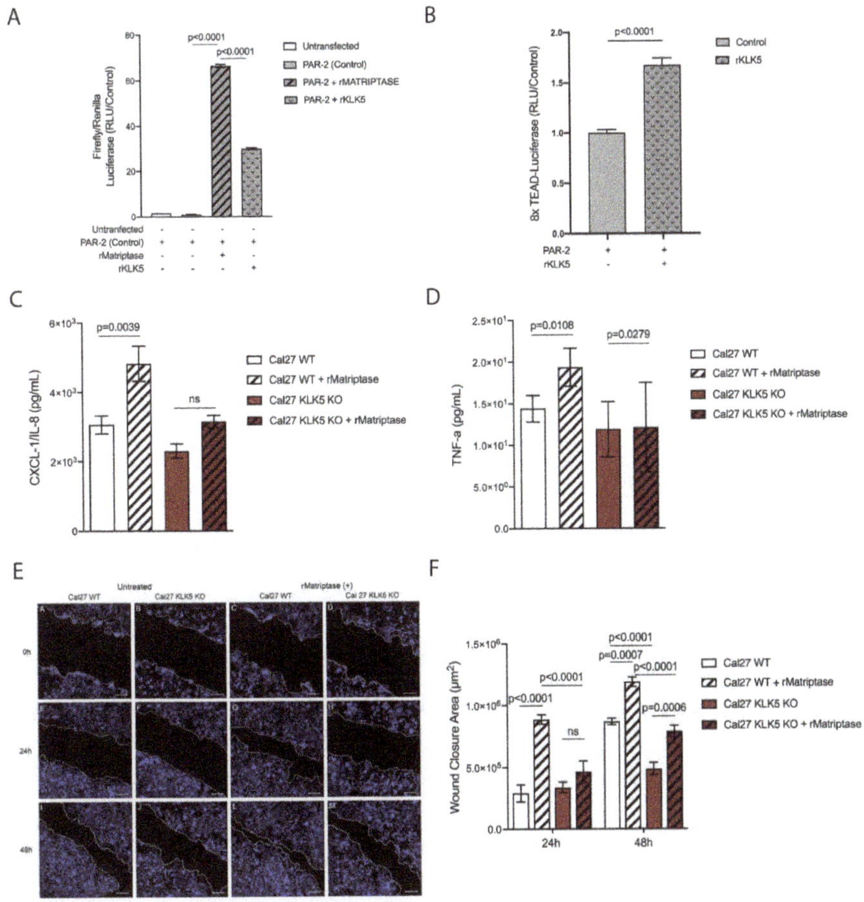

Figure 5. KLK5 activates YAP1-TAZ/TEAD transcription via PAR-2 and induces matriptase-mediated release of IL-8 and cell migration in OSCC cells. (A) KLK5-mediated PAR-2 activation analysis. HEK293T cells were transfected with pCDNA 3.1-PAR-2, pRL-Renilla luciferase and SRE-Firefly luciferase reporter plasmids and treated with hrKLK5 for 6 h, PAR-2 activation was measured by Luciferase activity. hrMatriptase was used as positive control for PAR-2 activation. *p*-values (One-Way ANOVA with Tukey's post-hoc test) are displayed in the graphs. (**B**) Transcriptional activity of TEAD was also measured by luciferase assay with a reporter containing tandem TEAD-binding sites in HEK293T cells transfected with pCDNA 3.1-PAR-2 plasmid and treated with hrKLK5 for 6 h. *p*-values (two-tailed unpaired parametric *t*-test) are displayed in the graphs. (**C**,**D**) WT and KLK5 KO OSCC Cal 27 cell lines were serum-starved for 2 h, treated or not with hrMatriptase for 24 h, and proinflammatory cytokine release was evaluated by ELISA. (**C**) hrMatriptase treatment stimulated the release of CXCL-1/IL-8 in WT cells but not in KLK5 KO cells. (**D**) hrMatriptase treatment stimulated the release of TNF-α in both WT and KLK5 KO cells, although in KO cells this effect was very discreet. *p*-values (One-Way ANOVA with Tukey's post-hoc test) are displayed in the graphs. (**E**,**F**) WT and KLK5 KO OSCC Cal 27 monolayers were serum-starved for 16 h, treated or not with hrMatriptase, scratched, and wound closure was evaluated for up to 48 h. (**E**) Representative images showing that hrMatriptase treatment induces wound closure in both WT and KLK5 KO cells (third and fourth columns) and this effect is partially inhibited in KLK5 KO cells (forth column) compared to WT cells (third column). Wound closure was also delayed in untreated KLK5 KO cells (second column) compared to WT cells (first column). (**F**) Quantification of the wound closure shows a delay in both matriptase treated (striped columns) and untreated (unstriped columns) KLK5 KO cells (red columns) when compared to WT cells (white columns) after 24 and 48 h. *p*-values (One-Way ANOVA with Tukey's post-hoc test) are displayed in the graphs. (**A–F**) Data are representative of two independent experiments and values are expressed in mean ± SEM.

4. Discussion

This study aimed to explore the inhibitory role for LEKTI in matriptase-dependent SCCs and investigate additional players operating in this pathway. We showed here that LEKTI expression is either absent or remarkably reduced in less differentiated, more aggressive OSCCs, while matriptase expression is prevalent among OSCCs of different grades. Protease to inhibitor unbalance has been associated with uncontrolled proteolysis and malignant disorders before [25,26], and this result is in agreement with recent findings from our group, where LEKTI was shown to be downregulated in OSCCs in comparison with premalignant lesions and normal oral mucosa, with increased KLK5/SPINK5 mRNA (protease/inhibitor) ratio being associated with worse prognosis [21]. Furthermore, SPINK5 expression levels have been found to be downregulated in head and neck squamous cell carcinomas (HNSCCs) [27–29].

Consisting with an inhibitory role for LEKTI in the development of SCCs, we show that co-expression of LEKTI with matriptase in basal keratinocytes partially reverts matriptase-mediated premalignant phenotype, as evidenced by reduced alopecia, hyperplasia, and myeloid cell recruitment. In line with the fact that the tumorigenic microenvironment of SCCs often displays reduced quantities of myeloid CD11c$^+$ dermal DCs, which indicates disruption in the DC immunostimulatory capacity [30,31], we showed a matriptase-dependent significant decrease in DCs. This was accompanied by increased macrophage infiltration, a phenotype rescued by co-expression of LEKTI in the skin of transgenic mice. Conspicuous influx of tumor-associated macrophages in SCC lesions can contribute to carcinogenesis and tumor growth [32]. More importantly, LEKTI expression also delayed the onset and progression of chemically induced carcinogenesis. Comparable inhibition was previously observed by co-expression of the matriptase cognate inhibitors hepatocyte growth factor inhibitors 1 and 2 (HAI-1 and HAI-2) [8,33].

In agreement with the fact that LEKTI inhibits matriptase-dependent skin desquamation in Netherton Syndrome by the direct inhibition of KLK5 [11] and our data shows that concomitant expression of LEKTI with matriptase rescues aberrant KLK5 expression. This indicates a role for KLK5 as a matriptase substrate. Dysregulated KLK5 has been implicated in a wide range of epithelial cancers [21,34–38]. Our data supports the existence of a matriptase-activated KLK5-dependent PAR-2 signaling axis in SCCs. PAR-2 is activated by several trypsin-like serine proteases, including matriptase and KLK5, signaling to various downstream pathways that modulate cell proliferation, migration and invasion, and cytokine production [10,21,36,39]. Regarding the potential mechanisms by which KLK5 can contribute to tumorigenesis, we show that KLK5 can induce PAR2 dependent activation of NFKB and Hippo-YAP1. Indeed, it has been shown that PAR-2 can lead to NFκB activation in inflammation and SCCs, and PAR-2 can be activated by KLK5 in OSCC [7,33,36,40,41]. In addition, PAR-2 activation by SLIGRL (PAR-2 peptide agonist) can result in YAP1 activation [42]. Hippo–YAP1/TEAD regulates organ size, tissue homeostasis, and tumorigenesis in mammals. Abnormal upregulation of this pathway occurs in many human malignancies and promotes tumor formation, progression, and metastasis [43,44].

Tumor invasion and metastasis are complex biological processes in which detachment and migration rely on matrix-degrading proteases. Knockdown of matriptase in endometrial cancer cells inhibits migration and invasion ability in vitro [45], and our recent work showed that CRISPR-mediated disruption of KLK5 blocks OSCC cell migration [21]. Our current results showed impaired matriptase-mediated migration in KLK5 KO OSCC cells, supporting a potential role of this pathway in cancer invasion and metastasis.

Taken together, our data support a model wherein matriptase activates PAR-2 via KLK5 and modulates SCC development and progression. This knowledge can contribute for the development of targeted therapeutics in HNSCC.

5. Conclusions

Our data identify a third signaling pathway for matriptase-dependent carcinogenesis in vivo, wherein matriptase activates PAR-2 via KLK5 and thus modulates SCC development and progression.

Supplementary Materials: The following are available online at https://www.mdpi.com/article/10.3390/cancers13174395/s1, Figure S1: Co-expression of LEKTI attenuates matriptase-mediated premalignant skin phenotype.

Author Contributions: E.Z.M.d.S.: Investigation, Methodology, Formal analysis, Writing original draft; T.F.d.C.F.-S., Y.Y., G.A.P., M.H.K., G.V.V., M.F.C., L.M.A.R.I., M.G.M.: Investigation, Formal analysis; M.G.A.: Investigation, Methodology, Formal analysis; C.K.d.F.: Investigation, Formal analysis, Validation; A.R.d.S., J.C.A.-F., V.L.D.B., R.I.-B.: Conceptualization, Methodology, Resources; K.U.S.: Conceptualization, Methodology, Writing—review and editing, Supervision, Formal analysis, Funding Acquisition. All authors have read and agreed to the published version of the manuscript.

Funding: This work was supported by the Fundaçao de Amparo a Pesquisa do Estado de São Paulo (São Paulo Research Foundation; FAPESP, 2014/06316-2) and Fundaçao de Apoio ao Ensino, Pesquisa e Assistencia do Hospital das Clinicas da Faculdade de Medicina de Ribeirao Preto da Universidade de São Paulo (FAEPA) research grants to K.U.S. E.Z.M.d.S.—post-doctoral fellowship FAPESP #: 2016/13228-8 and FAPESP BEPE fellowship # 2017/24730-9. M.G.A., M.H.K.—master's fellowship from CAPES (Brazilian Ministry of Education), G.V.V.—Ph.D. fellowship from CAPES. The funders had no role in study design, data collection, analysis, decision to publish, or manuscript preparation.

Institutional Review Board Statement: The study was conducted according to the guidelines of the Declaration of Helsinki and approved by the Ethics Committee on Human Research of Ribeirao Preto Clinical Hospital and Ribeirao Preto Medical School, University of São Paulo (protocol #: 50533515.6.0000.5440 approved on: 21 June 2016). All experiments involving mice were approved by the Ethics Committee on Animal Research of Ribeirao Preto Medical School, University of São Paulo (protocol #: 003/2015-1 approved on 29 June 2015) and are in accordance with the Guidelines of the Brazilian College of Animal Experimentation.

Informed Consent Statement: Only incisional biopsies were included in this study and patient consent was waived since the work was done using paraffin-embedded non-identified samples, which according to Brazilian laws, do not require patient consent forms.

Data Availability Statement: Data is contained within the article or supplementary material.

Acknowledgments: The authors are grateful to Vani Maria Alves, from the Department of Cell and Molecular Biology and Pathogenic Bioagents and Marcia Aparecida Oliva and Laura Midori Kawasse, from the Oncopathology Laboratory, Department of Pathology and Legal Medicine, both from FMRP-USP, Ribeirao Preto, SP, for excellent technical assistance. The authors also thank Thomas H. Bugge, from the National Institute of Dental and Craniofacial Research, National Institutes of Health for the technical support and for critically reading the manuscript.

Conflicts of Interest: The authors declare no potential conflict of interest.

References

1. Bray, F.; Ferlay, J.; Soerjomataram, I.; Siegel, R.L.; Torre, L.A.; Jemal, A. Global cancer statistics 2018: GLOBOCAN estimates of incidence and mortality worldwide for 36 cancers in 185 countries. *CA Cancer J. Clin.* **2018**, *68*, 394–424. [CrossRef]
2. Jemal, A.; Siegel, R.; Ward, E.; Hao, Y.; Xu, J.; Murray, T.; Thun, M.J. Cancer statistics, 2008. *CA Cancer J. Clin.* **2008**, *58*, 71–96. [CrossRef]
3. Molinolo, A.A.; Amornphimoltham, P.; Squarize, C.H.; Castilho, R.M.; Patel, V.; Gutkind, J.S. Dysregulated molecular networks in head and neck carcinogenesis. *Oral Oncol.* **2009**, *45*, 324–334. [CrossRef]
4. Forastiere, A.; Koch, W.; Trotti, A.; Sidransky, D. Head and neck cancer. *N. Engl. J. Med.* **2001**, *345*, 1890–1900. [CrossRef]
5. Bagan, J.V.; Scully, C. Recent advances in Oral Oncology 2007: Epidemiology, aetiopathogenesis, diagnosis and prognostication. *Oral Oncol.* **2008**, *44*, 103–108. [CrossRef] [PubMed]
6. Murray, A.S.; Varela, F.A.; List, K. Type II transmembrane serine proteases as potential targets for cancer therapy. *Biol. Chem.* **2016**, *397*, 815–826. [CrossRef]

7. Varela, F.; Hyland, T.; List, K. Physiological Functions and Role of Matriptase in Cancer. In *Extracellular Targeting of Cell Signaling in Cancer: Strategies Directed at MET and RON Receptor Tyrosine Kinase Pathways*, 1st ed.; Janetka, J.W., Benson, R.M., Eds.; John Wiley & Sons Ltd.: Hoboken, NJ, USA, 2018.
8. List, K.; Szabo, R.; Molinolo, A.; Sriuranpong, V.; Redeye, V.; Murdock, T.; Burke, B.; Nielsen, B.S.; Gutkind, J.S.; Bugge, T.H. Deregulated matriptase causes ras-independent multistage carcinogenesis and promotes ras-mediated malignant transformation. *Genes Dev.* **2005**, *19*, 1934–1950. [CrossRef]
9. Szabo, R.; Rasmussen, A.L.; Moyer, A.B.; Kosa, P.; Schafer, J.M.; Molinolo, A.A.; Gutkind, J.S.; Bugge, T.H. c-Met-induced epithelial carcinogenesis is initiated by the serine protease matriptase. *Oncogene* **2011**, *30*, 2003–2016. [CrossRef]
10. Sales, K.U.; Friis, S.; Konkel, J.E.; Godiksen, S.; Hatakeyama, M.; Hansen, K.K.; Rogatto, S.R.; Szabo, R.; Vogel, L.K.; Chen, W.; et al. Non-hematopoietic PAR-2 is essential for matriptase-driven pre-malignant progression and potentiation of ras-mediated squamous cell carcinogenesis. *Oncogene* **2015**, *34*, 346–356. [CrossRef]
11. Sales, K.U.; Masedunskas, A.; Bey, A.L.; Rasmussen, A.L.; Weigert, R.; List, K.; Szabo, R.; Overbeek, P.A.; Bugge, T.H. Matriptase initiates activation of epidermal pro-kallikrein and disease onset in a mouse model of Netherton syndrome. *Nat. Genet.* **2010**, *42*, 676–683. [CrossRef]
12. Komatsu, N.; Saijoh, K.; Toyama, T.; Ohka, R.; Otsuki, N.; Hussack, G.; Takehara, K.; Diamandis, E.P. Multiple tissue kallikrein mRNA and protein expression in normal skin and skin diseases. *Br. J. Dermatol.* **2005**, *153*, 274–281. [CrossRef]
13. Shaw, J.L.; Diamandis, E.P. Distribution of 15 human kallikreins in tissues and biological fluids. *Clin. Chem.* **2007**, *53*, 1423–1432. [CrossRef]
14. Yousef, G.M.; Diamandis, E.P. Tissue kallikreins: New players in normal and abnormal cell growth? *Thromb. Haemost.* **2003**, *90*, 7–16. [CrossRef]
15. Kryza, T.; Silva, M.L.; Loessner, D.; Heuzé-Vourc'h, N.; Clements, J.A. The kallikrein-related peptidase family: Dysregulation and functions during cancer progression. *Biochimie* **2016**, *122*, 283–299. [CrossRef]
16. Fortelny, N.; Cox, J.H.; Kappelhoff, R.; Starr, A.E.; Lange, P.F.; Pavlidis, P.; Overall, C.M. Network analyses reveal pervasive functional regulation between proteases in the human protease web. *PLoS Biol.* **2014**, *12*, e1001869. [CrossRef]
17. Association, W.M. World Medical Association Declaration of Helsinki: Ethical principles for medical research involving human subjects. *JAMA* **2013**, *310*, 2191–2194.
18. Murillas, R.; Larcher, F.; Conti, C.J.; Santos, M.; Ullrich, A.; Jorcano, J.L. Expression of a dominant negative mutant of epidermal growth factor receptor in the epidermis of transgenic mice elicits striking alterations in hair follicle development and skin structure. *EMBO J.* **1995**, *14*, 5216–5223. [CrossRef] [PubMed]
19. List, K.; Haudenschild, C.C.; Szabo, R.; Chen, W.; Wahl, S.M.; Swaim, W.; Engelholm, L.H.; Behrendt, N.; Bugge, T.H. Matriptase/MT-SP1 is required for postnatal survival, epidermal barrier function, hair follicle development, and thymic homeostasis. *Oncogene* **2002**, *21*, 3765–3779. [CrossRef] [PubMed]
20. Gioanni, J.; Fischel, J.L.; Lambert, J.C.; Demard, F.; Mazeau, C.; Zanghellini, E.; Ettore, F.; Formento, P.; Chauvel, P.; Lalanne, C.M.; et al. Two new human tumor cell lines derived from squamous cell carcinomas of the tongue: Establishment, characterization and response to cytotoxic treatment. *Eur. J. Cancer Clin. Oncol.* **1988**, *24*, 1445–1455. [CrossRef]
21. Alves, M.G.; Kodama, M.H.; da Silva, E.Z.M.; Gomes, B.B.M.; da Silva, R.A.A.; Vieira, G.V.; Alves, V.M.; da Fonseca, C.K.; Santana, A.C.; Cecílio, N.T.; et al. Relative expression of KLK5 to LEKTI is associated with aggressiveness of oral squamous cell carcinoma. *Transl. Oncol.* **2021**, *14*, 100970. [CrossRef]
22. Schindelin, J.; Arganda-Carreras, I.; Frise, E.; Kaynig, V.; Longair, M.; Pietzsch, T.; Preibisch, S.; Rueden, C.; Saalfeld, S.; Schmid, B.; et al. Fiji: An open-source platform for biological-image analysis. *Nat. Methods* **2012**, *9*, 676–682. [CrossRef]
23. Chang, Y.C.; Wu, J.W.; Wang, C.W.; Jang, A.C. Hippo Signaling-Mediated Mechanotransduction in Cell Movement and Cancer Metastasis. *Front. Mol. Biosci.* **2019**, *6*, 157. [CrossRef]
24. Yan, M.; Xu, Q.; Zhang, P.; Zhou, X.J.; Zhang, Z.Y.; Chen, W.T. Correlation of NF-kappaB signal pathway with tumor metastasis of human head and neck squamous cell carcinoma. *BMC Cancer* **2010**, *10*, 437. [CrossRef]
25. Vogel, L.K.; Sæbø, M.; Skjelbred, C.F.; Abell, K.; Pedersen, E.D.; Vogel, U.; Kure, E.H. The ratio of Matriptase/HAI-1 mRNA is higher in colorectal cancer adenomas and carcinomas than corresponding tissue from control individuals. *BMC Cancer* **2006**, *6*, 176. [CrossRef] [PubMed]
26. Kit, O.I.; Frantsiyants, E.; Kozlova, L.S.; Bandovkina, V.A.; Kolesnikov, E.N.; Kozhushko, M.A.; Kazieva, T.B.; Averkin, M.; Trifanov, V. Protease/protease inhibitor balance in blood plasma to predict postoperative complications in operated patients with pancreatic head cancer. *J. Clin. Oncol.* **2018**, *36*, 258. [CrossRef]
27. Leusink, F.K.; Van Diest, P.J.; Frank, M.H.; Broekhuizen, R.; Braunius, W.; Van Hooff, S.R.; Willems, S.M.; Koole, R. The Co-Expression of Kallikrein 5 and Kallikrein 7 Associates with Poor Survival in Non-HPV Oral Squamous-Cell Carcinoma. *Pathobiology* **2015**, *82*, 58–67. [CrossRef]
28. Gonzalez, H.E.; Gujrati, M.; Frederick, M.; Henderson, Y.; Arumugam, J.; Spring, P.W.; Mitsudo, K.; Kim, H.W.; Clayman, G.L. Identification of 9 genes differentially expressed in head and neck squamous cell carcinoma. *Arch. Otolaryngol. Head Neck Surg.* **2003**, *129*, 754–759. [CrossRef]
29. Ye, H.; Yu, T.; Temam, S.; Ziober, B.L.; Wang, J.; Schwartz, J.L.; Mao, L.; Wong, D.T.; Zhou, X. Transcriptomic dissection of tongue squamous cell carcinoma. *BMC Genom.* **2008**, *9*, 69. [CrossRef]

30. Bluth, M.J.; Zaba, L.C.; Moussai, D.; Suárez-Fariñas, M.; Kaporis, H.; Fan, L.; Pierson, K.C.; White, T.R.; Pitts-Kiefer, A.; Fuentes-Duculan, J.; et al. Myeloid dendritic cells from human cutaneous squamous cell carcinoma are poor stimulators of T-cell proliferation. *J. Investig. Dermatol.* **2009**, *129*, 2451–2462. [CrossRef]
31. Yanofsky, V.R.; Mitsui, H.; Felsen, D.; Carucci, J.A. Understanding dendritic cells and their role in cutaneous carcinoma and cancer immunotherapy. *Clin. Dev. Immunol.* **2013**, *2013*, 624123. [CrossRef] [PubMed]
32. Pettersen, J.S.; Fuentes-Duculan, J.; Suárez-Fariñas, M.; Pierson, K.C.; Pitts-Kiefer, A.; Fan, L.; Belkin, D.A.; Wang, C.Q.; Bhuvanendran, S.; Johnson-Huang, L.M.; et al. Tumor-associated macrophages in the cutaneous SCC microenvironment are heterogeneously activated. *J. Investig. Dermatol.* **2011**, *131*, 1322–1330. [CrossRef]
33. Sales, K.U.; Friis, S.; Abusleme, L.; Moutsopoulos, N.M.; Bugge, T.H. Matriptase promotes inflammatory cell accumulation and progression of established epidermal tumors. *Oncogene* **2015**, *34*, 4664–4672. [CrossRef] [PubMed]
34. Kim, H.; Scorilas, A.; Katsaros, D.; Yousef, G.M.; Massobrio, M.; Fracchioli, S.; Piccinno, R.; Gordini, G.; Diamandis, E.P. Human kallikrein gene 5 (KLK5) expression is an indicator of poor prognosis in ovarian cancer. *Br. J. Cancer* **2001**, *84*, 643–650. [CrossRef]
35. Papachristopoulou, G.; Malachias, A.; Devetzi, M.; Kamouza, E.; Scorilas, A.; Xynopoulos, D.; Talieri, M. Uncovering the clinical impact of kallikrein-related peptidase 5 (KLK5) mRNA expression in the colorectal adenoma-carcinoma sequence. *Clin. Chem. Lab. Med.* **2019**, *57*, 1251–1260. [CrossRef] [PubMed]
36. Johnson, J.J.; Miller, D.L.; Jiang, R.; Liu, Y.; Shi, Z.; Tarwater, L.; Williams, R.; Balsara, R.; Sauter, E.R.; Stack, M.S. Protease-activated Receptor-2 (PAR-2)-mediated Nf-κB Activation Suppresses Inflammation-associated Tumor Suppressor MicroRNAs in Oral Squamous Cell Carcinoma. *J. Biol. Chem.* **2016**, *291*, 6936–6945. [CrossRef] [PubMed]
37. Pettus, J.R.; Johnson, J.J.; Shi, Z.; Davis, J.W.; Koblinski, J.; Ghosh, S.; Liu, Y.; Ravosa, M.J.; Frazier, S.; Stack, M.S. Multiple kallikrein (KLK 5, 7, 8, and 10) expression in squamous cell carcinoma of the oral cavity. *Histol. Histopathol.* **2009**, *24*, 197–207.
38. Ma, H.; Hockla, A.; Mehner, C.; Coban, M.; Papo, N.; Radisky, D.C.; Radisky, E.S. PRSS3/Mesotrypsin and kallikrein-related peptidase 5 are associated with poor prognosis and contribute to tumor cell invasion and growth in lung adenocarcinoma. *Sci. Rep.* **2019**, *9*, 1844. [CrossRef]
39. Neville, B.W.; Day, T.A. Oral cancer and precancerous lesions. *CA Cancer J. Clin.* **2002**, *52*, 195–215. [CrossRef]
40. Oikonomopoulou, K.; Hansen, K.K.; Saifeddine, M.; Tea, I.; Blaber, M.; Blaber, S.I.; Scarisbrick, I.; Andrade-Gordon, P.; Cottrell, G.S.; Bunnett, N.W.; et al. Proteinase-activated receptors, targets for kallikrein signaling. *J. Biol. Chem.* **2006**, *281*, 32095–32112. [CrossRef]
41. Pawar, N.R.; Buzza, M.S.; Antalis, T.M. Membrane-Anchored Serine Proteases and Protease-Activated Receptor-2-Mediated Signaling: Co-Conspirators in Cancer Progression. *Cancer Res.* **2019**, *79*, 301–310. [CrossRef]
42. Mo, J.S.; Yu, F.X.; Gong, R.; Brown, J.H.; Guan, K.L. Regulation of the Hippo-YAP pathway by protease-activated receptors (PARs). *Genes Dev.* **2012**, *26*, 2138–2143. [CrossRef]
43. Maehama, T.; Nishio, M.; Otani, J.; Mak, T.W.; Suzuki, A. The role of Hippo-YAP signaling in squamous cell carcinomas. *Cancer Sci.* **2021**, *112*, 51–60. [CrossRef]
44. Nishio, M.; Maehama, T.; Goto, H.; Nakatani, K.; Kato, W.; Omori, H.; Miyachi, Y.; Togashi, H.; Shimono, Y.; Suzuki, A. Hippo vs. Crab: Tissue-specific functions of the mammalian Hippo pathway. *Genes Cells* **2017**, *22*, 6–31. [CrossRef] [PubMed]
45. Sun, P.; Xue, L.; Song, Y.; Mao, X.; Chen, L.; Dong, B.; Braicu, E.L.; Sehouli, J. Regulation of matriptase and HAI-1 system, a novel therapeutic target in human endometrial cancer cells. *Oncotarget* **2018**, *9*, 12682–12694. [CrossRef] [PubMed]

cancers

MDPI

Article

miRNAs as Biomarkers for Diagnosing and Predicting Survival of Head and Neck Squamous Cell Carcinoma Patients

Igor Piotrowski [1,2,3], Xiang Zhu [3], Tatiana Dandolini Saccon [4], Sarah Ashiqueali [3], Augusto Schneider [5], Allancer Divino de Carvalho Nunes [3], Sarah Noureddine [3], Agnieszka Sobecka [1,6], Wojciech Barczak [1,6], Mateusz Szewczyk [6,7], Wojciech Golusiński [6,7], Michal M. Masternak [3,6] and Paweł Golusiński [8,9,*]

1 Radiobiology Lab, Department of Medical Physics, Greater Poland Cancer Centre, 61-866 Poznan, Poland; igor.piotrowski@wco.pl (I.P.); agnieszka.sobecka@wco.pl (A.S.); wojciech.barczak@oncology.ox.ac.uk (W.B.)
2 Department of Electroradiology, Poznan University of Medical Sciences, ul. Garbary 15, 61-866 Poznan, Poland
3 Burnett School of Biomedical Sciences, College of Medicine, University of Central Florida, Orlando, FL 32827, USA; xiang.zhu@ucf.edu (X.Z.); sarah.ashiqueali@knights.ucf.edu (S.A.); anunes@umn.edu (A.D.d.C.N.); sarahnoureddine@knights.ucf.edu (S.N.); Michal.Masternak@ucf.edu (M.M.M.)
4 Centro de Desenvolvimento Tecnológico, Universidade Federal de Pelotas, Pelotas 96010-610, Brazil; tatiana.saccon@ufpel.ucf.edu
5 Faculdade de Nutrição, Universidade Federal de Pelotas, Pelotas 96010-610, Brazil; augusto.schneider@ufpel.edu.br
6 Department of Head and Neck Surgery, Poznan University of Medical Sciences, 61-701 Poznan, Poland; mateusz.szewczyk@wco.pl (M.S.); wojciech.golusinski@wco.pl (W.G.)
7 Department of Head and Neck Surgery, Greater Poland Cancer Centre, 61-866 Poznan, Poland
8 Department of Otolaryngology and Maxillofacial Surgery, University of Zielona Gora, 65-417 Zielona Gora, Poland
9 Department of Maxillofacial Surgery, Poznan University of Medical Sciences, 61-701 Poznan, Poland
* Correspondence: p.golusinski@cm.uz.zgora.pl

Citation: Piotrowski, I.; Zhu, X.; Saccon, T.D.; Ashiqueali, S.; Schneider, A.; de Carvalho Nunes, A.D.; Noureddine, S.; Sobecka, A.; Barczak, W.; Szewczyk, M.; et al. miRNAs as Biomarkers for Diagnosing and Predicting Survival of Head and Neck Squamous Cell Carcinoma Patients. *Cancers* **2021**, *13*, 3980. https://doi.org/10.3390/cancers13163980

Academic Editor: David Wong

Received: 7 July 2021
Accepted: 3 August 2021
Published: 6 August 2021

Publisher's Note: MDPI stays neutral with regard to jurisdictional claims in published maps and institutional affiliations.

Copyright: © 2021 by the authors. Licensee MDPI, Basel, Switzerland. This article is an open access article distributed under the terms and conditions of the Creative Commons Attribution (CC BY) license (https://creativecommons.org/licenses/by/4.0/).

Simple Summary: Head and Neck Squamous Cell Carcinoma (HNSCC) is the sixth most common cancer worldwide. It arises from the epithelium of the upper aerodigestive tract. Increasing evidence suggests that there is a significant role of microRNAs in HNSCC formation and progression. The aim of this study was to explore and compare the expression of HNSCC related miRNAs in tumor vs neighboring healthy tissue of HNSCC patients with tumors located in either the oral cavity, oropharynx, or larynx. Our results demonstrated that expression of these miRNAs was significantly different not only between healthy and tumor tissues, but also among tumor locations. Further analysis indicated that microRNA expression could be used to distinguish between tumor and healthy tissues, and prognose the overall survival of patients.

Abstract: Head and Neck Squamous Cell Carcinoma (HNSCC) is the sixth most common cancer worldwide. These tumors originate from epithelial cells of the upper aerodigestive tract. HNSCC tumors in different regions can have significantly different molecular characteristics. While many microRNAs (miRNAs) have been found to be involved in the regulation of the carcinogenesis and pathogenesis of HNSCC, new HNSCC related miRNAs are still being discovered. The aim of this study was to explore potential miRNA biomarkers that can be used to diagnose HNSCC and prognose survival of HNSCC patients. For this purpose, we chose a panel of 12 miRNAs: miR-146a-5p, miR-449a, miR-126-5p, miR-34a-5p, miR-34b-5p, miR-34c-5p, miR-217-5p, miR-378c, miR-6510-3p, miR-96-5p, miR-149-5p, and miR-133a-5p. Expression of these miRNAs was measured in tumor tissue and neighboring healthy tissue collected from patients diagnosed with HNSCC ($n = 79$) in either the oral cavity, oropharynx, or larynx. We observed a pattern of differentially expressed miRNAs at each of these cancer locations. Our study showed that some of these miRNAs, separately or in combination, could serve as biomarkers distinguishing between healthy and tumor tissue, and their expression correlated with patients' overall survival.

Keywords: head and neck squamous cell carcinoma; oral cancer; oropharyngeal cancer; laryngeal cancer; miRNA

1. Introduction

Head and Neck Squamous Cell Carcinoma (HNSCC) accounts for approximately 830,000 new cases and 430,000 deaths annually, making it the sixth most common cancer worldwide [1]. HNSCC includes tumors located in the oral cavity, larynx, nasopharynx, oropharynx, and hypopharynx, contributing to the heterogeneous nature of this disease. The most commonly recognized risk factors for oral and laryngeal cancer include tobacco use and alcohol consumption, whereas the oropharyngeal tumors are often linked to infection by human papillomavirus (HPV). Although some progress has been made in the diagnosis and therapy of HNSCC, the 5-year survival of these patients over the last couple of decades has improved modestly compared to the survival of patients diagnosed with other solid tumors [2]. Malignant transformation in HNSCC begins with hyperplasia of mucosal epithelial cells followed by dysplasia, formation of carcinoma in situ, and finally invasive carcinoma. Recognition of molecular markers that are responsible for carcinogenesis and progression of the disease is crucial for enhanced understanding of the mechanisms behind these processes and could assist in improved diagnosis and prediction of patients' outcomes.

MicroRNAs (miRNA) are non-coding, short (around 22 nucleotides in length), RNA molecules that regulate the expression of genes through binding to mRNAs, ultimately targeting them for degradation [3]. miRNAs are shown to regulate over 60% of human protein-coding genes [4], and are active in many physiological processes, including apoptosis, aging, and the cell cycle [5,6], as well as in the development of many pathological processes [7]. miRNAs also play an important role in carcinogenesis. Some research shows that miRNA expression profile in solid tumors is significantly altered from that of healthy tissue [8]. Since miRNAs are stably expressed in human tissues and are linked with the initiation and progression of cancer, they could serve as biomarkers for diagnosis, prediction of response to therapy and treatment outcome [9]. Research on HNSCC indicated that many miRNAs are differentially expressed in cancer tissue, with some classified as oncogenic, and others as tumor suppressors [10]. Although the roles of miRNAs in cancer are widely studied, novel cancer-related miRNAs are still being identified [11].

Since miRNAs play a crucial role in cancer development, many studies have investigated the application of miRNA expression analysis as a prognostic tool for HNSCC. Recently published data showed that expression of miRNAs in healthy and tumor tissue of oral squamous cell carcinoma can be associated with patient prognosis, and could be used to predict recurrence and patient survival [9,12]. Several studies demonstrated that expression of miRNAs in patients' serum could be used as a factor predicting patients' survival, disease progression, or risk of developing side effects [13–15]. Some researchers also pointed out that prognostic, cancer-related miRNAs present in the bloodstream likely originate from tumor cells [14,16]. Identifying a miRNA expression profile in HNSCC tumor is of remarkable importance for establishing pathways of disease development and progression, which could help to elucidate potential targets in novel therapeutic approaches.

To achieve this objective, we investigated the differential expression of miRNAs in HNSCC, analyzed the potential of these miRNAs as biomarkers distinguishing between healthy and tumor tissue, and correlated their expression with patient survival. We assessed matched tumor and neighboring healthy tissues collected from 79 patients with either oral cavity, oropharynx, or larynx HNSCC and we investigated 12 miRNAs of interest: miR-146a-5p, miR-449a, miR-126-5p, miR-34a-5p, miR-34b-5p, miR-34c-5p, miR-217-5p, miR-378c, miR-6510-3p, miR-96-5p, miR-149-5p, and miR-133a-5p. The panel was selected based on a literature search of their differential expression in HNSCC tissue [11,16–23]. Among them, miR-378c and miR-6510-3p were identified in our previous study via analysis of the Cancer Genome Atlas (TCGA) dataset of HNSCC cases [11]. The miRNA panel was chosen to include both miRNAs with oncogenic (i.e., miR-96-5p) and tumor suppressor (i.e., miR-217-5p) activity, as both types of miRNA can be associated with disease progression and outcome [20,21]. Although the miRNAs chosen for analysis were previously shown to play a role in HNSCC, their expression in tumor and healthy tissue in different HNSCC locations and association with patient survival have not been fully investigated.

2. Materials and Methods

2.1. Study Subjects

In this study, we recruited a total of 79 patients diagnosed with head and neck squamous cell carcinoma in either the oral cavity (n = 37), oropharynx (n = 9), or larynx (n = 33), between 2015 and 2017 at the Department of Head and Neck Surgery at The Greater Poland Cancer Centre in Poznan, Poland. Table 1 presents patients' characteristics. Staging was determined based on the TNM 7th edition by The Union for International Cancer Control (UICC). Exclusion criteria for participation in the study was defined as follows: patients diagnosed with more than one malignancy, patients previously undergoing other therapeutic modalities such as chemo- and radiotherapy, and HPV positive malignancies. The study protocol was approved by the Ethics Committee of the Poznan University of Medical Sciences (Decision No. 915/16), and written informed consent was provided by all participants.

Table 1. Clinical characteristics of the study group.

Characteristic	**Tumor Location** **Oral Cavity**	**Oropharynx**	**Larynx**
Number of subjects	37	9	33
Age (years, mean ± SD)	59.3 (±11.2)	52.9 (±12.1)	62.2 (±12.5)
Sex, n (%)			
Male	30 (81%)	5 (56%)	25 (76%)
Female	7 (19%)	4 (44%)	8 (24%)
T stage, n (%)			
T1-2	26 (70%)	6 (67%)	4 (12%)
T3-4	11 (30%)	3 (33%)	28 (85%)
NA	0 (0%)	0 (0%)	1 (3%)
N stage, n (%)			
N0	11 (30%)	0 (0%)	14 (42%)
N+	25 (68%)	8 (89%)	17 (52%)
NA	1 (3%)	1 (11%)	2 (6%)
M stage, n (%)			
M0	37 (100%)	9 (100%)	30 (91%)
NA	0 (0%)	0 (0%)	3 (9%)
Grading, n (%)			
G1	6 (16%)	0 (0%)	2 (6%)
G2	27 (73%)	2 (22%)	24 (73%)
G2/G3	1 (3%)	0 (0%)	1 (3%)
G3	3 (8%)	4 (44%)	5 (15%)
NA	0 (0%)	3 (33%)	1 (3%)

SD—standard deviation; T stage—size of tumor; N stage—spread to lymph nodes; M—distant metastasis; NA—data not available.

2.2. Experimental Methods

Patients were subjected to primary surgical treatment, during which cancer tissue as well as normal epithelium tissue (within 2 cm distance from the tumor margin) were collected. Both tissues were placed in cryovials on dry-ice directly following the surgical excision and subsequently transferred to −80 °C freezer, and later to liquid nitrogen for long-term storage.

The frozen tissue samples were homogenized with a mortar and pestle with Qiazol (Qiagen, Valencia, CA, USA). Total RNA was isolated using a commercial column purification system (miRNeasy Mini Kit, Qiagen, Valencia, CA, USA) with a DNase treatment (RNase-free DNase Set, Qiagen, Valencia, CA, USA) following manufacturer's instructions. RNA concentration and purity were measured spectrophotometrically (absorbance at 260, 230, and 280 nm) using Epoch™ Microplate Spectrophotometer (BioTek, Winooski, VT, USA).

To measure miRNA expression, 10 ng of total RNA were reverse-transcribed using TaqMan® Advanced miRNA cDNA Synthesis Kit (Applied Biosystems, Foster City, CA, USA), according to the manufacturer's instructions. For the RT-qPCR reaction a 1:10 dilution of cDNA template was prepared. RT-qPCR was performed using the Quantstudio™ 7 Flex System. We used TaqMan® Advanced miRNA Assays (Applied Biosystems, Foster City, CA, USA) with primers specific for selected miRNA: miR-146a-5p (478399_mir), miR-449a-5p (478561_mir), miR-126-5p (477888_mir), miR-34a-5p (478048_mir), miR-34b-5p (478050_mir), miR-34c-5p (478052_mir), miR-217-5p (478773_mir), miR-378c-5p (478864_mir), miR-6510-3p (480744_mir), miR-96-5p (478215_mir), miR-149-5p (477917_mir), miR-133a-5p (478706_mir). Supplemental Table S1 presents stem loops used for miRNA expression analysis. Each reaction was performed in two technical replicates. The $2^{-\Delta\Delta Ct}$ method was used to calculate and normalize expression of each miRNA, in which $-\Delta\Delta Ct$ was calculated as:

$$-\Delta\Delta Ct = \text{average}\ (\Delta Ct_{\text{healthy tissue}}) - \Delta Ct_{\text{tumor tissue}}, \quad (1)$$

where

$$\Delta Ct = Ct_{\text{target gene}} - Ct_{\text{housekeeping gene}}. \quad (2)$$

The housekeeping gene was miR-16-5p. In the above calculation, $-\Delta\Delta Ct > 0$ denotes upregulated expression, whereas $-\Delta\Delta Ct < 0$ denotes downregulated expression, relative to healthy tissue.

2.3. Statistical Analysis

Through RT-qPCR experiments, we obtained 1785 expression data points with an observed Ct of <40 and 135 non-detects without an assigned Ct value. We noticed that the non-detect rate in our RT-qPCR experiment increased with average ΔCt of each gene (Figure 1A), suggesting that the non-detects are not randomly distributed and imputation must be implemented. We first evaluated the bias of the non-detects imputed by a conventional method of setting the Ct of each non-detect to 40. The results showed that the bias caused by such an imputation method remarkably skewed the data towards low expression (Figure 1B). To deal with this issue and impute the missing expression data more appropriately, we then performed the expectation-maximization (EM) algorithm proposed by McCall et al. [24] using R package (HTqPCR, www.bioconductor.org/packages/release/bioc/html/HTqPCR.html, accessed on 11/16/2020) [25]. The results indicated that the EM algorithm produced unbiased imputation for the non-detects (Figure 1C).

Wilcoxon matched-pairs signed-ranks test was used to compare the expression of each miRNA in tumor tissue vs. in healthy tissue. Heatmap was produced to visualize the expression profile of each miRNA across tumor locations and tissue types. To evaluate the accuracy of each miRNA expression as a biomarker in distinguishing tumor from healthy tissue, we performed bootstrap logistic regression based on which ROC curves were produced, and sensitivity, specificity, and AUC were estimated. In addition, we performed stepwise logistic regression to explore whether combined miRNA expressions can be used as a more valid biomarker in the diagnosis of HNSCC. Finally, we assessed the prognosis value of each miRNA through survival analysis using the Kaplan–Meier survival curves. In this analysis, we first used R package (survminer) to find an optimal cut-off, splitting the expression of each miRNA into low and high expression. We then compared survival curves between low and high expression using logrank test. To further evaluate the prognosis of the selected miRNAs on overall survival of the HNSCC patients, we performed multiple Cox regression analysis with age at diagnosis, T stage, and N stage included as confounding variables. Due to the fact that the results of the small cohort of oropharyngeal cancer (n = 9) might be prone to bias, they were only presented in Supplementary Figures. Statistical analysis was performed using Stata MP software 14.1 (StataCorp LLC, College Station, TX, USA) and R packages. All tests were two-tailed and the p value < 0.05 was considered to be statistically significant.

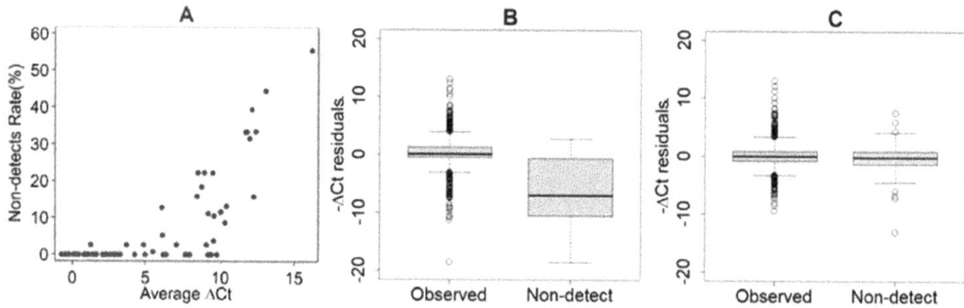

Figure 1. The scatter plot (**A**) indicating that the non-detect rate increases with average ΔCt by location, tissue type and gene. The boxplots showing the comparison of −ΔCt residues between observed expressions and non-detects, (**B**) if conventional imputation of setting the Ct of each non-detect to 40 was conducted, and (**C**) if expectation-maximization (EM) algorithm was used to estimate the Ct of each non-detect.

3. Results

3.1. *miRNA Expression*

First, we measured the expression of selected miRNAs in tumor tissue and neighboring healthy tissue of HNSCC patients. In oral cancer, we observed upregulation of miR-146a-5p, miR34b-5p, and miR34c-5p, and downregulation of miR-126-5p, miR378c-5p, miR-6510-3p, and miR-149-5p in tumor tissue compared with neighboring healthy tissue (Figure 2A). In laryngeal cancer, we observed upregulation of miR-146a-5p, miR-449a-5p, miR-34a-5p, miR-34b-5p, miR-34c-5p, and miR-96-5p and downregulation of miR-6510-3p (Figure 2A). The median values of miRNA expressions in tumor tissue relative to healthy tissue are presented in supplementary materials (Table S2). Clustering analysis of miRNA expression in each sample revealed that the expression of miRNAs strongly depended on the tissue type (tumor or healthy) and, to a lesser extent, on the tumor location (Figure 2B). Figure 2C presents a simple summary of miRNA expression analysis. The results of miRNA expression analysis for all three tumor locations (including oropharyngeal cancer) are presented in Supplementary Figure S1.

3.2. *miRNA Expression as a Biomarker to Distinguish Healthy Tissue from Tumor Tissue*

3.2.1. Expression of Individual miRNAs as a Biomarker

For the analysis of miRNAs as biomarkers, we considered expression of miRNAs showing both sensitivity and specificity ≥ 0.8 as strong candidates, and miRNAs showing both sensitivity and specificity ≥ 0.7 as potential, weaker candidates. For some miRNA expressions (miR-34c-5p and miR-149-5p in oral cancer, miR-34c-5p in laryngeal cancer) our analysis was not able to determine a unique threshold value, and ROC could not be accurately estimated.

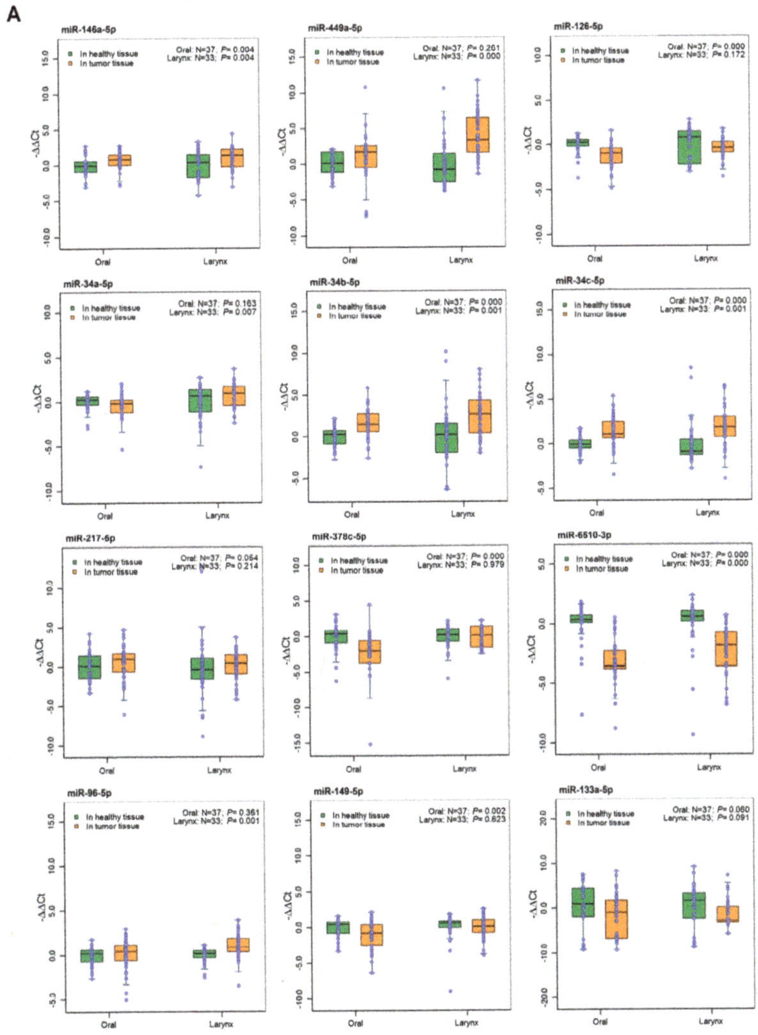

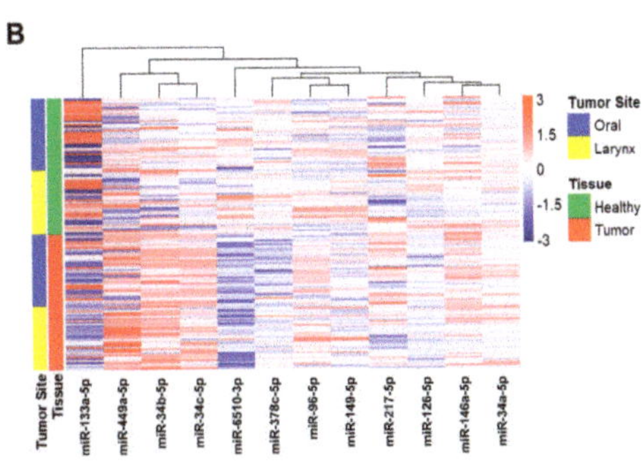

C

miRNA	Tumor site: Oral cavity	Tumor site: Larynx
miR-146a-5p	↑	↑
miR-449a-5p		↑
miR-126-5p	↓	
miR-34a-5p		↑
miR-34b-5p	↑	↑
miR-34c-5p	↑	↑
miR-378c-5p	↓	
miR-6510-3p	↓	↓
miR-96-5p		↑
miR-149-5p	↓	

Figure 2. Expression of miRNAs in tumor tissue and nearby healthy tissue of HNSCC patients. (**A**) Boxplots showing the comparison of expression of 12 miRNAs between tumor tissue and nearby healthy tissue in the oral cavity, and larynx. Expression was calculated using −ΔΔCt method; values

above 0 denote upregulation, and values below 0 denote downregulation of miRNA, compared with healthy tissue. Wilcoxon signed-rank test was performed. (**B**) Heatmap and dendrogram showing the overall expression profile of the 12 miRNAs in patient samples. Samples were grouped by tissue type and tumor location. The colors in the heatmap represent the value of $-\Delta\Delta Ct$ ranging from downregulation (dark blue) to upregulation (red). (**C**) Summary of findings on miRNA expressions as biomarkers in diagnosis of HNSCC ("↓" denotes downregulation and "↑" denotes upregulation).

For oral cancer, no miRNA expression reached the threshold sensitivity and specificity for a strong biomarker (Figure 3A). Expression of miR-378c-5p (sensitivity = 0.811, specificity = 0.703, ROC Area = 0.813, threshold = −0.3), and miR-6510-3p (sensitivity = 0.946, specificity = 0.784, ROC Area = 0.913, threshold = 0.08) demonstrated satisfying sensitivity and moderate specificity (indicating a higher risk of false positives) as biomarkers. The sensitivity and specificity of the expression of miR-126-5p (sensitivity = 0.784, specificity = 0.784, ROC Area = 0.790, threshold = −0.3) was close to 0.8, suggesting that this miRNA might be a candidate biomarker.

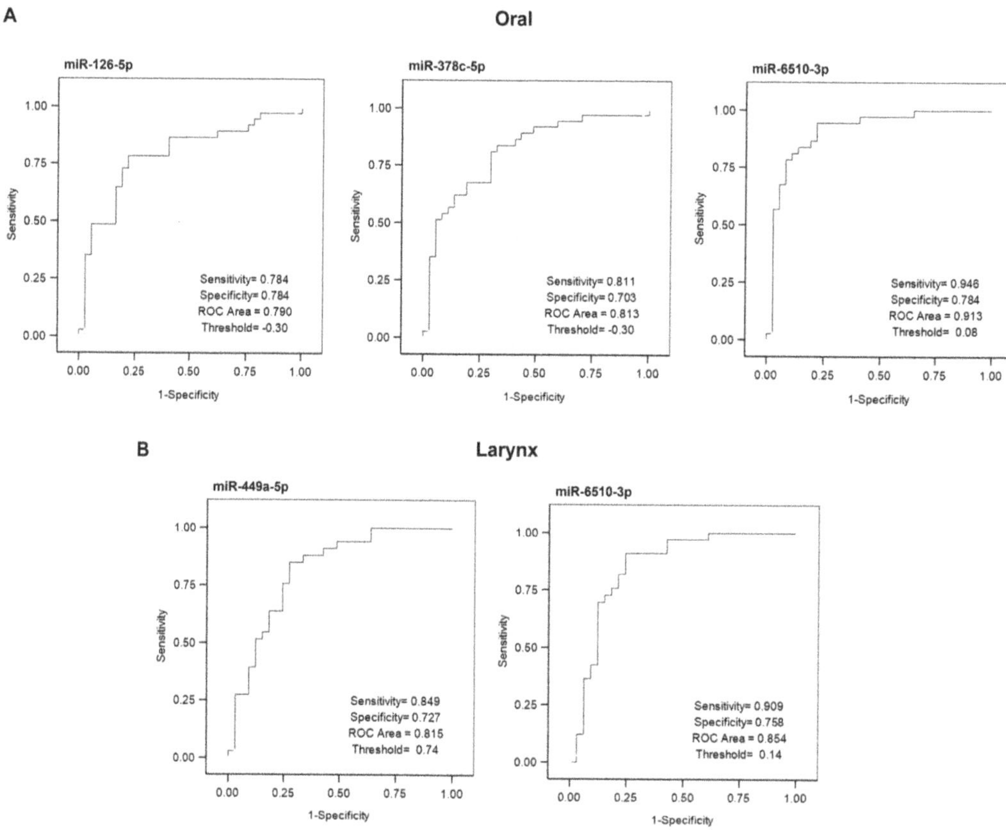

Figure 3. ROC curves showing miRNA biomarkers with potential in distinguishing tumor tissue from nearby healthy tissue in oral (**A**) and laryngeal (**B**) cancer.

For laryngeal cancer, no miRNA expression reached the threshold sensitivity and specificity for a strong biomarker (Figure 3B). Expression of miR-449a (sensitivity = 0.849, specificity = 0.727, ROC Area = 0.815, threshold = 0.74), and miR-6510-3p (sensitivity = 0.909, specificity = 0.758, ROC Area = 0.854, threshold = 0.14) presented as weaker biomarker candidates, with high sensitivity and moderate specificity (higher risk of false positives).

ROC curves for all miRNAs, analyzed in tissues collected form oral, oropharyngeal, and laryngeal cancer, are presented in Supplemental Figures S2–S4.

3.2.2. Combined Expression of miRNAs as a Biomarker

Since the expression of some miRNAs in neighboring healthy tissue compared to tumor tissue exhibited potential for use as biomarkers distinguishing between the two tissues, we decided to further investigate whether combined miRNA expression could be used with higher success. For oral cancer samples, we observed that the combined expression of miR-6510-3p and miR-34c-5p could be used as a biomarker with high success (sensitivity = 0.919, specificity = 0.946, ROC Area = 0.952) (Figure 4A). Similar analysis was performed for miRNA expression in laryngeal tumor and neighboring healthy tissue, where the combined expression of miR-449a-5p, miR-6510-3p, and miR-133a-5p showed strong ability to distinguish tumor tissues (sensitivity = 0.909, specificity = 0.879, ROC Area = 0.932) (Figure 4B).

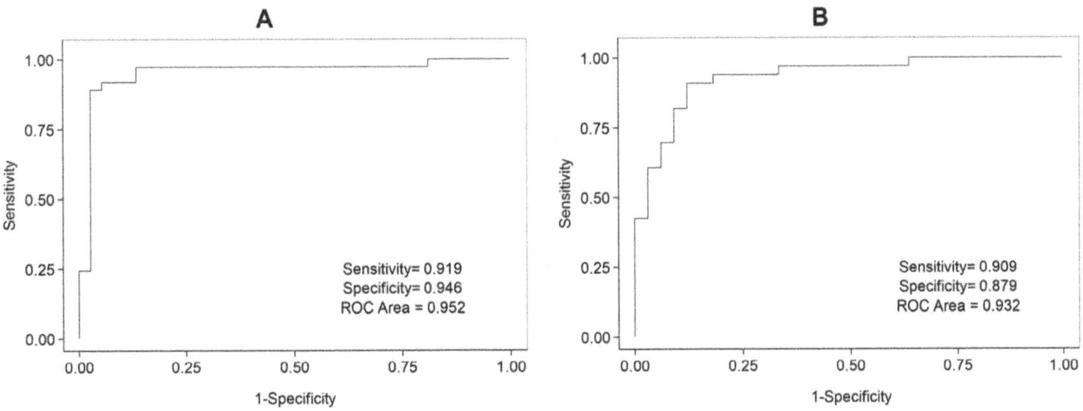

Figure 4. ROC showing accuracy of combined miRNA expression as a biomarker in distinguishing oral (**A**) and laryngeal (**B**) tumor tissue compared with neighboring healthy tissue. (**A**) Combined expression of miR-6510-3p and miR-34c-5p as a biomarker distinguishing oral tumor tissue from nearby healthy tissue with a threshold of −0.194 determined by the best model: logit(tumor) = −2.239 − 1.127(miR-6510-3p) + 1.392(miR-34c-5p). (**B**) Combined expression of miR-449a-5p, miR-6510-3p, and miR-149-5p as a biomarker in distinguishing laryngeal tumor tissue from nearby healthy tissue with a threshold of −0.212 determined by the best model: logit(tumor) = −1.855 + 0.438(miR-449a-5p) − 1.246(miR-6510-3p) + 1.473(miR-149-5p). In both models, if logit(tumor) score is greater than its corresponding threshold, the diagnosis is positive, otherwise the diagnosis is negative.

3.3. *Association of miRNA Expressions with Survival of HNSCC Patients*

Since expression of some of the selected miRNAs in tumor tissue was significantly different than in neighboring healthy tissue, we proposed that expression of these miRNAs could have a marked impact on patient survival. In oral cancer, low expression of miR-449a-5p (cutoff = −1.55), miR-34a-5p (cutoff = −0.70), miR-217-5p (cutoff = 2.73), miR-96-5p (cutoff = −0.10), and miR-133a-5p (cutoff = −7.53) in tumor tissue correlated with a significantly shorter survival (Figure 5A). Survival of patients with laryngeal cancer was significantly reduced in patients with high expression of miR-146a-5p (cutoff = 2.31) or low expression of miR-449a-5p (cutoff = 0.04) in tumor tissue (Figure 5B). Supplemental Figures S5–S7 present survival data for all miRNAs expressed in the tumor tissues collected from oral, oropharyngeal, and laryngeal cancer.

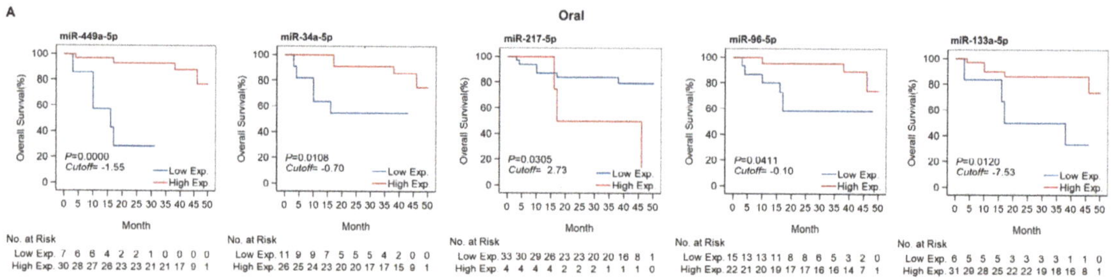

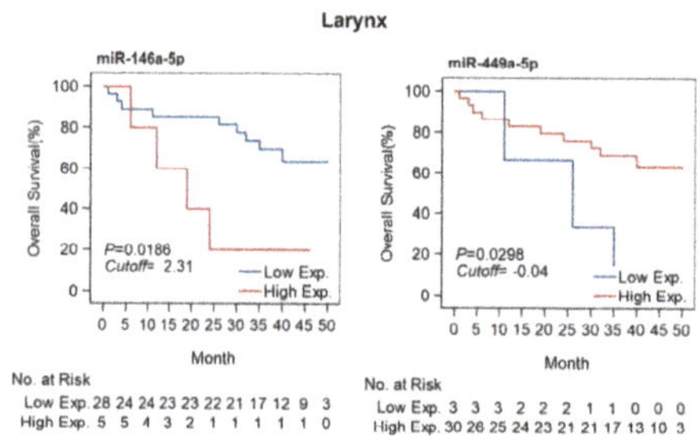

C

miRNA	Oral Cancer		Laryngeal Cancer	
	HR(SE)	P-value	HR(SE)	P-value
miR-146a-5p	0.364(0.401)	0.359	9.671(7.400)	0.003
miR-449a-5p	0.068(0.061)	0.003	0.388(0.291)	0.207
miR-126-5p	0.214(0.166)	0.046	0.881(0.836)	0.894
miR-34a-5p	0.155(0.131)	0.028	3.466(2.171)	0.047
miR-34b-5p	0.772(0.577)	0.729	6.761(4.674)	0.006
miR-34c-5p	-	-	0.736(0.467)	0.630
miR-217-5p	5.210(4.065)	0.034	2.012(1.140)	0.217
miR-378c-5p	3.803(3.294)	0.123	3.770(2.379)	0.035
miR-6510-3p	0.242(0.204)	0.092	0.914(0.971)	0.932
miR-96-5p	0.225(0.175)	0.055	0.155(0.224)	0.198
miR-149-5p	0.436(0.324)	0.265	2.183(2.311)	0.461
miR-133a-5p	0.153(0.118)	0.015	2.703(2.272)	0.237

Figure 5. Relation between miRNA expression and overall survival. (**A**,**B**) Kaplan Meier survival curves showing comparison of overall survival between low and high miRNA expression defined by optimal cutoff of $-\Delta\Delta Ct$ in oral (**A**) and laryngeal (**B**) cancers. (**C**) Summary of multiple Cox regression analysis for oral and laryngeal cancer investigating hazard ratio (HR) of high expression vs. low expression of each miRNA in separate survival models with age at diagnosis, T staging and N staging under control. HR < 1 denotes high miRNA expression at lower risk compared to low expression. HR > 1 denotes high miRNA expression at higher risk than low expression.

To verify whether relative miRNA expression in tumor tissue correlated with patient survival, we performed multiple Cox regression analysis, controlling for age at diagnosis, T stage, and N stage (Figure 5C). The analysis indicated that in oral cancer high expression of miR-449a-5p (HR = 0.068), miR-126-5p (HR = 0.214), miR-34a-5p (HR = 0.155), and miR-133a-5p (HR = 0.153) had a lower hazard, corresponding to a higher survival rate, and high expression of miR-217-5p (HR = 5.21) corresponded to a higher hazard and a lower survival rate. In laryngeal cancer, high expression of miR-146a-5p (HR = 9.671), miR-34a-5p (HR = 3.466), miR-34b-5p (HR = 6.761), and miR-378c-5p (HR = 3.77) was associated with a higher hazard and lower survival rate.

We also conducted multiple Cox regression analysis to investigate whether the combined miRNA expression was associated with patients' overall survival when controlled for age at diagnosis, T staging, and N staging. No significant association with age or staging was observed (Supplementary Table S3).

4. Discussion

Although the role of miRNAs in HNSCC development has been widely studied, and new miRNAs with significant oncogenic or cancer suppressor roles are being discovered, expression of these miRNAs in tumor and healthy tissue of patients with HNSCC at different sites has rarely been addressed. We investigated whether the differences in expression of miRNAs can be used as markers to distinguish between tumor and healthy tissue, and whether miRNA expression correlates with patient survival. Based on a literature search for miRNAs differentially expressed in HNSCC tissue [11,16–23], and on our previous study using deep sequencing on small cohorts of HNSCC patients [11], we selected a panel of 12 miRNAs for analysis: miR-146a-5p, miR-449a-5p, miR-126-5p, miR-34a-5p, miR-34b-5p, miR-34c-5p, miR-217-5p, miR-146a-5p, miR-378c-5p, miR-6510-3p, miR-96-5p, miR-149-5p, and miR-133a-5p. Expression of these miRNAs was measured in tumor and healthy tissue of 79 patients diagnosed with HNSCC located in either the oral cavity, oropharynx, or larynx.

It is noted that miR-146a-5p has previously been described in several types of cancer, including head and neck cancer [26]. Zhu et al. [27] observed upregulation of miR-146a-5p in oral cancer cell lines compared with normal keratinocyte cell line, and showed that miR-146a-5p inhibits NF-κB1 and reduces apoptotic cell death in oral cancer cell lines. Recent results show that expression of this miRNA in HNSCC is also connected with the status of human papillomavirus (HPV) infection [28]. Hung et al. also confirmed that this miRNA has an oncogenic and pro-metastatic effect in oral cancer by targeting genes involved in the NF-κB pathway [16]. The same authors also observed a high plasma level of miR-146a-5p in patients with oral cancer compared to controls, which decreased after surgery, suggesting that the miRNA present in plasma originated from the tumor. This indicates that miR-146a-5p might be a useful HNSCC marker in both tumor tissue and circulation. It is important to note that other studies point towards an anti-oncogenic effect of miR-146a-5p related to the targeting of SOX2 in oral cancer cell lines [29]. Only one study connected miR-146a-5p expression with laryngeal cancer and observed an increased risk of this cancer in individuals with the rs2910164 polymorphism in the miR-146a-5p gene [30]. In our study, we showed an upregulation of miR-146a-5p in oral and laryngeal tumor tissue, which was also observed by others [27]. Furthermore, our results indicate that laryngeal cancer patients with high expression of miR-146a-5p show worse overall survival, which is in line with its oncogenic role described in the literature.

It is noted that miR-449a-5p has been identified as a potential anti-oncogenic factor in several cancers [31,32]. Expression of miR-449a-5p was shown to be downregulated in tumor tissue compared with normal tissue in nasopharyngeal carcinoma [17]. Studies show that miR-449a has a suppressive effect on the progression, migration, and sphere formation of nasopharyngeal cancer cells [33,34]. In nasopharyngeal cancer, this miRNA was found to suppress LDH oncogenic activity [33]. Recently a decreased expression of miR-449a-5p was shown to correlate with positive nodal status in laryngeal cancer [35]. Interestingly, in our study, the expression of miR-449a-5p was significantly upregulated in laryngeal tumor tissue compared with healthy tissue. Downregulation of miR-449a-5p in oral and laryngeal tumor tissue compared with healthy tissue also correlated with reduced overall survival. Additionally, miR-449a-5p expression showed potential value as a biomarker for differentiating between healthy and cancerous tissue, with relatively high sensitivity and moderate specificity in laryngeal cancer.

The role of miR-126-5p has been described in different cancer types. Sasahira et al. observed its downregulation in oral tumor tissue relative to normal oral mucosa [18]. The same authors showed that decreased levels of miR-126-5p activate vascular endothelial growth factor (VEGF)-A and, consequently, prompt angiogenesis and lymphangiogenesis. Additionally, low expression of miR-126-5p in tumor tissue correlated with a more advanced stage of the disease, positive nodal status, as well as a poor disease-free survival rate in oral cancer patients. Low expression of miR-126-5p in oral squamous cell carcinoma has also been observed by others, with evidence of negative regulation of tumorigenesis via targeting of KRAS, as well as migration and invasion of cancer cells via downregulation of ADAM9 [36,37]. miR-126-5p was also detected in serum exosomes of oral carcinoma patients, and its level correlated with patient survival [38]. Our analysis showed a markedly downregulated expression of miR-126-5p in tumor tissue compared with healthy tissue in patients with oral tumors. Low expression of miR-126-5p was also associated with a lower survival rate. Interestingly, in esophageal cancer, this relationship might be opposite—patients with low tumor expression of miR-126 were characterized by worse prognosis than patients with high miR-126-5p expression [39]. This suggests that the location of HNSCC tumor might play a crucial role in the effect that miRNA expression has on the therapy outcome. While miR-126-5p is downregulated in tumor tissue compared to healthy tissue, it is upregulated in plasma of patients with oral cancer compared with healthy individuals, suggesting its potential as a biomarker [40]. Our analysis showed that miR-126-5p expression is a moderate biomarker in distinguishing between tumor and healthy tissue in oral cancer.

Similar to miR-126-5p, the dysregulation of miR-34 family expression has been observed in various types of cancer. miRNAs of the miR-34 family, and miR-34a in particular, are directly regulated by p53, which is a tumor suppressor protein [41]. The miR-34 family is involved in cell-cycle regulation, downregulation of epithelial-to-mesenchymal transition (EMT), suppression of stem-like phenotype, induction of apoptosis and senescence, and inhibition of glycolysis [41]. In a comprehensive meta-analysis, Li et al. concluded that miR-34a-5p is generally expressed at lower levels in HNSCC tumor tissue compared with control tissue [42]. Li et al. also suggested that the level of miR-34a measured in tumor tissue might serve as a clinical biomarker of the disease progression. Our data showed an upregulation of miR-34a-5p in laryngeal cancer tissue compared with healthy tissue, and no difference in oral cancer. An upregulation of miR-34a in tumor vs healthy tissue was also observed by Kalfert et al. only in oropharyngeal cancer, and not in laryngeal cancer [43]. Kalfert et al. underscored that the difference in expression of miR-34a supports the hypothesis of site-specific oncogenesis of HNSCC. Additionally, Kalfert et al. showed a positive correlation between p16 positivity, which is a surrogate marker for HPV infection, and miR-34a expression. As only oropharyngeal cancer cases showed p16 positivity, the miR-34a increase in this tumor location might be explained by p16 status. However, it cannot explain the increase in miR-34a-5p observed in our study, since HPV positivity was one of the exclusion criteria. In our study, oral cancer patients with high expression of miR-34a showed significantly better overall survival than patients with low miR-34a expression, which is in line with the tumor suppressor role of this miRNA. For laryngeal cancer the Cox analysis revealed that high expression of miR-34a-5p was also associated with lower survival rate.

Despite having a similar tumor suppressive role to miR-34a, miR-34b has been found to be upregulated in HNSCC tumor tissue compared with neighboring healthy tissue [44]. Roy et al. observed a significant upregulation of miR-34b expression in oral cancer tissue compared with normal tissue, however the normal tissue was collected from distinct healthy patients [45]. A significant upregulation of miR-34b-5p was also observed in both oral and laryngeal cases of HNSCC in our study. Ren et al. showed that expression of miR-34b was significantly downregulated in metastatic oral cancer tissue compared with nonmetastatic tissue, which is consistent with suppressive effect of miR-34b on EMT [46]. Our analysis indicated that in laryngeal cancer patients, higher expression of miR-34b-5p was associated with lower survival rate.

Both miR-34b and miR-34c are encoded by a common transcript, separate from miR-34a [47]. Severino et al. showed that, similar to miR-34b, miR-34c expression is upregulated in HNSCC tumor tissue compared with tumor-free margins [44]. However, other studies exhibited downregulated expression of this miRNA in laryngeal carcinoma tumor tissue compared with negative margins [48,49]. miR-34c acts as a tumor suppressor, with some studies displaying suppressive effect of this miRNA on cancer growth and invasiveness [48], and others showing significantly worse overall survival in patients with low expression of miR-34c in laryngeal tumor tissue [50]. Our results revealed that miR-34c-5p expression was significantly upregulated in tumor tissue compared with neighboring healthy tissue in both oral and laryngeal tumor tissue.

Research on the role of miR-217-5p in cancer revealed a potential suppressive effect in different cancer types [20]. In esophageal cancer, this miRNA was found to be repressed after exposure of cells to cigarette smoke condensate, which in turn resulted in overexpression of miR-217-5p's direct target, kallikrein 7 (KLK7), which induced growth and invasiveness of cancer cells [51]. Miao et al. demonstrated that the expression of miR-217 is significantly lower in laryngeal cancer tissue compared with adjacent paracarcinoma tissue, and also confirmed its suppressive effect on cancer migration, invasion, and proliferation, as well as its ability to induce apoptosis and G1 arrest in a laryngeal cancer cell line [20]. Our results indicated no regulation of miR-217-5p in oral or laryngeal cancer tissue compared with healthy tissue. Additionally, our results revealed a tendency of shorter overall survival of oral cancer patients with high expression of miR-217-5p. The inverse correlation of miR-217-5p expression with survival is contradictory to its anti-tumorigenic role.

Although miR-378c-5p has been recognized in some types of cancer, its role was not yet described in HNSCC. In colon cancer, expression of miR-378c is significantly downregulated in tumor tissue compared with normal colorectal mucosa tissue [52]. miR-378c was recognized as a potential early, progressively downregulated marker in carcinogenesis of esophageal carcinoma [53]. Results previously published by our group showed the differential expression of miR-378c in a TCGA dataset of HNSCC cases for the first time [11]. We demonstrated a significant downregulation of miR-378c in HNSCC tumor tissue compared with neighboring healthy tissue. Additionally, expression of miR-378c inversely correlated with the cancer T stage, suggesting a potential role in disease progression. In the current study, we confirmed the downregulation of miR-378c-5p in oral tumor tissue compared with healthy tissue. In laryngeal cancer, no significant regulation was observed. High expression of this miRNA in laryngeal tumor tissue was associated with lower survival rate. Expression of miR-378c-5p also displayed potential as a biomarker differentiating oral tumor tissue from healthy tissue with good sensitivity and moderate specificity.

The data on miR-6510-3p related to cancer are very scarce. Chen et al. found that this miRNA is significantly downregulated in oral lichen planus (OLP) buccal mucosa compared with normal mucosa from healthy individuals [54]. OLP is characterized by a tumor-like microenvironment and is associated with an increased risk of developing oral squamous cell carcinoma [55,56]. The downregulated expression of miR-6510-3p in OLP suggests that it might play a significant role in premalignant transformation in squamous cell carcinoma development. In a previous study, we demonstrated significant downregulation of miR-6510-3p expression in HNSCC tumor tissue compared with healthy tissue, as well as an inverse correlation between miR-6510-3p expression and the cancer T stage [11]. In this study, a strong downregulation of miR-6510-3p in tumor tissue compared to healthy tissue was also observed in both investigated tumor locations. Our study also showed that miR-6510-3p expression in oral and laryngeal cancer could serve as a biomarker distinguishing between healthy and tumor tissue, with very high sensitivity and specificity.

The role of miR-96-5p has previously been recognized in different types of cancer, including HNSCC. Wang et al. determined that this miRNA is significantly upregulated in oral tumor tissue compared with adjacent normal tissue, and confirmed that miR-96-5p induces proliferation, invasion, and EMT in oral cancer cells through reduction in FOXF2 expression [21]. Other studies on oral cancer cell lines suggested that miR-96-5p can induce chemo- and radio-resistance [57]. In our study, the increased expression of miR-96-5p in oral cancer tissue was not significant; however, we did observe a significantly shorter overall survival of patients with low expression of miR-96-5p, which seems contradictory to its pro-oncogenic role proposed by others. We also observed upregulated miR-96-5p expression in laryngeal cancer tissue. These results suggest that the change in expression of this miRNA might depend on the location of HNSCC.

The role of miR-149-5p has been described in several types of cancer including HNSCC [58,59]. The data on miR-149-5p expression in HNSCC tumor tissue suggests potential dependence on tumor location. Recent results indicate that in oral carcinoma cells miR-149-5p targets CDK6, an oncogene that regulates the cell cycle [60]. Tu et al. showed that the level of miR-149 was downregulated in HNSCC tumor tissue compared with non-cancerous tissue matched to the tumor location [22]. Additionally, expression of this miRNA in lymph node metastatic lesion was lower than both matched primary tumor and matched non-cancerous tissue and its low expression correlated with worse survival. This suggests that miR-149-5p might be an important factor in HNSCC progression. Similarly, research on tongue squamous cell carcinoma displayed downregulation of this miRNA in tumor tissue [61]. Interestingly, in oral cancer, the expression of miR-149-5p was significantly upregulated in tumor tissue compared with peripheral control tissue [62]. Our results showed that expression of miR-149-5p was significantly downregulated in oral tumor tissue, but no difference was observed in laryngeal cancer.

It is noted that miR-133a-5p has been recognized as a HNSCC suppressor inducing apoptosis and inhibiting migration of cancer cells [23,63]. Several studies showed down-regulation of miR-133a in HNSCC tumor tissue of various locations compared with normal adjacent tissues [63,64]. Recent research indicates that in oral carcinoma miR-133a-5p acts as a tumor suppressor, inducing apoptosis, and inhibiting proliferation, migration and invasion through regulation of Notch signaling pathway [65]. Others indicated that the suppressive effect of this miRNA in laryngeal cancer might be related to the targeting of CD47, a transmembrane protein widely expressed in tumor cells [66]. Our results also indicated a downregulation of miR-133a-5p expression in oral and laryngeal tumor tissue; however, the difference was not significant. The low expression of this miRNA in oral cancer patients also correlated with shorter overall survival.

We also analyzed whether miRNA combinations could be used as a biomarker. Our analysis allowed us to identify two sets of miRNAs in oral and laryngeal cancer. In oral cancer combined expression of miR-6510-3p and miR-34c-5p showed very high sensitivity and specificity in distinguishing between tumor and healthy tissue. Both miRNAs were previously found to have a suppressive role in HNSCC. In laryngeal cancer, the combined expression of miR-449a-5p, miR-6510-3p, and miR-149-5p presented as a valuable biomarker for distinguishing tumor and healthy tissue. Similar to the oral cancer miRNAs, these three biomarker miRNAs were shown to have tumor suppressive effects in HNSCC. Both sets were better in distinguishing the two tissues than any single miRNA expression separately. This suggests that the combined targets of these miRNAs can potentially accelerate cancer progression and malignancy, decreasing the chances of disease-free survival. However, in this study we were not able to demonstrate a correlation between expression of these miRNA sets and overall survival of patients.

Although our study reveals some new findings regarding the expression of selected miRNAs in tumor and healthy tissue of patients diagnosed with oral, oropharyngeal, or laryngeal cancer, it has some limitations. The analysis of oropharyngeal cancer was based on a small sample of patients (n = 9). Although we observed significant differences in miRNA expression in this group, the power to make a correct conclusion about survival and ROC analysis may be limited due to the small sample size. In the future, further investigations in a larger cohort are needed. In this study we focused on identifying miRNAs regulated differently in tumor tissue with neighboring healthy tissue. While this comparison might be used for prediction of treatment outcome, its value as a biomarker for cancer diagnosis is limited. Further analysis of miRNA expression should be performed on patients' serum and circulating tumor cells in patients' plasma, which could be used as a routine diagnostic tool.

5. Conclusions

In this study, we conducted analysis of the expression of several miRNAs in tumor and healthy tissue collected from patients diagnosed with HNSCC in the oral cavity, oropharynx, and larynx regions. Our results demonstrated that tumor tissue differs significantly in expression of miRNAs from healthy tissue and that the tumors of different locations differed in expression of some miRNAs. Additionally, the expression of miR-449-5p was the strongest predictor of overall survival in oral cancer, with several other miRNAs showing some correlation with survival. Analysis of combined expression of miRNAs presented two groups of miRNAs (miR6510-3p and miR-34c-5p for oral cancer, and miR449a-5p, miR-6510-3p, and miR-133a-5p for laryngeal cancer), which showed a very strong ability to distinguish between tumor tissue and neighboring healthy tissue. Future research on miRNAs should focus on an in-depth investigation of their mechanisms of action in HNSCC. Specifically, the mechanism related to the effects of miR-6510-3p expression in this tumor should be investigated, since its role in HNSCC has not yet been described.

Supplementary Materials: The following are available online at https://www.mdpi.com/article/10.3390/cancers13163980/s1, Figure S1: Boxplots showing the comparison of expression of 12 miRNAs between tumor tissue and nearby healthy tissue in the oral cavity, oropharynx, and larynx, Figure S2: ROC curves showing accuracy of gene expression (−ΔΔCt) of each miRNA as a biomarker in distinguishing oral tumor tissue from nearby healthy tissue, Figure S3: ROC curves showing accuracy of gene expression (−ΔΔCt) of each miRNA as a biomarker in distinguishing oropharyngeal tumor tissue from nearby healthy tissue, Figure S4: ROC curves showing accuracy of gene expression (−ΔΔCt) of each miRNA as a biomarker in distinguishing laryngeal tumor tissue from nearby healthy tissue, Figure S5: Kaplan Meier survival curves showing comparison of overall survival between low and high miRNA expression de-fined by optimal cutoff of −ΔΔCt in oral cancer, Figure S6: Kaplan Meier survival curves showing comparison of overall survival between low and high miRNA expression defined by optimal cutoff of −ΔΔCt in oropharyngeal cancer, Figure S7: Kaplan Meier survival curves showing comparison of overall survival between low and high miRNA expression defined by optimal cutoff of −ΔΔCt in laryngeal cancer, Table S1: Stem-loops used for miRNA expression analysis, Table S2: Median expression of miRNAs in tumor tissue relative to healthy tissue. Values calculated using −ΔΔCt method, Table S3: Summary of multiple Cox regression analysis investigating association of the combined miRNA expression (miRNAs detected in combined biomarker analysis) with overall survival of oral cancer and larynx cancer with age at diagnosis, T staging and N staging under control.

Author Contributions: Conceptualization, M.M.M. and P.G.; Methodology, X.Z., T.D.S., M.M.M. and P.G.; Software, X.Z.; Validation, I.P., T.D.S., S.A., A.D.d.C.N. and S.N.; Formal Analysis, I.P., X.Z. and A.S. (Augusto Schneider); Investigation, I.P., T.D.S., S.A. and A.S. (Agnieszka Sobecka); Resources, A.S. (Agnieszka Sobecka), W.B., M.S., W.G. and P.G.; Writing—Original Draft Preparation, I.P. and X.Z.; Writing—Review and Editing, I.P., X.Z., S.A., A.S. (Augusto Schneider), A.D.d.C.N., S.N., M.M.M. and P.G.; Visualization, I.P., X.Z. and A.S. (Augusto Schneider); Supervision, W.G., M.M.M. and P.G.; Project Administration, M.M.M. and P.G.; Funding Acquisition, M.M.M. All authors have read and agreed to the published version of the manuscript.

Funding: This research was funded by the fellowship from Polish National Agency for Academic Exchange, Walczak Scholarship Programme (grant no. PPN/WAL/2019/1/00024/U/00001) (I.P.).

Institutional Review Board Statement: The study was conducted according to the guidelines of the Declaration of Helsinki, and approved by the Institutional Ethics Committee of the Poznan University of Medical Sciences (Decision No. 915/16).

Informed Consent Statement: Informed consent was obtained from all subjects involved in the study.

Conflicts of Interest: The authors declare no conflict of interest.

References

1. Bray, F.; Ferlay, J.; Soerjomataram, I.; Siegel, R.L.; Torre, L.A.; Jemal, A. Global cancer statistics 2018: GLOBOCAN estimates of incidence and mortality worldwide for 36 cancers in 185 countries. *CA Cancer J. Clin.* **2018**, *68*, 394–424. [CrossRef]
2. Johnson, D.E.; Burtness, B.; Leemans, C.R.; Lui, V.W.Y.; Bauman, J.E.; Grandis, J.R. Head and neck squamous cell carcinoma. *Nat. Rev. Dis. Primers* **2020**, *6*, 92. [CrossRef] [PubMed]
3. Rana, T.M. Illuminating the silence: Understanding the structure and function of small RNAs. *Nat. Rev. Mol. Cell Biol.* **2007**, *8*, 23–36. [CrossRef] [PubMed]
4. Friedman, R.C.; Farh, K.K.; Burge, C.B.; Bartel, D.P. Most mammalian mRNAs are conserved targets of microRNAs. *Genome Res.* **2009**, *19*, 92–105. [CrossRef] [PubMed]
5. Victoria, B.; Nunez Lopez, Y.O.; Masternak, M.M. MicroRNAs and the metabolic hallmarks of aging. *Mol. Cell Endocrinol.* **2017**, *455*, 131–147. [CrossRef]
6. Li, N.; Long, B.; Han, W.; Yuan, S.; Wang, K. microRNAs: Important regulators of stem cells. *Stem Cell Res. Ther.* **2017**, *8*, 110. [CrossRef]
7. Li, Y.; Kowdley, K.V. MicroRNAs in common human diseases. *Genom. Proteom. Bioinform.* **2012**, *10*, 246–253. [CrossRef]
8. Volinia, S.; Calin, G.A.; Liu, C.G.; Ambs, S.; Cimmino, A.; Petrocca, F.; Visone, R.; Iorio, M.; Roldo, C.; Ferracin, M.; et al. A microRNA expression signature of human solid tumors defines cancer gene targets. *Proc. Natl. Acad. Sci. USA* **2006**, *103*, 2257–2261. [CrossRef] [PubMed]
9. Huang, Y.; Liu, Z.; Zhong, L.; Wen, Y.; Ye, Q.; Cao, D.; Li, P.; Liu, Y. Construction of an 11-microRNA-based signature and a prognostic nomogram to predict the overall survival of head and neck squamous cell carcinoma patients. *BMC Genom.* **2020**, *21*, 691. [CrossRef]

10. Chen, D.; Cabay, R.J.; Jin, Y.; Wang, A.; Lu, Y.; Shah-Khan, M.; Zhou, X. MicroRNA Deregulations in Head and Neck Squamous Cell Carcinomas. *J. Oral Maxillofac. Res.* **2013**, *4*, e2. [CrossRef]
11. Nunez Lopez, Y.O.; Victoria, B.; Golusinski, P.; Golusinski, W.; Masternak, M.M. Characteristic miRNA expression signature and random forest survival analysis identify potential cancer-driving miRNAs in a broad range of head and neck squamous cell carcinoma subtypes. *Rep. Pract. Oncol. Radiother.* **2018**, *23*, 6–20. [CrossRef] [PubMed]
12. Rajan, C.; Roshan, V.G.D.; Khan, I.; Manasa, V.G.; Himal, I.; Kattoor, J.; Thomas, S.; Kondaiah, P.; Kannan, S. miRNA expression profiling and emergence of new prognostic signature for oral squamous cell carcinoma. *Sci. Rep.* **2021**, *11*, 7298. [CrossRef] [PubMed]
13. Guo, L.; Cai, X.; Hu, W.; Hua, W.; Yan, W.; Lin, Y.; Yin, S.; Chen, Y. Expression and clinical significance of miRNA-145 and miRNA-218 in laryngeal cancer. *Oncol. Lett.* **2019**, *18*, 764–770. [CrossRef]
14. Shi, J.; Bao, X.; Liu, Z.; Zhang, Z.; Chen, W.; Xu, Q. Serum miR-626 and miR-5100 are Promising Prognosis Predictors for Oral Squamous Cell Carcinoma. *Theranostics* **2019**, *9*, 920–931. [CrossRef] [PubMed]
15. Powrozek, T.; Mlak, R.; Brzozowska, A.; Mazurek, M.; Golebiowski, P.; Malecka-Massalska, T. miRNA-130a Significantly Improves Accuracy of SGA Nutritional Assessment Tool in Prediction of Malnutrition and Cachexia in Radiotherapy-Treated Head and Neck Cancer Patients. *Cancers (Basel)* **2018**, *10*, 294. [CrossRef]
16. Hung, P.S.; Liu, C.J.; Chou, C.S.; Kao, S.Y.; Yang, C.C.; Chang, K.W.; Chiu, T.H.; Lin, S.C. miR-146a enhances the oncogenicity of oral carcinoma by concomitant targeting of the IRAK1, TRAF6 and NUMB genes. *PLoS ONE* **2013**, *8*, e79926. [CrossRef]
17. Li, T.; Chen, J.X.; Fu, X.P.; Yang, S.; Zhang, Z.; Chen, K.H.; Li, Y. microRNA expression profiling of nasopharyngeal carcinoma. *Oncol. Rep.* **2011**, *25*, 1353–1363. [CrossRef]
18. Sasahira, T.; Kurihara, M.; Bhawal, U.K.; Ueda, N.; Shimomoto, T.; Yamamoto, K.; Kirita, T.; Kuniyasu, H. Downregulation of miR-126 induces angiogenesis and lymphangiogenesis by activation of VEGF-A in oral cancer. *Br. J. Cancer* **2012**, *107*, 700–706. [CrossRef]
19. Wang, L.; Yu, J.; Xu, J.; Zheng, C.; Li, X.; Du, J. The analysis of microRNA-34 family expression in human cancer studies comparing cancer tissues with corresponding pericarcinous tissues. *Gene* **2015**, *554*, 1–8. [CrossRef]
20. Miao, S.; Mao, X.; Zhao, S.; Song, K.; Xiang, C.; Lv, Y.; Jiang, H.; Wang, L.; Li, B.; Yang, X.; et al. miR-217 inhibits laryngeal cancer metastasis by repressing AEG-1 and PD-L1 expression. *Oncotarget* **2017**, *8*, 62143–62153. [CrossRef]
21. Wang, H.; Ma, N.; Li, W.; Wang, Z. MicroRNA-96-5p promotes proliferation, invasion and EMT of oral carcinoma cells by directly targeting FOXF2. *Biol. Open* **2020**, *9*, bio049478. [CrossRef] [PubMed]
22. Tu, H.F.; Liu, C.J.; Chang, C.L.; Wang, P.W.; Kao, S.Y.; Yang, C.C.; Yu, E.H.; Lin, S.C.; Chang, K.W. The association between genetic polymorphism and the processing efficiency of miR-149 affects the prognosis of patients with head and neck squamous cell carcinoma. *PLoS ONE* **2012**, *7*, e51606. [CrossRef]
23. Nohata, N.; Hanazawa, T.; Kikkawa, N.; Mutallip, M.; Fujimura, L.; Yoshino, H.; Kawakami, K.; Chiyomaru, T.; Enokida, H.; Nakagawa, M.; et al. Caveolin-1 mediates tumor cell migration and invasion and its regulation by miR-133a in head and neck squamous cell carcinoma. *Int. J. Oncol.* **2011**, *38*, 209–217. [PubMed]
24. McCall, M.N.; McMurray, H.R.; Land, H.; Almudevar, A. On non-detects in qPCR data. *Bioinformatics* **2014**, *30*, 2310–2316. [CrossRef]
25. Dvinge, H.; Bertone, P. HTqPCR: High-throughput analysis and visualization of quantitative real-time PCR data in R. *Bioinformatics* **2009**, *25*, 3325–3326. [CrossRef]
26. Iacona, J.R.; Lutz, C.S. miR-146a-5p: Expression, regulation, and functions in cancer. *Wiley Interdiscip. Rev. RNA* **2019**, *10*, e1533. [CrossRef] [PubMed]
27. Zhu, F.Y.; Gan, C.W.; Wang, M.X.; Sun, B.C.; Li, F.J.; Qiu, Y.H.; Wang, K. MiR-146a-5p inhibits proliferation and promotes apoptosis of oral squamous cell carcinoma cells by regulating NF-kappaB signaling pathway. *Eur. Rev. Med. Pharmacol. Sci.* **2020**, *24*, 3717–3723. [CrossRef]
28. Emmett, S.E.; Stark, M.S.; Pandeya, N.; Panizza, B.; Whiteman, D.C.; Antonsson, A. MicroRNA expression is associated with human papillomavirus status and prognosis in mucosal head and neck squamous cell carcinomas. *Oral Oncol.* **2021**, *113*, 105136. [CrossRef]
29. Shi, Z.; Johnson, J.J.; Jiang, R.; Liu, Y.; Stack, M.S. Decrease of miR-146a is associated with the aggressiveness of human oral squamous cell carcinoma. *Arch Oral Biol.* **2015**, *60*, 1416–1427. [CrossRef]
30. Lin, D.; Dong, W.D.; Lu, M.P.; Xing, G.Q.; Dong, J.D.; Zhang, W.Q. The association between rs2910164 G>C polymorphism in pre-microRNA-146a and laryngeal cancer in Jiangsu Han population. *J. Otolaryngol. Ophthal. Shangdong Univ.* **2014**, *02*, 46–50.
31. Xu, B.; Zhang, X.; Wang, S.; Shi, B. MiR-449a suppresses cell migration and invasion by targeting PLAGL2 in breast cancer. *Pathol. Res. Pract.* **2018**, *214*, 790–795. [CrossRef]
32. Wu, D.; Liu, J.; Chen, J.; He, H.; Ma, H.; Lv, X. miR-449a Suppresses Tumor Growth, Migration, and Invasion in Non-Small Cell Lung Cancer by Targeting a HMGB1-Mediated NF-kappaB Signaling Pathway. *Oncol. Res.* **2019**, *27*, 227–235. [CrossRef] [PubMed]
33. Li, H.; Li, X.; Ge, X.; Jia, L.; Zhang, Z.; Fang, R.; Yang, J.; Liu, J.; Peng, S.; Zhou, M.; et al. MiR-34b-3 and miR-449a inhibit malignant progression of nasopharyngeal carcinoma by targeting lactate dehydrogenase A. *Oncotarget* **2016**, *7*, 54838–54851. [CrossRef] [PubMed]

34. Chan, L.S.; Lung, H.L.; Ngan, R.K.; Lee, A.W.; Tsao, S.W.; Lo, K.W.; Kahn, M.; Lung, M.L.; Wieser, R.; Mak, N.K. Role of miR-96/EVI1/miR-449a Axis in the Nasopharyngeal Carcinoma Cell Migration and Tumor Sphere Formation. *Int. J. Mol. Sci.* **2020**, *21*, 5495. [CrossRef] [PubMed]
35. Kawasaki, H.; Takeuchi, T.; Ricciardiello, F.; Lombardi, A.; Biganzoli, E.; Fornili, M.; De Bortoli, D.; Mesolella, M.; Cossu, A.M.; Scrima, M.; et al. Definition of miRNA Signatures of Nodal Metastasis in LCa: miR-449a Targets Notch Genes and Suppresses Cell Migration and Invasion. *Mol. Ther. Nucleic Acids* **2020**, *20*, 711–724. [CrossRef]
36. Han, J.; Wang, L.; Wang, X.; Li, K. Downregulation of Microrna-126 Contributes to Tumorigenesis of Squamous Tongue Cell Carcinoma via Targeting KRAS. *Med. Sci. Monit.* **2016**, *22*, 522–529. [CrossRef]
37. Qin, W.J.; Lv, L.H.; Zhang, M.; Zhou, X.; Liu, G.Q.; Lu, H.J. MiR-126 inhibits cell migration and invasion by targeting ADAM9 in oral squamous cell carcinoma. *Eur. Rev. Med. Pharmacol. Sci.* **2019**, *23*, 10324–10331. [CrossRef]
38. Chen, C.M.; Chu, T.H.; Chou, C.C.; Chien, C.Y.; Wang, J.S.; Huang, C.C. Exosome-derived microRNAs in oral squamous cell carcinomas impact disease prognosis. *Oral Oncol.* **2021**, *120*, 105402. [CrossRef]
39. Toxopeus, E.; Lynam-Lennon, N.; Biermann, K.; Dickens, G.; de Ruiter, P.E.; van Lanschot, J.; Reynolds, J.V.; Wijnhoven, B.; O'Sullivan, J.; van der Laan, L. Tumor microRNA-126 controls cell viability and associates with poor survival in patients with esophageal adenocarcinoma. *Exp. Biol. Med. (Maywood)* **2019**, *244*, 1210–1219. [CrossRef]
40. Tachibana, H.; Sho, R.; Takeda, Y.; Zhang, X.; Yoshida, Y.; Narimatsu, H.; Otani, K.; Ishikawa, S.; Fukao, A.; Asao, H.; et al. Circulating miR-223 in Oral Cancer: Its Potential as a Novel Diagnostic Biomarker and Therapeutic Target. *PLoS ONE* **2016**, *11*, e0159693. [CrossRef]
41. Hermeking, H. MicroRNAs in the p53 network: Micromanagement of tumour suppression. *Nat. Rev. Cancer* **2012**, *12*, 613–626. [CrossRef] [PubMed]
42. Li, J.; Liu, K.; Zhang, T.; Yang, Z.; Wang, R.; Chen, G.; Kang, M. A comprehensive investigation using meta-analysis and bioinformatics on miR-34a-5p expression and its potential role in head and neck squamous cell carcinoma. *Am. J. Transl. Res.* **2018**, *10*, 2246–2263.
43. Kalfert, D.; Pesta, M.; Kulda, V.; Topolcan, O.; Ryska, A.; Celakovsky, P.; Laco, J.; Ludvikova, M. MicroRNA profile in site-specific head and neck squamous cell cancer. *Anticancer Res.* **2015**, *35*, 2455–2463.
44. Severino, P.; Bruggemann, H.; Andreghetto, F.M.; Camps, C.; Klingbeil Mde, F.; de Pereira, W.O.; Soares, R.M.; Moyses, R.; Wunsch-Filho, V.; Mathor, M.B.; et al. MicroRNA expression profile in head and neck cancer: HOX-cluster embedded microRNA-196a and microRNA-10b dysregulation implicated in cell proliferation. *BMC Cancer* **2013**, *13*, 533. [CrossRef]
45. Roy, R.; Singh, R.; Chattopadhyay, E.; Ray, A.; Sarkar, N.; Aich, R.; Paul, R.R.; Pal, M.; Roy, B. MicroRNA and target gene expression based clustering of oral cancer, precancer and normal tissues. *Gene* **2016**, *593*, 58–63. [CrossRef] [PubMed]
46. Ren, Z.H.; Wu, K.; Yang, R.; Liu, Z.Q.; Cao, W. Differential expression of matrix metalloproteinases and miRNAs in the metastasis of oral squamous cell carcinoma. *BMC Oral Health* **2020**, *20*, 24. [CrossRef] [PubMed]
47. Hermeking, H. The miR-34 family in cancer and apoptosis. *Cell Death Differ.* **2010**, *17*, 193–199. [CrossRef] [PubMed]
48. Cai, K.M.; Bao, X.L.; Kong, X.H.; Jinag, W.; Mao, M.R.; Chu, J.S.; Huang, Y.J.; Zhao, X.J. Hsa-miR-34c suppresses growth and invasion of human laryngeal carcinoma cells via targeting c-Met. *Int. J. Mol. Med.* **2010**, *25*, 565–571. [CrossRef]
49. Liu, M.; Wu, H.; Liu, T.; Li, Y.; Wang, F.; Wan, H.; Li, X.; Tang, H. Regulation of the cell cycle gene, BTG2, by miR-21 in human laryngeal carcinoma. *Cell Res.* **2009**, *19*, 828–837. [CrossRef]
50. Re, M.; Ceka, A.; Rubini, C.; Ferrante, L.; Zizzi, A.; Gioacchini, F.M.; Tulli, M.; Spazzafumo, L.; Sellari-Franceschini, S.; Procopio, A.D.; et al. MicroRNA-34c-5p is related to recurrence in laryngeal squamous cell carcinoma. *Laryngoscope* **2015**, *125*, E306–E312. [CrossRef]
51. Xi, S.; Inchauste, S.; Guo, H.; Shan, J.; Xiao, Z.; Xu, H.; Miettenen, M.; Zhang, M.R.; Hong, J.A.; Raiji, M.T.; et al. Cigarette smoke mediates epigenetic repression of miR-217 during esophageal adenocarcinogenesis. *Oncogene* **2015**, *34*, 5548–5559. [CrossRef] [PubMed]
52. Gungormez, C.; Gumushan Aktas, H.; Dilsiz, N.; Borazan, E. Novel miRNAs as potential biomarkers in stage II colon cancer: Microarray analysis. *Mol. Biol. Rep.* **2019**, *46*, 4175–4183. [CrossRef]
53. Slaby, O.; Srovnal, J.; Radova, L.; Gregar, J.; Juracek, J.; Luzna, P.; Svoboda, M.; Hajduch, M.; Ehrmann, J. Dynamic changes in microRNA expression profiles reflect progression of Barrett's esophagus to esophageal adenocarcinoma. *Carcinogenesis* **2015**, *36*, 521–527. [CrossRef] [PubMed]
54. Chen, J.; Du, G.; Wang, Y.; Shi, L.; Mi, J.; Tang, G. Integrative analysis of mRNA and miRNA expression profiles in oral lichen planus: Preliminary results. *Oral Surg. Oral Med. Oral Pathol. Oral Radiol.* **2017**, *124*, 390–402 e317. [CrossRef] [PubMed]
55. Aghbari, S.M.H.; Abushouk, A.I.; Attia, A.; Elmaraezy, A.; Menshawy, A.; Ahmed, M.S.; Elsaadany, B.A.; Ahmed, E.M. Malignant transformation of oral lichen planus and oral lichenoid lesions: A meta-analysis of 20095 patient data. *Oral Oncol.* **2017**, *68*, 92–102. [CrossRef]
56. Peng, Q.; Zhang, J.; Ye, X.; Zhou, G. Tumor-like microenvironment in oral lichen planus: Evidence of malignant transformation? *Expert Rev. Clin. Immunol.* **2017**, *13*, 635–643. [CrossRef]
57. Vahabi, M.; Pulito, C.; Sacconi, A.; Donzelli, S.; D'Andrea, M.; Manciocco, V.; Pellini, R.; Paci, P.; Sanguineti, G.; Strigari, L.; et al. miR-96-5p targets PTEN expression affecting radio-chemosensitivity of HNSCC cells. *J. Exp. Clin. Cancer Res.* **2019**, *38*, 141. [CrossRef]

58. Ke, Y.; Zhao, W.; Xiong, J.; Cao, R. miR-149 Inhibits Non-Small-Cell Lung Cancer Cells EMT by Targeting FOXM1. *Biochem. Res. Int.* **2013**, *2013*, 506731. [CrossRef]
59. Xu, Y.; Lin, Y.P.; Yang, D.; Zhang, G.; Zhou, H.F. Clinical Significance of miR-149 in the Survival of Patients with Laryngeal Squamous Cell Carcinoma. *BioMed Res. Int.* **2016**, *2016*, 8561251. [CrossRef] [PubMed]
60. Lv, T.; Liu, H.; Wu, Y.; Huang, W. Knockdown of lncRNA DLEU1 inhibits the tumorigenesis of oral squamous cell carcinoma via regulation of miR1495p/CDK6 axis. *Mol. Med. Rep.* **2021**, *23*, 1–11. [CrossRef]
61. Lai, H.; Xu, G.; Meng, H.; Zhu, H. Association of SP1 rs1353058818 and STAT3 rs1053004 gene polymorphisms with human tongue squamous cell carcinoma. *Biosci. Rep.* **2019**, *39*, BSR20190955. [CrossRef] [PubMed]
62. Tandon, D.; Dewangan, J.; Srivastava, S.; Garg, V.K.; Rath, S.K. miRNA genetic variants: As potential diagnostic biomarkers for oral cancer. *Pathol. Res. Pract.* **2018**, *214*, 281–289. [CrossRef]
63. Mutallip, M.; Nohata, N.; Hanazawa, T.; Kikkawa, N.; Horiguchi, S.; Fujimura, L.; Kawakami, K.; Chiyomaru, T.; Enokida, H.; Nakagawa, M.; et al. Glutathione S-transferase P1 (GSTP1) suppresses cell apoptosis and its regulation by miR-133alpha in head and neck squamous cell carcinoma (HNSCC). *Int. J. Mol. Med.* **2011**, *27*, 345–352. [CrossRef] [PubMed]
64. Chattopadhyay, E.; Singh, R.; Ray, A.; Roy, R.; De Sarkar, N.; Paul, R.R.; Pal, M.; Aich, R.; Roy, B. Expression deregulation of mir31 and CXCL12 in two types of oral precancers and cancer: Importance in progression of precancer and cancer. *Sci. Rep.* **2016**, *6*, 32735. [CrossRef]
65. Liu, W.; Shi, X.; Wang, B. microRNA-133a exerts tumor suppressive role in oral squamous cell carcinoma through the Notch signaling pathway via downregulation of CTBP2. *Cancer Gene Ther.* **2021**, 1–11. [CrossRef]
66. Li, H.; Wang, Y.; Li, Y.Z. MicroRNA-133a suppresses the proliferation, migration, and invasion of laryngeal carcinoma cells by targeting CD47. *Tumour Biol.* **2016**, *37*, 16103–16113. [CrossRef] [PubMed]

cancers

Article

Piperlongumine, a Potent Anticancer Phytotherapeutic, Induces Cell Cycle Arrest and Apoptosis In Vitro and In Vivo through the ROS/Akt Pathway in Human Thyroid Cancer Cells

Fang-Ping Kung [1,†], Yun-Ping Lim [2,3,4,†], Wen-Ying Chao [5], Yi-Sheng Zhang [6], Hui-I Yu [1], Tsai-Sung Tai [1], Chieh-Hsiang Lu [1], Shu-Hsin Chen [6], Yi-Zhen Li [6], Pei-Wen Zhao [6], Yu-Pei Yen [1] and Ying-Ray Lee [7,*]

1 Division of Endocrinology and Metabolism, Department of Internal Medicine, Ditmanson Medical Foundation Chia-Yi Christian Hospital, Chiayi 60002, Taiwan; 07266@cych.org.tw (F.-P.K.); 04490@cych.org.tw (H.-I.Y.); 04015@cych.org.tw (T.-S.T.); 02602@cych.org.tw (C.-H.L.); 15159@cych.org.tw (Y.-P.Y.)
2 Department of Pharmacy, College of Pharmacy, China Medical University, Taichung 406040, Taiwan; limyp@mail.cmu.edu.tw
3 Department of Internal Medicine, China Medical University Hospital, Taichung 404332, Taiwan
4 Department of Medical Research, China Medical University Hospital, Taichung 404332, Taiwan
5 Department of Nursing, Min-Hwei College of Health Care Management, Tainan 73658, Taiwan; A129@o365.mhchcm.edu.tw
6 Department of Medical Research, Ditmanson Medical Foundation Chia-Yi Christian Hospital, Chiayi 60002, Taiwan; s1030722@alumni.ncyu.edu.tw (Y.-S.Z.); 10472@cych.org.tw (S.-H.C.); 10862@cych.org.tw (Y.-Z.L.); 14421@cych.org.tw (P.-W.Z.)
7 Department of Microbiology and Immunology, School of Medicine, College of Medicine, Kaohsiung Medical University, Kaohsiung 80708, Taiwan
* Correspondence: yingray@kmu.edu.tw
† These authors contributed equally to this work.

Citation: Kung, F.-P.; Lim, Y.-P.; Chao, W.-Y.; Zhang, Y.-S.; Yu, H.-I.; Tai, T.-S.; Lu, C.-H.; Chen, S.-H.; Li, Y.-Z.; Zhao, P.-W.; et al. Piperlongumine, a Potent Anticancer Phytotherapeutic, Induces Cell Cycle Arrest and Apoptosis In Vitro and In Vivo through the ROS/Akt Pathway in Human Thyroid Cancer Cells. *Cancers* **2021**, *13*, 4266. https://doi.org/10.3390/cancers13174266

Academic Editors: David Wong, Felisbina Queiroga and Bruno Cogliati

Received: 15 July 2021
Accepted: 20 August 2021
Published: 24 August 2021

Publisher's Note: MDPI stays neutral with regard to jurisdictional claims in published maps and institutional affiliations.

Copyright: © 2021 by the authors. Licensee MDPI, Basel, Switzerland. This article is an open access article distributed under the terms and conditions of the Creative Commons Attribution (CC BY) license (https://creativecommons.org/licenses/by/4.0/).

Simple Summary: There is no effective treatment currently available for patients with anaplastic, recurrent papillary, or follicular thyroid cancers. Reactive oxygen species (ROS) are believed to hold promise as a new therapeutic strategy for multiple human cancers. However, studies on ROS inducers for human thyroid cancer treatment are scarce. This study assesses the anticancer activity and the detailed downstream mechanisms of piperlongumine, a ROS inducer, in human thyroid cancer cells. We demonstrate that piperlongumine inhibits cell proliferation, regulates the cell cycle, and induces cellular apoptosis in various types of human thyroid cancer cells. The antihuman thyroid cancer activity of piperlongumine was through ROS induction, and it further suppressed the downstream Akt signaling pathway to elevate mitochondria-dependent apoptosis. A mouse xenograft study demonstrated that piperlongumine was safe and could inhibit tumorigenesis in vivo. The present study provides strong evidence that piperlongumine can be used as a therapeutic candidate for human thyroid cancers.

Abstract: Thyroid cancer (TC) is the most common endocrine malignancy, and its global incidence has steadily increased over the past 15 years. TC is broadly divided into well-differentiated, poorly differentiated, and undifferentiated types, depending on the histological and clinical parameters. Thus far, there are no effective treatments for undifferentiated thyroid cancers or advanced and recurrent cancer. Therefore, the development of an effective therapeutic is urgently needed for such patients. Piperlongumine (PL) is a naturally occurring small molecule derived from long pepper; it is selectively toxic to cancer cells by generating reactive oxygen species (ROS). In this study, we demonstrate the potential anticancer activity of PL in four TC cell lines. For this purpose, we cultured TC cell lines and analyzed the following parameters: Cell viability, colony formation, cell cycle, apoptosis, and cellular ROS induction. PL modulated the cell cycle, induced apoptosis, and suppressed tumorigenesis in TC cell lines in a dose- and time-dependent manner through ROS induction. Meanwhile, an intrinsic caspase-dependent apoptosis pathway was observed in the TC cells under PL treatment. The activation of Erk and the suppression of the Akt/mTOR pathways through ROS induction were seen in cells treated with PL. PL-mediated apoptosis in TC cells was through the ROS-Akt pathway. Finally, the anticancer effect and safety of PL were also demonstrated in vivo. Our findings indicate that PL exhibits antitumor activity and has the potential for use as a chemotherapeutic agent against TC. This is the first study to show the sensitivity of TC cell lines to PL.

Keywords: novel therapeutic strategy; safe anticancer treatment; anaplastic thyroid cancer; recurrent thyroid cancer; effective treatment

1. Introduction

Thyroid cancer (TC) is the most common malignancy of the endocrine organs, and its incidence has steadily increased in the past 15 years [1,2]. Most tumors (>95% of TCs) are derived from follicular cells, whereas, some tumors (medullary TC) are derived from C cells. Follicular cell-derived TCs are broadly divided into well-differentiated (DTC), poorly differentiated (PDTC), and anaplastic (ATC) types, depending on the histological and clinical parameters. DTC includes papillary TC (PTC) and follicular TC (FTC); most DTCs have a slowly progressive course and a normally favorable prognosis. Although the initiated treatment in most patients with PTC and FTC is effective, 10–15% of patients have tumor recurrence. By contrast, ATC is rare; this type is highly aggressive and 100% lethal within one year after diagnosis [3,4]. Currently, there is no effective treatment for ATC, because the patients fail to respond to the available radiotherapy and chemotherapeutic agents [5]. Therefore, the development of an effective therapeutic for ATC and recurrent DTC and PDTC is urgently needed.

Piperlongumine (PL), also known as piplartine, is an amide alkaloid isolated from long pepper, *Piper longum* L. [6]. PL exerts extensive biological activities, including antiplatelet, antimicrobial, antiangiogenetic, antidiabetic, antidepressant, antiatherosclerotic, neuroprotective, and anticancer properties [7]. PL also exerts antitumor activity in selectively affected transformed cell types in lymphoma [8], melanoma [9], glioblastoma [10], oral [11], head and neck [12], lung [13], breast [14], liver [15], cholangiocarcinoma [16], renal [17], pancreatic [18], gastric [19], colon [20], bladder [21], and prostate [22] cancers, without affecting normal cells. The anticancer activities occur through the p38/JNK, MAPK, and NF-κB pathways and by the induction of high levels of reactive oxygen species (ROS) [16,23,24]. PL also causes cell death through both caspase-dependent apoptosis and necrosis and induces the downregulation of Bcl2 expression and the activation of caspase-3, poly (ADP-ribose) polymerase (PARP), and JNK [16,24]. PL has shown antitumor activity in several whole-animal models, and it is highly safe when used in vivo [25,26]. However, no study has provided evidence for the effects of PL against TC. In the present study, we aimed to evaluate the therapeutic effects of PL in human multidrug-resistant PTC, FTC, and ATC cells. Furthermore, the anticancer mechanisms of PL were also determined in these cells, and a mouse xenograft model was used to confirm the therapeutic property of PL in vivo.

2. Results

2.1. *Piperlongumine Inhibits Cell Proliferation and Colony Formation of Human Thyroid Cancer Cells*

To investigate the effects of PL in human TCs, we performed the cell viability test using the CCK-8 assay on four human TC cell lines (IHH-4, WRO, 8505c, and KMH-2; see Section 4.1) treated with various concentrations of PL (0, 1, 2, 3, 4, and 5 μM) for 24 and 48 h. Cell growth was inhibited by PL treatment in each cell line in a dose- and time-dependent manner (Figure 1A), suggesting that PL could suppress cell growth in various types of human TC cells. The IC_{50} values of PL in these cells are shown in Table 1. Among them, KMH-2 was more sensitive to PL treatment, and distinctly, WRO was more resistant when incubated with PL. In addition, to assess whether PL could suppress cellular proliferation, an in vitro colony formation assay was conducted, and the results showed that PL could significantly reduce colony formation in a dosage-dependent manner (Figure 1B). Moreover,

IHH-4 cells were more sensitive than the other cell lines in the PL anticolony formation analysis (Figure 1B). Altogether, we demonstrated that PL is a potential anticancer agent against multiple human TCs. Further, PL doses of 2 and 3 μM were selected as effective doses for subsequent experiments owing to their abilities to suppress IHH-4, 8505c, and KMH-2 cells below 50% of the control.

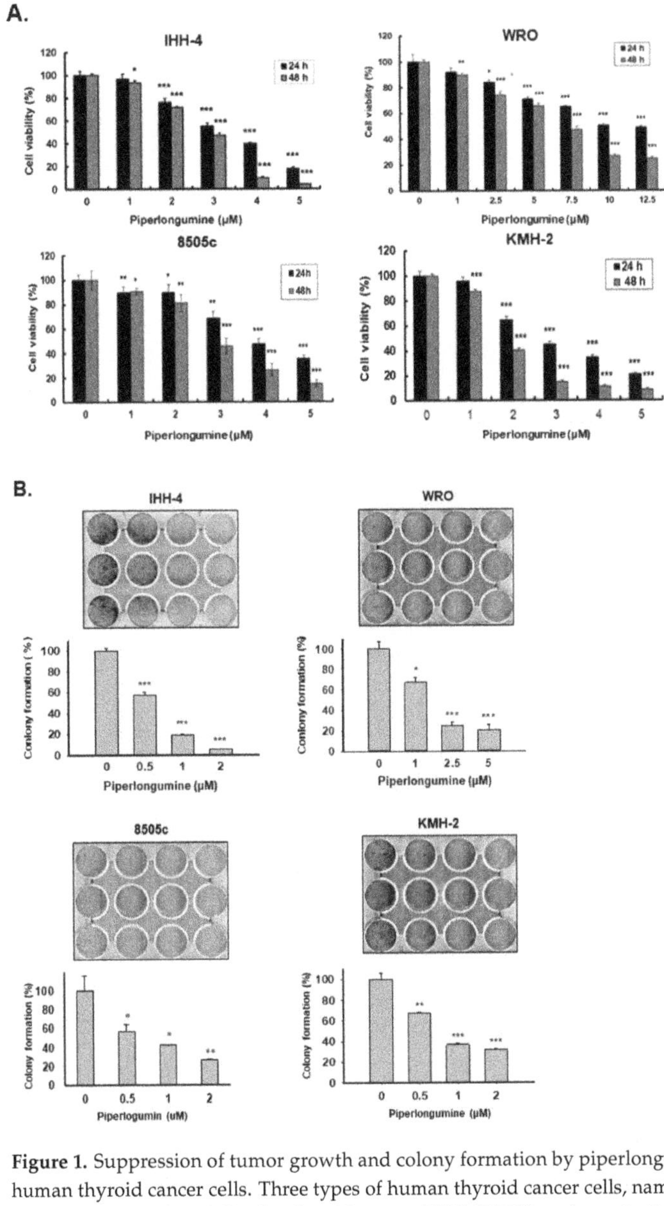

Figure 1. Suppression of tumor growth and colony formation by piperlongumine (PL) in vitro in human thyroid cancer cells. Three types of human thyroid cancer cells, namely, papillary thyroid cancer (PTC) (IHH-4), follicular thyroid cancer (FTC) (WRO), and anaplastic thyroid cancer (ATC) (8505c and KMH-2), were incubated with PL, and the (**A**) cellular viability and (**B**) colony formation were examined. Dimethyl sulfoxide (DMSO) was used as a negative control. Three independent experiments of cellular viability were conducted. Compared with the control group, * indicates $p < 0.05$, ** indicates $p < 0.01$, *** indicates $p < 0.001$.

Table 1. IC_{50} values of piperlongumine in human thyroid cancer cells.

	Time (h)	Cell Line			
		IHH-4	WRO	8505c	KMH-2
Piperlongumine (μM)	24	3.2	12.52	3.3	2.4
	48	2.8	5.58	2.8	1.7

2.2. PL Induces G2/M Phase Cell Cycle Arrest and Cellular Apoptosis through the Intrinsic Caspase-Dependent Pathway in Human Thyroid Cancer Cells

Because PL treatment reduced the viability of TC cells (Figure 1), we further examined whether growth inhibition by PL led to cell cycle alteration and apoptosis. Cell cycle analysis indicated that a large cell population was in the G2/M phase with 6–12 h treatment in all the TC cell lines after PL treatment (Figure 2), suggesting that PL could promote cell cycle arrest at the G2/M phase in human TCs. Cellular apoptosis was also observed to occur in a dosage-dependent manner in all the cells treated with PL (Figure 3A). Moreover, the expression or the activation of caspases, such as caspase-8, caspase-9, and caspase-3, and PARP were examined. As shown in Figure 3A, caspase-9, caspase-3, and PARP were activated due to PL treatment in a dose-dependent manner. However, caspase-8 was not activated in PL-mediated apoptosis. These data suggest that PL induced cellular apoptosis in KMH-2 and IHH-4 cells through an intrinsic pathway, and the extrinsic pathway was not involved.

To unravel whether caspase activation is involved in PL-mediated cellular apoptosis in IHH-4, 8505c, and KMH-2 cells, the Z-VAD-FMK was used. The activation of caspase-3 and PARP were significantly reversed in the cells with pre-treatment of Z-VAD-FMK (Figure 3B), confirming that PL can induce a caspase activation in human TC cells. Furthermore, PL-induced cellular apoptosis was significantly reversed in the cells pre-incubated with Z-VAD-FMK (Figure 3C), demonstrating that caspase-dependent apoptosis occurred in IHH-4, 8505c, and KMH-2 cells when treated with PL. Altogether, our findings demonstrate that PL inhibited cell growth and induced apoptosis in human TC cell lines through an intrinsic caspase-dependent pathway.

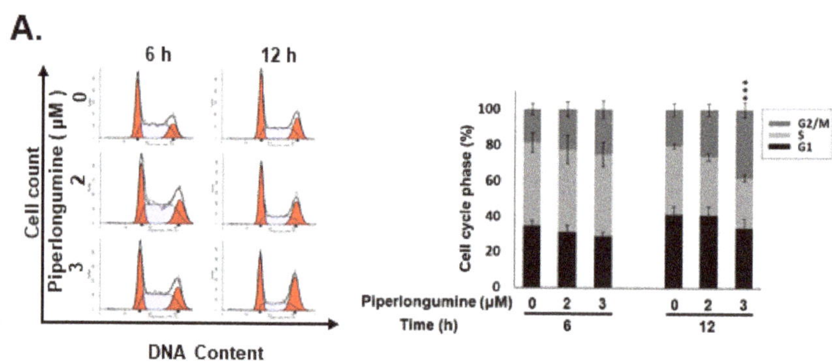

Figure 2. *Cont.*

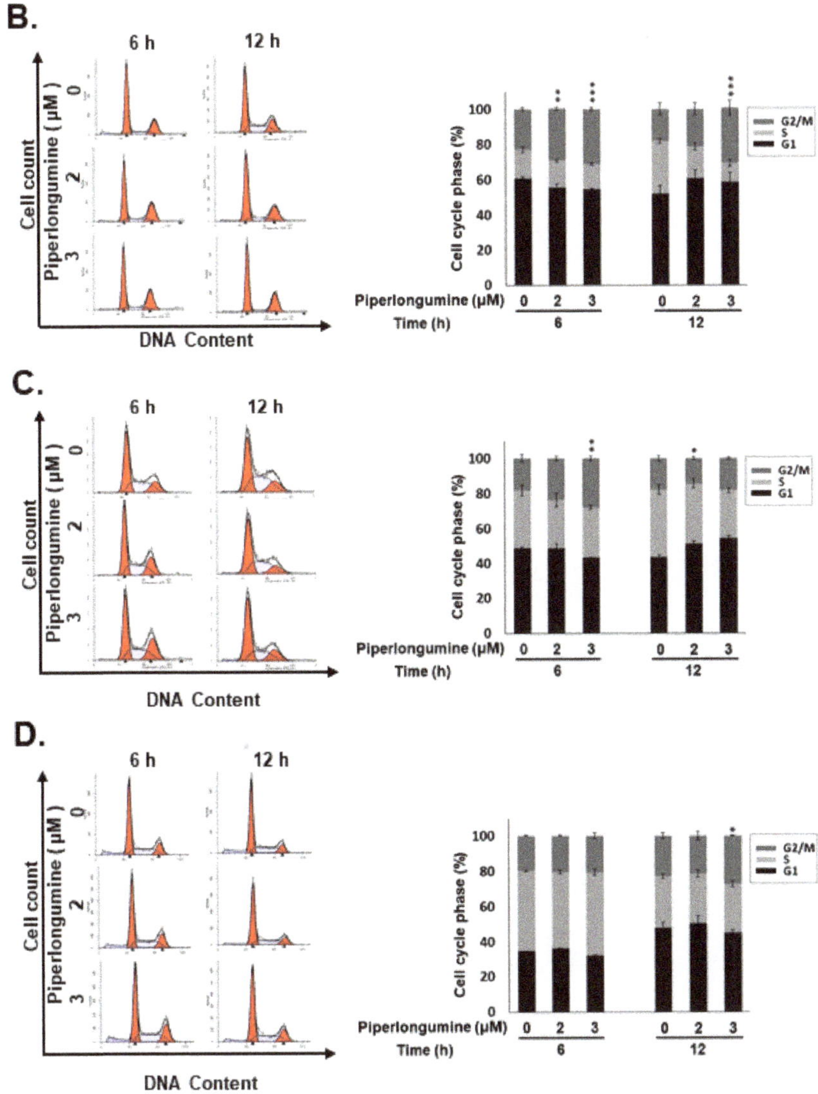

Figure 2. Induction of cell cycle arrest at the G2/M phase by piperlongumine (PL) in human thyroid cancer cells. Human thyroid cancer cells, namely, (**A**) IHH-4, (**B**) WRO, (**C**) 8505c, and (**D**) KMH-2 cells, were incubated with a control medium or PL, and the cell cycle was examined with fluorescence-activated cell sorting (FACS) flow cytometry. Dimethyl sulfoxide (DMSO) was used as a negative control. Three independent experiments were conducted. * indicated comparing with the untreated group. * indicates $p < 0.05$. ** indicates $p < 0.01$. *** indicates $p < 0.001$.

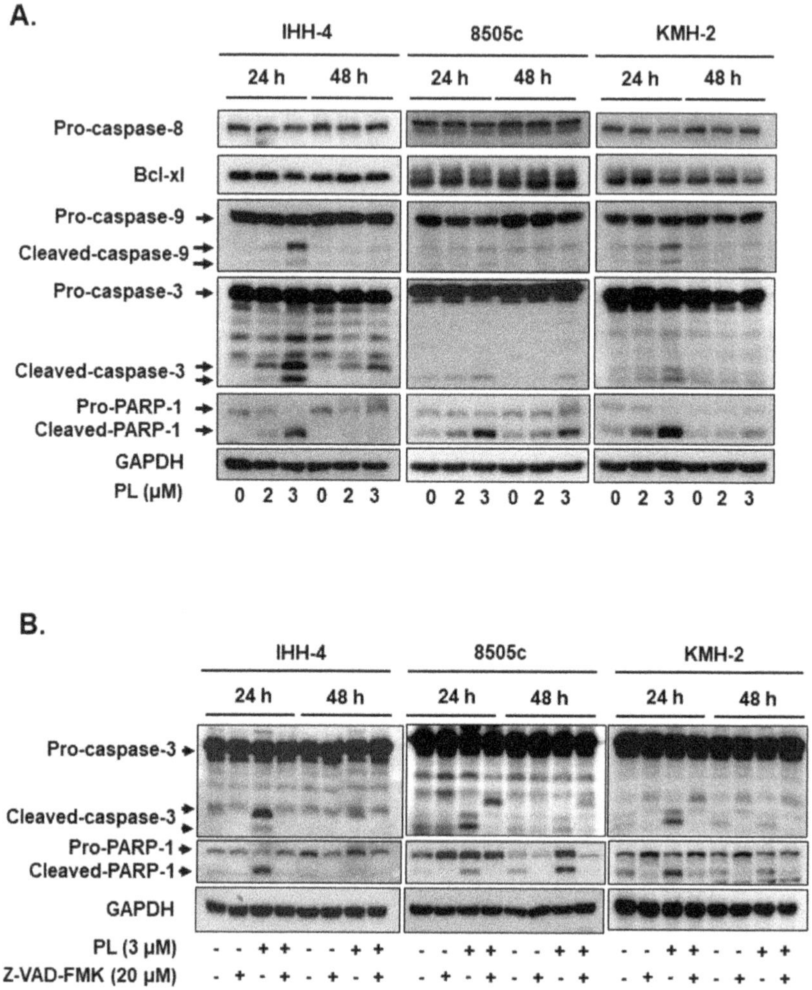

Figure 3. *Cont.*

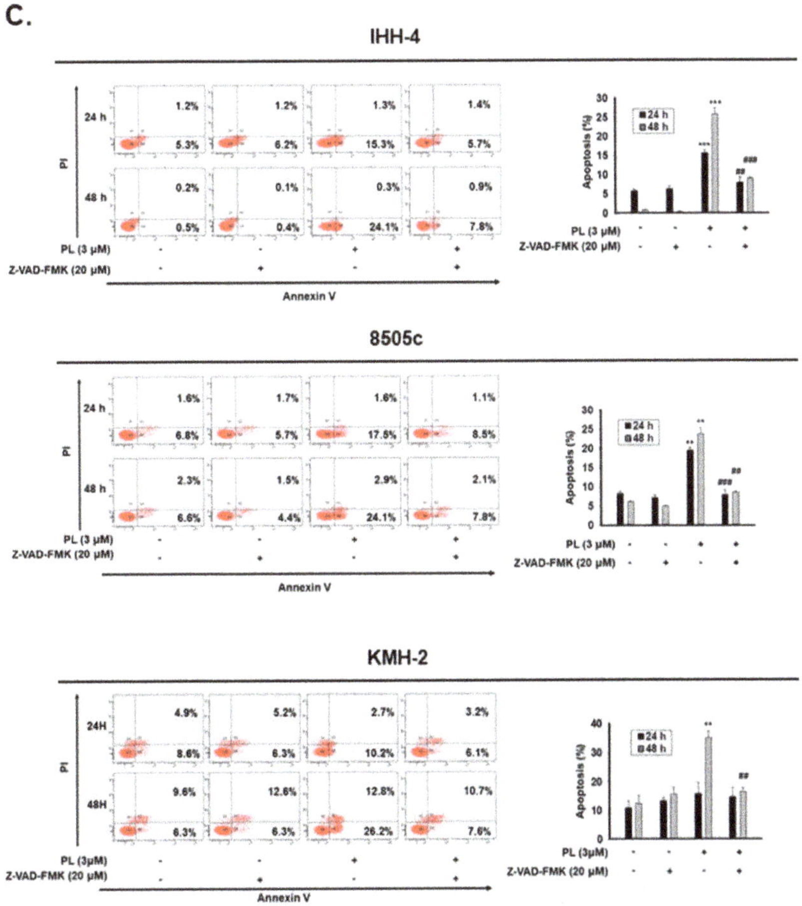

Figure 3. Activation of mitochondria-dependent cellular apoptosis by piperlongumine (PL) in human thyroid cancer cells. (**A**) Three human thyroid cancer cells were incubated with PL, and the expression of caspases and poly (ADP-ribose) polymerase (PARP) was examined by western blotting. (**B**) Z-VAD-FMK was used to suppress PL-mediated caspase-dependent apoptosis, and the activation of caspase and PARP was assessed by western blots, and (**C**) the cellular apoptosis was determined with flow cytometry. Dimethyl sulfoxide (DMSO) was used to be a negative control. GAPDH was used as a loading control. Three independent experiments were conducted. The data include western blotting, and the flow cytometry dot plot was shown with a representative experiment. * indicated comparing with the untreated group and # showed comparing with PL treated group. ** and ## indicates $p < 0.01$. *** and ### indicates $p < 0.001$.

2.3. ROS Induction Is Involved in PL-Mediated Antitumor Behavior in Human Thyroid Cancer Cells

PL exerts anticancer activity in various types of cancers by promoting the induction of ROS [27]. In the present study, the induction of ROS in IHH-4, 8505c, and KMH-2 cells under PL treatment was also determined (Figure 4A). Pre-treatment with N-acetylcysteine (NAC; a selective ROS scavenger) significantly reduced PL-mediated ROS activation (Figure 4A). Moreover, the suppression of PL-induced ROS using NAC showed significant reversal of cytotoxic effects of PL in IHH-4, 8505c, and KMH-2 cells, according to cellular viability and colony formation (Figure 4B,C). In addition, cells pre-treated with NAC showed complete reversal of PL-induced cell cycle progression at the G2/M phase (Figure 4D).

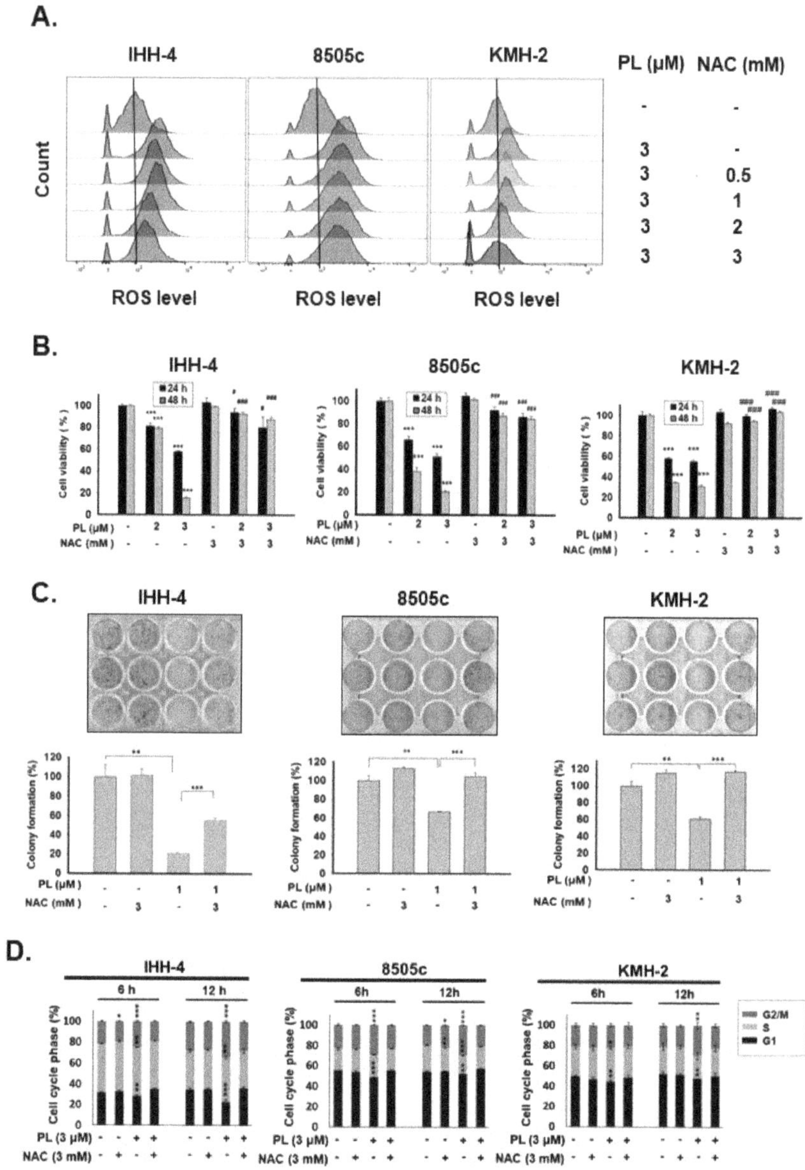

Figure 4. Modulation of cell growth, tumorigenesis, and cell cycle by piperlongumine (PL) by inducing reactive oxygen species (ROS) in human thyroid cancer cells. Three human thyroid cancer cells were incubated with PL, and NAC was used to inhibit ROS activation. (**A**) The expression of ROS was determined with fluorescence-activated cell sorting (FACS) flow cytometry. (**B**) Cellular viability and (**C**) colony formation were examined. (**D**) Cell cycle regulation was determined by FACS flow cytometry. Dimethyl sulfoxide (DMSO) was used as a negative control. ROS expression in (**A**) was confirmed with two independent experiments, and a representative experiment was shown. The cell viability, colony formation, and cell cycle were obtained in three independent experiments. * indicated comparing with the untreated group and # showed comparing with PL treated group. * and # indicates $p < 0.05$. ** indicates $p < 0.01$. *** and ### indicates $p < 0.001$.

Furthermore, the role of ROS in PL-mediated cellular apoptosis in IHH-4, 8505c, and KMH-2 cells was also evaluated. The activation of caspase-9, caspase-3, and PARP were determined in these cells under PL treatment; however, the activation of caspases and PARP in these cells after incubation with PL was reduced in the presence of NAC (Figure 5A). Moreover, the number of apoptotic cells significantly increased under PL treatment, but these effects were blocked by pre-treatment with NAC (Figure 5B). Taken together, these findings indicate that PL induces TC apoptosis and cell cycle arrest, as well as in vitro tumorigenesis, by generating and inducing intracellular ROS.

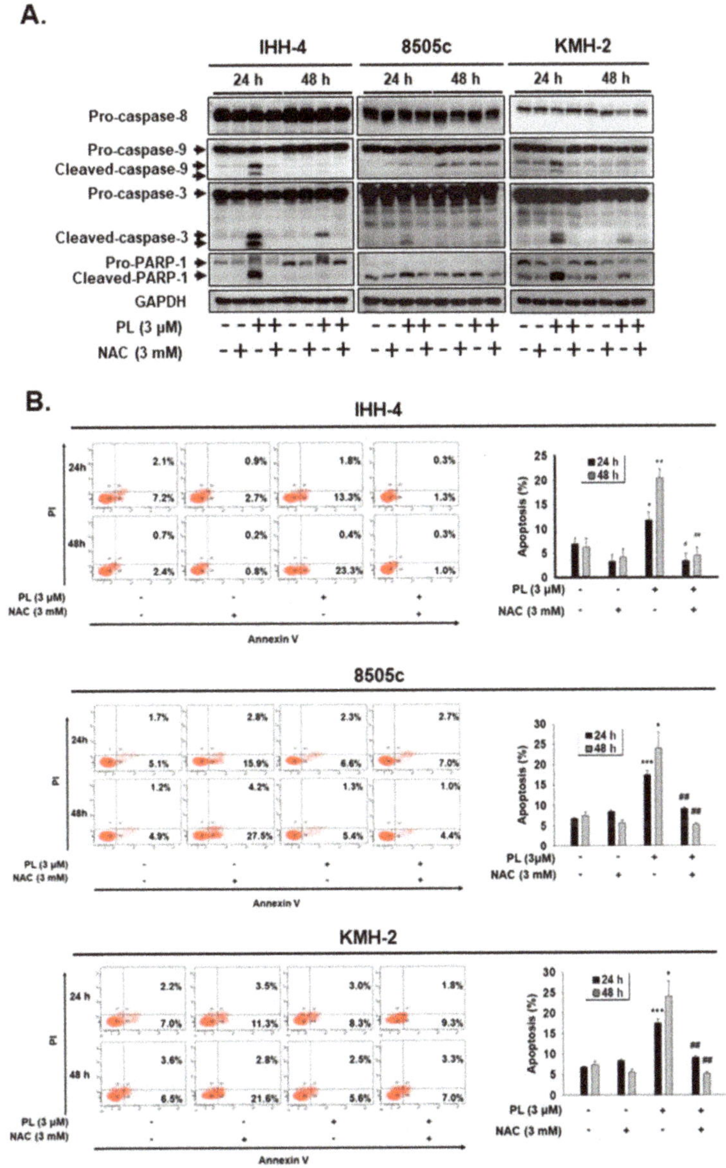

Figure 5. Piperlongumine (PL)-mediated cellular apoptosis was regulated by inducing reactive oxygen species (ROS) in human thyroid cancer cells. Three human thyroid cancer cells were treated with PL;

(**A**) the activation of caspase and poly (ADP-ribose) polymerase (PARP) and (**B**) cellular apoptosis were investigated by western blotting and fluorescence-activated cell sorting (FACS) flow cytometry. NAC was used to suppress PL-induced ROS. DMSO was used as a negative control. GAPDH was used as a loading control. Three independent experiments each for western blotting and flow cytometry were conducted, and western blots and the flow cytometry dot plots of a representative experiment are shown. * indicated comparing with the untreated group and # showed comparing with PL treated group. * and # indicates $p < 0.05$. ** and ## indicates $p < 0.01$. *** indicates $p < 0.001$.

2.4. PL Induces Cellular Apoptosis through the ROS-Modulated Akt Pathway

ROS is involved in apoptosis regulation by regulating Erk, JNKs, p38, and through the Akt/mTOR pathways in many types of cancers [28,29]. Thus, we further assessed whether Erk, JNK, p38, and Akt activation are involved in regulating the apoptotic effect caused by PL. PL increased p-ERK expression, but decreased the expression of p-p38, p-Akt, and p-mTOR in KMH-2 cells (Figure 6A). Because ROS activation was found to occur in these cells under PL treatment (Figure 4), herein, the relationship of ROS with Erk, p38, and Akt pathways was further examined. The activation of Erk and the suppression of the Akt pathway under PL treatment were found to be reversed by NAC pre-treatment (Figure 6B). However, p38 activation in PL-treated cells showed no significant difference without or with co-incubation with NAC (Figure 6B). This finding suggested that PL-mediated ROS activation may be the upstream pathway of the Erk and Akt pathways. PL-mediated ROS activation could induce TC apoptosis and regulate the activation of the Erk and Akt pathways. We further evaluated the activation of caspase-3 and PARP in cells treated with PL in combination with or without PD98059 (an inhibitor of Erk) and perifosine (an inhibitor of Akt). We demonstrated that reducing PL-mediated Erk activation could not reverse PL-mediated cellular apoptosis (Figure 6C), and an increase in PL-mediated Akt inactivation could promote PL-mediated cellular apoptosis (Figure 6D). In addition, a dominant active Akt construct was transfected to reverse PL-suppressed Akt activation in KMH-2 cells, and the PL-mediated activation of caspase-3 and PARP was found to be inhibited (Figure 7A). Moreover, the PL-mediated cellular apoptosis was significantly reduced in KMH-2 cells in the presence of constitutive Akt activation (Figure 7B). These findings demonstrate that PL treatment in TC cells can activate ROS, reduce Akt signaling, and finally, induce cellular apoptosis.

2.5. PL Reduces Tumor Growth and Induces Tumor Cell Apoptosis In Vivo

PL displays a good antitumor behavior in various human cancers [23]. We demonstrated that PL exerts anticancer activities, such as cell cycle arrest, growth inhibition, and apoptosis induction, and suppresses colony formation in vitro. In the present study, a mouse xenograft model was used to evaluate the anti-TC activity and the safety of PL in vivo. Nude mice received a subcutaneous injection of IHH-4 cells into the right flank and intraperitoneal injection of PL at various doses. Tumorigenesis was examined every two days, and the bodyweight of mice was also recorded. The tumor volumes in the mice treated with PL (10 mg/kg) were significantly lower than those in the control groups (Figure 8A,B). Moreover, the tumor weight in the PL (10 mg/kg) treatment group significantly decreased when compared with that in the control group (Figure 8C). To address the safety of PL in this study, we assessed the bodyweight and the pathological phenomena of mouse tissues by (hematoxylin and eosin) HE staining. The bodyweight of mice showed no significant difference between the control and PL treatment groups. Furthermore, no pathological findings in the liver and kidney were detected in the control or PL treatment groups (Figure 8D). In addition, there was no significant infiltration of immune cells in any of the tissues (Figure 8D). Finally, tumor cell apoptosis in vivo was assessed with the TUNEL assay, which revealed that PL treatment could induce tumor apoptosis in a dose-dependent manner in vivo (Figure 8E). These data demonstrate that PL exhibits good therapeutic action against human TCs and is safe for treatment in vivo.

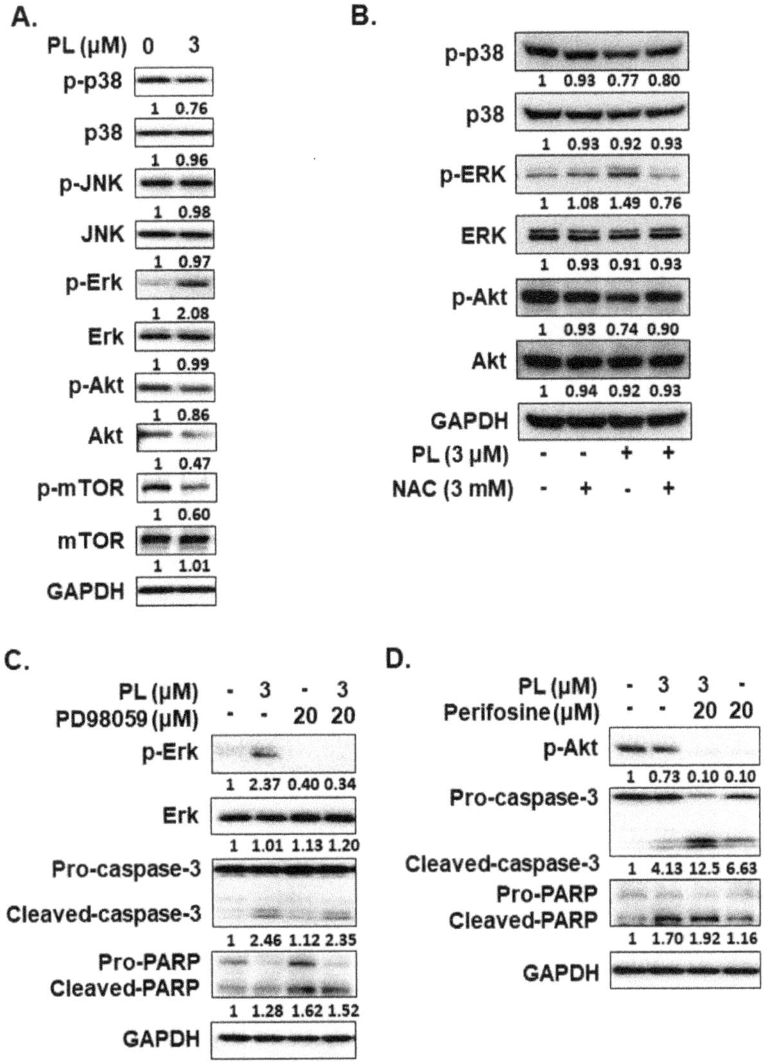

Figure 6. Regulation of the activation of the Erk and Akt signaling pathways by piperlongumine (PL) through reactive oxygen species (ROS) induction. KMH-2 cells were used to evaluate the regulation of signaling pathways under PL treatment. (**A**) The expression and the activation of Erk, JNK, p38, Akt, and mTOR were examined by western blotting in PL-treated cells. (**B**) NAC was used to block PL-mediated ROS induction and p38 activation, and the resulting expression of Erk and Akt was examined by western blotting. (**C**) PD98059 was used to suppress PL-mediated Erk activation, and (**D**) perifosine was used to promote PL-mediated Akt activation, and the activation of caspase-3 and PARP was examined by western blotting. DMSO was used as a negative control. GAPDH was used as a loading control. Three independent experiments were conducted, and a representative experiment is shown.

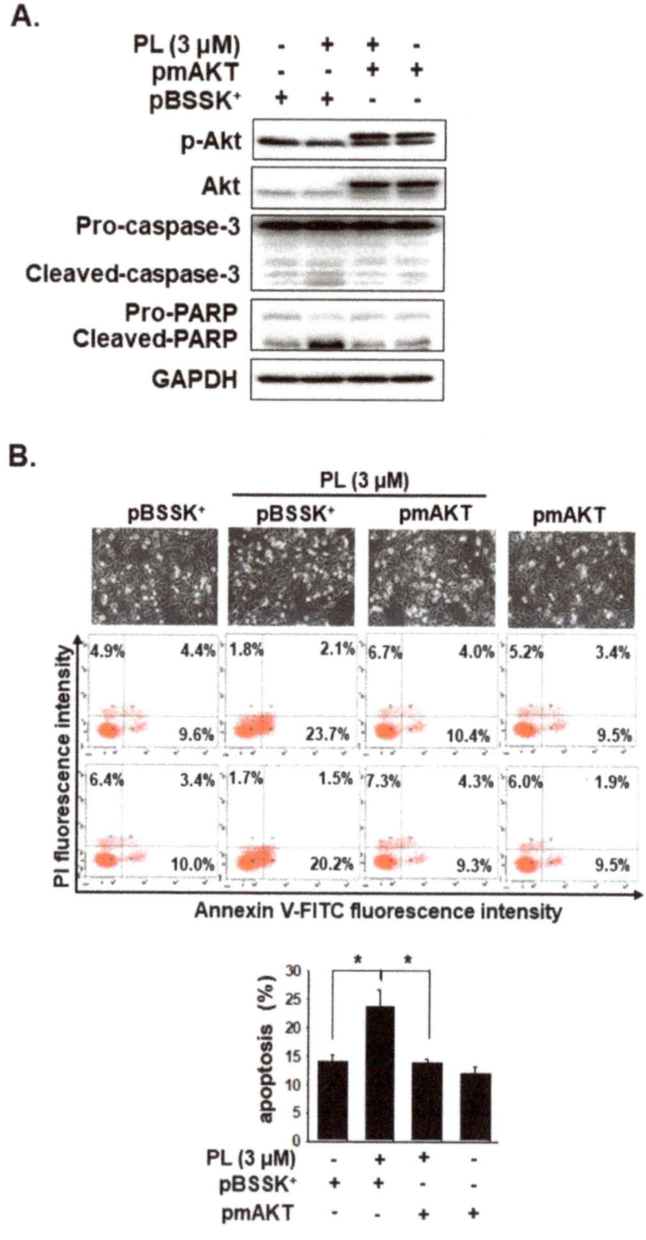

Figure 7. Piperlongumine (PL) induced cellular apoptosis through the Akt signaling pathway. KMH-2 cells were used to verify the involvement of the Akt signaling pathway in PL-caused apoptosis. (**A**) KMH-2 cells were transfected with or without constitutively active Akt construct, and the expression and the activation of Akt, caspase-3, and PARP were investigated by western blots after incubation with PL for 24 h. (**B**) Cellular apoptosis was investigated by flow cytometry. pBSSK$^+$ was used as a negative control of transfection. pmAKT was a constitutively active form of the ATK construct. DMSO was used as a negative control. GAPDH was used as a loading control. Three independent experiments each for western blotting and flow cytometry were conducted, and the western blots and flow cytometry dot plots of a representative experiment are shown. * indicates $p < 0.05$.

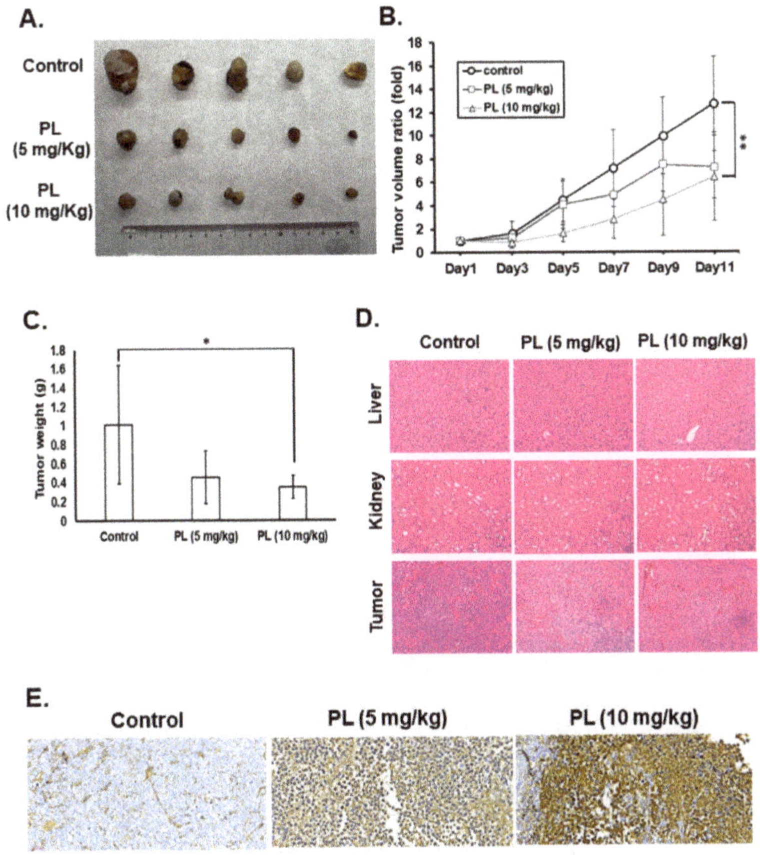

Figure 8. Piperlongumine significantly inhibited human thyroid cancer cell growth in vivo. IHH-4-bearing mice were treated with either dimethyl sulfoxide (DMSO) (n = 5) or PL (n = 5/group). (**A**) Representative images of excised tumors from each group. (**B**) Tumor volume and (**C**) tumor weight were measured to reflect tumor growth in vivo. (**D**) Hematoxylin and eosin staining to determine cellular pathology and immune cell infiltration. 200× magnification. (**E**) Apoptosis in tumors was measured by terminal deoxynucleotidyl transferase-mediated nick end labeling (TUNEL) staining, and the brown staining indicates apoptotic cells. 40× magnification. DMSO was used as a negative control. * indicates $p < 0.05$. ** indicates $p < 0.01$.

3. Discussion

TC predominantly affects women worldwide; among all TCs, ATC and recurrent PTC and FTC display a highly aggressive nature. Currently, no effective treatments have been reported. Recently, it is growing interest to develop therapeutic compounds from natural products. Our findings indicate that PL induces a ROS-mediated mechanism, which perturbs numerous cellular signaling pathways and finally elevates apoptosis in TC cells. This finding is based on two crucial findings. Primary, ROS accumulation happened in PL-incubated cells. Second, blocking of ROS with NAC considerably suppressed all PL-induced effects, including the activation of caspases, reducing apoptotic gene expression, and inhibition of cell survival signaling.

Activation of either the intrinsic apoptosis pathway or the extrinsic apoptosis pathway may cause the activation of effectors of various caspases, leading to cellular apoptosis. Caspases can be classified according to their function and activation as initiator caspases (caspases 2, 8, and 9) or executioner caspases (caspases 3, 6, and 7) [30]. Our results indicate that PL generates ROS and activates caspase-9 and caspase-3. However, caspase-8 is not involved. Caspase-9 is well known to be one of the key factors for the intrinsic pathway. These mechanisms were initiated by the generation of ROS, due to PL treatment. As shown in our study, pre-treatment with NAC could fully inhibit this cascade of events (Figure 5). Caspase-mediated pathways related to apoptosis were induced by PL in these cells, as confirmed by the addition of the Z-VAD-FMK pan-caspase inhibitor, which completely suppressed cell death induced by PL (Figure 3). Moreover, overexpression of Bcl-$_{XL}$ and Bcl-2 can act against the cytotoxic effect of chemotherapeutic drugs, suggesting a potential role of these proteins in TC resistance to drug-induced cytotoxicity [31]. In this study, we also determined the decrease in Bcl-$_{XL}$ expression in KMH-2 cells under PL treatment (Figure 3A), suggesting that PL treatment in human TC cells can induce cellular apoptosis through the intrinsic mitochondrial death pathway.

Previous research has indicated that PL is highly selective in targeting cancer cells, but not healthy cells [7]. Furthermore, a good safety profile has been observed in previous in vivo toxicological testing, showing high absorption through the gastrointestinal tract and >50% of bioavailability after oral administration of PL in mice [32]. The effects of PL on various molecular targets involved in cancer development and progression have been reported, focusing on its low toxicity and advanced pharmacokinetic features. However, owing to the lack of nanomolar potency and less soluble in water of PL, its applicability is presently limited. Pharmaceutical chemists may be able to derive compounds characterized by improved anticancer activity and appropriate drug-like physicochemical parameters. In addition, the use of modern drug delivery systems may improve the efficacy and solve issues related to water solubility [33]. Recently, great efforts have been made in to optimize PL activity through physicochemical parameters, pharmacokinetics, and safety, making PL a good candidate for future anticancer therapy.

Standard treatment for TC usually includes surgery, thyroid-stimulating hormone suppressive therapy, and radioactive irradiation (RAI) for the ablation of TC remnants [34]. However, approximately 5% of TC patients have metastasis, and there is currently a lack of effective drugs and limited efficacy of RAI for this condition. Several naturally occurring products, such as flavonoids, resveratrol, quercetin, catechins, myricetin, apigenin, and curcumin, have been investigated as having the potential to slow down or inhibit dedifferentiation and cancer progression [35]. Some human studies have also described the beneficial effects of plant-derived products with the potential for TC management and have found that different classes of flavonoids may have distinct effects in determining TC risk [36].

ROS-independent molecular targets of PL, such as PI3K/Akt/mTOR inhibition, have been reported [37]. PL also has anti-invasive properties, such as inhibiting the expression of Twist, N-cadherin, p120-ctn/vimentin/N-cadherin complex, IL-6, IL-8, MMP-9, and ICAM-1 [38]. PL also shows antiangiogenic effects by decreasing the expression of vascular endothelial growth factor protein [39]. Furthermore, PL works in concert with existing breast and prostate cancer treatments by increasing the efficacy of the chemotherapeutics 5-fluorouracil, cisplatin, doxorubicin, paclitaxel, and curcumin, increasing intracellular ROS levels, and enhancing the radiosensitivity of cancer cells [38,40,41]. ROS generation is strictly regulated in normal physiology; however, dysregulated ROS activation can cause cellular damage [42]. Because there is usually a higher level of ROS production in cancer cells than in normal cells, once intracellular ROS production is increased, cancer cells are more sensitive than normal cells, resulting in cellular damage [43,44]. Furthermore, exogenously generated ROS-induced oxidative stress is considered an effective therapy is owing to its selective effects against cancer cells, and PL has been reported to induce ROS in cancer cells, but not in normal cell lines [45,46]. Although enhancing intracellular ROS has been believed to be a new therapeutic strategy for multiple human cancers, related studies in TC remain scarce.

In this study, we found that PL significantly inhibited tumor growth in four TC cancer cell lines through G2/M phase arrest, in agreement with the findings of previous reports [12,19]. PL-induced ROS generation led to TC cell apoptosis by activating the Akt signaling pathway, thereby initiating caspase and PARP activation (Figure 7). Because these effects were fully abolished by the addition of the ROS scavenger NAC, as well as Akt inducer and dominant active Akt transfection, these results indicate that the antitumor activity of PL was primarily due to ROS generation (Figures 3–7). Prolonged exposure to PL for 48 h enhanced the number of apoptotic cells, and the effect was completely reversed by co-exposure to NAC (Figure 3C). Inhibition of cell survival signaling by the Akt/mTOR pathway was also targeted by PL. In previous studies, PL has been shown to inhibit cell growth in various cancer cell lines, with an IC_{50} value of less than 10 μM, in agreement with the small micromolar doses observed here [12–15,19].

Human TC cell lines are frequently used for TC studies, with various lines originating from differentiated and undifferentiated human thyroid tumors. They have maintained hallmarks of cancer cells and are pure, genetically identical, and easily cultured, and can be genetically operated [47]. The cell line originates from clinical specimens and is generated by the selection of the resistance or rapidly proliferating cells during passages. The IHH-4 line used in this study was derived from a patient with cervical lymph node metastasis; it possesses multidrug resistance in PTC, accounting for approximately 80% of all TCs reported. The KMH-2 and 8505c lines are derived from ATC, which is undifferentiated, grows rapidly in elderly patients, and is not sensitive to most therapies [48]; as such, it is often fatal. PL showed great inhibitory effects against all three cell lines, suggesting broad applicability in TC treatment. In addition, WRO cells seemed more resistant to PL treatment than IHH-4, KMH-2, and 8505c cells (Figure 1). Because the PL-mediated anticancer effect in this study is through ROS and downstream Akt signaling (Figure 9), we also detected ROS induction in WRO cells under PL treatment. We demonstrated that higher doses of PL are required for raising ROS levels in WRO cells than in IHH-4, KMH-2, and 8505c cells (Figures 4A and S1).

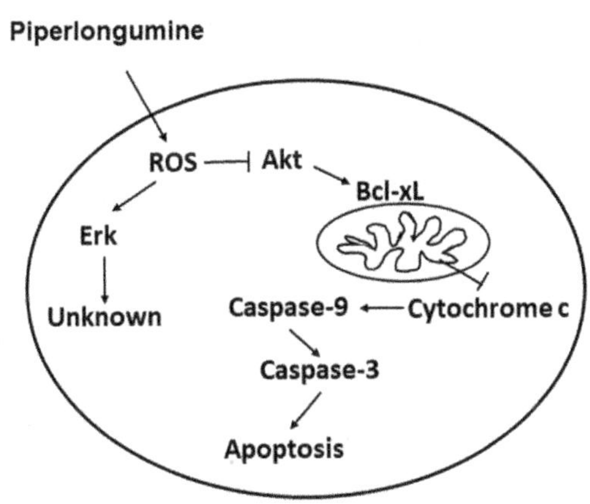

Figure 9. Illustration of the mechanisms involved in piperlongumine-mediated cellular apoptosis.

Cell proliferation and cell survival are modulated with an intricate network, such as signaling pathways; controlling one or two factors of them may not be enough to manage malignant growth. Because of the increased metabolism in malignant cells, ROS accumulation is common, and can be used as a therapeutic target for malignant cells. In the present study, we demonstrated that PL successfully induces ROS overexpression in TC cells to interfere with cell survival mechanisms and achieve apoptosis induction.

4. Materials and Methods

4.1. Cell lines and Cell Culture

The human multidrug-resistant PTC cell line (IHH-4) and ATC cell lines (KMH-2 and 8505c) were purchased from the Japan Collection of Research Bioresources Cell Bank (JCRB). The FTC cell line (WRO) was provided by Prof. Jen-Der Lin [49]. IHH-4 and KMH-2 cells were cultured with Dulbecco's minimal essential medium (DMEM) + Roswell Park Memorial Institute (RPMI) (1:1) medium (Gibco, Gaithersburg, MD, USA); 8505c cells were maintained with MEM (Gibco); and WRO cells were incubated in RPMI 1640 medium (GIBCO) supplemented with 10% fetal bovine serum (Biological Industries, Kibbutz Beit Haemek, Israel), at 37 °C in a 5% CO_2 incubator. The passage ranges for the cell lines used in this study were p05–p20 post-purchase.

4.2. Cell Viability Assay

The cells (5×10^3 cells/well) were seeded in a 96-well cell culture dish, and the plate containing the previous medium was incubated. After overnight incubation, the cells were administrated with the control medium (containing 0.01% DMSO) or PL (Cayman, MI, USA). After administration for 24 and 48 h, cell survival was detected with a CCK-8 assay kit (Enzo Life Sciences, Farmingdale, NY, USA). Three independent assays were performed.

4.3. Colony Formation Assay

The cells were cultured in 6-well plates (10^3 cells/well), and the plates were cultured at 37 °C with the previous medium. And the cells were administrated with DMSO or PL. After 12 days of incubation, the colony of the cells and the colony morphology were detected under 10% crystal violet (Sigma-Aldrich, St. Louis, MO, USA) staining. The colony size and the colony number were examined.

4.4. Cell Cycle Analysis

Cells (1×10^6 cells/dish) were cultured in a 10 cm culture plate. After overnight cell attachment, the cells were administrated with DMSO or PL with various durations. The process for cell cycle analysis was referred to as previously reported [4,50,51]. Three independent assays were performed.

4.5. Cell Death Analysis

Cells were treated with DMSO or PL for various durations. The process for determining apoptosis with flow cytometry was referred to as previously reported [52]. The protein expressions involved in the mechanisms that PL-mediated apoptosis were examined with Western blots. caspase-3, caspase-8, caspase-9, anti-PARP, and anti-GAPDH antibodies were purchased from Cell Signaling Inc. (Danvers, MA, USA). GAPDH was used as a loading control. Z-VAD-FMK, a pan-caspase inhibitor, was purchased from ApexBio Technology LLC (Houston, TX, USA). The inhibitor of ROS, NAC, was purchased from Cayman. Moreover, the MAPK and Akt signaling pathways, including the total forms and phosphorylated forms of ERK, JNK, p38, Akt, and mTOR (all antibodies were purchased from Cell Signaling), were examined in the cells after PL treatment. PD98058 and perifosine, inhibitors of Erk and Akt, respectively, were purchased from LC Laboratories (Woburn, MA, USA). The plasmid, including $pBSSK^+$ and constitutively active human Akt1 construct (pmAkT), were used as previously reported [50]. Three independent assays were performed.

4.6. Measurement of Cellular ROS

Cells (5×10^5/dish) were treated with DMSO or PL for various durations. The cells were stained with dichlorofluorescin diacetate (DCFH-DA, 10 μM; Sigma-Aldrich) at 37 °C for 30 min; they were then washed three times with ice-cold phosphate-buffered saline (PBS). The fluorescence expressed by cells was measured by flow cytometry (Becton

Dickinson, San Diego, CA, USA). In the ROS blocking experiments, cells were pre-treated with NAC for 2 h.

4.7. Western Blotting

Cells were treated with DMSO or PL for various durations and were lysed with the M-PER^{TM} protein extraction reagent (Thermo Fisher Scientific Inc., Rockford, IL, USA) with a 0.1% protease inhibitor cocktail. The sample containing total protein was loaded, using sodium dodecyl sulfate-polyacrylamide gel electrophoresis gels to separate proteins and then transferred the proteins on polyvinylidene fluoride membranes. The proteins were detected after blocking with a primary antibody and a secondary antibody. The experimental process could refer to our previous paper [53]. All original western blots are included in Figures S2–S75.

4.8. Xenograft Study

Six-week-old female nude mice (BALB/cAnN.Cg-Foxn1nu/CrlNarl) were purchased from the National Laboratory Animal Center. Mice were maintained in the animal facility of National Chiayi University, Chiayi, Taiwan. Because PTC is common in TC patients, we used a multidrug-resistant PTC cell line (IHH-4) to examine the anticancer activity of PL in vivo. IHH-4 cells (2×10^6 cells/mice) were subcutaneously injected into the nude mice in the right flank. When the tumor growth was less than 1 mm^3, the mice were divided into three groups of five animals each, and were then administrated with DMSO or PL. In the experimental group, mice received injections with PL (containing 5 and 10 mg/kg) on days 1, 3, 5, 7, 9, and 11. The mice received an injection of DMSO to be the control group. The tumor size was also measured on these days. The tumor volume was estimated using the following equation: $L \times S^2/2$, where L is the longest diameter, and S is the shortest diameter). After treatment for 11 days, mice were sacrificed, and tumor weights were determined. The experimental protocol complied with Taiwan's Animal Protection Act and was approved by the Laboratory Animal Care and Use Committee of the National Chiayi University (IACUC Approval No. 102021).

4.9. Immunohistochemical Staining and TUNEL Assay of Tissue Sections

The tumor tissues were fixed with formalin (Sigma-Aldrich) at 4 °C for 24 h, and soaked with 100% alcohol for 5 min. Specimens were embedded with paraffin after incubation with xylene. Finally, the specimens were sectioned into 4 μm slices. The specimens were stained with HE (Sigma-Aldrich) for 15 min at room temperature, and the stained specimens were visualized under a microscope. In addition, the specimen slices were stained by the terminal deoxynucleotidyl transferase-mediated nick end labeling (TUNEL) assay with the DeadEnd™ Colorimetric Apoptosis Detection System (Promega, Madison, WI, USA) referring to the manufacturer's instructions. Briefly, the slides were incubated by equilibration buffer for 10 min, and treated with proteinase K for 10 min. The samples were washed with PBS and treated with TdT enzyme at 37 °C for 1 h. The slides were then incubated with horseradish peroxidase-labeled streptavidin, and detected with diaminobenzidine. Finally, the images of the specimens were examined under a microscope.

4.10. Statistical Analysis

Data are shown as the mean ± SD in the experiments. Differences between the control group and test groups were analyzed with a one-way analysis of variance and Fisher's least significant difference test. In the in vivo mice study, the Mann–Whitney *U*-test was used. Statistical significance was defined at $p < 0.05$, < 0.01, or < 0.001 in the statistical tests.

5. Conclusions

Our findings demonstrate that PL exhibits antitumor activity against TC cell lines, including ATC and multiple drug-resistant PTC and FTC cells, by inhibiting cell proliferation, enhancing G2/M phase arrest, and promoting apoptosis. In addition, PL can trigger TC cell death through intrinsic cellular apoptosis by ROS induction and inhibition of downstream Akt signaling (Figure 9). This is the first study to report that TC cell lines are sensitive to PL. The study also investigated the anti-TC mechanisms of PL in detail, which suggested that it is a potential chemotherapeutic agent for TC treatment.

Supplementary Materials: The following are available online at https://www.mdpi.com/article/10.3390/cancers13174266/s1, Figure S1: Modulation of cellular ROS with piperlongumine (PL) treatment in WRO cells., Figures S2–S75: Original western blots.

Author Contributions: F.-P.K., Y.-P.L., and Y.-R.L. designed and performed the research and wrote the paper. Y.-S.Z., S.-H.C., Y.-Z.L., P.-W.Z., and Y.-P.Y. performed the in vitro and in vivo experiments. W.-Y.C., H.-I.Y., T.-S.T., and C.-H.L. contributed to interpreting the results in the manuscript. Y.-R.L. conceived the original idea and contributed to the final version of the manuscript. All authors have read and agreed to the published version of the manuscript.

Funding: This study was supported by the Ministry of Science and Technology, Taiwan, R.O.C. (MOST109-2314-B-705-001 and MOST110-2320-B-039-016-MY3); Kaohsiung Medical University, Kaohsiung, Taiwan (KMU-Q110011A); China Medical University, Taichung, Taiwan (CMU109-MF-92); Chiayi Christian Hospital, Chiayi, Taiwan (R109-024); and Min-Hwei College of Health Care Management, Tainan, Taiwan (MHCHCM109001).

Institutional Review Board Statement: The study was conducted according to the guidelines of the Laboratory Animal Care and Use Committee of the National Chiayi University, and approved by the Laboratory Animal Care and Use Committee of the National Chiayi University (No. 102021).

Informed Consent Statement: Not applicable.

Data Availability Statement: Data is contained within the article or supplementary material.

Acknowledgments: The authors thank the Laboratory Animal Center of National Chiayi University, Chiayi, Taiwan, for providing the facilities and animal care required for this study, and thank LiTzung Biotechnology, Kaohsiung, Taiwan, for providing pathological assistance for this study.

Conflicts of Interest: The authors declare no conflict of interest.

References

1. Hodgson, N.C.; Button, J.; Solorzano, C.C. Thyroid Cancer: Is the Incidence Still Increasing? *Ann. Surg. Oncol.* **2004**, *11*, 1093–1097. [CrossRef]
2. Vanderpump, M.P. The Epidemiology of Thyroid Disease. *Br. Med. Bull.* **2011**, *99*, 39–51. [CrossRef]
3. Kebebew, E.; Greenspan, F.S.; Clark, O.H.; Woeber, K.A.; McMillan, A. Anaplastic Thyroid Carcinoma. Treatment Outcome and Prognostic Factors. *Cancer* **2005**, *103*, 1330–1335. [CrossRef] [PubMed]
4. Hua, S.C.; Chang, T.C.; Chen, H.R.; Lu, C.H.; Liu, Y.W.; Chen, S.H.; Yu, H.I.; Chang, Y.P.; Lee, Y.R. Reversine, a 2,6-Disubstituted Purine, as an Anti-cancer Agent in Differentiated and Undifferentiated Thyroid Cancer Cells. *Pharm. Res.* **2012**, *29*, 1990–2005. [CrossRef] [PubMed]
5. Shaha, A.R. Implications of Prognostic Factors and Risk Groups in the Management of Differentiated Thyroid Cancer. *Laryngoscope* **2004**, *114*, 393–402. [CrossRef] [PubMed]
6. Chatterjee, A.; Dutta, C.P. Alkaloids of Piper longum Linn. I. Structure and Synthesis of Piperlongumine and Piperlonguminine. *Tetrahedron* **1967**, *23*, 1769–1781. [CrossRef]
7. Bezerra, D.P.; Pessoa, C.; de Moraes, M.O.; Saker-Neto, N.; Silveira, E.R.; Costa-Lotufo, L.V. Overview of the Therapeutic Potential of Piplartine (Piperlongumine). *Eur. J. Pharm. Sci.* **2013**, *48*, 453–463. [CrossRef] [PubMed]
8. Han, S.S.; Son, D.J.; Yun, H.; Kamberos, N.L.; Janz, S. Piperlongumine Inhibits Proliferation and Survival of Burkitt Lymphoma In Vitro. *Leuk. Res.* **2013**, *37*, 146–154. [CrossRef]
9. Song, X.; Gao, T.; Lei, Q.; Zhang, L.; Yao, Y.; Xiong, J. Piperlongumine Induces Apoptosis in Human Melanoma Cells Via Reactive Oxygen Species Mediated Mitochondria Disruption. *Nutr. Cancer* **2018**, *70*, 502–511. [CrossRef]
10. Liu, J.M.; Pan, F.; Li, L.; Liu, Q.R.; Chen, Y.; Xiong, X.X.; Cheng, K.; Yu, S.B.; Shi, Z.; Yu, A.C.; et al. Piperlongumine Selectively Kills Glioblastoma Multiforme Cells via Reactive Oxygen Species Accumulation Dependent JNK and p38 Activation. *Biochem. Biophys. Res. Commun.* **2013**, *437*, 87–93. [CrossRef] [PubMed]
11. Chen, S.Y.; Liu, G.H.; Chao, W.Y.; Shi, C.S.; Lin, C.Y.; Lim, Y.P.; Lu, C.H.; Lai, P.Y.; Chen, H.R.; Lee, Y.R. Piperlongumine Suppresses Proliferation of Human Oral Squamous Cell Carcinoma through Cell Cycle Arrest, Apoptosis and Senescence. *Int. J. Mol. Sci.* **2016**, *17*, 616. [CrossRef]

12. Roh, J.L.; Kim, E.H.; Park, J.Y.; Kim, J.W.; Kwon, M.; Lee, B.H. Piperlongumine Selectively Kills Cancer Cells and Increases Cisplatin Antitumor Activity in Head and Neck Cancer. *Oncotarget* **2014**, *5*, 9227–9238. [CrossRef]
13. Zheng, J.; Son, D.J.; Gu, S.M.; Woo, J.R.; Ham, Y.W.; Lee, H.P.; Kim, W.J.; Jung, J.K.; Hong, J.T. Piperlongumine Inhibits Lung Tumor Growth via Inhibition of Nuclear Factor Kappa B signaling Pathway. *Sci. Rep.* **2016**, *6*, 26357. [CrossRef] [PubMed]
14. Lee, H.N.; Jin, H.O.; Park, J.A.; Kim, J.H.; Kim, J.Y.; Kim, B.; Kim, W.; Hong, S.E.; Lee, Y.H.; Chang, Y.H.; et al. Heme Oxygenase-1 Determines the Differential Response of Breast Cancer and Normal Cells to Piperlongumine. *Mol. Cells* **2015**, *38*, 327–335. [CrossRef]
15. Chen, Y.; Liu, J.M.; Xiong, X.X.; Qiu, X.Y.; Pan, F.; Liu, D.; Lan, S.J.; Jin, S.; Yu, S.B.; Chen, X.Q. Piperlongumine Selectively Kills Hepatocellular Carcinoma Cells and Preferentially Inhibits their Invasion via ROS-ER-MAPKs-CHOP. *Oncotarget* **2015**, *6*, 6406–6421. [CrossRef]
16. Thongsom, S.; Suginta, W.; Lee, K.J.; Choe, H.; Talabnin, C. Piperlongumine Induces G2/M Phase Arrest and Apoptosis in Cholangiocarcinoma Cells through the ROS-JNK-ERK Signaling Pathway. *Apoptosis* **2017**, *22*, 1473–1484. [CrossRef] [PubMed]
17. Golovine, K.; Makhov, P.; Naito, S.; Raiyani, H.; Tomaszewski, J.; Mehrazin, R.; Tulin, A.; Kutikov, A.; Uzzo, R.G.; Kolenko, V.M. Piperlongumine and its Analogs Down-regulate Expression of c-Met in Renal Cell Carcinoma. *Cancer Biol. Ther.* **2015**, *16*, 743–749. [CrossRef]
18. Mohammad, J.; Singh, R.R.; Riggle, C.; Haugrud, B.; Abdalla, M.Y.; Reindl, K.M. JNK Inhibition Blocks Piperlongumine-Induced Cell Death and Transcriptional Activation of Heme Oxygenase-1 in Pancreatic Cancer Cells. *Apoptosis* **2019**, *24*, 730–744. [CrossRef] [PubMed]
19. Duan, C.; Zhang, B.; Deng, C.; Cao, Y.; Zhou, F.; Wu, L.; Chen, M.; Shen, S.; Xu, G.; Zhang, S.; et al. Piperlongumine Induces Gastric Cancer Cell Apoptosis and G2/M cell Cycle Arrest both In Vitro and In Vivo. *Tumour Biol.* **2016**, *37*, 10793–10804. [CrossRef]
20. Kumar, S.; Agnihotri, N. Piperlongumine, a Piper Alkaloid Targets Ras/PI3K/Akt/mTOR Signaling Axis to Inhibit Tumor Cell Growth and Proliferation in DMH/DSS Induced Experimental Colon Cancer. *Biomed. Pharmacother.* **2019**, *109*, 1462–1477. [CrossRef]
21. Liu, D.; Qiu, X.Y.; Wu, X.; Hu, D.X.; Li, C.Y.; Yu, S.B.; Pan, F.; Chen, X.Q. Piperlongumine Suppresses Bladder Cancer Invasion via Inhibiting Epithelial Mesenchymal Transition and F-actin Reorganization. *Biochem. Biophys. Res. Commun.* **2017**, *494*, 165–172. [CrossRef] [PubMed]
22. Oblad, R.; Doughty, H.; Lawson, J.; Christensen, M.; Kenealey, J. Application of Mixture Design Response Surface Methodology for Combination Chemotherapy in PC-3 Human Prostate Cancer Cells. *Mol. Pharmacol.* **2018**, *94*, 907–916. [CrossRef] [PubMed]
23. Tripathi, S.K.; Biswal, B.K. Piperlongumine, a Potent Anticancer Phytotherapeutic: Perspectives on Contemporary Status and Future Possibilities as an Anticancer Agent. *Pharmacol. Res.* **2020**, *156*, 104772. [CrossRef] [PubMed]
24. Li, W.; Wen, C.; Bai, H.; Wang, X.; Zhang, X.; Huang, L.; Yang, X.; Iwamoto, A.; Liu, H. JNK Signaling Pathway is Involved in Piperlongumine-mediated Apoptosis in Human Colorectal Cancer HCT116 Cells. *Onco.l Lett.* **2015**, *10*, 709–715. [CrossRef]
25. Bezerra, D.P.; Castro, F.O.; Alves, A.P.; Pessoa, C.; Moraes, M.O.; Silveira, E.R.; Lima, M.A.; Elmiro, F.J.; Costa-Lotufo, L.V. In Vivo Growth-inhibition of Sarcoma 180 by Piplartine and Piperine, Two Alkaloid Amides from Piper. *Braz. J. Med. Biol. Res.* **2006**, *39*, 801–807. [CrossRef]
26. Bezerra, D.P.; de Castro, F.O.; Alves, A.P.; Pessoa, C.; de Moraes, M.O.; Silveira, E.R.; Lima, M.A.; Elmiro, F.J.; de Alencar, N.M.; Mesquita, R.O.; et al. In Vitro and In Vivo Antitumor Effect of 5-FU Combined with Piplartine and Piperine. *J. Appl Toxicol.* **2008**, *28*, 156–163. [CrossRef]
27. Mohler, H.; Pfirrmann, R.W.; Frei, K. Redox-directed Cancer Therapeutics: Taurolidine and Piperlongumine as Broadly Effective Antineoplastic Agents (Review). *Int. J. Oncol.* **2014**, *45*, 1329–1336. [CrossRef]
28. Son, Y.; Cheong, Y.K.; Kim, N.H.; Chung, H.T.; Kang, D.G.; Pae, H.O. Mitogen-Activated Protein Kinases and Reactive Oxygen Species: How Can ROS Activate MAPK Pathways? *J. Signal. Transduct.* **2011**, *2011*, 792639. [CrossRef]
29. Zhao, Y.; Hu, X.; Liu, Y.; Dong, S.; Wen, Z.; He, W.; Zhang, S.; Huang, Q.; Shi, M. ROS Signaling under Metabolic Stress: Cross-Talk between AMPK and AKT Pathway. *Mol. Cancer* **2017**, *16*, 79. [CrossRef] [PubMed]
30. McArthur, K.; Kile, B.T. Apoptotic Caspases: Multiple or Mistaken Identities? *Trends Cell Biol.* **2018**, *28*, 475–493. [CrossRef]
31. Stassi, G.; Todaro, M.; Zerilli, M.; Ricci-Vitiani, L.; Di Liberto, D.; Patti, M.; Florena, A.; Di Gaudio, F.; Di Gesu, G.; De Maria, R. Thyroid Cancer Resistance to Chemotherapeutic Drugs Via Autocrine Production of Interleukin-4 and Interleukin-10. *Cancer Res.* **2003**, *63*, 6784–6790. [PubMed]
32. Bezerra, D.P.; Pessoa, C.; Moraes, M.O.; Alencar, N.M.; Mesquita, R.O.; Lima, M.W.; Alves, A.P.; Pessoa, O.D.; Chaves, J.H.; Silveira, E.R.; et al. In Vivo Growth Inhibition of Sarcoma 180 by Piperlonguminine, an Alkaloid Amide from the Piper Species. *J. Appl. Toxicol.* **2008**, *28*, 599–607. [CrossRef] [PubMed]
33. Liu, Y.; Li, Q.Y.; Wang, Y.P.; Liu, Y.M.; Liu, B.; Liu, M.M.; Liu, B.M. Spectroscopic Investigation of the Anticancer Alkaloid Piperlongumine Binding to Human Serum Albumin from the Viewpoint of Drug Delivery. *Luminescence* **2018**, *33*, 305–311. [CrossRef] [PubMed]
34. Fallahi, P.; Mazzi, V.; Vita, R.; Ferrari, S.M.; Materazzi, G.; Galleri, D.; Benvenga, S.; Miccoli, P.; Antonelli, A. New Therapies for Dedifferentiated Papillary Thyroid Cancer. *Int. J. Mol. Sci.* **2015**, *16*, 6153–6182. [CrossRef]
35. Pistollato, F.; Masias, M.; Agudo, P.; Giampieri, F.; Battino, M. Effects of Phytochemicals on Thyroid Function and Their Possible Role in Thyroid Disease. *Ann. N. Y. Acad. Sci.* **2019**, *1443*, 3–19. [CrossRef] [PubMed]

36. Xiao, Q.; Park, Y.; Hollenbeck, A.R.; Kitahara, C.M. Dietary Flavonoid Intake and Thyroid Cancer Risk in the NIH-AARP Diet and Health Study. *Cancer Epidemiol. Biomarkers Prev.* **2014**, *23*, 1102–1108. [CrossRef]
37. Wang, F.; Mao, Y.; You, Q.; Hua, D.; Cai, D. Piperlongumine Induces Apoptosis and Autophagy in Human Lung Cancer Cells through Inhibition of PI3K/Akt/mTOR Pathway. *Int. J. Immunopathol. Pharmacol.* **2015**, *28*, 362–373. [CrossRef]
38. Piska, K.; Gunia-Krzyzak, A.; Koczurkiewicz, P.; Wojcik-Pszczola, K.; Pekala, E. Piperlongumine (Piplartine) as a Lead Compound for Anticancer Agents-Synthesis and Properties of Analogues: A Mini-review. *Eur. J. Med. Chem.* **2018**, *156*, 13–20. [CrossRef] [PubMed]
39. Liu, Y.; Chang, Y.; Yang, C.; Sang, Z.; Yang, T.; Ang, W.; Ye, W.; Wei, Y.; Gong, C.; Luo, Y. Biodegradable Nanoassemblies of Piperlongumine Display Enhanced Anti-angiogenesis and Anti-tumor Activities. *Nanoscale* **2014**, *6*, 4325–4337. [CrossRef]
40. Piska, K.; Koczurkiewicz, P.; Wnuk, D.; Karnas, E.; Bucki, A.; Wojcik-Pszczola, K.; Jamrozik, M.; Michalik, M.; Kolaczkowski, M.; Pekala, E. Synergistic anticancer activity of doxorubicin and piperlongumine on DU-145 Prostate Cancer Cells-The Involvement of Carbonyl Reductase 1 Inhibition. *Chem. Biol. Interact.* **2019**, *300*, 40–48. [CrossRef]
41. Yao, J.X.; Yao, Z.F.; Li, Z.F.; Liu, Y.B. Radio-sensitization by Piper Longumine of Human Breast Adenoma MDA-MB-231 Cells In Vitro. *Asian Pac. J. Cancer Prev.* **2014**, *15*, 3211–3217. [CrossRef] [PubMed]
42. Cao, P.; Xia, Y.; He, W.; Zhang, T.; Hong, L.; Zheng, P.; Shen, X.; Liang, G.; Cui, R.; Zou, P. Enhancement of Oxaliplatin-induced Colon Cancer Cell Apoptosis by Alantolactone, a Natural Product Inducer of ROS. *Int. J. Biol. Sci.* **2019**, *15*, 1676–1684. [CrossRef] [PubMed]
43. Circu, M.L.; Aw, T.Y. Reactive Oxygen Species, Cellular Redox Systems, and Apoptosis. *Free Radic. Biol. Med.* **2010**, *48*, 749–762. [CrossRef] [PubMed]
44. Nogueira, V.; Hay, N. Molecular Pathways: Reactive Oxygen Species Homeostasis in Cancer Cells and Implications for Cancer Therapy. *Clin. Cancer Res.* **2013**, *19*, 4309–4314. [CrossRef] [PubMed]
45. Xiong, X.X.; Liu, J.M.; Qiu, X.Y.; Pan, F.; Yu, S.B.; Chen, X.Q. Piperlongumine Induces Apoptotic and Autophagic Death of the Primary Myeloid Leukemia Cells from Patients via Activation of ROS-p38/JNK Pathways. *Acta. Pharmacol. Sin.* **2015**, *36*, 362–374. [CrossRef]
46. Liu, J.; Wang, Z. Increased Oxidative Stress as a Selective Anticancer Therapy. *Oxid. Med. Cell Longev.* **2015**, *2015*, 294303. [CrossRef]
47. van Staveren, W.C.; Solis, D.Y.; Hebrant, A.; Detours, V.; Dumont, J.E.; Maenhaut, C. Human Cancer Cell Lines: Experimental Models for Cancer Cells in Situ? For Cancer Stem Cells? *Biochim. Biophys. Acta.* **2009**, *1795*, 92–103. [CrossRef]
48. Chintakuntlawar, A.V.; Foote, R.L.; Kasperbauer, J.L.; Bible, K.C. Diagnosis and Management of Anaplastic Thyroid Cancer. *Endocrinol. Metab. Clin. North. Am.* **2019**, *48*, 269–284. [CrossRef]
49. Lu, C.H.; Liu, Y.W.; Hua, S.C.; Yu, H.I.; Chang, Y.P.; Lee, Y.R. Autophagy Induction of Reversine on Human Follicular Thyroid Cancer Cells. *Biomed. Pharmacother.* **2012**, *66*, 642–647. [CrossRef]
50. Lee, Y.R.; Chen, S.H.; Lin, C.Y.; Chao, W.Y.; Lim, Y.P.; Yu, H.I.; Lu, C.H. In Vitro Antitumor Activity of Aloperine on Human Thyroid Cancer Cells through Caspase-Dependent Apoptosis. *Int. J. Mol. Sci.* **2018**, *19*, 312. [CrossRef]
51. Lu, C.H.; Chen, S.H.; Chang, Y.S.; Liu, Y.W.; Wu, J.Y.; Lim, Y.P.; Yu, H.I.; Lee, Y.R. Honokiol, a Potential Therapeutic Agent, Induces Cell Cycle Arrest and Program Cell Death In Vitro and In Vivo in Human Thyroid Cancer Cells. *Pharmacol. Res.* **2017**, *115*, 288–298. [CrossRef] [PubMed]
52. Chang, J.M.; Kam, K.H.; Chao, W.Y.; Zhao, P.W.; Chen, S.H.; Chung, H.C.; Li, Y.Z.; Wu, J.Y.; Lee, Y.R. Berberine Derivatives Suppress Cellular Proliferation and Tumorigenesis In Vitro in Human Non-Small-Cell Lung Cancer Cells. *Int. J. Mol. Sci.* **2020**, *21*, 4218. [CrossRef] [PubMed]
53. Huang, K.J.; Kuo, C.H.; Chen, S.H.; Lin, C.Y.; Lee, Y.R. Honokiol Inhibits In Vitro and In Vivo Growth of Oral Squamous Cell Carcinoma through Induction of Apoptosis, Cell Cycle Arrest and Autophagy. *J. Cell Mol. Med.* **2018**, *22*, 1894–1908. [CrossRef] [PubMed]

cancers

MDPI

Article

Intratumoral (Poly-ICLC) Therapy for Dogs with Advanced Cancers: First Report on Clinical Effectiveness, Quality of Life, and Adverse Events

Alessandra Rinah Nogueira Voges [1,†], Rodrigo Ubukata [1,†], Karina Velloso Braga Yazbek [2], Otávia Luisa Caballero [3], Andres Mario Salazar [4], Cristina de Oliveira Massoco [5] and Maria Lucia Zaidan Dagli [5,*]

1 PROVET, Veterinary Oncology Hospital, São Paulo 04081-004, Brazil; alessandra.rinah@yahoo.com.br (A.R.N.V.); ubukata@gmail.com (R.U.)
2 All Care Vet, São Paulo 04077-003, Brazil; kayazbek@yahoo.com.br
3 Orygen Biotecnologia, Ltd.a, São Paulo 04544-150, Brazil; ocaballero@orygen.com.br
4 Oncovir, Inc., Washington, DC 20008, USA; asalazar@oncovir.com
5 Department of Pathology, School of Veterinary Medicine and Animal Science, University of São Paulo, São Paulo 05508-270, Brazil; cmassoco@usp.br
* Correspondence: mlzdagli@usp.br; Tel.: +55-11-3091-7712
† Current Professional Address: E+ Especialidades Veterinárias, São Paulo 04078-013, Brazil.

Simple Summary: Although polyinosinic-polycytidylic acid-poly-l-lysine carboxymethylcellulose (poly-ICLC) is widely used as a standalone agent for treating human cancers, there are no reports on its use for treating canine cancers. We aimed to investigate the clinical efficacy, quality of life, and adverse events of poly-ICLC treatment in dogs with advanced cancers. Our results showed that intratumoral poly-ICLC therapy was well tolerated in dogs with advanced cancers, with clinical benefit and improved quality of life scores observed in some dogs. Our findings suggested that patients with lower tumor burden may benefit more from this treatment.

Abstract: Polyinosinic-polycytidylic acid-poly-l-lysine carboxymethylcellulose (poly-ICLC) is a synthetic double-stranded viral RNA analog widely tested as a component of human therapeutic cancer vaccines and as a standalone agent for treating human cancers. However, there are no reports on the use of poly-ICLC for treating canine cancers. This study aimed to investigate the clinical efficacy, quality of life (QL), and adverse events of poly-ICLC treatment in dogs with advanced cancers. The treatment protocol consisted of weekly intratumoral doses of poly-ICLC. The canine patients underwent clinical, laboratory, and imaging tests, and their owners answered weekly QL questionnaires. Fourteen canine patients with different types of spontaneous advanced tumors were enrolled. Most dogs had received prior conventional therapies. Five dogs received at least 12 doses of poly-ICLC: the injected tumor was stable in three dogs, there was a partial response in one, and the injected tumor significantly enlarged in the other. The QL scoring remained stable or increased in most cases. Mild adverse events related to poly-ICLC were observed in 10 of the 14 patients. The data showed that intratumoral poly-ICLC therapy was well tolerated in dogs with advanced cancers, with clinical benefit and improved QL scores observed in some dogs.

Keywords: cancer; dog; therapy; quality of life; viral RNA

Citation: Voges, A.R.N.; Ubukata, R.; Yazbek, K.V.B.; Caballero, O.L.; Salazar, A.M.; Massoco, C.d.O.; Dagli, M.L.Z. Intratumoral (Poly-ICLC) Therapy for Dogs with Advanced Cancers: First Report on Clinical Effectiveness, Quality of Life, and Adverse Events. *Cancers* **2021**, *13*, 2237. https://doi.org/10.3390/cancers13092237

Academic Editors: Felisbina Queiroga and Bruno Cogliati

Received: 9 March 2021
Accepted: 31 March 2021
Published: 7 May 2021

Publisher's Note: MDPI stays neutral with regard to jurisdictional claims in published maps and institutional affiliations.

Copyright: © 2021 by the authors. Licensee MDPI, Basel, Switzerland. This article is an open access article distributed under the terms and conditions of the Creative Commons Attribution (CC BY) license (https://creativecommons.org/licenses/by/4.0/).

1. Introduction

Poly-ICLC (polyinosinic-polycytidyl acid, stabilized with poly-L-lysine and carboxymethylcellulose), or Hiltonol® is a synthetic double-stranded viral RNA analog that mimics a danger signal by acting as a pattern recognition receptor agonist [1,2]. Poly-ICLC was initially developed as an interferon inducer, but the current evidence indicates broader biological effects, including specific antitumor and antiviral activities [3,4]. Although Poly-IC and Poly-ICLC have similar effects, as both are recognized by the cytosolic RNA helicase MDA-5 and by TLR3, in order to improve Poly-IC stability to endogenous RNAses in humans, a modified version complexed with poly-Lysine and carboxymethyl cellulose (poly-ICLC) was developed and used in innumerous clinical trials [1,5].

Poly-ICLC activates multiple elements of innate and adaptive immunity, including induction of interferons (IFNs), cytokines and proinflammatory chemokines, maturation of dendritic cells, natural killer cell cytotoxicity, and specific T cell responses, through Toll-like receptor 3 (TLR3) and melanoma differentiation-associated protein 5 (MDA5) [6–8]. The exact interaction between dsRNA, IFN, and IFN-inducible systems has not been fully elucidated, but the role of dsRNAs, such as Poly-ICLC may be bimodal: starting with the induction of IFN-related genes and the expression of dsRNA-dependent systems, such as 2′5′OAS, PKR, TLR3, RIG I, MDA5, and probably others, followed by their activation by dsRNA [3,8,9].

Poly-ICLC has been widely tested as a component of therapeutic vaccines for human cancers, including melanoma, gastrointestinal carcinoma, squamous cell carcinoma, gliomas, and multiple myelomas [10–14]. In addition, it has been experimentally employed as a standalone agent for treating established tumors [15–21].

Cancer is a growing cause of death in canines worldwide [22,23]. Although the prevalence rates of different types of cancer vary, it is estimated that up to 1 in 4 dogs will develop cancer at some point in their life and that almost 50% of dogs over 10 years of age will die from cancer-related problems [23]. The choice of therapeutic protocols for treating canine cancers depends on the histological type and stage of the disease, the general condition of the animal, and the owner's finances, since many treatments can be very expensive. Standard approaches such as surgery, chemotherapy, and radiotherapy are ineffective in many cases.

The concept of human cancer treatment has changed in recent years, with the immune response's modulation against tumor cells proving to be an effective therapeutic strategy in an increasing number of cancer types [24–27]. The field of veterinary cancer immunotherapy has also advanced in the past few decades, but with few approved therapies [28,29]. Nevertheless, cancer immunotherapies are expected to expand the availability of veterinary oncology treatment options in the future.

Poly-ICLC has been investigated in a series of clinical studies in humans, alone or as an adjuvant in cancer vaccines (10–20). As a standalone therapy, when applied intratumorally, the tumor itself is the antigen source, resulting in "autovaccination". These studies have indicated the clinical safety of using this product in humans, with evidence of efficacy against cancer. Therefore, we investigated whether poly-ICLC treatment could be a therapeutic alternative for canine cancer patients. To our knowledge, there are no previous reports on the use of poly-ICLC for treating canine cancers.

This study aimed to evaluate the clinical efficacy, safety, and quality of life that resulted from intratumoral injections of Poly-ICLC in dogs with advanced cancers.

2. Materials and Methods

2.1. Patient Eligibility

All the owners of the dogs included in this study signed informed consent forms.

Canine patients with histologically confirmed unresectable tumors of the following types: carcinomas, adenocarcinomas, sarcomas, and lymphomas were considered for the trial. Patients were required to present at least one site with an accessible primary tumor that could be easily injected with poly-ICLC. The lesion had to measure at least 10 mm in the longest diameter and the patient was required to present acceptable renal, hepatic and hematological functions. Owners had to be willing to comply with all the study requirements.

Patients were excluded if they presented, in addition to the tumor, clinically significant unstable medical conditions that could compromise the patient's safety, severe hepatic impairment, defined as alanine aminotransferase or aspartate aminotransferase > 3 times the upper limit of normal (ULN), total bilirubin > 2 times ULN, or kidney disease classified by the IRIS classification higher than grade II.

2.2. Treatment Plan

The study was conducted at the PROVET–Veterinary Oncology Hospital and ALL CARE Vet Hospital in São Paulo, Brazil. As there are no previous reports in the literature of intratumoral Poly-ICLC applications in dogs and taking into account a previous publication of lethality of an intravenous (IV) dose of 1 mg of Poly-ICLC/kg of body weight in 1 of 3 adult dogs [30], the decision was made to use approximately 1/10 of the IV dose for safety reasons. In human clinical trials in which Poly-ICLC was used intratumorally, the frequency of application ranged from one to three times a week [18,20]. We proposed to administer one injection per week to increase compliance with the treatment. The treatment plan consisted of weekly intratumoral injections of poly-ICLC for 12 weeks. Poly-ICLC was administered according to the animal's weight as follows: animals from 2–5 kg received 0.25 mg, from 5–10 kg received 0.5 mg, and >10 kg received 1 mg. Half of the respective dose was administered in the first week of treatment. If well tolerated, the complete dose was administered between the second and twelfth weeks.

Poly-ICLC was injected at one or more points within a single target lesion throughout the treatment. Analgesics using tramadol hydrochloride and dipyrone were allowed and used as necessary. The use of corticosteroids was avoided as much as possible, unless the inflammatory response was intense and the veterinarian deemed it necessary. No other antitumor therapies were used during the study period. Upon completing the 12-week treatment period, all canine participants continued receiving poly-ICLC on a compassionate basis for as long as the veterinarian judged that the patient benefited from the treatment.

2.3. Evaluation Schedule

During screening for inclusion in the study, the canine patients underwent clinical and laboratory evaluations (hematological, blood biochemistry, and urinalysis), a lesion biopsy, tumor staging by ultrasound, radiography, and computed tomography, and immune-inflammatory evaluation (immunophenotyping, complete blood count-based inflammatory score, and immunological evaluation of the tumor biopsy through immunohistochemistry tests). During all visits, immediately prior to the poly-ICLC injection, the patients underwent clinical evaluation and physical examination, including caliper measurement of tumor diameter. In addition, the owners completed the weekly QL questionnaire.

Following the sixth week of treatment and up to ten days after the end of the experimental treatment, the imaging exams and the laboratory evaluations were repeated. A tumor biopsy was taken for histological and immunological evaluation after week 12 of the treatment.

2.4. Disease Assessment—Response to Therapy

The response to therapy was determined for the injected (target) lesions and noninjected (or nontarget) lesions. The tumor responses were defined as stable disease (SD), partial response (PR), or complete response (CR) according to the canine response evaluation criteria in solid tumors v1.0 [31]. The tumors were assessed using the one-dimensional measurement of the injected lesion with a caliper. In addition, noninjected lesions were assessed using a computed tomography scan at week 12 of the treatment.

Adverse events were classified according to veterinary cooperative oncology group—common terminology criteria for adverse events (VCOG, 2016) [32]. Toxicity was fully evaluated throughout the study duration and included blood hematologic and biochemical panels, including hematologic, liver, and kidney function.

2.5. Quality of Life Scoring

The QL assessment was performed for all dogs during poly-ICLC treatment and consisted of a questionnaire answered by the dog's owner. The questionnaire consisted of

12 questions with a choice of four answers [33]. Responses to each question ranged from zero to three points, resulting in up to 36 points. A score of zero was considered the worst QL, and 36 was the best. The questions addressed behavior, interaction with the owner, assessment of pain, appetite, sleep disturbances, and signs of vomiting, diarrhea, urinary incontinence, or repletion.

3. Results

3.1. Patients' Characteristics

The characteristics of the canine patients are presented in Table 1. Nine females and five males from different breeds (7 Mongrels, 1 Cocker Spaniel, 1 Labrador, 1 Teckel, 1 Maltese, 1 Shih-Tsu, and 1 Bulldog) aged 3–16 years and presenting different types of spontaneous tumors were enrolled in the study.

Table 1. Characteristics of the canine patients.

Dog	Breed	Age (Years)	Gender	Primary Tumor	Location	TNM	WHO Stage	Metastasis	Prior Therapies
1	Mongrel	16	Male *	STS	Jaw	T3N1M1	IV	Olfactory bulb	Surgery, chemotherapy (4 Dx)
2	Mongrel	3	Male *	STS	Thorax	T4N1M1	IV	LN, lungs	None
3	Cocker spaniel	13	Female *	STS	Abdominal region	T4N1M1	IV	LN, lungs, liver, spleen	Surgery, ECT
4	Mongrel	8	Female *	STS	Mandible	T3N1M0	III	LN	Surgery, chemotherapy (5 Dx)
5	Mongrel	16	Female *	STS	Thorax	T4N1M1	IV	Lungs, liver	None
6	Labrador	11	Female	Histiocytic sarcoma	Thorax	T3N1M1	IV	LN, lungs, skin, liver, spleen, intestinal serosa, myocardium, esophagus	None
7	Teckel	12	Male *	Carcinoma	Dorsal to the left eye	T2N0M0	II	none	Radiotherapy
8	Maltese	13	Female *	Carcinoma	Left perianal region	T3N1M0	III	LN	Chemotherapy) (RTKi)
9	Mongrel	15	Female *	Adenocarcinoma	Breast, inguinal region and inner face of the right pelvic limb	T4N2M1	IV	LN, skin, liver	Surgery, chemotherapy (4 Dx, 1 Cb)
10	Mongrel	12	Female *	Adenocarcinoma	Paravertebral	T3N2M1	IV	Subcutaneous, lungs	None
11	Shih-tzu	15	Female *	Adenocarcinoma	Breast, vulva	T3N1M1	IV	LN, liver, skin	Surgery, chemotherapy (4 Cb)
12	Mongrel	9	Female *	Adenocarcinoma	Breast	T4N2M1	IV	LN, lungs, skin, liver, spleen	Surgery, chemotherapy (5 Dx, 6 Cb)
13	Bulldog	11	Male *	Multicentric NHL	Lymphoid tissue	stage IV	IV	Liver, Spleen	Chemotherapy (Vc, Cc, Dx, Pr)
14	Mongrel	10	Male *	Multicentric NHL	Lymphoid tissue	stage IV	IV	Liver, Spleen	Chemotherapy (1 Vc, 1 Cc, 1 Dx)

[1] STS, soft tissue sarcoma; NHL, non-Hodgkin's lymphoma; LN, lymph node; TNM (Tumor, Lymph node, and Metastasis); WHO, World Health Organization. ECT, eletrochemotherapy; Dx, doxorubicin; Cb, carboplatin; Vc, Vincristine; Cc, cyclophosphamide; RTKi, toceranib (receptor tyrosine kinase inhibitor). * Castrated/Spayed.

There were five cases of soft tissue sarcomas, one histiocytic sarcoma, two carcinomas, four adenocarcinomas, and two multicentric non-Hodgkin lymphomas. Most cases were in the World Health Organization (WHO) stage IV (11/14 cases), 2/14 cases were in stage III, and 1/14 cases was stage II. Most tumors (13/14) had metastasized, mostly to the lymph nodes, lungs, liver, and spleen. Prior therapies included surgery (6/14), chemotherapy (8/14), radiotherapy (1/14), and electrochemotherapy (1/14). Two of the 14 cases had no prior treatment.

3.2. Response to Therapy

The target lesions' responses, injected with Poly-ICLC, are presented in Table 2 and Figure 1.

Table 2. Response to treatment of target lesions injected with poly-ICLC. Dogs were separated according to the number of weeks of treatment.

Dog	Treatment Duration (Weeks)	Dose of poly-ICLC (mg)	BOR	Survival (Days)	Survival after poly-ICLC (Days)	Concurrent Disease	Status
4	30	1.00	PD	411	259	None	alive
7	22	0.50	SD	346	206	MVI	alive
8	31	0.25	PR	520	252	CKD, MVI, TVI	dead (kidney and heart failure)
9	21	1.00	SD	470	181	None	Dead (euthanasia)
11	21	0.50	SD	2108	178	MVI, TVI	alive
10	8	1.00	SD	123	133	None	dead (euthanasia)
6	6	1.00	PD	133	112	None	dead (euthanasia)
5	3	0.50	NE	555	77	MVI	alive
13	3	1.00	PD	218	24	MVI, TVI	Dead (respiratory failure)
1	2	0.50	NE	607	46	MVI, TVI	Dead (seizure)
14	2	1.00	PD	100	24	None	Dead (respiratory failure)
2	1	0.25	PD	127	13	None	Dead (probable cause?)
12	1	0.25	PD	443	6	MVI	Dead (euthanasia)

BOR, Best Overall Response; SD, stable disease; PD, progressive disease; CKD, chronic kidney disease; MVI, mitral valve insufficiency; TVI, tricuspid valve insufficiency; NE, Not Evaluable. According to RECIST (Response Evaluation Criteria in Solid Tumours), a requirement for SD is that it should be met at least once no less than 6–8 weeks after the first dose of trial treatment/baseline assessment, otherwise the best response will be Not Evaluable (NE).

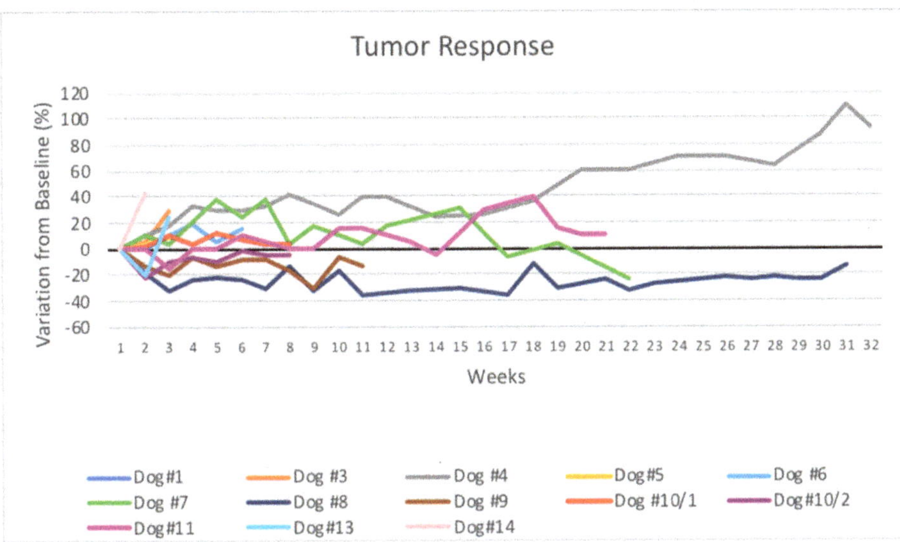

Figure 1. Spider plot of tumor burden changes during intratumoral Poly-ICLC therapy in 12 patients. The longest diameters of the injected lesions are demonstrated as changes from baseline of one lesion from each dog, except from Dog #10 that had two lesions injected with Poly-ICLC.

In the canine patients that completed the course of treatment of 12 weeks or more, 3/5 exhibited less than a 20% increase in the size of the target lesion (SD), 1/5 exhibited at least a 30% reduction in tumor diameter (PR), and one presented more than a 20% increase in tumor size (PD).

Three of the five animals were alive 21 weeks after the start of poly-ICLC treatment, with just one being euthanized because of eventual progressive disease and another animal died due to a preexisting cardiac condition. Thus, in most canine patients that received the full proposed treatment course, a prolonged stabilization of the injected tumor was achieved. An adenocarcinoma patient that left the study after eight weeks for reasons unrelated to the disease (dog 10) exhibited stability of the injected lesion confirmed by CT measurement at week nine. The majority of cases received less than the full course of treatment due to either progressive disease or other issues that precluded continuation in the study. The three patients that presented stages II and III (dogs 4, 7, and 8) had the best outcomes in the study. The disease remained stable in patient 4 for 12 weeks before progressing. Patient 7 is still being treated on a compassionate basis, and although the tumor increased in size in the first month of treatment, it remained practically stable for the following five weeks before starting to decrease. Patient 8 presented PR in the injected lesion and died of diseases unrelated to the tumor 31 weeks after starting the treatment.

The remaining patients that were treated with poly-ICLC for six weeks or less had progressive disease (6/8 cases). In these cases, the treatment was generally unable to restrict tumor growth, possibly due to its more advanced stage at the start of treatment. Notably, the initial size of the injected tumors of those patients that were unable to complete the full course of treatment were, in all cases except dog 13, more than twice as large as the tumors of the patients that were able to complete the treatment. The baseline mean of the size of the injected tumors in the dogs that received the full course of treatment was 3.64 cm, whereas in the dogs that received six or fewer injections, the mean baseline tumor size was 9.86 cm.

For quantifying the effects of poly-ICLC on nontarget lesions, five patients that had at least two measurements of the lesions using computed tomography were analyzed (Table 3). Most metastatic lesions increased in size, and new metastases were identified in three of the five cases. Overall, the nontarget lesions progressed in four patients, whereas one patient presented SD. Thus, even in the cases where there was an apparent control of the target tumor, there was no abscopal effect.

Table 3. Response to treatment in nontarget lesions, not injected with poly-ICLC.

Dog	Treatment Duration (Weeks)	Baseline Tumor Burden Nontarget Lesions	Final Tumor Burden Nontarget Lesions	BOR Nontarget Lesions	BOR Target and Nontarget Lesions
4	30	LN: 1 enlarged (3.8 cm)	LN: 1 enlarged (3.9 cm)	SD	SD
8	31	LN: 2 enlarged (4.6 cm and 1.9 cm)	LN: 3 enlarged (4.5, 2.1, 1.4 cm)	PD	PD
9	21	Liver: 2 nodules (4.6 cm and 2.9 cm)	Liver: 2 nodules (4.7 and 3,3 cm) + another Spleen: 2 nodules (0.9 and 2.2 cm)	PD	PD
10	6	Lungs: unquantifiable nodules	Lungs: unquantifiable nodules	PD	PD
11	21	Liver: 1 nodule (5.1 cm) LN: 3 enlarged (1.7, 0.9, 0.5 cm)	Liver: 4 nodules (7.0, 1.0, 0.3, 0.7 cm) LNs: 4 enlarged (2.1, 1.0, 1.5, 1.2 cm) Stomach: 1 nodule (1.2 cm)	PD	PD

BOR, best overall response; SD, stable disease; PD, progressive disease; LN, lymph nodes.

3.3. *Quality of Life Scoring*

The variation of the QL scores in relation to the baseline measurements of canine patients that received one, two, or more poly-ICLC injections are presented in Figure 2. The QL score remained stable during the treatment for most dogs. The only dog that presented partial tumor response (dog 8) presented an increase in QL scores throughout the treatment period that varied from 8% to 37% in relation to the baseline measurement. Before starting the treatment, this dog had difficulties defecating, probably due to the perianal lesion and lymph node enlargements, which resolved after three doses of poly-ICLC. In this patient, analgesics could be reduced and were removed from the treatment protocol after nine weeks of treatment with poly-ICLC.

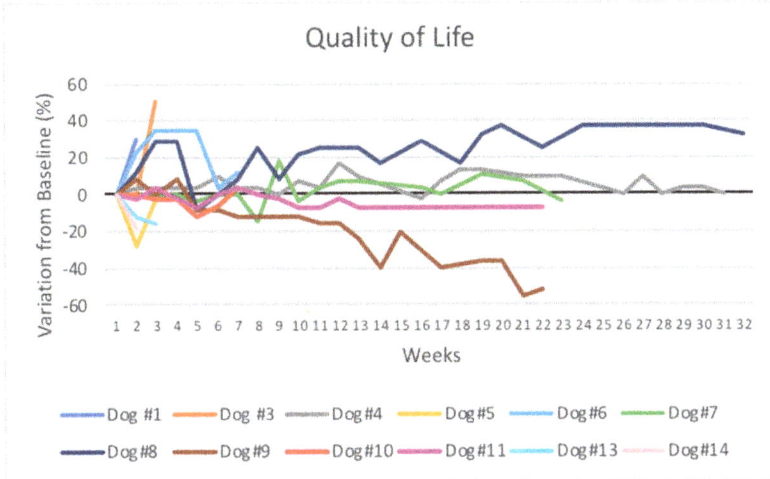

Figure 2. Spider plot of quality of life scores during intratumoral poly-ICLC therapy in 12 patients. The quality of life scores are demonstrated as changes from baseline.

One patient (dog 9) presented a marked decrease in QL scores that reached 56% at week 21. This patient presented a very large subcutaneous lesion that evolved with an important inflammation of the subcutaneous tissue of the caudal and ventral abdominal region and edema of the hind paw. Treatment with corticosteroids decreased the inflammation, but the disease progressed with many nodules in the inguinal and the inner face of the right pelvic limb. The patient was euthanized 181 days after the initiation of treatment due to progressive disease and marked QL decrease.

A combination chart with quality of life and tumor response is shown in Figure 3. There was no obvious correlation between measured tumor response and quality of life.

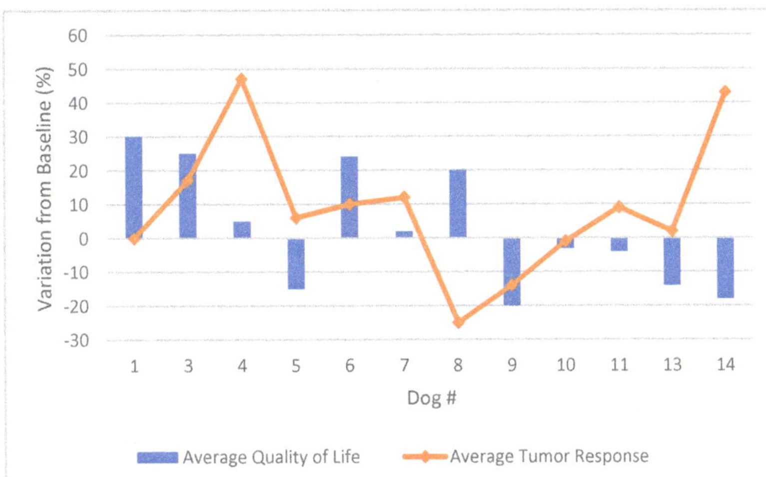

Figure 3. Combination chart of the average change from baseline QL scores and tumor response scores during intratumoral poly-ICLC therapy in 12 patients.

3.4. Adverse Events

The adverse events related and unrelated to poly-ICLC treatment are presented in Table 4. All 14 dogs that received at least one intratumoral dose of poly-ICLC were included in this assessment. Adverse events related to poly-ICLC were mild and included grade 1 lethargy and fatigue in half of the dogs. Allergic/hypersensitivity reactions were observed in three dogs: one had a grade 1 reaction, and two had grade 2 reactions. There was no discontinuation of the treatment due to adverse events.

Table 4. Adverse events related and not related to poly-ICLC treatment.

Adverse Events	Grade 1	Grade 2	Grade 3	Grade 4	Grade 5	TOTAL
Related to poly-ICLC						
Lethargy/fatigue	7/14	0	0	0	0	7/14
Allergic reaction/ hypersensitivity	1/14	2/14	0	0	0	3/14
Not related to poly-ICLC						
Skin ulceration	0	2/14	2/14	0	0	4/14
Seizure	0	0	0	1/14	1/14	2/14
Dyspnea	0	0	0	0	3/14	3/14
Increased alkaline phosphatase	1/6	2/6	2/6	0	0	5/6
Increased ALT	0	2/6	2/6	0	0	4/6
Increased BUN	2/6	1/6	0	0	0	3/6
Increased creatinine	1/6	0	0	0	0	1/6
Decreased hemoglobin	5/6	1/6	0	0	0	6/6
Cystitis	1/6	0	0	0	0	1/6

4. Discussion

This study represents the first report on the use of poly-ICLC for treating canine cancers, in which a strategy of therapeutic intratumoral injections of the dsRNA viral mimic and TLR agonist, poly-ICLC was tested. This strategy was well tolerated and generated clinical benefit in some patients.

There is considerable interest in developing immunotherapies for canine cancers [34–36]. Poly-ICLC represents a highly practical potential option for this aim. Poly-ICLC, a synthetic complex of carboxymethylcellulose, polyinosinic-polycytidylic acid, and poly-L-lysine, can stimulate cytotoxic cytokine release and increase the tumoricidal activities of various immunohematopoietic cells by inducing interferon-gamma production [18]. This phenomenon has been extensively investigated in humans as a standalone therapy or combined with vaccine antigens.

Salazar and colleagues [15] first investigated in 1996 the use of poly-ICLC for the long-term treatment of human cancers at a dose similar to that used in our study. This was a pilot study for the treatment of malignant gliomas and showed prolonged quality of survival with tumor stabilization or regression. Several trials have been conducted since then, including poly-ICLC as the therapeutic immunostimulant for different types of human cancers (10–20).

The present trial involved 14 dogs, of which five received the planned 12 weekly doses of poly-ICLC. Three of these dogs presented stable disease, and one showed partial response. Notably, the quality of life of the patients receiving long-term poly-ICLC remained stable or improved throughout the planned treatment.

The owners reported important improvements in the quality of life of their dogs. Considering the advanced neoplasm of these animals, we consider the achievement of at least three months of stable disease with a good quality of life to be clinically relevant, although randomized placebo control studies would be required to confirm this clinical effect. The best outcome shown in the three patients that presented less advanced disease (WHO stages II and III) may signify that patients with lower tumor burden may receive increased benefit from this treatment.

Nine animals received from one to six doses of poly-ICLC due to their poor clinical condition in most cases. All nine patients were in WHO stage IV and had metastasis.

All 14 dogs included in our study were evaluated for adverse events. Regardless of the type of cancer, the WHO stage, and number of doses received, the adverse events of intratumoral poly-ICLC injections were mild and consisted only of lethargy/fatigue similar to that described in human studies [18]. Low-grade allergy/hypersensitivity reactions were observed in two animals.

Given the advanced stage of the canine patients' tumors included in our study and the multiple mechanisms for immune suppression in oncologic patients, using a single immune stimulator may not be enough for inducing robust anticancer immunity. Recently, Kyi et al. [18] tested poly-ICLC in patients with recurrent metastatic disease therapy (head and neck squamous cell cancer and melanoma) in which prior systemic treatment had failed. Patients received two intratumoral treatment cycles, followed by intramuscular boosters biweekly for seven weeks, with a one-week rest period. One patient completed two IT treatment cycles and achieved clinical benefit (SD and progression-free survival of six months). Poly-ICLC was well tolerated in patients with solid cancer and generated local and systemic immune responses, as evident in the patient achieving clinical benefit. These results were similar to those obtained in our study.

In a recent study [37], it was shown that intratumoral injection of poly-ICLC was significantly less effective in inducing tumor T cell infiltration and controlling growth of tumors in mice as compared with systemic (intravenous or intramuscular) administration. Systemically administered Poly-ICLC resulted in the enhancement of T cell infiltrates into solid mouse tumors and correlated with a substantial therapeutic effect. Intramuscular or combined IT and IM routes Poly-ICLC treatment may increase the therapeutic benefit in these patients.

5. Conclusions

The goal of any trial to test new cancer therapies is to show that its use can improve the patient's condition without causing adverse events. Overall, it can be concluded from our initial study that poly-ICLC is well tolerated in dogs with different types of advanced cancers, exhibits clinical efficacy in injected lesions and improves the quality of life in some cases. This justifies further trials of poly-ICLC for verifying its benefits for dogs with specific cancers and the inclusion of intramuscular application to achieve an enhanced systemic response.

Author Contributions: Conceptualization, O.L.C., A.M.S., K.V.B.Y., C.d.O.M. and M.L.Z.D.; methodology, A.R.N.V., R.U., K.V.B.Y., C.d.O.M.; software, O.L.C., M.L.Z.D.; validation, O.L.C. and M.L.Z.D.; formal analysis, O.L.C., M.L.Z.D. investigation, A.R.N.V., R.U., K.V.B.Y., C.d.O.M.; resources, O.L.C.; data curation, A.R.N.V., R.U., K.V.B.Y., C.d.O.M.; writing—original draft preparation, M.L.Z.D.; writing—review and editing, O.L.C., A.M.S.; visualization, A.R.N.V.; supervision, O.L.C., C.d.O.M., M.L.Z.D.; project administration, C.d.O.M.; funding acquisition, O.L.C. All authors have read and agreed to the published version of the manuscript.

Funding: This project was sponsored by "Financiadora de Estudos e Projetos (FINEP): PEI 0069/15" and by Orygen Biotecnologia Ltd.a.

Institutional Review Board Statement: The study was conducted according to the guidelines of the Declaration of Helsinki, and approved by the Ethics Committee on the use of animals of the School of Veterinary Medicine and Animal Science of the University of São Paulo (FMVZ-USP), Brazil (Process Number 4.036.310.817)).

Informed Consent Statement: All the dogs' owners included in this study signed informed consent forms.

Data Availability Statement: Data are available upon request.

Acknowledgments: The authors are grateful to Andrew Simpson for help in design of the trial, preparation and critical review of the manuscript.

Conflicts of Interest: The authors declare no conflict of interest. The funders had no role in the design of the study; in the collection, analyses, or interpretation of data; in the writing of the manuscript, but approved the decision to publish the results.

References

1. Martins, K.A.; Bavari, S.; Salazar, A.M. Vaccine adjuvant uses of poly-IC and derivatives. *Expert Rev. Vaccines* **2015**, *14*, 447–459. [CrossRef] [PubMed]
2. Sultan, H.; Salazar, A.M.; Celis, E. Poly-ICLC, a multi-functional immune modulator for treating cancer. *Semin. Immunol.* **2020**, *49*, 101414. [CrossRef] [PubMed]
3. Sammons, M.L.; Stephen, E.L.; Levy, H.B.; Baron, S.; Hilmas, D.E. Interferon induction in cynomolgus and rhesus monkey after repeated doses of a modified polyriboinosinic-polyribocytidylic acid complex. *Antimicrob. Agents Chemother.* **1977**, *11*, 80–83. [CrossRef] [PubMed]
4. Caskey, M.; Lefebvre, F.; Filali-Mouhim, A.; Cameron, M.J.; Goulet, J.P.; Haddad, E.K.; Breton, G.; Trumpfheller, C.; Pollak, S.; Shimeliovich, I.; et al. Synthetic double-stranded RNA induces innate immune responses similar to a live viral vaccine in humans. *J. Exp. Med.* **2011**, *208*, 2357–2366. [CrossRef]
5. Levy, H.B.; Baer, G.; Baron, S.; Buckler, C.E.; Gibbs, C.J.; Iadarola, M.J.; London, W.T.; Rice, J. A modified polyriboinosinic-polyribocytidylic acid complex that induces interferon in primates. *J. Infect. Dis.* **1975**, *132*, 434–439. [CrossRef]
6. Sultan, H.; Wu, J.; Kumai, T.; Salazar, A.M.; Celis, E. Role of MDA5 and interferon-I in dendritic cells for T cell expansion by antitumor peptide vaccines in mice. *Cancer Immunol. Immunother.* **2018**, *67*, 1091–1103. [CrossRef] [PubMed]
7. Barral, P.M.; Sarkar, D.; Su, Z.; Barber, G.N.; DeSalle, R.; Racaniello, V.R.; Fisher, P.B. Functions of the cytoplasmic RNA sensors Rig-I And MDA-5: Key regulators of innate immunity. *Pharmacol. Ther.* **2009**, *124*, 219–234. [CrossRef]
8. Matsumoto, M.; Seya, T. Tlr3: Interferon induction by double-stranded RNA including poly (I:C). *Adv. Drug Deliv. Rev.* **2008**, *60*, 805–812. [CrossRef]
9. Kato, H.; Takeuchi, O.; Sato, S.; Yoneyama, M.; Yamamoto, M.; Matsui, K.; Uematsu, S.; Jung, A.; Kawai, T.; Ishii, K.J.; et al. Differential roles of MDA5 and RIG-I helicases in the recognition of RNA viruses. *Nature* **2006**, *441*, 101–115. [CrossRef]
10. Takeoka, T.; Nagase, H.; Kurose, K.; Ohue, Y.; Yamasaki, M.; Takiguchi, S.; Sato, E.; Isobe, M.; Kanazawa, T.; Matsumoto, M.; et al. NY-ESO-1 Protein Cancer Vaccine with Poly-ICLC and OK-432: Rapid and Strong Induction of NY-ESO-1-specific Immune Responses by Poly-ICLC. *J. Immunother.* **2017**, *40*, 140–147. [CrossRef] [PubMed]
11. Dillon, P.M.; Petroni, G.R.; Smolkin, M.E.; Brenin, D.R.; Chianese-Bullock, K.A.; Smith, K.T.; Olson, W.C.; Fanous, I.S.; Nail, C.J.; Brenin, C.M.; et al. A pilot study of the immunogenicity of a 9-peptide breast cancer vaccine plus poly-ICLC in early stage breast cancer. *J. Immunother. Cancer* **2017**, *5*, 92. [CrossRef] [PubMed]
12. Hilf, N.; Kuttruff-Coqui, S.; Frenzel, K.; Bukur, V.; Stevanović, S.; Gouttefangeas, C.; Platten, M.; Tabatabai, G.; Dutoit, V.; van der Burg, S.H.; et al. Actively personalized vaccination trial for newly diagnosed glioblastoma. *Nature* **2019**, *565*, 240–245, Erratum in **2019**, *566*, E13. [CrossRef] [PubMed]
13. Migliorini, D.; Dutoit, V.; Allard, M.; Grandjean Hallez, N.; Marinari, E.; Widmer, V.; Philippin, G.; Corlazzoli, F.; Gustave, R.; Kreutzfeldt, M.; et al. Phase I/II trial testing safety and immunogenicity of the multipeptide IMA950/poly-ICLC vaccine in newly diagnosed adult malignant astrocytoma patients. *Neuro Oncol.* **2019**, *21*, 923–933. [CrossRef] [PubMed]
14. Ott, P.A.; Hu, Z.; Keskin, D.B.; Shukla, S.A.; Sun, J.; Bozym, D.J.; Zhang, W.; Luoma, A.; Giobbie-Hurder, A.; Peter, L.; et al. An immunogenic personal neoantigen vaccine for patients with melanoma. *Nature* **2017**, *547*, 217–221, Erratum in **2018**, *555*, 402. [CrossRef]
15. Salazar, A.M.; Hilton, B.L.; Steven, O.; Meir, K.; Barbara, S.; Douglas, B.; Hernando, M.; Norman, M.; Karen, S.; Daniel, D.; et al. Long-term treatment of malignant gliomas with intramuscularly administered polyinosinic-polycytidylic acid stabilized with polylysine and carboxymethylcellulose: An open pilot study. *Neurosurgery* **1996**, *38*, 1096–1103. [CrossRef] [PubMed]
16. Giantonio, B.J.; Hochster, H.; Blum, R.; Wiernik, P.H.; Hudes, G.R.; Kirkwood, J.; Trump, D.; Oken, M.M. Toxicity and response evaluation of the interferon inducer poly ICLC administered at low dose in advanced renal carcinoma and relapsed or refractory lymphoma: A report of two clinical trials of the Eastern Cooperative Oncology Group. *Invest. New Drugs.* **2001**, *19*, 89–92. [CrossRef]
17. Butowski, N.; Chang, S.M.; Junck, L.; DeAngelis, L.M.; Abrey, L.; Fink, K.; Cloughesy, T.; Lamborn, K.R.; Salazar, A.M.; Prados, M.D. A phase II clinical trial of poly-ICLC with radiation for adult patients with newly diagnosed supratentorial glioblastoma: A North American Brain Tumor Consortium (NABTC01-05). *J. Neurooncol.* **2009**, *91*, 175–182. [CrossRef]
18. Kyi, C.; Roudko, I.; Sabado, R.; Saenger, Y.; Loging, W.; Mandeli, J.; Thin, T.H.; Lehrer, D.; Donovan, M.; Posner, M.; et al. Therapeutic immune modulation against solid cancers with intratumoral poly-ICLC: A pilot trial. *Clin. Cancer Res.* **2018**, *24*, 4937–4948. [CrossRef]
19. de la Torre, A.N.; Contractor, S.; Castaneda, I.; Cathcart, C.S.; Razdan, D.; Klyde, D.; Kisza, P.; Gonzales, S.F.; Salazar, A.M. A Phase I trial using local regional treatment, nonlethal irradiation, intratumoral and systemic polyinosinic-polycytidylic acid polylysine carboxymethylcellulose to treat liver cancer: In search of the abscopal effect. *J. Hepatocell. Carcinoma.* **2017**, *4*, 111–121. [CrossRef] [PubMed]
20. Hammerich, L.; Marron, T.U.; Upadhyay, R.; Svensson-Arvelund, J.; Dhainaut, M.; Hussein, S.; Zhan, Y.; Ostrowski, D.; Yellin, M.; Marsh, H.; et al. Systemic clinical tumor regressions and potentiation of PD1 blockade with in situ vaccination. *Nat. Med.* **2019**, *25*, 814–824. [CrossRef]
21. Hartman, L.L.; Crawford, J.R.; Makale, M.T.; Milburn, M.; Joshi, S.; Salazar, A.M.; Hasenauer, B.; VandenBerg, S.R.; MacDonald, T.J.; Durden, D.L. Pediatric phase II trials of poly-ICLC in the management of newly diagnosed and recurrent brain tumors. *J. Pediatr. Hematol. Oncol.* **2014**, *36*, 451–457. [CrossRef] [PubMed]

22. Adams, V.J.; Evans, K.M.; Sampson, J.; Wood, J.L.N. Methods and mortality results of a health survey of purebred dogs in the UK. *J. Small Anim. Pract.* **2010**, *51*, 512–524. [CrossRef] [PubMed]
23. Schiffman, J.D.; Breen, M. Comparative oncology: What dogs and other species can teach us about humans with cancer. *Philos. Trans. R. Soc. Lond. B Biol. Sci.* **2015**, *370*, 20140231. [CrossRef] [PubMed]
24. Mellman, I.; Coukos, G.; Dranoff, G. Cancer immunotherapy comes of age. *Nature* **2011**, *480*, 480–489. [CrossRef] [PubMed]
25. Wolchok, J. Putting the Immunologic Brakes on Cancer. *Cell* **2018**, *175*, 1452–1454. [CrossRef]
26. Ribas, A.; Wolchok, J.D. Cancer immunotherapy using checkpoint blockade. *Science* **2018**, *359*, 1350–1355. [CrossRef]
27. Khalil, D.N.; Smith, E.L.; Brentjens, R.J.; Wolchok, J.D. The future of cancer treatment: Immunomodulation, CARs and combination immunotherapy. *Nat. Rev. Clin. Oncol.* **2016**, *13*, 273–290, Erratum in **2016**, *13*, 394. [CrossRef]
28. Killick, D.R.; Stell, A.J.; Catchpole, B. Immunotherapy for canine cancer—Is it time to go back to the future? *J. Small Anim. Pract.* **2015**, *56*, 229–241. [CrossRef]
29. Bergman, P.J. Cancer Immunotherapies. *Vet. Clin. N. Am. Small Anim. Pract.* **2019**, *49*, 881–902. [CrossRef]
30. Smith, J.S.; Yager, P.A.; Baer, G.M. Minimal interferon induction in dogs, using a modified polyriboinosinic-polyribocytidylic acid complex. *Am. J. Vet. Res.* **1980**, *41*, 1833–1835.
31. Nguyen, S.M.; Thamm, D.H.; Vail, D.M.; London, C.A. Response evaluation criteria for solid tumours in dogs (v1.0): A Veterinary Cooperative Oncology Group (VCOG) consensus document. *Vet. Comp. Oncol.* **2013**, *13*, 176–183. [CrossRef]
32. Veterinary Cooperative Oncology Group-Common Terminology. Criteria for Adverse Events (VCOG-CTCAE) following Chemotherapy or Biological Antineoplastic Therapy in Dogs and Cats V1.1. *Vet. Comp. Oncol.* **2016**, *14*, 417–446. [CrossRef]
33. Yazbek, K.V.; Fantoni, D.T. Validity of a health-related quality-of-life scale for dogs with signs of pain secondary to cancer. *J. Am. Vet. Med. Assoc.* **2005**, *226*, 1354–1358. [CrossRef] [PubMed]
34. Thacker, E.L. Immunomodulators, immunostimulants, and immunotherapies in small animal veterinary medicine. *Vet. Clin. N. Am. Small Anim. Pract.* **2010**, *40*, 473–483. [CrossRef] [PubMed]
35. Regan, D.; Guth, A.; Coy, J.; Dow, S. Cancer immunotherapy in veterinary medicine: Current options and new developments. *Vet. J.* **2016**, *207*, 20–28. [CrossRef] [PubMed]
36. Bergman, P.J. Veterinary oncology immunotherapies. *Vet. Clin. N. Am. Small Anim. Pract.* **2018**, *48*, 257–277. [CrossRef] [PubMed]
37. Sultan, H.; Wu, J.; Fesenkova, V.I.; Fan, A.E.; Addis, D.; Salazar, A.M.; Celis, E. Poly-IC enhances the effectiveness of cancer immunotherapy by promoting T cell tumor infiltration. *J. Immunother. Cancer* **2020**, *8*. [CrossRef]

cancers

MDPI

Review

The Nervous System as a Regulator of Cancer Hallmarks: Insights into Therapeutic Implications

Karla V. Torres-Juárez [1], Felisbina Luisa Queiroga [2,3,4,5,*] and Laura P. Romero-Romero [1,*]

1 Department of Pathology, Faculty of Veterinary Medicine and Zootechnics, National Autonomous University of Mexico (UNAM), CDMX 04510, Mexico
2 Department of Veterinary Sciences, University of Trás-os-Montes and Alto Douro, 5000-801 Vila Real, Portugal
3 Animal and Veterinary Research Centre (CECAV), University of Trás-os-Montes and Alto Douro, 5000-801 Vila Real, Portugal
4 Center for the Study of Animal Sciences, CECA-ICETA, University of Porto, 4099-002 Porto, Portugal
5 Associate Laboratory for Animal and Veterinary Science (AL4AnimalS), 5000-801 Vila Real, Portugal
* Correspondence: fqueirog@utad.pt (F.L.Q.); lromeror@unam.mx (L.P.R.-R.); Tel.: +35-19-1782-6982 (F.L.Q.)

Simple Summary: The nervous system communicates with the whole organism, regulating several physiological pathways. The modification of nerve activity could deregulate the state of cellular and tissue homeostasis which could drive cancer development. This paper provides the current state of knowledge, in an evidence-oriented manner, that the nervous system is able to participate in the carcinogenesis process by inducing biochemical, physiological, and cellular modifications involved in the hallmarks of cancer.

Abstract: The involvement of the nervous system in the development of cancer is controversial. Several authors have shown opinions and conflicting evidence that support the early effect of the nervous system on the carcinogenic process. For about a century, research has not been enough, questions remain open, ideas are not discarded, and although more research is still needed to answer all the questions, there is now enough evidence to support the theories and give hope of finding one more possible form of treatment. It is clear that malignant neoplasms have endogenous characteristics that allow them to establish and progress. Some of these characteristics known as hallmarks of cancer, are damage mechanisms in the pathology but necessary during other physiological processes which show some nerve dependence. The nervous system communicates with the whole organism, regulating physiological processes necessary to respond to external stimuli and for the maintenance of homeostasis. The modification of nerve activity could generate an overload and deregulate the state of cellular and tissue homeostasis; this could drive cancer development. In this review, we will address the issue in an evidence-oriented manner that supports that the nervous system is able to participate in the initial and progressive process of carcinogenesis by inducing biochemical, physiological, and cellular modifications involved in the hallmarks of cancer.

Keywords: nervous system; cancer hallmarks; tumor innervation; neurotumoral communication; immunity neuromodulation

Citation: Torres-Juárez, K.V.; Queiroga, F.L.; Romero-Romero, L.P. The Nervous System as a Regulator of Cancer Hallmarks: Insights into Therapeutic Implications. *Cancers* **2022**, *14*, 4372. https://doi.org/10.3390/cancers14184372

Academic Editor: Guido Rindi

Received: 27 July 2022
Accepted: 6 September 2022
Published: 8 September 2022

Copyright: © 2022 by the authors. Licensee MDPI, Basel, Switzerland. This article is an open access article distributed under the terms and conditions of the Creative Commons Attribution (CC BY) license (https://creativecommons.org/licenses/by/4.0/).

1. A Bit of History

The neurobiology of cancer is not a new concept. For a long time, it was claimed that tumors were devoid of innervation, but in the nineteenth century, Young [1], through histological preparations of impregnation with methylene blue, managed to observe nerve fibers in the parenchyma of different tumor types, such as mammary carcinomas and small round cell sarcomas. At that time, the accepted belief was that the nerves found in neoplastic tissues were there due to their entrapment from their pre-existence in healthy tissue. The transformation of healthy tissue and the accelerated cellular proliferation of the malignant cells would cause the invasion of adjacent tissues and thus confine the nerves within the neoplastic tissue. This being the case, when nerves are deprived of their function and invaded by neoplastic cells, they would eventually atrophy and disappear. In 1933, Ryrie [2] explained different theories about the anatomical and physiological relationship of nerve fibers in a neoplasm. His work highlights the thought of the existence of nerve fibers in tumors, the possibility that these are newly formed, perineural invasion (PI), fibrosis after damage as a result of PI and neural regeneration similar to peripheral damage within the neoplastic tissue. However, he concluded that there is no reason to assume any trophic influence on malignant cells [2]. In 1949, Shapiro published the results of his research regarding the relationship between nerves and tumor cells, and their possible physiological role in mesotheliomas and carcinomas in the anterior chamber of the eye, transplanted in mice and rabbits, respectively. In his conclusion, he supports the suggestion that the variation in the growth rate of tumors may be due to vasomotor changes caused by the alteration in the innervation present in each tumor [3].

On the other hand, during the 1950s, some studies were conducted to show the relationship between psychosomatic problems and the development of the disease. The organism is considered as a whole, so psychosocial problems may be more associated with the cause than with the effect of the disease, which is why it is suggested that the personality structure could play a role in the pathogenesis of cancer in individuals predisposed [4]. Psychological therapy, until then, was little used as a palliative practice in cancer patients and when exploring the benefit of intensive deep psychotherapy, a significant reduction in the size of the tumors presented by the patients was reported, when recognizing and working on traumatic events of their past [5]. On the contrary, it has been seen that patients suffering from cancer and manifesting problems of depression, anxiety, chronic stress, or social isolation, have poor prognoses, which include an accelerated progression of the disease, higher rates of recurrences, and low response to treatments, compared to those patients who do not have these disorders [6,7]. In other studies, when examining the association of social stress with relapse and mortality rates in some hormone-sensitive or lifestyle-related tumors, these have been found to be increased mainly in patients with breast [6], colorectal, pancreatic [8], and gastric tumors [7], a fact that is currently being corroborated in experimental works in vivo [9,10].

Until now, the role of the emotional factors in the carcinogenesis process is not entirely clear, however, the balance seems to tilt towards active participation in the cancer initiation and progression [11,12].

Some authors support that cancer is a disease of cellular differentiation, communication and tissue organization [13,14], and the environmental factor is described as one of the most important elements in the progression stage. The emotional state, in many cases, depends on the environment and has the ability to generate important biological modifications which can cause the prolonged release of molecules that could stimulate the formation of a tumor due to the signals it transmits and the processes it activates [10]. In addition, modifications or interruptions in the interactions between the nervous system and tissues are likely to transform the microenvironment and create the appropriate conditions for tumor development [9].

Over the years, cancer research has taken a reductionist approach, neglecting the complex functioning of a complete organism with all its component parts. Until now, the cancer study model has excluded important mechanisms, such as cellular communication through gap junctions, stem cells, mechanical damage [13,15], dysbiosis [16,17], ion channels [18–20], and the neurobiology of the oncological process.

2. Emotional State, Personality and Cancer

Several studies support that people who present different stress-generating situations, such as a hostile family environment, a compromised economic situation or conditions of neglect or isolation, present accelerated progression of the disease with an ineffective response to treatment [6,21]. On the other hand, in different studies in which patients, in addition to pharmacological treatment, receive group and/or individual emotional therapy, coupled with physical exercise, or another type of social support and in some cases nutritional control, it was possible to observe an evident improvement, obtaining favorable results with a better response to treatments, prolonging disease-free survival, and notably improving quality of life [22,23]. Similarly, it has been proposed that character and personality, which define how we face adversity, could predispose to the development of breast cancer in women [24]. Studies in animals have produced results that support the fact that type of behavior at an early age causes differences in the production of endogenous glucocorticoids associated with the variation in the natural function of the Hypothalamic Pituitary Adrenal (HPA) axis, causing oscillation in the function of the immune system and generating individuals potentially susceptible to specific processes of disease at the end of life [25]. Therefore, if personality and character can predispose to the appearance of diseases due to the variations it causes in the functioning of the immune system, generally causing a suppressive effect, this could be the basis of the individual differences in the disease stages.

3. Participation of the Nervous System in Physiological Proliferative Processes

In support of the concept that the nervous system could play a central role in the pathophysiology of cancer, there is evidence that in healthy tissues, nerves and some of their products regulate various mechanisms deeply involved in the carcinogenesis process, such as: control stem cell proliferation [26,27], angiogenesis [28], cell migration [29], cell translocation [30], cell differentiation [31], recruitment and activation of immune cells [32], and energy metabolism [33]. All of them are necessary mechanisms in proliferative physiological processes such as embryogenesis, healing or epimorphic regeneration. In addition, these normal physiological proliferative processes, are nerve-dependent. When nerves are absent, they alter, delay, or these processes simply do not occur [27,34].

In the case of angiogenesis, it is known that nerves and blood vessels are closely related in structure and organization, both during embryonic development as in response to damage, and both depend on bidirectional signals for self-regulation [35]. In various tumors, it has been possible to identify areas with intratumoral nerve fibers, located by immunohistochemistry [36–39]. These fibers are frequently found on the periphery of tumors, some in the stroma and others in a perivascular manner, however, some blood vessels are devoid of innervation, in our opinion this could cause failure in cellular communication between both elements. Consequently, the angiogenic process would be deregulated and this could partially explain the irregular formation of blood vessels and their heterogeneity within a tumor [40].

4. Evidence of the Involvement of the Nervous System in Cancer

Research on the neurobiology of cancer is so far scant. However, today there is greater interest and acceptance due to the evidence that is now available. For instance, epidemiological studies report a correlation between neuroactive therapies (benzodiazepines, lithium, or tricyclic antidepressants) with a reduced risk of cancer [41]. There is also an inverse correlation between neurodegenerative diseases and cancer. A good example is the case of patients with Parkinson's disease or Alzheimer's disease, where a low risk of cancer was identified, both before and after diagnosis [42].

Cognitive damage is a frequent event in cancer patients, and in some cases, diffuse brain damage has been identified through imaging studies. It is not clear whether this damage can occur as an adverse effect of anti-cancer therapies or neurodegeneration due to oxidative damage, inflammation and/or age effect associated with the oncologic disease [43]. Another interesting finding is the one that has been seen in patients with spinal cord damage and prostate denervation, who have a lower incidence of prostate cancer [44].

Other studies in patients with breast cancer have reported that women who use beta-blockers incidentally for some cardiovascular disease, before and/or during the year of breast cancer diagnosis, show a reduction in progression and mortality or remain free of the disease for a longer period, compared to patients who did not undergo beta-blocker treatment [45–47]. It is relevant to highlight that the use of beta-blockers seems to be more efficient in triple-negative breast tumors, which represents an option to be studied as a therapeutic target for this subtype of breast tumor, which has a highly aggressive biological behavior and so far, and does not show fully efficient treatment options [47].

The parallel benefit of the use of beta-blockers has been observed in complementary treatments in patients with other types of cancer, such as melanoma [48], pancreatic [49] and lung cancer [50]. In addition, it has been experimentally observed that treatments with beta-blockers in conjunction with other therapies, such as chemotherapeutics [51], immunotherapies with αPD-1 [48], or metformin [52], experimentally prevent the tumor growth and its progression. On the contrary, the action of catecholamines, when using adrenergic agonists such as isoproterenol, increases the occurrence of metastases to distant tissues compared to control tissues [9].

The development of metastases, via the hematogenous and/or lymphatic pathways, has long been considered the most relevant mechanism in malignancy and the leading cause of mortality in cancer patients. More recently, the neural pathway of metastasis was described and called perineural invasion (PI). Perineural invasion is considered a strong factor indicative of malignancy and has been associated with the worst prognosis [53].

Infiltration of neural tissues has been observed in various types of tumors, such as pancreatic [54,55], squamous cell carcinomas [56], and colorectal carcinoma [57–59], among others. Although the mechanisms by which this phenomenon occurs are not entirely clear, in vitro studies have shown that nervous and tumor tissue can have mutual tropism and exert chemoattraction, since the nervous tissue releases signaling molecules that guide tumor cells towards them [60,61].

5. Tumor Innervation

The process of forming new nerve fibers is called axonogenesis. The mechanisms by which tumors become innervated are still to be clarified, but for this new formation to occur, the confluence of various stimuli that include growth factors, neurotrophins, cytocines, and axonal guide molecules is necessary, resulting in the differentiation of neural precursors and their development or elongation. Various factors related to the development and activation of the nervous system have been found to be highly expressed in tumors, such as pro-NGF, NGF, BDNF, NT, VEGF, FGF, TGFs, semaphorins, and netrins [41,62–64]. These neurotrophic factors could be the inducers of the new formation of nerve fibers within tumor tissues. It is known that they can be produced by the same tumor cells and/or inflammatory cells, exert autocrine or paracrine stimulation, and activate signaling pathways through their action on TrkA and p75NTR receptors. The activation of these receptors leads to the stimulation of cell proliferation and cell survival [65].

The possible ability of tumors to stimulate their own innervation as occurs with angiogenesis and lymphangiogenesis, has been defined as neoneurogenesis [64,66]. Although the presence of nerve fibers has been described in various types of tumors, neoformation has only been described in prostate cancer [67]. Hyperinervation occurs in pancreatic tumors that could be partially dependent on β2ADR signaling [49]. Interestingly, a recent study has explained the presence of prostate tumor cells that presented expression patterns similar to those found in the subventricular zone (SVZ) and when isolated they were able to proliferate and differentiate into neurons, thus suggesting that the adrenergic neoformations observed in tumors may originate in situ from progenitor cells that could reach the tumor from neurogenic areas of the brain [68]. However, other substances could influence the modification in the number of nerve fibers present in tumors, for example, we can find physiological nerve plasticity in reproductive organs due to variations in circulating hormonal concentrations [69], therefore, it could be possible that some tumors stimulated by hormones such as breast tumors, may suffer nerve plasticity due to the presence of hormones in their microenvironment.

6. Nerve Diversity in Varied Tumor Types

One question about tumor innervation is, if it is similar to the innervation in the primary tissue where it originated, or if the tumor innervation is totally independent and established according to the needs of the new neoplastic tissue.

The prostate and pancreas are highly innervated organs, and their function responds to a large extent to nervous stimuli, likewise, the colon presents high nerve density and it is in primary tumors of these organs that the innervation of neoplastic tissues has been studied the most, seeking their presence and effect [49,67].

Perivascular denervation of sensory fibers (NPY, VIP, SP) and adrenergic fibers of blood vessels present in the stroma and the submucosa adjacent to the tumor have been observed in colon cancer studies. The greatest loss occurred in TH^+ and NPY^+ fibers from tumors that showed a higher degree of malignancy, which raises potential temporary modifications [39]. It is suggested that nerve degeneration is the result of the release of tumor factors and that it could lead to bowel dysfunction [39,70]. In a study with a large number of samples (n = 236) using tissue array slides immunostained for protein gene product 9.5 (PGP9.5) to identify nerve tissue, Albo et al. [71] observed PGP9.5+ nerves in the stroma of colon adenocarcinomas in 67% of the cases, being 11% considered as having high nerve density (>20 nerves per hpf). Moreover, rectal tumors showed a high degree of innervation (73%), of which 20% showed high nerve density. Patients with high nerve density had a lower 5-year survival rate in comparison with those who did not show the presence of nerves (40% vs. 86%) and high recurrence and median survival of 2.8 years vs. 9.7 years. For this reason, nerve density was considered a powerful independent prognostic factor, of greater relevance than observing positive lymph nodes [71].

In the case of breast tumors, at first, the presence of nerve fibers was reported only at the junction between healthy tissue and neoplastic tissue, without evidence of perivascular innervation, and associated with well-differentiated tumors [36]. Twenty years later, in another study, nerve fibers were observed in the tumor stroma and around blood vessels; high nerve density was then associated with tumors with a higher degree of malignancy [37]. Although both studies present contradictory results, they cannot be considered determining or misleading, since reports and information on the matter are still scarce.

In a recent study, the association between perineural invasion (PNI) and locoregional recurrence (LRR) accounting for age, tumor size, nodal involvement, estrogen receptor, progesterone receptor, HER2 status, histologic tumor grade, presence of lymphovascular invasion (LVI), and receipt of chemotherapy and/or radiation among a broadly representative cohort of breast cancer patients was evaluated [72]. Their analyses revealed that PNI represents a significant risk factor associated with LRR following the definitive treatment of invasive breast cancer, and must be considered among other clinicopathologic features, such as young age, lymphovascular invasion, and high-grade tumor [72].

The presence of peripheral innervation has been found in other types of tumors, as well as the presence of neurotrophic factors that could benefit them, such is the case of ameloblastomas [73], papillary thyroid carcinoma [74], and among others, some benign tumors such as cutaneous neurofibromas, where a greater number of nerve fibers was found in the evaluated neoplastic tissues, compared to the control tissue [75]. For the most part, the abundance of nerves in newly formed tissues is associated with malignancy and the nerve phenotypes mainly reported correspond to adrenergic innervation, although it is important to note that there is nerve variation in the different histological classifications of primary tumors of the same cell line.

7. Neurotumoral Communication

The interaction of neural elements with tumor cells is defined as neuro-neoplastic synapses. The nerves in the tumor parenchyma, through the release of neurotransmitters, can interact with membrane receptors on tumor cells and other types of cells present in the stroma, for example, endothelial cells, leukocytes, and fibroblasts [76]. Neurotumoral synapses can also occur in CNS tumors, such as gliomas. Gliomas form neurotumoral synapses through a presynaptic neuron and a postsynaptic tumor cell, and these synapses can be as functional as synapses formed by neurons. Additionally, the presence of AMPA receptors (AMPARs) in glioma cells of the postsynaptic region was also reported [77]. AMPARs are ionotropic glutamate receptors that are expressed throughout the CNS and regulate most of the fast-excitatory transmission. The changes that occur at AMPARs synapses lead to different forms of synaptic plasticity that trigger some of the important mechanisms involved in learning and memory formation [78].

Neurotumoral communication could occur through two pathways, the humoral and the neural, with an important function of the peripheral autonomic and sensory nerves [79]. Although it is mentioned that the hematogenous pathway does not turn out to be the main provider for the delivery of neural stimuli towards a tumor [80], the transport of stem cells and neural progenitors from the subventricular area of the brain, capable of migrating and colonizing selectively a tumor and metastasis sites, occurs through blood vessels. These progenitor cells can then differentiate into adrenergic neurons [65], and therefore, the innervation of a tumor can occur from within itself. On the other hand, some neurotransmitters may not originate from nerve cells and be synthesized in situ from other stromal cells, such as lymphocytes, macrophages, and neutrophils, since these have the necessary machinery for synthesis, release, and inactivation of catecholamines, in fact, the phagocytic system is considered a diffuse adrenergic organ [81].

The release of neurotransmitters in tumors can occur as a consequence of a multidirectional communication between the NS, the immune system, and tumor cells [80]. For example, cytokines released by inflammatory and/or tumor cells could cross the blood-brain barrier or transmit signals to the brain through afferent nerves, inducing structural and functional changes in the brain [82], which could trigger a homeostatic imbalance and influence the development of a neoplastic process. In this way, the central nervous system could not only promote the progression of the disease, but also monitor the processes of carcinogenesis initiation. Peripheral innervation would also intervene in an important way in this communication, this being the means for the stimuli to reach their destination, in addition, peripheral innervation controls various cellular phenomena similar to those observed in tumors, so it would not be strange that innervation assumed a central role in the pro and/or anti-carcinogenic mechanisms in the different stages of the disease (Figure 1).

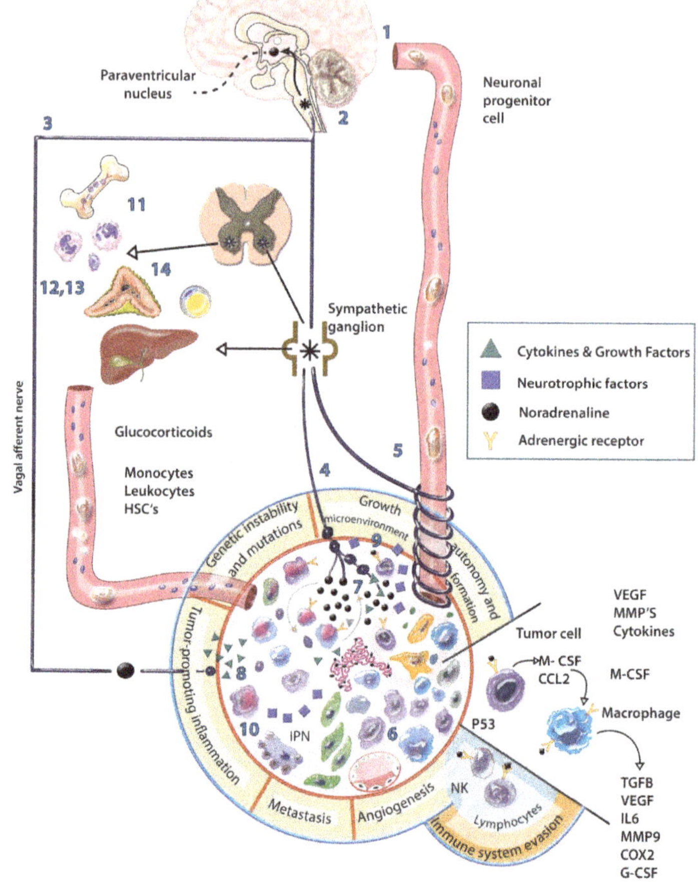

Figure 1. Physiological and neural modulation of the tumor. The communication of the nervous system (NS) with the tumor can be direct, through the hematogenous (1) or neural route (2), or indirect through the stimulation of other organs (3); this communication could be bidirectional. Nerve fibers can be present in the periphery of the tumor, perivascularly, even in the stroma (4), although they are generally scarce, the fibers present dilations that could amplify the signal several nanometers away, activating membrane receptors of various cell types present in the stroma. The direct influence of the NS on the tumor involves: the migration of progenitor cells through the blood from the paraventricular nucleus (PVN) to the tumor (5), to subsequently differentiate into nerve cells; increased infiltration by macrophages and MDSC cells in the tumor (6); the release of neural products in the tumor (7); the increased expression of COX-2, CSF-1, TGFb, MMP9, VEGF, VCAM1 and the decrease in IFNb by adrenergic activation of macrophages within the tumor (8); the release of neurotrophic factors such as NGF, BDNF and FGF by tumor cells that can induce neurite outgrowth (9) and perineural invasion (10). Indirectly, adrenergic stimulation is capable of: mobilizing bone marrow cells for subsequent recruitment into the tumor stroma (11); decreasing the mitotic activity and differentiation of lymphocytes and the proliferation and cytotoxicity of NK cells (12); promoting the trafficking of phagocytic cells (13); mobilization of energy from adipose tissue and liver (14) for later use by tumor cells. In addition, according to the physiology of a healthy individual, the increase in circulating glucocorticoids induced by stress causes a decrease in insulin levels, which results in an increase in circulating glucose generating a protumoral effect. * Sympathetic neurons.

8. Neuromodulation of Immunity

Both the Nervous System and the Immune System share soluble mediators and receptors in common, for example, leukocytes have receptors for neurotransmitters and neurons have Toll-like receptors (TLR), this is how communication occurs between systems and the brain can detect when an inflammatory process is occurring. In general, we know that the SNS promotes the humoral immune response and inhibits the cellular immune response by decreasing the cytotoxic function of T lymphocytes and NK cells [83]. The general conception of the functioning of the SNS and the PNS is that they have antagonistic effects, but in fact, they work together. During physiological processes, both exert coordinated and dependent functions on one another for the maintenance of homeostasis [84].

In the inflammatory process, sensory nerves pick up inflammatory signals during the innate immune response and release neurotransmitters and neuropeptides such as SP, and CGRP in response to damage. The release of pro-inflammatory mediators is set off and the inflammatory response is triggered, which includes vasodilation, vascular permeability, and pain. During this process, signals produced can be felt by the CNS, through the uptake of cytokines, prostaglandins, and damage molecules that can cross the blood-brain barrier, or by receiving an afferent signal captured by the vagus nerve [84].

Subsequently, through an efferent signal, the peripheral ganglia release acetylcholine to stimulate peripheral nerves that release noradrenaline, this adrenergic activity in T lymphocytes causes them to release acetylcholine and by cholinergic activation, the production of TNFα, IL-1β and IL-12 is modulated in macrophages [83]. T lymphocytes will now produce INFγ, achieving an anti-inflammatory effect [84] for the modulation of the inflammatory process, bringing the tissue to a homeostatic balance. Some populations of NK cells also have the ability to synthesize acetylcholine, either by neural stimulation or by an inflammatory process, overexpressing genes related to cytokines and chemokines involved in the regulation of the immune response and chemotaxis during the inflammatory process, and underexpress genes related to cytotoxicity. The local effect of positive NK cholinecetyltransferase (ChAT+) cells is the decrease in the migration of infiltrating monocytes/macrophages and the presence of pro-inflammatory cytokines [83]. Another anti-inflammatory mechanism, which is regulated by the vagus nerve, is the release of dopamine in the adrenal glands; this effect has been observed in experimental sepsis models stimulated with electroacupuncture, which manages to modulate the production of catecholamines [85].

Adrenergic stimuli, on the other hand, influence the migration of immune cells locally in the initial phase of inflammation, regulated by α-receptors. In the late stages, adrenergic signals have an anti-inflammatory impact and are regulated by β-receptors. Noradrenaline has anti-inflammatory effects at high concentrations and, conversely, at low concentrations, it exerts a pro-inflammatory effect [85].

Both macrophages and precursor monocytes express β-adrenergic receptors. The stimulation of these receptors in macrophages produces the synthesis of different cytokines such as IL-6, IL-10, and TNFα. It also modulates the organization and dynamics of actin and the deformation of macrophages to induce phagocytosis and migration [86] (Figure 1).

9. Intrinsic Properties of Malignant Tumors and Their Mechanisms Related to the Nervous System

In 2000, Hanahan and Weinberg published the first hallmarks of cancer. In 2011, the same authors added another four elements to the model, which already considered the interactions of cancer cells with the stroma and recognized the latter as an active participant in the process of tumorigenesis. Today, 14 essential characteristics have been described that add up to the formation and progression of cancer [87–89]. The nervous system, directly or indirectly, can influence some of these necessary elements and other factors that have been found relevant in the development of cancer, for example, the modulation of stem cells, the stimulation of ion channels or the alteration of the microbiota [90]. Below, we describe how the nervous system could influence the main mechanisms involved in the formation and malignancy of neoplastic tissues (Figure 2).

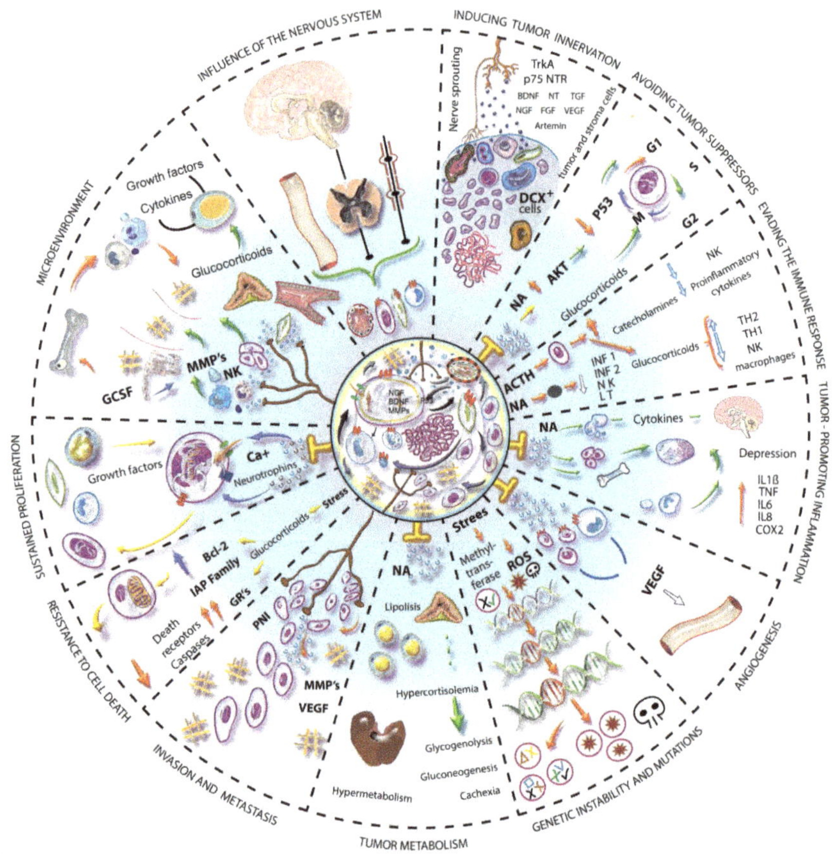

Figure 2. Influence of the nervous system in hallmarks of cancer. The nervous system by its pleiotropic effect can influence different mechanisms involved in the hallmarks of cancer. Communication occurs via hematogenous or neural pathways and stimulates different types of cells inside and outside the tumor. INDUCING TUMOR INNERVATION. Tumor and stromal cells release various neurotrophic factors that induce peripheral nerves to innervate the tumor. Furthermore, DCX^+ cells can differentiate and generate tumor innervations. AVOIDING TUMOR SUPPRESSORS. The adrenergic stimulus causes an increase in cAMP and activation of AKT, which results in the ubiquitination of p53 and its subsequent degradation, which will cause the accumulation of DNA damage and the maintenance of cell division. EVADING THE IMMUNE RESPONSE. Both the HPA axis and the SNA regulate the response to stress, the consequent release of glucocorticoids and catecholamines have an immunosuppressive effect, suppress the production of proinflammatory cytokines and improve the production of anti-inflammatory cytokines. TUMOR-PROMOTING INFLAMMATION. The mobilization of monocytes towards the blood circulation depends on the local adrenergic activity of the bone marrow, in addition, CSF-1 increases its expression by adrenergic activity in macrophages and favors its recruitment in the tumor. Proinflammatory cytokines can activate the HPA axis. Cytokines released by chronic stress promote depression-like behavior by disrupting the synthesis and signal transduction of neurotransmitters. ANGIOGENESIS. Adrenergic stimulation and chronic stress favor the expression of proangiogenic factors and some sympathetic nerve fibers can release NPY. GENETIC INSTABILITY AND MUTATION. Stress-related hormones can cause DNA damage and promote cell transformation through the production of ROS,

expression of DNA damage genes, and decreased DNA repair. The decrease in methyltransferase due to stress favors gene expression and facilitates cell transformation and genotypic diversity. TUMOR METABOLISM. Adrenergic stimulation produces energy mobilization from the liver, increases glycogenolysis and gluconeogenesis in hepatocytes, also inhibits insulin secretion from the pancreas, stimulates glucagon, and also increases lipolysis. These events favor the availability of energy for the tumor. INVASION AND METASTASIS. The modification of the microenvironment and the extracellular matrix, the formation of blood vessels and the presence of M2 macrophages, are important events in the development of metastases that can be induced by adrenergic activation. A perineural invasion is a form of metastasis in which communication occurs between the nerve fiber and tumor cells. RESISTANCE TO CELL DEATH. Glucocorticoids have an antiapoptotic effect and give survival signals to epithelial cells since they favor the expression of BCL-2 and IAP family proteins and prevent cell death. SUSTAINED PROLIFERATION. The neurotrophins that are released by tumor cells stimulate the nerve fibers that correspond to the release of neurotransmitters to activate the receptors of the cells present in the tumor, where the adrenergic stimulus can activate calcium channels and the mobilization of calcium activates IGF receptors that promote proliferation and tumorigenesis. In addition, the expression of neurotrophins by binding to their receptor induces proliferation and survival, an example of this is the binding of BDNF/TrkB. MICROENVIRONMENT. The microenvironment of a tumor is dynamic, resulting from extensive communication between tumor and stromal cells, inducing each other. The microenvironment is as important as the proliferation of tumor cells, it could even be more powerful due to the transformation capacity it has on epithelial cells. The nervous system may participate in the formation of the microenvironment of a tumor in a pleiotropic manner.

9.1. Evasion of Tumor Suppressors, Genetic Instability and Mutation

Stress has been associated with the progression and malignancy of cancer. However, some studies suggest that it can also facilitate genetic instability and mutation being involved in the early stages of the carcinogenic process [91].

The 8-hydroxydeoxyguanosine (8-OH-dG) is a marker of oxidative DNA damage, capable of causing mutagenesis and carcinogenesis, which occurs as a result of exposure to carcinogens and other hazardous materials. By examining the formation of 8-OH-dG in peripheral blood leukocytes in workers of a manufacturing corporation, it was determined that 8-OH-dG levels were associated with the perception of psychological stress and workload, being considered a risk factor for cancer, particularly in women [92].

It is well known that the formation of reactive oxygen species (ROS) causes DNA damage. The prolonged serum elevation of glucocorticoid levels can negatively influence mitochondrial function leading to mitochondrial damage with a negative impact on cellular metabolism and consequent inflammatory reaction [92]. All these observations have been drastically reduced through the administration of diazepam prior to a stressor stimulus and antipsychotic drugs such as olanzapine [90], which reduced the perception of stress in the individuals. This fact supports the idea that stress could be involved in preliminary steps for the initiation of the carcinogenic process by the alteration of cellular metabolism, tissue stress, consequent inflammation, the formation of ROS, and the consequent transformation of the healthy microenvironment in a pro-tumoral microenvironment.

In other studies, the possible effects of stress on tumor initiation have been studied. During the exposure of 3T3 fibroblasts to cortisol and noradrenaline for periods longer than 10 min, it was possible to observe a modification in the expression of genes related to the signaling pathways of DNA damage; also, the repair capacity and the regulation of the cell cycle were negatively affected, the apoptosis of the damaged cells was avoided, which eventually led to the transformation of cells. The effect of exposure to noradrenaline caused greater damage compared to glucocorticoids, damage that managed to be blocked with the use of propranolol or a glucocorticoid receptor antagonist [91]. These mechanisms could certainly contribute to the formation of precancerous lesions.

Another effect of adrenergic signaling that could be significantly involved in the cancer initiation stage is decreased p53 function, which occurs through β-arrestin-1-mediated AKT activation and Mdm2 phosphorylation and activation, leading to the nuclear export, ubiquitination, and degradation of the p53 tumor suppressor. Both, adrenergic stimulation and hypothalamic-pituitary-adrenal HPA axis stimulation can synergistically boost Mdm2 function and decrease p53 levels [93]. Stress also causes the decrease in methyltransferase, which would also affect the repair capacity of DNA [94].

All this suggests that the NS could be involved in chromosomal instability and evasion of tumor suppressors during the initial stages of tumorigenesis and that stress could also participate in the initiation process. For this reason, the emotional state can be considered a risk factor for cancer development and progression [91,93].

9.2. Evasion of the Immune System

The activation of the HPA axis occurs via afferent or by stimulation of inflammatory mediators and counteracts the effects of inflammation by the effect of glucocorticoids released from the adrenal cortex [95]. It is possible that this immunosuppressive activation could participate in the cell transformation process, where it is necessary for cells to escape cytotoxic destruction for the progressive acquisition of malignant characteristics. Glucocorticoids inhibit the availability and function of lymphocytes, NK cells and macrophages; regulate gene transcription and gene expression of mediators, receptors, adhesion molecules, and cytokines. The cytokines from which their synthesis is mainly inhibited are the IL-1, IL-2, IL-3, IL-4, IL-5, IL-6, IL-8, TNFα, IFN-γ, GM-CSF. It is also reported that glucocorticoids can increase the transcription of βARs in different tissues [96–98], which triggers vasoconstriction by adrenergic stimulation.

In various studies, it has been observed that adrenergic stimulation has an immunosuppressive effect, for example, through T cell treatments with noradrenaline, epinephrine, or adrenergic agonists (such as isoproterenol or terbutaline). In vitro inhibition of the differentiation of naïve T lymphocytes to Th1 cells was also observed, by affecting early events involved in the initiation of proliferation, which may be explained in part by decreased IL-2 production by activated T cells. Likewise, the production of INF-γ in Th1 cells is decreased [99]. NK cells also express adrenergic receptors, the adrenergic activation of these cells inhibits their number and cytotoxicity. This fact could be verified in an experimental study by administering propranolol to mice stressed by sleep deprivation, increasing the number of NK cells and their cytotoxicity in vitro with melanoma cells [98]. Adrenergic signaling also promotes the activation and proliferation of myeloid-derived suppressor cells (MDSCs) and it has been proven that the use of propranolol can reverse these effects [100].

9.3. Tumor Associated Inflammation

The inflammatory cells that colonize the tumor microenvironment provide molecules that help maintain and support neoplastic cells. Molecules such as EGF, VEGF, FGF2, metalloproteinases, cytokines, and chymosins, make this particular characteristic an orchestrator of much of the rest of the essential features involved during carcinogenesis [101]. Some tumor cells through oncogenic activation are self-sufficient in the production of pro-inflammatory mediators such as cytokines, chymosins, prostaglandins and chemotactic factors, such as GM-CSF and G-CSF that favor the recruitment of inflammatory cells in the tumor and therefore the progression of solid tumors [102]. The immune system communicates in an important way with the nervous system, and this has the ability to influence pro-inflammatory and anti-inflammatory effects, which can undoubtedly favor the development of tumors.

Macrophages, endothelial cells, smooth muscle cells, and immune cells such as B and Th1 lymphocytes express β-adrenergic receptors. SNS affects clonal expansion, cytokine production, and target cell receptors expression; modifying the balance between the cellular and humoral immune response, increasing or inhibiting the response and mobilization of immunocytes. In macrophages, the response to adrenergic stimulation depends on the receptor that is activated. β-AR stimulation inhibits the activation of Ca^{2+}-dependent K^+ channels while α1-AR stimulation activates these Ca^{2+}-dependent K^+ channels. The mobilization of intracellular Ca^{2+} by adrenergic activation is important for the modulation of the phagocytic activity of macrophages and thus it can be explained that α-AR receptors regulate the improvement of phagocytic activity while β2-AR stimulation suppresses it [99,103]. In malignant tumors, myeloid cells and tumor-associated macrophages develop an immunosuppressed phenotype, which is regulated in part by TGFβ activity, and the stimulation of β-AR decreases the production of IL-6, IL-1β, IL-18, and increase in COX2 and CSF-1, the latter favoring the recruitment of macrophages in the tumor microenvironment [9].

The inflammatory process associated with a tumor participates not only in the recruitment and infiltration of differentiated inflammatory cells into the tumor parenchyma, but also in the mobilization of hematopoietic and progenitor stem cells (HSPCs) from the bone marrow for subsequent recruitment. The egress of HSPCs is regulated by G-CSF, which induces peripheral β-adrenergic signaling and allows the decrease in the chemokine CXCL12 in the bone, this decrease is what allows HSPCs mobilization [26,30]. Damage to hematopoietic regeneration and HSPC mobilization can occur due to sympathetic neuropathy in the bone marrow [95].

In some tumors, it has been possible to observe the joint participation of neural and immune elements. In induced hepatic tumors in rats, it was observed that Kupffer cells stimulated with LPS + noradrenaline significantly increase the expression of IL-6 and TGF-β compared to single induction with LPS. In addition, the blockade of the α1-ARs receptors caused the decrease in the expression of mRNA of CD90 [104], a protein associated with a higher incidence of metastases to distant organs [105], and a decrease in CD133, a marker of cancer progression [106]. Moreover, the activation of Kupffer cells and increase in cytokine levels was observed, which was related to the high density of nerve fibers present in the tumor [107]. So, it is possible that sympathetic innervation promotes carcinogenesis through the regulation of a pro-oncogenic inflammatory microenvironment.

9.4. Angiogenesis

One of the most relevant mechanisms for cancer growth and the one responsible for the formation of one of the main pathways for the development of metastases is angiogenesis. Angiogenesis supports the maintenance and nutrition of tumor cells and helps the formation of the microenvironment by altering the extracellular matrix. Angiogenic factors are released mainly by tumor and endothelial cells; however, some sympathetic nerve fibers may release NPY, which is a potent angiogenic factor [108]. VEGF is the transcendental protein that promotes the proliferation of endothelial cells for the formation of new blood vessels. Chronic stress adrenergic activation induces VEGF expression in endothelial cells and in some types of tumor cells through the ADRβ2-PKA-CREB pathway [109,110]. CREB-induced HDAC2 has also been seen to epigenetically inhibit the antiangiogenic factor TSP1 in prostate cancer models [111]. Multiple cells in the tumor stroma can secrete proangiogenic factors, such is the case of macrophages that, by adrenergic stimulation, can express VEGF, MMP9 [9], and G-CSF, which is also capable of mobilizing endothelial progenitor cells from the bone marrow [102].

Dopamine is one of the various endogenous regulators of angiogenesis which can modulate microvascular hyperpermeability, angiogenesis, and probably tumor growth. Receptors for both dopamine and serotonin have been found in various tumor types such as colon, ovaries, breast, kidney, and pancreas. Dopamine, by occupying the D2 receptors, can inhibit angiogenesis, preventing the phosphorylation of VEGFR-2 from endothelial cells in tumor tissue and blocking the effect of VEGF, generating an antiangiogenic effect [112]. Furthermore, some in vitro studies demonstrated that Dopamine D1 receptor agonists suppress the proliferation of osteosarcoma OS732 cells, through the downregulation of the ERK1/2 and P13K-Akt pathways [113,114]. These data indicate that dopamine may have a protective effect. In malignant tumors, the existence of low concentrations of dopamine has been described. On the other hand, serotonin is associated with angiogenic development in malignant tumors [113].

Perivascular nerve fibers have been described in breast tumors [37], colon [71], pancreas, prostate, ovarian tumors, and melanoma, among others [115]. Vascular malformations have been identified in different tumor types and it was observed that they have smooth muscle around them [40]. From our perspective, it is possible that those vascular malformations could be innervated and provide vasomotor stimuli.

9.5. Invasion and Metastasis

As mentioned above, the nerve fibers associated with a tumor can favor invasion and metastasis through the IPN, stimulate the mesenchymal epithelium transition, the synthesis of metalloproteinases, and confer positive survival signals to tumor cells, these mechanisms provide the morphology required for cell mobilization [110,111].

The immune system in its communication with the NS participates in this process and an example of this is the induction of the expression of pro-metastatic genes and genes characteristic of the M2 phenotype, such as TGFβ, in addition to the reduction in the expression of the IFN geneβ in macrophages located within the tumor microenvironment, in response to stress [9]. Additionally, β2ADR activation has a direct effect on tumor cells because the downstream signaling activates Src, regulated by PKA, and thus improves cell migration and invasion and growth of ovarian tumors [116]. Moreover, the β2ADR-cAMP-PKA pathway is involved in the activation of FAK, a focal adhesion kinase, implicated in cellular motility through the rearrangement of the cytoskeleton, increasing the phosphorylation of paxillin to improve the reassembly of F-actin and the remodeling of the extracellular matrix through the release of MMPs. Increased FAK has been observed in patients with metastatic prostate cancer and high levels of depression [117]. Similarly, there is evidence of reduction in metastases in experimental studies and in patients with various types of cancer, who have received treatment with selective and non-selective adrenergic antagonists, in a single treatment or combined with adjuvants, where, in addition, it has been found that the main regulator of these processes is β2ADR [52,110,118,119], however, it has also been reported that β2ADR agonists may have the same effect [120].

The NS can also be involved in conditions associated with metastasis which occurs during the neoplastic process. Such is the case of the common metastatic lesions to the bone that occur in breast cancer, where the invasion of tumor cells favors bone remodeling, causes osteolytic lesions, and induces outbreaks of sensory nerve fibers in the periosteum, in addition to histological disorganization and consequently bone pain [121]. Similar results were obtained in prostate cancer studies, where, in addition to sensory fibers, an increase in the density and disorganization of adrenergic fibers in bone tissue colonized by tumor cells, was observed [122]. In both studies, treatment with the anti-NGF antibody (mAb911) was administered, which was able to reduce nerve outbreaks and the consequent pain. One of the explanations for why breast cancer cells cause frequent metastases to bone is because sympathetic nerves abundantly innervate bone tissue and release noradrenaline to regulate remodeling and homeostasis. Osteoclasts express βARs and their stimulation causes the expression of RANKL, an osteoclastogenic cytokine with promigratory properties to cancer cells [123]. It should be considered that the lymphoid organs are also mainly innervated by the SNS, and it is where a large part of the invasion by tumor cells occurs, so it would be interesting to know if there is a predilection to areas where prometastatic colonization can be directed by neural influence.

The participation of the NS in cancer metastasis is not contemplated in the mechanisms accepted to date. However, the influence of the NS can occur through several regulatory mechanisms that could explain the frequent metastases to sites with no apparent explanation for their occurrence by the hematogenous route, such is the case of frequent metastases to bone (a tissue highly innervated) in different tumor types [124]. In our opinion, there is a need to consider the inclusion of NS as a regulator of metastasis; however, more research is needed.

10. Conclusions and Perspectives

For many years, cancer has been one of the most studied diseases worldwide; however, to date there is no cure for this clinical condition, so it is possible that some pieces of the puzzle are still missing in the study model. It is necessary to make a change in the approach from which it has been seen for so many years and to be able to take a turn in the understanding of this disease.

As a result of several investigations, the need to expand and generate new knowledge in the cancer field has been widely recognized, but also in other areas that have been discarded for a long time and that could give interesting answers to integrate theories not very well accepted in their beginnings. One example of this is the Tissue Organization Field Theory (TOFT), where it is proposed that carcinogenesis is a relationship problem where reciprocal interactions occur between various cell types of the morpho-genetic field, and it is tissue disorganization and not autonomous uncontrollable cell proliferation that is the causal defect of cancer [125]. It is a reality that cancer is more than a proliferative disease, we have cellular and tissue diversity, a stroma, a microenvironment, diversity of substances, factors and molecules, all functioning at the same time for the benefit of the new tissue. It is also a fact that the nervous system has a pleiotropic effect and that is why it may be able to modulate multiple processes that occur during cancer, which is why further research is required to determine its degree of participation in the different stages and tumor strains.

Because the neurobiology of cancer has so far been little explored, but with solid evidence of its participation in the processes of carcinogenesis, the insertion of the nervous system in the study model is considered of high relevance. The plasticity and nerve remodeling of solid tumors, through various events such as axonogenesis, hyperinnervation or denervation at specific times of their development, activate or suppress the mechanism according to the requirements of the neoplasm, either for its formation or maintenance, similar to how it occurs in organogenesis. It is to be expected that the variation in their formation depends on the stage, grade and organ in which the tumors are located; there may be other factors that determine the timing, type, and density of nerve formation, such as metabolism, or the internal and external environment to which it is exposed. Finally, this information might answer what its biological function is in the process of carcinogenesis and how it could be used for the benefit of patients.

The nervous system represents a promising option for the understanding and, therefore, future treatment of cancer due to the pleiotropic effect it presents since it could attack different mechanisms at the same time. First, it is necessary to know what type of tumors are innervated and which are not, the phenotype of innervation present or absent, the anatomical location, its function or effect to which they are associated, and finally, understand the integration of how the association with tumor cells, their microenvironment and the organism occurs. It is relevant to visualize the possibility of generating future therapeutic targets, prognostic and/or early biomarkers of disease, and understand the possible etiopathogenesis of cancer contemplating all its parts.

With all the data collected in the multiple investigations carried out around the world which have allowed a deepening of the knowledge and understanding of its biology, it can be assumed with certainty that the nervous system has relevant participation in different stages of carcinogenesis by providing survival signals to tumor cells, promoting a favorable microenvironment for cell proliferation, and facilitating the migration of tumor cells, although the mechanisms are not yet fully defined.

Cancer is a multisystem, multistage, multimechanism disease, so the deregulation of different biological-systemic processes that can impact the carcinogenic process must be considered. For the study of the neurobiology of cancer it is important not to forget the complexity of organisms and reject reductionism, contemplating that living beings are complex, multicellular organisms, with multiple internal and external agents influencing them, although there are genes or proteins that alone can generate disease and are importantly involved in the development of various pathologies, including cancer; the extensive complexity of the neoplastic process does not allow any of its parts to be excluded for study, understanding and possible prevention and/or cure. The knowledge we have about the interactions between the nervous system and tumors is still limited, interest and research in this branch of cancer biology are increasing and although there are currently no definitive conclusions, a promising future is envisioned.

Author Contributions: Conceptualization, K.V.T.-J.; writing—original draft preparation, K.V.T.-J. and L.P.R.-R.; writing—review and editing, F.L.Q. and L.P.R.-R. All authors have read and agreed to the published version of the manuscript.

Funding: This work was financed by the Support Program for Technological Research and Innovation Projects of the UNAM (PAPIIT-UNAM) under the project IN221729; and the National Funds (FCT/MCTES, Fundação para a Ciência e a Tecnologia and Ministério da Ciência, Tecnologia e Ensino Superior) under the project UIDB/00211/2020. Authors also want to thank the support received by project UIDB/CVT/00772/2020, from FCT/MCTES.

Acknowledgments: The figures were created by K.V.T.-J. The authors thank Hugo Miranda for the excellent design.

Conflicts of Interest: The authors declare no conflict of interest.

References

1. Young, H.H. On the Presence of Nerves in Tumors and of Other Structures in Them as Revealed by a Modification of Ehrlich's Method of "Vital Staining" with Methylene Blue. *J. Exp. Med.* **1897**, *2*, 1–12. [CrossRef] [PubMed]
2. Ryrie, G.M. On the significance of nerve fibres in human malignant neoplasms. *J. Pathol. Bacteriol.* **1933**, *36*, 13–18. [CrossRef]
3. Shapiro, D.M.; Warren, S. Cancer Innervation. *Cancer Res.* **1949**, *9*, 707–711.
4. Tarlau, M.; Smalheiser, I. Personality patterns in patients with malignant tumors of the breast and cervix; an exploratory study. *Psychosom. Med.* **1951**, *13*, 117–121. [CrossRef] [PubMed]
5. Leshan, L.L.; Gassmann, M.L. Some Observations on Psychotherapy with Patients Suffering from Neoplastic Disease. *Am. J. Psychother.* **1958**, *12*, 723–734. [CrossRef]
6. Hinzey, A.; Gaudier-Diaz, M.M.; Lustberg, M.B.; DeVries, A.C. Breast cancer and social environment: Getting by with a little help from our friends. *Breast Cancer Res.* **2016**, *18*, 54. [CrossRef]
7. Kim, G.M.; Kim, S.J.; Song, S.K.; Kim, H.R.; Kang, B.D.; Noh, S.H.; Chung, H.C.; Kim, K.R.; Rha, S.Y. Prevalence and prognostic implications of psychological distress in patients with gastric cancer. *BMC Cancer* **2017**, *17*, 283. [CrossRef]
8. Batty, G.D.; Russ, T.C.; MacBeath, M.; Stamatakis, E.; Kivimäki, M. Psychological distress in relation to site specific cancer mortality: Pooling of unpublished data from 16 prospective cohort studies. *BMJ* **2017**, *356*, j108. [CrossRef]
9. Sloan, E.K.; Priceman, S.J.; Cox, B.F.; Yu, S.; Pimentel, M.A.; Tangkanangnukul, V.; Arevalo, J.M.; Morizono, K.; Karanikolas, B.D.; Wu, L.; et al. Sympathetic nervous system induces a metastatic switch in primary breast cancer. *Cancer Res.* **2010**, *70*, 7042–7052. [CrossRef]
10. Azpiroz, A.; De Miguel, Z.; Fano, E.; Vegas, O. Relations between different coping strategies for social stress, tumor development and neuroendocrine and immune activity in male mice. *Brain Behav. Immun.* **2008**, *22*, 690–698. [CrossRef]
11. Fox, B.H. The role of psychological factors in cancer incidence and prognosis. *Oncology* **1995**, *9*, 245–253, discussion 253–256. [PubMed]
12. Chida, Y.; Hamer, M.; Wardle, J.; Steptoe, A. Do stress-related psychosocial factors contribute to cancer incidence and survival? *Nat. Clin. Pract. Oncol.* **2008**, *5*, 466–475. [CrossRef] [PubMed]
13. Trosko, J.E.; Chang, C.C.; Upham, B.L.; Tai, M.H. Ignored hallmarks of carcinogenesis: Stem Cells and Cell-Cell communication. *Ann. N. Y. Acad. Sci.* **2004**, *1028*, 192–201. [CrossRef]
14. Soto, A.; Sonnenschein, C. The tissue organization field theory of cancer: A testable replacement for the somatic mutation theory. *Bioessays* **2011**, *33*, 332–340. [CrossRef]
15. Levin, M. Morphogenetic fields in embryogenesis, regeneration, and cancer: Non-local control of complex patterning. *BioSystems* **2012**, *109*, 243–261. [CrossRef]
16. Xuan, C.; Shamonki, J.M.; Chung, A.; Dinome, M.L.; Chung, M.; Sieling, P.A.; Lee, D.J. Microbial dysbiosis is associated with human breast cancer. *PLoS ONE* **2014**, *9*, e83744. [CrossRef]
17. Schwabe, R.F.; Jobin, C. The microbiome and cancer. *Nat. Rev. Cancer* **2013**, *13*, 800–812. [CrossRef]
18. Lobikin, M.; Chernet, B.; Lobo, D.; Levin, M. Resting Potential, Oncogene-induced Tumorigenesis, and Metastasis: The Bioelectric Basis of Cancer in vivo. *Phys. Biol.* **2012**, *9*, 065002. [CrossRef]
19. Prevarskaya, N.; Skryma, R.; Shuba, Y. Ion channels and the hallmarks of cancer. *Trends Mol. Med.* **2010**, *16*, 107–121. [CrossRef]
20. Huber, S.M. Oncochannels. *Cell Calcium.* **2013**, *53*, 241–255. [CrossRef]

21. Kruk, J.; Aboul-Enein, B.H.; Bernstein, J.; Gronostaj, M. Psychological Stress and Cellular Aging in Cancer: A Meta-Analysis. *Oxid. Med. Cell. Longev.* **2019**, *2019*, 1270397. [CrossRef] [PubMed]
22. Betof, A.S.; Dewhirst, M.W.; Jones, L.W. Effects and potential mechanisms of excercise training on cancer progression: A translational perspective. *Brain Behav. Immun.* **2013**, *30*, S75–S87. [CrossRef] [PubMed]
23. Adam, A.; Koranteng, F. Availability, accessibility, and impact of social support on breast cancer treatment among breast cancer patients in Kumasi, Ghana: A qualitative study. *PLoS ONE* **2020**, *15*, e0231691. [CrossRef] [PubMed]
24. Montes-Nogueira, I.; Gutiérrez-Ospina, G.; Romo-González, T. Towards a Psychoneuroimmunendocrine Hypothesis of Breast Cancer. *Adv. Neuroimmune Biol.* **2017**, *6*, 153–160. [CrossRef]
25. Cavigelli, S.A.; Bennett, J.M.; Michael, K.C.; Cousino, L. Female temperament, tumor development and life span: Relation to glucocorticoid and tumor necrosis factor α levels in rats. *Brain Behav. Immun.* **2008**, 727–735. [CrossRef]
26. Heidt, T.; Sager, H.B.; Courties, G.; Dutta, P.; Iwamoto, Y.; Zaltsman, A.; von Zur Muhlen, C.; Bode, C.; Fricchione, G.L.; Denninger, J.; et al. Chronic variable stress activates hematopoietic stem cells. *Nat. Med.* **2014**, *20*, 754–758. [CrossRef]
27. Martínez-Martínez, E.; Galván-Hernández, C.I.; Toscano-Márquez, B.; Gutiérrez-Ospina, G. Modulatory Role of Sensory Innervation on Hair Follicle Stem Cell Progeny during Wound Healing of the Rat Skin. *PLoS ONE* **2012**, *7*, e36421. [CrossRef]
28. Sharifpanah, F.; Saliu, F.; Bekhite, M.M.; Wartenberg, M.; Sauer, H. β-Adrenergic receptor antagonists inhibit vasculogenesis of embryonic stem cells by downregulation of nitric oxide generation and interference with VEGF signalling. *Cell Tissue Res.* **2014**, *358*, 443–452. [CrossRef]
29. Spiegel, A.; Shivtiel, S.; Kalinkovich, A.; Ludin, A.; Netzer, N.; Goichberg, P.; Azaria, Y.; Resnick, I.; Hardan, I.; Ben-Hur, H.; et al. Catecholaminergic neurotransmitters regulate migration and repopulation of immature human CD34+ cells through Wnt signaling. *Nat. Immunol.* **2007**, *8*, 1123–1131. [CrossRef]
30. Katayama, Y.; Battista, M.; Kao, W.M.; Hidalgo, A.; Peired, A.J.; Thomas, S.A.; Frenette, P.S. Signals from the sympathetic nervous system regulate hematopoietic stem cell egress from bone marrow. *Cell* **2006**, *124*, 407–421. [CrossRef]
31. Liang, W.; Zhuo, X.; Tang, Z. Calcitonin gene-related peptide stimulates proliferation and osteogenic differentiation of osteoporotic rat-derived bone mesenchymal stem cells. *Mol. Cell. Biochem.* **2015**, *402*, 101–110. [CrossRef]
32. Powell, N.D.; Sloan, E.K.; Bailey, M.T.; Arevalo, J.M.; Miller, G.E.; Chen, E.; Kobor, M.S.; Reader, B.F.; Sheridan, J.F.; Cole, S.W. Social stress up-regulates inflammatory gene expression in the leukocyte transcriptome via β-adrenergic induction of myelopoiesis. *Proc. Natl. Acad. Sci. USA* **2013**, *110*, 16574–16579. [CrossRef]
33. Collins, S.; Sarzani, R.; Bordicchia, M. Coordinate control of adipose 'browning' and energy expenditure by β-adrenergic and natriuretic peptide signalling. *Int. J. Obes. Suppl.* **2014**, *4* (Suppl. 1), S17–S20. [CrossRef]
34. Boilly, B.; Faulknerx, S.; Jobling, P.; Hondermarck, H. Nerve dependence: From regeneration to cancer. *Cancer Cell* **2017**, *31*, 342–354. [CrossRef]
35. Makita, T.; Sucov, H.M.; Gariepy, C.E.; Yanagisawa, M.; Ginty, D.D. Endothelins are vascular-derived axonal guidance cues for developing sympathetic neurons. *Nature* **2008**, *452*, 759–763. [CrossRef]
36. Mitchell, B.S.; Schumacher, U.; Stauber, V.V.; Kaiserling, E. Are breast tumours innervated? Immunohistological investigations using antibodies against the neuronal marker protein gene product 9.5 (PGP 9.5) in benign and malignant breast lesions. *Eur. J. Cancer* **1994**, *30*, 1100–1103. [CrossRef]
37. Zhao, Q.; Yang, Y.; Liang, X.; Du, G.; Liu, L.; Lu, L.; Dong, J.; Han, H.; Zhang, G. The clinicopathological significance of neurogenesis in breast cancer. *BMC Cancer* **2014**, *14*, 484. [CrossRef]
38. Lü, S.H.; Zhou, Y.; Que, H.P.; Liu, S.J. Peptidergic innervation of human esophageal and cardiac carcinoma. *World J. Gastroenterol.* **2003**, *9*, 399–403. [CrossRef]
39. Chamary, V.L.; Robson, T.; Loizidou, M.; Boulos, P.B.; Burnstock, G. Progressive loss of perivascular nerves adjacent to colorectal cancer. *Eur. J. Surg. Oncol.* **2000**, *26*, 588–593. [CrossRef]
40. Nagy, J.A.; Chang, S.H.; Shih, S.C.; Dvorak, A.M.; Dvorak, H.F. Heterogeneity of the tumor vasculature. *Semin. Thromb. Hemost.* **2010**, *36*, 321–331. [CrossRef]
41. Saloman, J.L.; Albers, K.M.; Rhim, A.D.; Davis, B.M. Can Stopping Nerves, Stop Cancer? *Trends Neurosci.* **2016**, *39*, 880–889. [CrossRef]
42. Driver, J.A. Inverse association between cancer and neurodegenerative disease: Review of the epidemiologic and biological evidence. *Biogerontology* **2014**, *15*, 547–557. [CrossRef]
43. Wefel, J.S.; Vardy, J.; Ahles, T.; Schagen, S.B. International Cognition and Cancer Task Force recommendations to harmonise studies of cognitive function in patients with cancer. *Lancet Oncol.* **2011**, *12*, 703–708. [CrossRef]
44. Rutledge, A.; Jobling, P.; Walker, M.M.; Denham, J.W.; Hondermarck, H. Spinal Cord Injuries and Nerve Dependence in Prostate Cancer. *Trends Cancer* **2017**, *3*, 812–815. [CrossRef]
45. Raimondi, S.; Botteri, E.; Munzone, E.; Cipolla, C.; Rotmensz, N.; DeCensi, A.; Gandini, S. Use of beta-blockers, angiotensin-converting enzyme inhibitors and angiotensin receptor blockers and breast cancer survival: Systematic review and meta-analysis. *Int. J. Cancer* **2016**, *139*, 212–219. [CrossRef]
46. Barron, T.I.; Connolly, R.M.; Sharp, L.; Bennett, K.; Visvanathan, K. Beta blockers and breast cancer mortality: A population-based study. *J. Clin. Oncol.* **2011**, *29*, 2635–2644. [CrossRef]

47. Melhem-Bertrandt, A.; Chavez-Macgregor, M.; Lei, X.; Brown, E.N.; Lee, R.T.; Meric-Bernstam, F.; Sood, A.K.; Conzen, S.D.; Hortobagyi, G.N.; Gonzalez-Angulo, A.M. Beta-blocker use is associated with improved relapse-free survival in patients with triple-negative breast cancer. *J. Clin. Oncol.* **2011**, *29*, 2645–2652. [CrossRef]
48. Kokolus, K.M.; Zhang, Y.; Sivik, J.M.; Schmeck, C.; Zhu, J.; Repasky, E.A.; Drabick, J.J.; Schell, T.D. Beta blocker use correlates with better overall survival in metastatic melanoma patients and improves the efficacy of immunotherapies in mice. *Oncoimmunology* **2017**, *7*, e1405205. [CrossRef]
49. Renz, B.W.; Takahashi, R.; Tanaka, T.; Macchini, M.; Hayakawa, Y.; Dantes, Z.; Maurer, H.C.; Chen, X.; Jiang, Z.; Westphalen, C.B.; et al. β2 Adrenergic-Neurotrophin Feedforward Loop Promotes Pancreatic Cancer. *Cancer Cell* **2018**, *33*, 75–90.e7, Erratum in: *Cancer Cell.* **2018**, *34*, 863–867. [CrossRef]
50. Nilsson, M.B.; Sun, H.; Diao, L.; Tong, P.; Liu, D.; Li, L.; Fan, Y.; Poteete, A.; Lim, S.O.; Howells, K.; et al. Stress hormones promote EGFR inhibitor resistance in NSCLC: Implications for combinations with β-blockers. *Sci. Transl. Med.* **2017**, *9*, eaao4307. [CrossRef]
51. Pasquier, E.; Street, J.; Pouchy, C.; Carre, M.; Gifford, A.J.; Murray, J.; Norris, M.D.; Trahair, T.; Andre, N.; Kavallaris, M. β-blockers increase response to chemotherapy via direct antitumour and anti-angiogenic mechanisms in neuroblastoma. *Br. J. Cancer* **2013**, *108*, 2485–2494. [CrossRef]
52. Rico, M.; Baglioni, M.; Bondarenko, M.; Laluce, N.C.; Rozados, V.; André, N.; Carré, M.; Scharovsky, O.G.; Menacho Márquez, M. Metformin and propranolol combination prevents cancer progression and metastasis in different breast cancer models. *Oncotarget* **2017**, *8*, 2874–2889. [CrossRef]
53. Marchesi, F.; Piemonti, L.; Mantovani, A.; Allavena, P. Molecular mechanisms of perineural invasion, a forgotten pathway of dissemination and metastasis. *Cytokine Growth Factor Rev.* **2010**, *21*, 77–82. [CrossRef]
54. Chen, J.W.; Bhandari, M.; Astill, D.S.; Wilson, T.G.; Kow, L.; Brooke-Smith, M.; Toouli, J.; Padbury, R.T. Predicting patient survival after pancreaticoduodenectomy for malignancy: Histopathological criteria based on perineural infiltration and lymphovascular invasion. *HPB J.* **2010**, *12*, 101–108. [CrossRef]
55. Yao, J.; Li, W.Y.; Li, S.G.; Feng, X.S.; Gao, S.G. Midkine promotes perineural invasion in human pancreatic cancer. *WJG* **2014**, *20*, 3018–3024. [CrossRef]
56. Qian, X.; Nguyen, D.T.; Dong, Y.; Sinikovic, B.; Kaufmann, A.M.; Myers, J.N.; Albers, A.E.; Graviss, E.A. Prognostic Score Predicts Survival in HPV-Negative Head and Neck Squamous Cell Cancer Patients. *Int. J. Biol. Sci.* **2019**, *15*, 1336–1344. [CrossRef]
57. Seefeld, P.H.; Bargen, J.A. The spread of carcinoma of the rectum: Invasion of lymphatics, veins and nerves. *Ann. Surg.* **1943**, *118*, 76–90. [CrossRef]
58. Chablani, P.; Nguyen, P.; Pan, X.; Robinson, A.; Walston, S.; Wu, C.; Frankel, W.L.; Chen, W.; Bekaii-Saab, T.; Chakravarti, A.; et al. Perineural Invasion Predicts for Distant Metastasis in Locally Advanced Rectal Cancer Treated with Neoadjuvant Chemoradiation and Surgery. *Am. J. Clin. Oncol.* **2017**, *40*, 561–568. [CrossRef]
59. Mirkin, K.A.; Hollenbeak, C.S.; Mohamed, A.; Jia, Y.; El-Deiry, W.S.; Messaris, E. Impact of perineural invasion on survival in node negative colon cancer. *Cancer Biol. Ther.* **2017**, *18*, 740–745. [CrossRef]
60. Ceyhan, G.O.; Demir, I.E.; Altintas, B.; Rauch, U.; Thiel, G.; Müller, M.W.; Giese, N.A.; Friess, H.; Schäfer, K.H. Neural invasion in pancreatic cancer: A mutual tropism between neurons and cancer cells. *Biochem. Biophys. Res. Commun.* **2008**, *374*, 442–447. [CrossRef]
61. Liebig, C.; Ayala, G.; Wilks, J.A.; Berger, D.H.; Albo, D. Perineural invasion in cancer: A review of the literature. *Cancer* **2009**, *115*, 3379–3391. [CrossRef]
62. Magnon, C. Role of the autonomic nervous system in tumorigenesis and metastasis. *Mol. Cell. Oncol.* **2015**, *2*, e975643. [CrossRef]
63. Chakravarthy, R.; Mnich, K.; Gorman, A.M. Nerve growth factor (NGF)-mediated regulation of p75 (NTR) expression contributes to chemotherapeutic resistance in triple negative breast cancer cells. *Biochem. Biophys. Res. Commun.* **2016**, *478*, 1541–1547. [CrossRef]
64. Mancino, M.; Ametller, E.; Gascón, P.; Almendro, V. The neuronal influence on tumor progression. *Biochim. Et Biophys. Acta* **2011**, *1816*, 105–118. [CrossRef]
65. Bradshaw, R.A.; Pundavela, J.; Biarc, J.; Chalkley, R.J.; Burlingame, A.L.; Hondermarck, H. NGF and ProNGF: Regulation of neuronal and neoplastic responses through receptor signaling. *Adv. Biol. Regul.* **2015**, *58*, 16–27. [CrossRef]
66. Entschladen, F.; Palm, D.; Lang, K.; Drell, T.L., 4th; Zaenker, K.S. Neoneurogenesis: Tumors may initiate their own innervation by the release of neurotrophic factors in analogy to lymphangiogenesis and neoangiogenesis. *Med. Hypotheses* **2006**, *67*, 33–35. [CrossRef]
67. Magnon, C.; Hall, S.J.; Lin, J.; Xue, X.; Gerber, L.; Freedland, S.J.; Frenette, P.S. Autonomic nerve development contributes to prostate cancer progression. *Science* **2013**, *341*, 1236361. [CrossRef]
68. Mauffrey, P.; Tchitchek, N.; Barroca, V.; Bemelmans, A.P.; Firlej, V.; Allory, Y.; Roméo, P.H.; Magnon, C. Progenitors from the central nervous system drive neurogenesis in cancer. *Nature* **2019**, *569*, 672–678. [CrossRef]
69. Blacklock, A.D.; Smith, P.G. Estrogen increases calcitonin gene-related peptide-immunoreactive sensory innervation of rat mammary gland. *J. Neurobiol.* **2004**, *59*, 192–204. [CrossRef]
70. Godlewski, J.; Łakomy, I.M. Changes in vasoactive intestinal peptide, pituitary adenylate cyclase-activating polypeptide and neuropeptide Y-ergic structures of the enteric nervous system in the carcinoma of the human large intestine. *Folia Histochem. Cytobiol.* **2010**, *48*, 208–216. [CrossRef]

71. Albo, D.; Akay, C.L.; Marshall, C.L.; Wilks, J.A.; Verstovsek, G.; Liu, H.; Agarwal, N.; Berger, D.H.; Ayala, G.E. Neurogenesis in colorectal cancer is a marker of aggressive tumor behavior and poor outcomes. *Cancer* **2011**, *117*, 4834–4845. [CrossRef]
72. Narayan, P.; Flynn, J.; Zhang, Z.; Gillespie, E.F.; Mueller, B.; Xu, A.J.; Cuaron, J.; McCormick, B.; Khan, A.J.; Cahlon, O.; et al. Perineural invasion as a risk factor for locoregional recurrence of invasive breast cancer. *Sci. Rep.* **2021**, *11*, 12781. [CrossRef]
73. Pagella, P.; Catón, J.; Meisel, C.T.; Mitsiadis, T.A. Ameloblastomas Exhibit Stem Cell Potential, Possess Neurotrophic Properties, and Establish Connections with Trigeminal Neurons. *Cells* **2020**, *9*, 644. [CrossRef] [PubMed]
74. Rowe, C.W.; Dill, T.; Griffin, N.; Jobling, P.; Faulkner, S.; Paul, J.W.; King, S.; Smith, R.; Hondermarck, H. Innervation of papillary thyroid cancer and its association with extra-thyroidal invasion. *Sci. Rep.* **2020**, *10*, 1539. [CrossRef] [PubMed]
75. Friedrich, R.E.; Behrendt, C.A.; Glatzel, M.; Hagel, C. Vascular Innervation in Benign Neurofibromas of Patients with Neurofibromatosis Type 1. *Anticancer Res.* **2015**, *35*, 6509–6516.
76. Palm, D.; Entschladen, F. Neoneurogenesis and the neuro-neoplastic synapse. *Neuronal Act. Tumor Tissue* **2007**, *39*, 91–98. [CrossRef]
77. Wang, H.; Zheng, Q.; Lu, Z.; Wang, L.; Ding, L.; Xia, L.; Zhang, H.; Wang, M.; Chen, Y.; Li, G. Role of the nervous system in cancers: A review. *Cell Death Discov.* **2021**, *7*, 76. [CrossRef]
78. Díaz-Alonso, J.; Nicoll, R.A. AMPA receptor trafficking and LTP: Carboxy-termini, amino-termini and TARPs. *Neuropharmacology* **2021**, *197*, 108710. [CrossRef]
79. Mravec, B.; Gidron, Y.; Hulin, I. Neurobiology of cancer: Interactions between nervous, endocrine and immune systems as a base for monitoring and modulating the tumorigenesis by the brain. *Semin. Cancer Biol.* **2008**, *18*, 150–163. [CrossRef]
80. Cole, S.W.; Nagaraja, A.S.; Lutgendorf, S.K.; Green, P.A.; Sood, A.K. Sympathetic nervous system regulation of the tumour microenvironment. *Nat. Rev. Cancer* **2015**, *15*, 563–572. [CrossRef]
81. Flierl, M.A.; Rittirsch, D.; Nadeau, B.A.; Chen, A.J.; Sarma, J.V.; Zetoune, F.S.; McGuire, S.R.; List, R.P.; Day, D.E.; Hoesel, L.M.; et al. Phagocyte-derived catecholamines enhance acute inflammatory injury. *Nature* **2007**, *449*, 721–725. [CrossRef] [PubMed]
82. Köhler, C.A.; Freitas, T.H.; Maes, M.; de Andrade, N.Q.; Liu, C.S.; Fernandes, B.S.; Stubbs, B.; Solmi, M.; Veronese, N.; Herrmann, N.; et al. Peripheral cytokine and chemokine alterations in depression: A meta-analysis of 82 studies. *Acta Psych. Scand.* **2017**, *135*, 373–387. [CrossRef] [PubMed]
83. Jiang, W.; Li, D.; Han, R.; Zhang, C.; Jin, W.N.; Wood, K.; Liu, Q.; Shi, F.D.; Hao, J. Acetylcholine-producing NK cells attenuate CNS inflammation via modulation of infiltrating monocytes/macrophages. *Proc. Natl. Acad. Sci. USA* **2017**, *114*, E6202–E6211. [CrossRef]
84. Rosas-Ballina, M.; Olofsson, P.S.; Ochani, M.; Valdés-Ferrer, S.I.; Levine, Y.A.; Reardon, C.; Tusche, M.W.; Pavlov, V.A.; Andersson, U.; Chavan, S.; et al. Acetylcholine-synthesizing T cells relay neural signals in a vagus nerve circuit. *Science* **2011**, *334*, 98–101. [CrossRef]
85. Torres-Rosas, R.; Yehia, G.; Peña, G.; Mishra, P.; del Rocio Thompson-Bonilla, M.; Moreno-Eutimio, M.A.; Arriaga-Pizano, L.A.; Isibasi, A.; Ulloa, L. Dopamine mediates vagal modulation of the immune system by electroacupuncture. *Nat. Med.* **2014**, *20*, 291–295. [CrossRef]
86. Kim, T.H.; Ly, C.; Christodoulides, A.; Nowell, C.J.; Gunning, P.W.; Sloan, E.K.; Rowat, A.C. Stress hormone signaling through β-adrenergic receptors regulates macrophage mechanotype and function. *FASEB* **2019**, *33*, 3997–4006. [CrossRef] [PubMed]
87. Hanahan, D.; Weinberg, R.A. The hallmarks of cancer. *Cell* **2000**, *100*, 57–70. [CrossRef]
88. Hanahan, D.; Weinberg, R.A. Hallmarks of cancer: The next generation. *Cell* **2011**, *144*, 646–674. [CrossRef]
89. Hanahan, D. Hallmarks of Cancer: New Dimensions. *Cancer Discov.* **2022**, *12*, 31–46. [CrossRef]
90. De, R.; Mazumder, S.; Sarkar, S.; Debsharma, S.; Siddiqui, A.A.; Saha, S.J.; Banerjee, C.; Nag, S.; Saha, D.; Bandyopadhyay, U. Acute mental stress induces mitochondrial bioenergetic crisis and hyper-fission along with aberrant mitophagy in the gut mucosa in rodent model of stress-related mucosal disease. *Free Radic. Biol. Med.* **2017**, *113*, 424–438. [CrossRef]
91. Flint, M.S.; Baum, A.; Chambers, W.H.; Jenkins, F.J. Induction of DNA damage, alteration of DNA repair and transcriptional activation by stress hormones. *Psychoneuroendocrinology* **2007**, *32*, 470–479. [CrossRef] [PubMed]
92. Irie, M.; Asami, S.; Nagata, S.; Miyata, M.; Kasai, H. Relationships between perceived workload, stress and oxidative DNA damage. *Int. Arch. Occup. Environ. Health* **2001**, *74*, 153–157. [CrossRef] [PubMed]
93. Hara, M.R.; Sachs, B.D.; Caron, M.G.; Lefkowitz, R.J. Pharmacological blockade of a β(2)AR-β-arrestin-1 signaling cascade prevents the accumulation of DNA damage in a behavioral stress model. *Cell Cycle* **2013**, *12*, 219–224. [CrossRef] [PubMed]
94. Glaser, R.; Thorn, B.E.; Tarr, K.L.; Kiecolt-Glaser, J.K.; D'Ambrosio, S.M. Effects of stress on methyltransferase synthesis: An important DNA repair enzyme. *Health Psychol.* **1985**, *4*, 403–412. [CrossRef]
95. Hanoun, M.; Maryanovich, M.; Arnal-Estapé, A.; Frenette, P.S. Neural regulation of hematopoiesis, inflammation, and cancer. *Neuron* **2015**, *86*, 360–373. [CrossRef] [PubMed]
96. Collins, S.; Caron, M.G.; Lefkowitz, R.J. Beta-adrenergic receptors in hamster smooth muscle cells are transcriptionally regulated by glucocorticoids. *J. Biol. Chem.* **1988**, *263*, 9067–9070. [CrossRef]
97. Mak, J.C.; Nishikawa, M.; Barnes, P.J. Glucocorticosteroids increase beta 2-adrenergic receptor transcription in human lung. *Am. J. Physiol.* **1995**, *268 Pt 1*, L41–L46. [CrossRef]
98. De Lorenzo, B.H.; de Oliveira Marchioro, L.; Greco, C.R.; Suchecki, D. Sleep-deprivation reduces NK cell number and function mediated by β-adrenergic signalling. *Psychoneuroendocrinology* **2015**, *57*, 134–143. [CrossRef]

99. Bellinger, D.L.; Millar, B.A.; Perez, S.; Carter, J.; Wood, C.; ThyagaRajan, S.; Molinaro, C.; Lubahn, C.; Lorton, D. Sympathetic modulation of immunity: Relevance to disease. *Cell. Immunol.* **2008**, *252*, 27–56. [CrossRef]
100. Mohammadpour, H.; MacDonald, C.R.; Qiao, G.; Chen, M.; Dong, B.; Hylander, B.L.; McCarthy, P.L.; Abrams, S.I.; Repasky, E.A. β2 adrenergic receptor-mediated signaling regulates the immunosuppressive potential of myeloid-derived suppressor cells. *J. Clin. Investig.* **2019**, *129*, 5537–5552. [CrossRef]
101. Raposo, T.P.; Arias-Pulido, H.; Chaher, N.; Fiering, S.N.; Argyle, D.J.; Prada, J.; Pires, I.; Queiroga, F.L. Comparative aspects of canine and human inflammatory breast cancer. *Semin. Oncol.* **2017**, *44*, 288–300. [CrossRef] [PubMed]
102. Carvalho, M.I.; Silva-Carvalho, R.; Pires, I.; Prada, J.; Bianchini, R.; Jensen-Jarolim, E.; Queiroga, F.L. A Comparative Approach of Tumor-Associated Inflammation in Mammary Cancer between Humans and Dogs. *BioMed Res. Int.* **2016**, *2016*, 4917387. [CrossRef] [PubMed]
103. Aliper, A.M.; Frieden-Korovkina, V.P.; Buzdin, A.; Roumiantsev, S.A.; Zhavoronkov, A. A role for G-CSF and GM-CSF in nonmyeloid cancers. *Cancer Med.* **2014**, *3*, 737–746. [CrossRef]
104. Bujak, J.K.; Kosmala, D.; Szopa, I.M.; Majchrzak, K.; Bednarczyk, P. Inflammation, Cancer and Immunity-Implication of TRPV1 Channel. *Front. Oncol.* **2019**, *9*, 1087. [CrossRef]
105. Sauzay, C.; Voutetakis, K.; Chatziioannou, A.; Chevet, E.; Avril, T. CD90/Thy-1, a Cancer-Associated Cell Surface Signaling Molecule. *Front. Cell Dev. Biol.* **2019**, *7*, 66. [CrossRef]
106. Glumac, P.M.; LeBeau, A.M. The role of CD133 in cancer: A concise review. *Clin. Transl. Med.* **2018**, *7*, 18. [CrossRef]
107. Huan, H.B.; Wen, X.D.; Chen, X.J.; Wu, L.; Wu, L.L.; Zhang, L.; Yang, D.P.; Zhang, X.; Bie, P.; Qian, C.; et al. Sympathetic nervous system promotes hepatocarcinogenesis by modulating inflammation through activation of alpha1-adrenergic receptors of Kupffer cells. *Brain Behav. Immun.* **2017**, *59*, 118–134. [CrossRef]
108. Horvathova, L.; Padova, A.; Tillinger, A.; Osacka, J.; Bizik, J.; Mravec, B. Sympathectomy reduces tumor weight and affects expression of tumor-related genes in melanoma tissue in the mouse. *Stress* **2016**, *19*, 528–534. [CrossRef]
109. Zhang, X.; Zhang, Y.; He, Z.; Yin, K.; Li, B.; Zhang, L.; Xu, Z. Chronic stress promotes gastric cancer progression and metastasis: An essential role for ADRB2. *Cell Death Dis.* **2019**, *10*, 788. [CrossRef]
110. Zhao, Y.; Li, W. Beta-adrenergic signaling on neuroendocrine differentiation, angiogenesis, and metastasis in prostate cancer progression. *Asian J. Androl.* **2019**, *21*, 253–259. [CrossRef]
111. Kilpatrick, L.E.; Alcobia, D.C.; White, C.W.; Peach, C.J.; Glenn, J.R.; Zimmerman, K.; Kondrashov, A.; Pfleger, K.; Ohana, R.F.; Robers, M.B.; et al. Complex Formation between VEGFR2 and the β_2-Adrenoceptor. *Cell Chem. Biol.* **2019**, *26*, 830–841.e9. [CrossRef] [PubMed]
112. Basu, S.; Sarkar, C.; Chakroborty, D.; Nagy, J.; Mitra, R.B.; Dasgupta, P.S.; Mukhopadhyay, D. Ablation of peripheral dopaminergic nerves stimulates malignant tumor growth by inducing vascular permeability factor/vascular endothelial growth factor-mediated angiogenesis. *Cancer Res.* **2004**, *64*, 5551–5555. [CrossRef]
113. Peters, M.; Meijer, C.; Fehrmann, R.; Walenkamp, A.; Kema, I.P.; de Vries, E.; Hollema, H.; Oosting, S.F. Serotonin and Dopamine Receptor Expression in Solid Tumours Including Rare Cancers. *Pathol. Oncol. Res.* **2020**, *26*, 1539–1547. [CrossRef]
114. Gao, J.; Zhang, C.; Gao, F.; Li, H. The effect and mechanism of dopamine D1 receptors on the proliferation of osteosarcoma cells. *Mol. Cell. Biochem.* **2017**, *430*, 31–36. [CrossRef] [PubMed]
115. Wang, W.; Li, L.; Chen, N.; Niu, C.; Li, Z.; Hu, J.; Cui, J. Nerves in the Tumor Microenvironment: Origin and Effects. *Front. Cell Dev. Biol.* **2020**, *8*, 601738. [CrossRef]
116. Armaiz-Pena, G.N.; Allen, J.K.; Cruz, A.; Stone, R.L.; Nick, A.M.; Lin, Y.G.; Han, L.Y.; Mangala, L.S.; Villares, G.J.; Vivas-Mejia, P.; et al. Src activation by β-adrenoreceptors is a key switch for tumour metastasis. *Nat. Commun.* **2013**, *4*, 1403. [CrossRef]
117. Cheng, Y.; Gao, X.H.; Li, X.J.; Cao, Q.H.; Zhao, D.D.; Zhou, J.R.; Wu, H.X.; Wang, Y.; You, L.J.; Yang, H.B.; et al. Depression promotes prostate cancer invasion and metastasis via a sympathetic-cAMP-FAK signaling pathway. *Oncogene* **2018**, *37*, 2953–2966. [CrossRef]
118. Choy, C.; Raytis, J.L.; Smith, D.D.; Duenas, M.; Neman, J.; Jandial, R.; Lew, M.W. Inhibition of β2-adrenergic receptor reduces triple-negative breast cancer brain metastases: The potential benefit of perioperative β-blockade. *Oncol. Rep.* **2016**, *35*, 3135–3142. [CrossRef]
119. Shaashua, L.; Shabat-Simon, M.; Haldar, R.; Matzner, P.; Zmora, O.; Shabtai, M.; Sharon, E.; Allweis, T.; Barshack, I.; Hayman, L.; et al. Perioperative COX-2 and β-Adrenergic Blockade Improves Metastatic Biomarkers in Breast Cancer Patients in a Phase-II Randomized Trial. *Clin. Cancer Res.* **2017**, *23*, 4651–4661. [CrossRef]
120. Rivero, E.M.; Piñero, C.P.; Gargiulo, L.; Entschladen, F.; Zänker, K.; Bruzzone, A.; Lüthy, I.A. The β 2-Adrenergic Agonist Salbutamol Inhibits Migration, Invasion and Metastasis of the Human Breast Cancer MDA-MB- 231 Cell Line. *Curr. Cancer Drug Targets* **2017**, *17*, 756–766. [CrossRef]
121. Bloom, A.P.; Jimenez-Andrade, J.M.; Taylor, R.N.; Castañeda-Corral, G.; Kaczmarska, M.J.; Freeman, K.T.; Coughlin, K.A.; Ghilardi, J.R.; Kuskowski, M.A.; Mantyh, P.W. Breast cancer-induced bone remodeling, skeletal pain, and sprouting of sensory nerve fibers. *J. Pain* **2011**, *12*, 698–711. [CrossRef] [PubMed]
122. Jimenez-Andrade, J.M.; Ghilardi, J.R.; Castañeda-Corral, G.; Kuskowski, M.A.; Mantyh, P.W. Preventive or late administration of anti-NGF therapy attenuates tumor-induced nerve sprouting, neuroma formation, and cancer pain. *Pain* **2011**, *152*, 2564–2574. [CrossRef]

123. Elefteriou, F. Role of sympathetic nerves in the establishment of metastatic breast cancer cells in bone. *J. Bone Oncol.* **2016**, *5*, 132–134. [CrossRef] [PubMed]
124. Welch, D.R.; Hurst, D.R. Defining the Hallmarks of Metastasis. *Cancer Res.* **2019**, *79*, 3011–3027. [CrossRef] [PubMed]
125. Sonnenschein, C.; Soto, A.M. Over a century of cancer research: Inconvenient truths and promising leads. *PLoS Biol.* **2020**, *18*, e3000670. [CrossRef]

cancers

Review

In Vivo and In Vitro Models of Hepatocellular Carcinoma: Current Strategies for Translational Modeling

Guilherme Ribeiro Romualdo [1,2,3,†], Kaat Leroy [4,†], Cícero Júlio Silva Costa [1], Gabriel Bacil Prata [2,3], Bart Vanderborght [5,6], Tereza Cristina da Silva [1], Luís Fernando Barbisan [2], Wellington Andraus [7], Lindsey Devisscher [6], Niels Olsen Saraiva Câmara [8], Mathieu Vinken [4,‡] and Bruno Cogliati [1,*,‡]

1 Department of Pathology, School of Veterinary Medicine and Animal Science, University of São Paulo (USP), São Paulo 05508-270, Brazil; romualdo.gr15@gmail.com (G.R.R.); cicerocosta@usp.br (C.J.S.C.); terezacs678@gmail.com (T.C.d.S.)
2 Department of Structural and Functional Biology, Biosciences Institute, São Paulo State University (UNESP), Botucatu 18618-689, Brazil; gabriel.bacil@unesp.br (G.B.P.); luis.barbisan@unesp.br (L.F.B.)
3 Department of Pathology, Botucatu Medical School, São Paulo State University (UNESP), Botucatu 18618-687, Brazil
4 Department of Pharmaceutical and Pharmacological Sciences, Vrije Universiteit Brussel, 1090 Brussels, Belgium; Kaat.Leroy@vub.be (K.L.); Mathieu.Vinken@vub.be (M.V.)
5 Gut-Liver Immunopharmacology Unit, Basic and Applied Medical Sciences, Liver Research Center Ghent, Faculty of Medicine and Health Sciences, Ghent University, 9000 Ghent, Belgium; Bart.Vanderborght@ugent.be
6 Hepatology Research Unit, Internal Medicine and Paediatrics, Liver Research Center Ghent, Faculty of Medicine and Health Sciences, Ghent University, 9000 Ghent, Belgium; LINDSEY.DEVISSCHER@ugent.be
7 Department of Gastroenterology, Clinics Hospital, School of Medicine, University of São Paulo (HC-FMUSP), São Paulo 05403-000, Brazil; wellingtonandraus@gmail.com
8 Department of Immunology, Institute of Biomedical Sciences IV, University of São Paulo (USP), São Paulo 05508-000, Brazil; niels@icb.usp.br
* Correspondence: bcogliati@usp.br; Tel.: +55-1130911200; Fax: +55-1130917829
† Authors share first authorship.
‡ Authors share corresponding authorship.

Simple Summary: Hepatocellular carcinoma (HCC) is a highly incident and deadly malignant neoplasia, and only a few anti-HCC drugs are currently available. Thus, the development of HCC models has become essential for both basic and translational research, improving the understanding of HCC pathophysiology and molecular landscape. The present paper provides a state-of-the-art overview of in vivo and in vitro models used for translational modeling of HCC, focusing on their molecular hallmarks. Our paper depicts the key features, advantages and disadvantages of the main bioassays available, shedding light on standard HCC model choice.

Abstract: Hepatocellular carcinoma (HCC) is the sixth most common cancer worldwide and the third leading cause of cancer-related death globally. HCC is a complex multistep disease and usually emerges in the setting of chronic liver diseases. The molecular pathogenesis of HCC varies according to the etiology, mainly caused by chronic hepatitis B and C virus infections, chronic alcohol consumption, aflatoxin-contaminated food, and non-alcoholic fatty liver disease associated with metabolic syndrome or *diabetes mellitus*. The establishment of HCC models has become essential for both basic and translational research to improve our understanding of the pathophysiology and unravel new molecular drivers of this disease. The ideal model should recapitulate key events observed during hepatocarcinogenesis and HCC progression in view of establishing effective diagnostic and therapeutic strategies to be translated into clinical practice. Despite considerable efforts currently devoted to liver cancer research, only a few anti-HCC drugs are available, and patient prognosis and survival are still poor. The present paper provides a state-of-the-art overview of in vivo and in vitro models used for translational modeling of HCC with a specific focus on their key molecular hallmarks.

Citation: Romualdo, G.R.; Leroy, K.; Costa, C.J.S.; Prata, G.B.; Vanderborght, B.; da Silva, T.C.; Barbisan, L.F.; Andraus, W.; Devisscher, L.; Câmara, N.O.S.; et al. In Vivo and In Vitro Models of Hepatocellular Carcinoma: Current Strategies for Translational Modeling. *Cancers* **2021**, *13*, 5583. https://doi.org/10.3390/cancers13215583

Academic Editor: Luca Di Tommaso

Received: 13 September 2021
Accepted: 4 November 2021
Published: 8 November 2021

Publisher's Note: MDPI stays neutral with regard to jurisdictional claims in published maps and institutional affiliations.

Copyright: © 2021 by the authors. Licensee MDPI, Basel, Switzerland. This article is an open access article distributed under the terms and conditions of the Creative Commons Attribution (CC BY) license (https://creativecommons.org/licenses/by/4.0/).

Keywords: hepatocarcinogenesis; liver cancer; animal model; cell culture; gene mutation; epigenetic alteration; translational research

1. Hepatocellular Carcinoma: Worldwide Trends and Mechanisms

1.1. Epidemiology and Contributing Factors

Liver cancers, mainly represented by hepatocellular carcinoma (HCC), accounted for about 840,000 incident cases and 780,000 deaths in 2018 [1]. HCC corresponds to approximately 78% of all hepatobiliary malignancies, being the sixth most incident neoplasm and the third leading cause of cancer-related deaths worldwide [2]. HCC has a poor prognosis, displaying an average survival of 11 months and a survival rate of 49–63%, 19–29%, and 17% after 1, 3, and 5 years of diagnosis, respectively [3,4]. Over 90% of HCC cases occur in a fibrotic or cirrhotic background, which is considered the main risk factor [1,5]. Moreover, populational data on HCC display two important features: geographical and gender disparities. Standardized incidence rates (cases or deaths per 100,000 people) in Asian and African continents are ~2-fold higher than in Europe and North America [1]. While most HCC cases globally are caused by chronic hepatitis B and/or C virus (HBV/HCV) infections (44–56% to HBV and 20–21% to HCV), lifestyle-related risk factors are fast-growing populational attributable factors for this malignancy in western HCC patients. Non-alcoholic fatty liver disease (NAFLD) is closely associated with metabolic syndrome and *diabetes mellitus*, which are independently linked to as many as 16% of HCC cases worldwide [6]. Excessive and chronic alcohol intake leading to alcoholic liver disease (ALD) is another important risk factor associated with 26% of HCC cases, standing only behind HBV infection. In Central Asia and Central Sub-Saharan Africa, HBV and HCV chronic infections are indeed the most prominent risk factors, responsible for 57–60% and 41–50% of all cases, respectively [6]. On the other hand, in Central Europe and North America, ALD and NAFLD-related metabolic syndromes are the most prominent ones, linked to 30–32% and 20–24% of all cases, respectively [6,7]. Since some authors consider chronic viral infections as the most important risk factors for HCC development, HBV/HCV-related HCC attributable fraction may in part explain the geographical disparity feature. Another important epidemiological feature is the marked male disparity (two to three-fold higher in males), whose mechanisms may involve the predominance of risk factors in men and the promoting/protective roles of sex hormones [2]. The influence of dietary factors on HCC emergence is not fully understood, but many epidemiological studies point to a marked protective effect of coffee consumption [8,9].

1.2. Hepatocarcinogenesis

Hepatocarcinogenesis represents a complex multistep process in which successively more aberrant monoclonal populations of hepatocytes evolve [10]. The pro-inflammatory and pro-fibrotic microenvironment forms the ideal background for the emergence of numerous human hepatocarcinogenesis-promoting genetic and epigenetic abnormalities [11,12]. Many cancer driver pathways have been repeatedly altered in HCC according to the distinct genotoxic insults and etiologies, allowing the classification of HCC in molecular and/or immune subclasses [13]. To unveil the main molecular alterations involved in HCC, The Cancer Genome Atlas Research Network (TCGA) has performed the first large-scale multiplatform analysis of HCC, including the evaluation of somatic mutations, DNA methylation, gene, protein, and microRNA (miRNA) expressions [14]. Further, Llovet et al. [13] recently segregated HCCs into two major morphological/pathophysiological/molecular phenotypes: proliferation and non-proliferation classes. The proliferation class is more aggressive and poorly differentiated, frequently related to HBV-related etiology. The non-proliferation phenotype is less aggressive, well-to-moderately differentiated, and linked to HCV, alcohol, and NASH-related causes. Telomerase reverse transcriptase (TERT) promoter mutations are the most common mutations in all HCCs analyzed (44%), frequently observed in both phenotypes and in co-occurrence with CDKN2A (p16) hypermethylation (53%), which is more common in the non-proliferation class. Upregulation of TERT and downregulation of CDKN2A enables the immortalization cancer hallmark. The activation of the Wnt/β-catenin pathway, conferring sustained proliferation hallmark, was also frequently featured in both phenotypes, as inactivating tumor suppressor AXIN1 (8%) and activating oncogene CTNNB1 (27%) mutations were observed in proliferation and non-proliferation classes, respectively. HBV-related proliferation class is also associated with the activation of key proliferation pathways, as PI3K–AKT–mTOR, RAS–MAPK, MET, and IGF. TP53 mutations (31%), conferring "evasion of growth suppressors", "genomic instability and mutation", and "resistance to cell death" cancer hallmarks, were frequently observed in proliferation class, also in keeping with global DNA hypomethylation signature [13–15]. As Wnt/β-catenin and TP53 pathways or TERT are altered in ~77% of HCCs, these dominant molecular drivers are key molecular therapeutic targets and remain undruggable [13,14].

In light of the unknown HCC molecular landscape and the urgent need for novel preventive and therapeutic strategies, the establishment of HCC models has become essential for both basic and translational research. Recently, with the continuous emergence of precision and personalized medicine, standardized and personalized HCC models are warranted. To achieve these requirements, the model should recapitulate key pathophysiological and molecular events observed during hepatocarcinogenesis in view of being effectively translated into clinical practice. Considering the current myriad of HCC models in the literature, we provide a comprehensive overview of the main in vivo and in vitro bioassays applied for HCC modelling, depicting their key molecular hallmarks.

2. In Vivo Models of HCC

2.1. Syngeneic and Xenograft Mouse Models

Syngeneic and xenograft experimental models are based on the injection or implantation of HCC cell lines or patient-derived xenograft (PDX) in either extrahepatic (heterotopic) or intrahepatic (orthotopic) microenvironments. In the syngeneic mouse models, injection of a murine HCC cell line enables the evaluation of molecular characteristics and tumor growth in a microenvironment of immunocompetent animals [16]. The xenografts mouse models comprise injection of human HCC cells or transplantation of fresh PDX into immunodeficient animals, such as non-obese diabetic/severe combined immunodeficiency (NOD-*scid*) and athymic Balb/c nude mice, delivering a translational model of HCC that recapitulates some of the relevant genetic alterations, i.e., TP53, FGFR1, and KRAS mutations [16–19]. The *scid*-mutated mice are leucopenic and have a compromised function of B and T lymphocytes, while NOD-*scid* mice feature both impairment of leucocyte activity and diminished activity of natural killer cells and innate immunity, allowing them to be easily grafted [20,21]. To establish a translational model to evaluate HCC, the PDX mouse model underwent improvements, and humanized mice, which will be further reviewed (see 2.5 Humanized mouse models), have been developed [17]. These kinds of features make syngeneic and xenograft mouse models widely employed in pre-clinical approaches of new treatment protocols and adequate to unveil molecular traits and pathological aspects similar to HCC patients. However, it is still uncertain whether morphologic, genomic, and molecular aspects of engrafted HCC tumors remain similar to samples obtained from patients [16].

To establish a reliable orthotopic PDX model, an early study by Sun et al. [22] evidenced that surgically removed HCC samples, further implanted into BALB/c nude mice and selected according to its invasive potential, resemble translational features regarding morphological aspects and increased alpha-fetoprotein expression (Table 1). Besides, the LCI-D20 model showed take rates of 100% and transplantability through mouse generations, as well as spontaneous liver, lung, and lymph nodes metastasis after 6–24 weeks of protocol. Regarding the metastatic potential, Genda et al. [23] yielded a PDX model with an orthotopic injection of Li7 and KYN-2 cells into *scid*-mutated mice and showed their metastatic potential, with 50% of the engrafted animals showing intrahepatic micrometastasis after 6 weeks. Besides, in vivo and in vitro assays unravel an underlying p160ROCK-dependent mechanism in the metastatic activity of Li7 cells by suppressing Rho signaling (Table 1) [23]. Likewise, the PDX mouse model with HCC samples obtained by needle biopsies provides a striking similarity to the original biopsies by upregulating molecular pathways related to hypoxia, cell cycle progression, and epithelial-to-mesenchymal transition, even after at least 6 retransplantations into NOD-*scid* mice (Table 1) [24]. A NOD-*scid* mouse model displaying an impairment in the interleukin 2 receptor tends to increase HCC engrafted, making it a reliable model to evaluate the tumoral behavior alterations in a human immune microenvironment [25]. Indeed, the tumoral microenvironment associated with the immune background enhances tumoral growth, suggesting that HCC exerts a survival strategy of modulating immune checkpoints and attenuating cytotoxic T cell activity. Hence, the plasma levels of pro-inflammatory cytokines, such as tumor necrosis factor-α and human interferon-γ, increase (0–4 weeks of protocol) followed by a marked decrease (4–8 weeks of protocol), mimicking the HCC survival behavior in HCC patients [24]. Although these models do not resemble the whole landscape of tumoral-immune dynamics, HCC establishment requires a short experimental time (compared to chemical and diet-induced models, for example), maintaining key features of the derived tumor. In this scenario, the PDX mouse models represent a promising and translational strategy for discovering new drug therapies and the pivotal molecular mechanisms underlying the HCC development since this model resembles some of the genomic, morphological, immunological, and microenvironmental tumor characteristics observed in patients.

Table 1. Summary of some of the patient-derived xenograft HCC protocols in mice.

Model	Procedure	Animal (Species, Strain, Age)	Timepoints and Incidence of Lesions	References
Ectopic implantation of human HCC in mouse;	- Samples of human HCC were sectioned and inoculated on the dorsum of the mice; a region pretreated with anti-asialo GM1;	Female Balb/c athymic nude mice at 4-week-old;	- 1st and 2nd generation: tumor growth at the implantation site with 100% (6/6) of transplantability and size averaged 1.0 cm in diameter at 4th week; - 3rd generation: spontaneous liver metastasis at 12th week; - 4th generation: 100% (5/5) of significant liver metastases in the 4th week; - 8th week: all animals with transplanted tumors died with important metastases	[26]
LCI-D20: orthotopic Implantation of human HCC	- Tissue fragments measuring 2 mm^3 of HCC from 30 human patients were implanted in the left hepatic lobe of the mice for 6 to 24 weeks;	Male BALB/cA nude mice at 4 to 6-week-old	- At 2nd week: initial liver metastasis with tumor colonies around the site of origin; - At 3rd: metastasis in mesenteric and iliac lymph nodes, hepatic hilum, mesentery and diaphragm; micrometastasis in pulmonary vessels; - At 6th week: generalized liver metastasis with vascular micrometastasis; micrometastasis in the lung parenchyma;	[22]

Table 1. *Cont.*

Model	Procedure	Animal (Species, Strain, Age)	Timepoints and Incidence of Lesions	References
HCC cells	- Different human HCC cells were cultured and suspended in saline solution at 1.108 cells/mL; afterward, 20 µL of cell suspension was injected into the mice's liver subserosa.	Male SCID at 6-week-old;	- At 6–7th week: the necropsy of the implanted animals was performed or before that period, if they showed signs of stress; - At 7th week: 2 of 12 cell lines had formed tumors only in the muscle; 5 of 12 cell lines had formed tumors in both muscles and lives; other 5 cells line did not form a tumor; - Intrahepatic metastases observed in 2 of 12 explored cells line (Li7; KYN-2), with the presence of neoplastic thrombi and new neoplastic sites distinct from the original or in the lobe that did not undergo implantation.	[23]
HCC-LY5 and HCC-LY10: Ectopic and orthotopic Implantation of human HCC in mice;	- Ectopic procedure: tissue fragments measuring 2 mm^3 of HCC from human patients were transplanted in the subcutaneous tissue of the right flank of the mice. Tumor growth was measured once a week by palpation; when the tumor mass reached 10–15 mm, the tumor was removed, reimplanted in other mice, three more times; - Orthotopic procedure: fragments of ectopic models were implanted in the left lobe of nude mice;	NOD/SCID male and female mice and T cell-immunodeficient BALB/c-nu/nu mice at 6 to 8-week-old	- Ectopic implantation: - At 4th week: the necropsy of the implanted animals was performed. - 5 of 24 human HCC samples were transplantable (20.83%); - The growth rate of the tumor and the growth time increased according to the advance of the passages. Variable growth over 7–6 weeks (tumors with 2.7 mm^3 to 7.2 mm^3); - Orthotopic implantation: - At 6th week: the necropsy of the implanted animals was performed. - The rate of tumor formation was 100% (8/8).	[27]

Table 1. *Cont.*

Model	Procedure	Animal (Species, Strain, Age)	Timepoints and Incidence of Lesions	References
Ectopic implantation from human HCC needle biopsies in mice	- 10 human HCC needle biopsies from patients were transplanted into the subcutaneous tissue of the mice	Nonobese, diabetic/severe combined immunodeficiency gamma-c mice at 10-week-old;	- 11 PDX models were established with 10 human HCC needle biopsies. - 4th to 28th week after implantation, it was the time necessary to observe tumor growth; - PDX subcutaneous injection of a biopsy cell suspension has a slow growth compared to intact tumor architecture; - Retransplanted tumors showed a shortened lag phase until the onset of tumor growth compared to the xenograft tumor derived from the biopsy tissue;	[24]

2.2. Chemical-Induced Rodent Models

2.2.1. Diethylnitrosamine (DEN)

Diethylnitrosamine (DEN or DENA, PubChem CID:5921), which is also known as N-nitrosodiethylamine (NDEA), is the most prominent and widely applied xenobiotic in chemically induced models of HCC. Although the daily human ingestion of total N-nitrosamines usually occurs in low microgram (μg) ranges, reaching 0.5 to 1.0 μg/day, DEN holds the "Group 2A: probably carcinogenic to humans" classification according to the International Agency for Research on Cancer [28–30]. Both volatile and non-volatile nitrosamines account for human exposure, mostly through oral and respiratory routes, as they can be detected (>0.1 μg/kg) in tobacco smoke, food additives, and cured or smoked meat products as either naturally occurring compounds or formed after food processing [29–32]. There is plenty of in vivo evidence showing that DEN bio-activation occurs primarily in the liver (by the hepatocytes), mostly mediated by cytochrome P450 (CYP) 2E1. Thus, the constitutive activity of this cytochrome is strongly correlated to DEN-related outcomes on tumorigenesis as incidence and severity (number and size) in rodents [33,34]. DEN undergoes alpha-hydroxylation and dealkylation reactions, thereby producing the unstable ethyl diazonium hydroxide molecule that may generate highly reactive carbonium ions, oxygen (ROS) and nitrogen (RNS) species [35]. These highly reactive metabolites may bind to different biomolecules, including DNA and proteins. DNA alkylation or oxidation induced by DEN—such as the formation of O^6-ethylguanine and O^4- and O^2-ethylthymine adducts mainly in centrilobular (zone 3) hepatocytes - may contribute to genomic instability, DNA damage, mutation, and tumor initiation [35–38]. Oxidative damage in proteins, such as conversion of protein thiol (-SH) groups to disulfides, is also featured in the liver after DEN exposure and may have direct implications on protein function and cell signaling [39]. As DEN was found to be a complete carcinogen in classical rodent bioassays (*i.e.*, a chemical that can induce HCC development without the association of secondary chemical or surgical procedure as a promoter), this N-nitrosamine was widely applied as an "initiating agent" in the past few years within a myriad of protocols in mice and rats (Table 2). This chemical literally "initiates" the hepatocarcinogenic process by the production of a stable, heritable mutational change in the target cell (hepatocytes). Although it is not yet clear if this genomic alteration activates or inactivates one (or more than one) oncogene or tumor suppressor gene at the level of a single hepatocyte, it is mostly accepted that chemically induced preneoplastic lesions and HCC itself may clonally expand from this single DEN- "initiated" hepatocyte [40]. Nonetheless, when administered in drinking water or single or few non-necrogenic intraperitoneal (i.p.) injections to juvenile/adult mice, a long latency time is necessary to achieve a high burden of neoplastic lesions (Table 2).

Table 2. Summary of some of the chemically induced hepatocarcinogenesis protocols in rats and mice strains.

Model	Procedure	Animal (Species, Strain, Age)	Timepoints and Incidence of Lesions	References
DEN	Single i.p., 90 mg/kg b.w.	Juvenile (5 weeks) C3H/He, DBA/2 and C57BL/6 mice (male)	- Adenomas: 0–20% at week 24, 10–50% at 36, and 20–50% at weeks 36–52 (strain-dependent) - Carcinomas: 0% at week 24, 0% at 36, and 0–40% at weeks 36–52 (strain-dependent)	[41]
	Multiple i.p. 1.5 or 3 mg/kg b.w. for 1 week (4×/week)	Juvenile (6 weeks) B6C3F1 and C3AF1 mice (male)	- Adenomas: 3–8% at weeks 100–120 (strain- and dose-dependent) - Carcinomas: 16–22% at weeks 100–120 (strain- and dose-dependent)	[42]
	Drinking water 15 mg/L for 3 weeks	Juvenile (4 weeks) B6C3F1 mice (male)	- Adenomas: 100% at week 24	[43]
	Multiple i.p. 25, 50, or 75 mg/kg b.w., for 4 or 8 weeks (1×/week)	Juvenile (4 weeks) C57BL/6 mice (male)	- Preneoplastic foci: 44–100% at week 33 (dose-dependent) - Adenomas: 0–33% at week 33 (dose-dependent) - Carcinomas: 0% at week 33	[44]
	Single i.p. 2.5, 10, 25 or 50 mg/kg b.w.	Infant (2 weeks) BALB/c mice (male)	- Adenomas: 7–87% at week 24 and 10–100% at week 40 (dose-dependent) - Carcinomas: 5–10% at both weeks at week 40 (dose-dependent)	- [45]
	Single i.p. 5 mg/kg b.w.	Infant (2 weeks) C3H/HeJ, B6C3F1 and C57BL mice (male)	- Adenomas: 90–100% at week 28 (strain-dependent)	[46]
	Single i.p. 1 mg/kg b.w.	Infant (2 weeks) C3H/HeJ, B6C3F1 and C57BL mice (male)	- Preneoplastic foci: 25–56% at week 22 and 46–100% at week 48 (strain-dependent) - Adenomas + Carcinomas: 25–67% at week 22 and 77–100% at week 48 (strain-dependent)	[47]
	Single i.p. 1 mg/kg b.w.	Infant (2 weeks) B6C3F1 mice (male)	- Preneoplastic foci: 75% at week 17 and 100% at week 22 - Adenomas: 0% at week 17 and 12.5% at week 22 - Carcinomas: 0% at both weeks 17 and 22	[48]
	Multiple i.p. 20 (1 dose), 30 (1 dose) and 50 mg/kg (6 doses) b.w., for 8 weeks (1×/week)	Infant (2 weeks) C57BL/6 mice (male and female)	- Adenomas + Carcinomas: 100% at week 24 - Obs.: Liver Fibrosis	[49]
	Gavage 80 mg/kg b.w. (weeks ~6–7)	Adult (~6–7 weeks) Sprague-Dawley rats (male)	- Carcinomas: ~5–20% at week 48 and 70	[50]

Table 2. *Cont.*

Model	Procedure	Animal (Species, Strain, Age)	Timepoints and Incidence of Lesions	References
	Single i.p., 200 mg/kg b.w. (week 6)	Juvenile (4 weeks) F344 rats (male)	- Preneoplastic foci: 100% at week 42 - Adenomas: 7% at week 42 - Carcinomas: 0% at week 42	[51]
	Multiple i.p. 70 mg/kg b.w., for 10 weeks (1×/week)	Adult (6 weeks) Sprague-Dawley rats (male)	- Carcinomas: 100% at week 20 - Obs.: Liver Cirrhosis	[52]
	Multiple i.p. 30 mg/kg b.w. for 11 weeks (2×/week)	Juvenile (4–5 weeks) Sprague-Dawley rats (male)		[53]
	Multiple gavage 70 mg/kg b.w. for 14 weeks (1×/week)	Juvenile (4–5 weeks) Wistar rats (male)	- Carcinomas: 80% at week 30 and 100% at week 34 - Lung metastasis: 20% at week 34 - Obs.: Liver Cirrhosis	[54]
DEN and CCl_4	- DEN: Single i.p. 10 mg/kg b.w. (week 2) - CCl_4: Multiple i.p. 0.25 to 1.50 mg/kg b.w. for 8 weeks (3×/week)	Infant (2 weeks) C3H/HeJ mice (male)	- Preneoplastic foci: 100% at week 17 - Adenomas: 100% at week 17 - Carcinomas: 12.5% at weeks 17	[55]
	- DEN: Single i.p. 10 mg/kg b.w. (week 2) - CCl_4: Multiple i.p. 0.2 mL/kg b.w. for 9 or 14 weeks (2×/week)	Infant (2 weeks) B6C3F1 mice (male)	- Preneoplastic foci: 100% at week 17 and 25% at week 22 - Adenomas: 37.5% at week 17 and 100% at week 22 - Carcinomas: 20% at week 17 and 50% at week 22	[48]
	- DEN: Single i.p. 200 mg/kg b.w. (week 4–5) - CCl_4: Multiple gavage 0.5 mL/kg b.w. for 6 weeks (3×/week)	Juvenile (4–5 weeks) F344 rats (male)	- Preneoplastic foci: ~81% at week 16 - Adenomas: 100% at week 16 - Carcinomas: 73% at week 16	[56]
CCl_4	- CCl_4: Multiple i.p. 0.2 mL/kg b.w. for 9 or 14 weeks (2×/week)	Infant (2 weeks) B6C3F1 mice (male)	- Preneoplastic foci: 0% at week 17 and 12.5% at week 22 - Adenomas: 0% at week 17 and 12.5% at week 22 - Carcinomas: 0% at week 17 and 25% at week 22	[48]
DEN and TAA	- DEN: Multiple i.p. 20 (1 dose), 30 (1 dose) and 50 mg/kg (6 doses) b.w., for 8 weeks (1×/week) - TAA: Multiple i.p. 300 mg/kg b.w., for 4 or 8 weeks (2×/week)	Infant (2 weeks) C57BL/6 mice (male and female)	- Adenomas + Carcinomas: 100% at week 24 (for both doses)	[49]
	- DEN: Single i.p. 200 mg/kg b.w. (week 6) - TAA: Multiple i.p. 200 mg/kg b.w. for 24 weeks (2×/week) (cycles of 3 weeks of administration and 1 week of rest)	Adult (6 weeks) Wistar rats (male)	- Preneoplastic foci: 100% at week 26 - Adenomas: 30% at week 26 - Carcinomas: 10% at week 26	[57]

Table 2. *Cont.*

Model	Procedure	Animal (Species, Strain, Age)	Timepoints and Incidence of Lesions	References
DEN, TAA and PB	- DEN: Single i.p. 200 mg/kg b.w. (week 6) - PB: 0.05% diet for 1 week - TAA: Drinking water 0.03% for 9, 10, 20 or 30 weeks	Adult (6 weeks) F344 rats (male)	- Adenomas: 0% at week 9, 16.7% at week 20, 100% at weeks 30 and 40 - Carcinomas: 0% at weeks 9 and 20, 25% at week 30, and 75% at week 40	[58]
DEN and PB	- DEN: Single i.p., 90 mg/kg b.w. (week 5) - PB: Drinking water 0.05% (week 7) for 17, 31 or 45 weeks	Juvenile (5 weeks) C3H/He, DBA/2 and C57BL/6 mice (male)	- Adenomas: 10–90% at week 24, 50–60% at 36, and 0–30% at weeks 36–52 (strain-dependent) - Carcinomas: 0% at week 24, 0–40% at 36, and 0–100% at weeks 36–52 (strain-dependent)	[41]
	- DEN: Gavage 80 mg/kg b.w. (weeks ~6–7) - PB: Drinking water 0.025, 0.05 or 0.1% (weeks ~7–8) for ~48 or ~70 weeks	Adult (~6–7 weeks) Sprague-Dawley rats (male)	- Carcinomas: ~5–20% at week 48 and 50–60% at week 70	[50]
	- DEN: Single i.p., 200 mg/kg b.w. (week 6) - PB: Drinking water 0.05% (week 7) for 36 weeks	Juvenile (4 weeks) F344 rats (male)	- Preneoplastic foci: 100% at week 42 - Adenomas: 64% at week 42 - Carcinomas: 50% at week 42	[51]
DEN, 2-AAF and PH	- DEN: Single i.p., 200 mg/kg b.w. (week 1) - 2-AAF: diet 0.02% for 2 weeks (weeks 3 and 4) - PH: 67% (week 3)	Adult Fischer 344 rats (male)	- Carcinomas: 68–71% at week 32, 75% at week 40 - Metastasis: ~3–4% at week 32, ~4% at week 40	[59]

i.p. = intraperitoneal; b.w. = body weight. 2-AAF: 2-acetylaminofluorene; CCl_4: carbon tetrachloride; DEN: diethylnitrosamine; PB: phenobarbital; PH: partial hepatectomy; TAA: thioacetamide.

Mindful of these findings, the studies of Vesselinovitch et al. shed light on the kinetics of using neonatal mice instead of juvenile/adult rodents. The main advantage of using neonatal model protocols in mice, also known as the "infant model", is the hepatic postnatal development context [42,60]. Compared to the adult liver, hepatocyte proliferation rates are higher in the liver of neonatal mice [61]. Thus, when given at low doses ranging from 1 to 50 mg/kg body weight to neonatal mice at 15–20 postnatal days, the pro-proliferative hepatic context promotes the clonal expansion of DEN-initiated hepatocytes, ultimately favoring hepatocellular (pre)neoplastic lesion development and shortening the time for HCC emergence compared to juvenile/adult animals. According to the findings of Vesselinovitch et al. [42], mice display a progressively lower HCC incidence as the age at DEN administration increased from neonatal (46–69%) to juvenile/adult mice (9–10%) in a strain- and dose-dependent manner. However, the fact that the latency time for HCC development following neonatal DEN administration remains long inspired the use of different types of promoters (i.e., substances or procedures that enhance tumorigenicity when administered after a carcinogen) and the establishment of multi-stage protocols. The features of these combined chemical and/or surgical procedures will be discussed in Sections 2.2.2–2.2.5. In multiple weekly administrations in mice and rats (Table 2), DEN also acts as a hepatotoxicant by causing damage and necrosis. These cellular processes trigger a progressive inflammatory response that may lead to extracellular matrix (ECM) accumulation, leading to fibrosis or cirrhosis (protocol-dependent) [49,52]. The chronic pro-inflammatory context, resulting in elevated levels of hepatomitogen cytokines, may promote clonal expansion of DEN-initiated hepatocytes by paracrine signaling [62], increasing the burden of neoplastic lesions in a shorter time (100% of animals at 20–24 weeks post-initiation) (Table 2), an effect similarly obtained by using 2-stage protocols with fibrogenic promoters, such as carbon tetrachloride (CCl_4) and thioacetamide (TAA) [49,52,53].

Concerning the early molecular alterations caused by non-fibrogenic and subnecrogenic DEN administration in the liver, Watanabe et al. [63] revealed some biologically relevant mRNA networks both 4h and 28 days post-initiation in mice. Most of these genes showed a dose-dependent increase after 4 h, but not after 28 days. At both time points, genes were associated with cancer (i.e., *Fos*, *Jun*, and *Myc* oncogenes), cell cycle arrest, and cell death (i.e., *Bax*, *Cdkn1a*, *CCng1*, and *Gadd45*) gene expression. Sequentially, the first and smallest morphologically recognizable lesion in chemically induced models of hepatocarcinogenesis in rodents are the preneoplastic foci, also called altered hepatocyte foci, AHF. In general, foci present clear phenotypical variations and are usually classified as basophilic, eosinophilic, or clear cell foci according to the tinctorial characteristic of most hepatocytes in Hematoxylin and Eosin (HE)-stained sections [55]. These phenotypes seem not to occur at random, considering that the cell lineages that originate from these lesions are theorized to undergo a "metabolic turnover". At first, DEN increases insulin growth factor 2 (IGF-2) levels, and IGF-2 downstream signaling decreases glucose-6-phosphatase (G6Pase) activity, promoting the emergence of glycogen storage phenotypes (eosinophilic and clear cell). The strong eosinophilia may result from the enhanced smooth endoplasmic reticulum (ER), peroxisome, or mitochondria. IGF signaling also promotes the Ras/Raf mitogen-activated signaling cascade, enhancing cell proliferation. Progressively, foci shift from anabolic to catabolic glucose metabolism to fuel cell proliferation, giving rise to the basophilic phenotype [64–66]. Along with the deregulated energetics hallmark, some AHF display *Hras* (10% of G6Pase-negative foci) and *Braf* (80–90%) oncogene mutations, which may provide a proliferative and growth advantage to these foci as late-stage neoplastic lesions also display these molecular alterations in higher frequency [67,68] (Figure 1). In this respect, *Braf* mutations are proposed to induce ERK1/Akt hyperphosphorylation and the induction of pro-survival/pro-proliferative complement component C5/C5a in basophilic foci [68] (Figure 1). For these reasons, AHF are generally considered putative preneoplastic lesions in chemically induced models, although the importance of morphologically similar lesions (glycogen-storing foci and small-cell change) is not completely understood in human hepatocarcinogenesis [65]. The molecular events that explain the stepwise progression of AHF to HCC are not fully unveiled, but recent findings indicate that some hepatocytes of DEN-induced AHF presenting oncogenic dephosphorylation of CCAAT/enhancer-binding protein alpha (C/EBPα) acquire a "stemness" feature, being classified as potential tumor-initiating hepatocyte (PTIH) [69]. Similar events were also described in the early and late stages of aggressive human HCC, suggesting that the preneoplastic foci with PTIHs are the origin of mouse HCC [69] (Figure 1).

In medium-term post-initiation timepoints (22–24 weeks), DEN has minimal effect on global miRNA expression and methylation profile in the liver, as only 8 miRNAs were upregulated and global/gene-specific methylation remained unaltered [70,71] (Figure 1). In more advanced stages, in a recent genome-wide investigation of stochastic point mutations, a high burden of potential coding alterations was observed in tumors (benign and malignant) harvested at 24–40 weeks post DEN initiation in C3H mice [38]. More than 80% of DEN-induced tumors had an activating hotspot mutation in either *Hras* or *Braf*, and around 20% of samples carried an activating mutation in *Egfr*. In addition, truncating mutations of gene suppressor *Apc* were exclusive to HCCs (21%). These alterations were considered putative oncogenic drivers of HCC in the DEN-induced model, as they may lead to the constitutive activation of Ras/Raf/MEK/ERK and Wnt/β-catenin signaling pathways, deregulating cell proliferation, growth, and survival processes. The downregulation of tumor suppressor miR-144–3p, as observed in human HCC, may also be accounted for Ras/Raf/MEK/ERK pathway activation in DEN-induced HCC as this miRNA downregulates *Egfr* [72] (Figure 1).

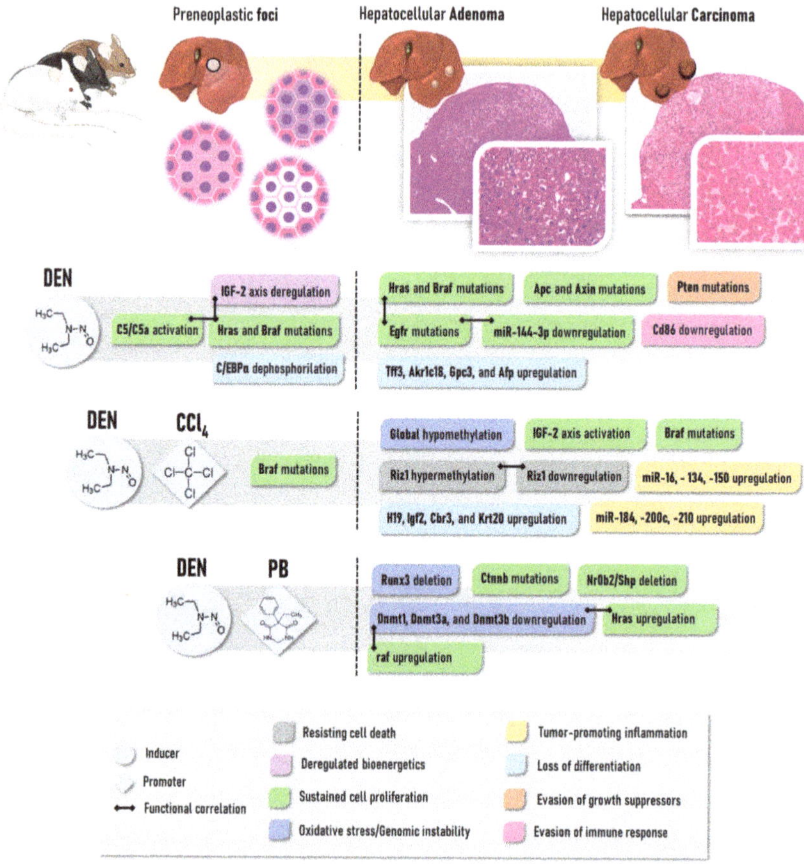

Figure 1. General depiction of the main molecular alterations and functional hallmarks involved in the development of preneoplastic (foci) and neoplastic (adenomas and carcinomas) lesions in widely-applied chemically induced models in mice. Strain-related and protocol-related variations should be considered. At late stages, molecular alterations are usually screened in a pool of neoplastic alterations, not considering if they are benign or malignant. CCl_4: carbon tetrachloride; DEN: diethylnitrosamine; PB: phenobarbital. The figure was composed with the aid of illustrations from the SMART-servier Medical Art available at https://smart.servier.com/ (accessed on 15 January 2021).

It is noteworthy that, as the occurrence of some mutations increased from benign to malignant tumors, also considering that some HCCs had a "nodule-in-nodule" morphological appearance at late stages, it has been hypothesized that the stepwise progression from benign tumors to HCC, similarly to the corresponding human disease. Another mutational profile also addressed activating *Braf* mutations in 89% of DEN-induced tumors carrying the V637E substitution, equivalent to the human V600E BRAF mutation. Of note, activating mutations in *Pik3ca* and inactivating mutations in the tumor suppressor *Pten* (PIK3CA inhibitor), involved in cell growth and angiogenesis, were also observed in 16% of tumors. Although common in the corresponding human disease, *Tp53*, *Tert*, and *Ctnnb1* mutations were not observed. Concerning the Wnt/β-catenin pathway, 5% of tumors displayed inactivating mutations in the tumor suppressor *Axin1* gene [73]. The transcriptomic profile of DEN-induced tumors sampled 44 weeks post-initiation (benign/malignant) also evidenced differential expressions of fetal/neonatal genes, such as *Tff3*, *Akr1c18*, *Gpc3*,

Afp, and *Abcd2*, which are involved in robust physiological proliferative responses of undifferentiated cells [74]. Compared to the other 9 genetically engineered mouse models, DEN-induced tumors in mice showed markedly lower expression of *Cd86*, which is an immune-checkpoint stimulator, a feature predictive of poor prognosis concerning immunotherapy strategies [75] (Figure 1). Of note, comparative analysis revealed that the transcriptomic profile of DEN-induced HCC is similar to the poorer survival group of human HCCs [76].

Ultimately, regarding the molecular alterations observed in models of multiple necrogenic DEN administrations, Liu et al. [77] characterized gene expression profiles in rats during the progression from liver cirrhosis to malignant lesions, also comprising adenomas, early and late HCCs. Compared to the cirrhotic stage, transcriptomic changes in late HCCs were increased by 32–46%, as 999 and 906 mRNAs were up- and downregulated, respectively. Interestingly, all stages shared 349 upregulated and 345 downregulated genes, which were mainly associated with fat metabolism (*Scd2*, *Fap4*, and *Fabp5*, upregulated), oxidative stress (*Akr1b7*, *Akr1b8*, and *Aldh3a1*, upregulated), anti-oxidant defense (many members of glutathione axis, as *Gstm3*), ECM synthesis (*Itga6*, *Lamc1*, *Col1a1*, and *Spp1*), cell growth, proliferation and migration (upregulation of many annexin isoform-coding genes, such as *Anxa1*, *Anxa2*, *Anxa3*, *Anxa5*, and *Anxa7*).

As further presented (Sections 2.2.2–2.2.8), there is a myriad of protocols applied in chemical-induced models (e.g., different chemical compounds, doses, frequencies of administration, etc.), and rat/mouse strains used (less or more susceptible), resulting in a clear methodological heterogeneity, and in the absence of a standard model. In general, regardless of the chemically induced protocol chosen, the models depicted in Section 2.2, mainly those induced by DEN, have been widely applied for the screening of predisposing and chemopreventive agents [57,78–83].

2.2.2. Carbon Tetrachloride (CCl_4)

In chemically induced rodent models, another widely applied xenobiotic is CCl_4 (PubChem CID:5943). This haloalkane, which is usually administered in multiple intragastrical or intraperitoneal doses, is considered a promoter in 2-stage hepatocarcinogenesis models after DEN initiation (Table 2). CCl_4 is metabolized in the hepatocytes by CYP2E1 to form the highly reactive oxygen trichloromethyl ($^*CCl_3$) and trichloromethyl peroxyl ($^*OOCCl_3$) radicals that promote lipid/protein damage and hepatocyte death, triggering an inflammatory response [84]. Oxidative stress, cell death, and inflammatory mediators are the stimuli for hepatic stellate cell (HSC) activation and collagen synthesis, ultimately leading to liver fibrosis and cirrhosis (a scenario that is absent in models using single or some DEN administrations) [5,85]. The establishment of a CCl_4-induced pro-inflammatory and pro-fibrogenic background is thought to promote the clonal expansion of DEN-initiated hepatocytes, increasing the incidence of adenomas and HCC by 87.5% and 50%, respectively, compared to mice receiving only DEN at 22 weeks post-initiation [48]. A similar increase in neoplastic lesion burden is also observed in rats, suggesting a CCl_4-mediated acceleration of HCC development [56,86] (Table 2).

Although *Braf* mutations are dependent on the genotoxic mechanism of DEN, which is absent in the CCl_4 regimen, Yamamoto et al. [68] showed that these alterations are maintained in (pre)neoplastic lesions induced by the DEN/CCl_4 protocol, suggesting the potential importance of this oncogene in tumors arising in a fibrotic context as well (Figure 1). Different from models using initiating non-fibrogenic DEN protocols, epigenetic alterations are key events in DEN/CCl_4-induced models of fibrosis-associated hepatocarcinogenesis. In this scenario, adenomas and carcinomas feature global DNA hypomethylation and decreased histone 3 lysine 9 trimethylation (H3K9me3), which are indicators of genomic instability. Furthermore, HCCs present promoter hypermethylation and functional downregulation of tumor suppressor *Riz1*, which was associated with accelerated tumor burden. Some of these alterations were also observed in the fibrotic tissue surrounding the lesions while absent in non-fibrotic tissue in DEN-initiated animals [70]

(Figure 1). In medium-term post-initiation timepoints (22 weeks), the DEN/CCl_4 protocol led to a distinct profile of 25 upregulated oncogenic and pro-fibrotic miRNAs, which are associated with proliferation, apoptosis, inflammation, and fibrosis functional networks, and thus also correlated with the increased neoplastic lesion burden [71] (Figure 1). Tumors arising from the CCl_4-induced fibrotic background also showed deregulated expression of oncofetal genes, such as the upregulation of *H19*, *Igf2*, *Cbr3*, and *Krt20* compared to DEN-induced tumors (Figure 1). In particular, continuous activation of the IGF-2-mediated axis in both tumors and surrounding fibrotic parenchyma, which is only observed in the early stages of mice submitted to the DEN protocol, mediates excessive hepatocyte proliferative stimuli following CCl_4-induced chronic liver injury, which could contribute to the increased (pre)neoplastic lesion burden [66,74]. Even though CCl_4 is routinely applied as a promoter by establishing necrogenic, inflammatory, and fibrotic responses, some protocols use this haloalkane as a complete carcinogen, as some hepatocytes are initiated by adduct formation between *CCl_3 radical and DNA while presenting decreased neoplastic lesion burden compared to DEN and DEN/CCl_4 protocols (Table 2) [48,74,84].

2.2.3. Thioacetamide (TAA)

TAA (PubChem CID 2723949) multiple i.p. injections or medium-long term administration in drinking water mimics chronic liver damage, fueling the development of DEN-induced (pre)neoplastic liver lesions in an inflammatory scenario in rodents (Table 2). TAA undergoes metabolic activation by CYP2E1 in the liver, generating S-oxide (TASO) and S, S-dioxide (TASO(2)) reactive compounds that sequentially exert amine lipids, protein damage, cell death, inflammatory response, HSC activation, excessive ECM synthesis, and fibrosis/cirrhosis in a protocol-dependent manner [87]. Most of the relevant histopathological and mechanistic data on DEN/TAA-induced hepatocarcinogenesis is derived from rat models. In short- and medium-term experiments, the screening of glutathione-S-transferase pi (GST-P)-positive foci by immunohistochemistry, which is not detected in normal liver, is widely applied and well-accepted in rat models. Placental GST-P is a long applied and accurate marker for the identification of putative preneoplastic lesions, as classical findings demonstrated that known hepatocarcinogens and hepatopromoters enhance the induction of GST-P+ foci, while non-hepatocarcinogens and non-hepatopromoters do not. In addition, late-stage neoplastic lesions, as liver adenomas and carcinomas, feature increased GST-P expression as well [88,89]. Noteworthy, TAA administration after single DEN administration increased the number and liver area occupied by GST-P+ foci by 5- and 10-fold compared to animals that were only initiated by DEN [90].

In the early stages of hepatocarcinogenesis, TAA promotion deregulated the expression of many G1/S and G2/M proteins, of which expression either increased or decreased, contributing to the clonal expansion of hepatocytes populations featuring checkpoint disruption and genomic instability in GST-P+ foci (Figure 2). Epigenetic alterations may be involved in these TAA-induced promoting mechanisms, such as the exon 2 of Cdkn2a featured hypermethylation, which was not found in animals submitted only to DEN initiation [90,91]. These early cell cycle alterations in GST-P+ foci may contribute to neoplastic lesion emergence since 50% of DEN/TAA-induced poorly differentiated HCCs display hypermethylation of exon 1 of *Cdkn2a* (Figure 2). The degradation (hyperphosphorylation) of tumor suppressor Retinoblastoma protein (pRb) (Figure 2), which has a pivotal role in the negative control of the cell cycle, is progressively increased in DEN/TAA-induced liver adenomas and carcinomas, while absent in early fibrotic stages [58]. More recently, Mizukami et al. [92] showed that TAA promotion might decrease the expression of TMEM70 and UBE2E2, involved in oxidative phosphorylation and cell cycling, in GST-P+ lesions by hypermethylation (Figure 2). Findings indicated that these alterations were acquired in early preneoplastic (foci) and increased in late neoplastic stages (adenomas and carcinomas) [92].

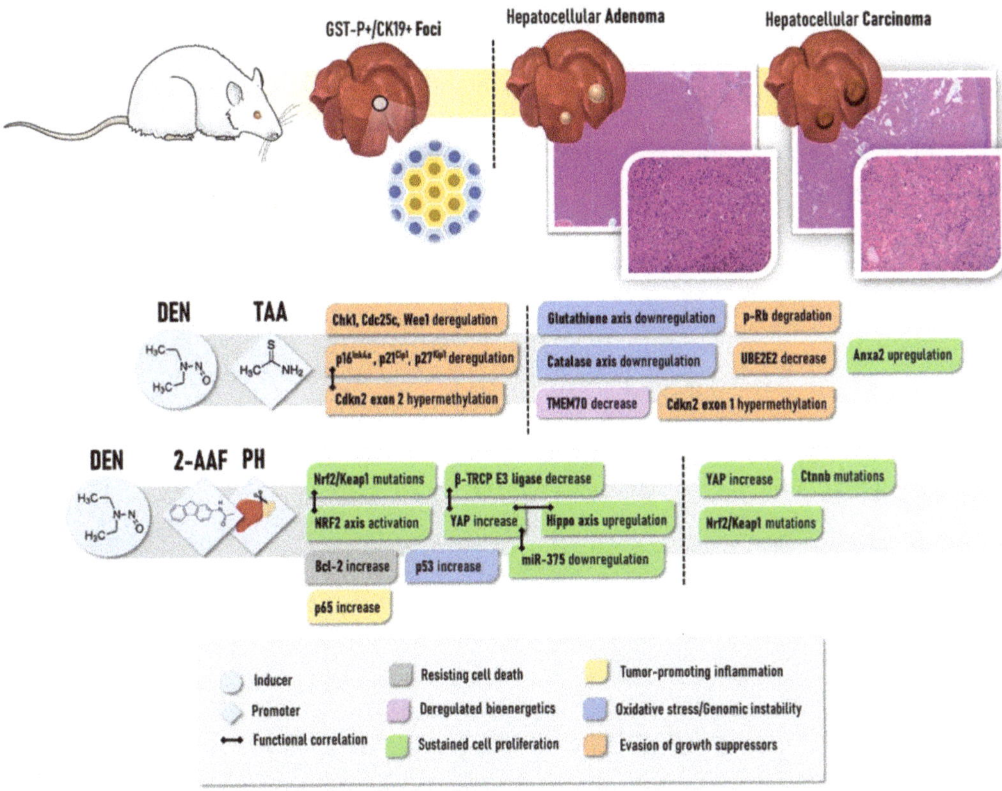

Figure 2. General depiction of the main molecular alterations and functional hallmarks involved in the development of preneoplastic (foci) and neoplastic (adenomas and carcinomas) lesions in widely-applied chemically induced models in rats. Strain-related and protocol-related variations should be considered. At late stages, molecular alterations are usually screened in a pool of neoplastic alterations, not considering if they are benign or malignant. 2-AAF: 2-acetylaminefluorene; DEN: diethylnitrosamine; PH: partial hepatectomy; TAA: thioacetamide. The figure was composed with the aid of illustrations from the SMART-servier Medical Art available at https://smart.servier.com/ (accessed on 15 January 2021).

Deregulation of the antioxidant axis, leading to increased oxidative stress, is also proposed to have key roles during DEN/TAA-induced hepatocarcinogenesis in rats [57]. Repeated administration of TAA depletes different anti-oxidant systems, decreasing total glutathione content and mRNA/activity of catalase, glutathione-S-transferase, and glutathione peroxidase [57,93]. Interestingly, the upregulation of Anxa2 was also seen in the liver of DEN/TAA-induced cirrhosis/hepatocarcinogenesis at 26 weeks post-initiation [57] (Figure 2). A hierarchical cluster analysis revealed that neoplastic lesions arising from both CCl_4-induced and TAA-induced cirrhotic backgrounds in mice had similar mRNA expression profiles, sharing the selective activation of the IGF-2 pathway in comparison to the tumors that emerged from the non-cirrhotic scenario [74]. Like CCl_4, repeated treatment with TAA is proposed to have initiating potential, while not as pronounced as DEN, as many DNA damage-inducible genes are upregulated, and (pre)neoplastic lesions (protocol-dependent) are observed in response to different TAA regimens in both rats and mice [74,94].

2.2.4. Phenobarbital (PB)

Phenobarbital (PubChem CID 4763) has been used for several decades as a promoter of 2-stage hepatocarcinogenesis models in rats and mice [41,45,50,95–97]. This non-genotoxic barbiturate is usually given in low doses in drinking water or diet after the administration of an initiating carcinogen like DEN (Table 2). In contrast to multiple DEN administration and CCl_4 or TAA regimens, PB promotion leads to a non-fibrogenic hepatic event. While the mechanisms regarding PB-related promotion are not fully elucidated, the hepatic context of PB administration positively selects hepatocytes harboring the activating mutations of the *Ctnnb1* gene, which lead to activation of Wnt/β-catenin signaling in about 80% of neoplastic lesions [98] (Figure 1). Moreover, most preneoplastic foci and neoplastic lesions induced by DEN/PB protocol present strong eosinophilia, whereas the common DEN-induced basophilic AHF is not as frequent [99]. For this reason, some authors denominate PB as a "tumor selector" or "selective promoter" rather than a classical promoter [100], considering that *Hras* mutations, which are frequent in DEN-induced tumors, are infrequent in DEN/PB-induced tumors. Moreover, *Ctnnb1* mutations are absent in protocols using only DEN as tumor initiator [38,100].

PB administration induces CYP450 enzymes, increasing the metabolic capacity of hepatocytes, which could increase the bio-activation of hepatotoxic drugs, thus enhancing their genotoxic/cytotoxic effects [101]. Some 2-stage rat protocols apply 1-week-long 0.05% PB interventions after DEN initiation and preceding promoter administrations, such as TAA (Table 2) [58]. It is suggested that the activation of the nuclear constitutive active/androstane receptor (CAR), which is involved in the induction of CYP450 enzymes, is essential for liver tumor promotion by PB in mice since CAR knockout (KO) mice led to the absence of (pre)neoplastic lesions in DEN/PB-induced protocol [102]. Furthermore, the sex-dependent interplay between CAR and β-catenin, being pronounced in male mice, may regulate enzyme induction and hepatocyte proliferation [103], which could explain the outgrowth of HCC with predominant eosinophilic phenotype and activated β-catenin signaling. More recently, Aleksic et al. [37] found that chromosomal instability may precede the outgrowth of *Ctnnb*-mutated hepatocytes. At early tumorigenesis stages, 29% of neoplastic alterations had chromosomal gains and/or losses, which increased in late stages, as 92% of tumors harbored these alterations. Among those, the loss of distal chromosome 4q, including the tumor suppressors Runx3 and Nr0b2/Shp, was an early and persistent event during DEN/PB-induced hepatocarcinogenesis (Figure 1). In contrast, Ctnnb occurred at high frequency only at late stages. In addition, PB and other chemicals have been shown to block gap junctional intercellular communication (GJIC) to exert their promotional activity. Although all mechanisms are not fully understood, Moennikes et al. [104] demonstrated that functional connexin 32 (Cx32) is required for tumor promotion by PB, considering that Cx32 null mice did not feature marked increases in size, volume, and/or the number of (pre)neoplastic lesions in response to PB promotion compared to Cx32 wild-type mice. In contrast, Cx26 KO mice have only minor effects on DEN/PB-induced mouse hepatocarcinogenesis [105].

Regarding the protocols (Table 2), PB administration after different DEN initiation protocols leads to 3–5-fold and 4–6-fold increases in the number of GST-P+ preneoplastic foci and HCCs in different rat strains in a concentration-dependent and time-dependent manner, respectively, compared to DEN counterparts [50,97]. The PB promotion effects in mice depend on the timing of DEN initiation. When given to mice submitted to DEN initiation at 2 weeks of age, PB did not alter (C3H/HeJ) or paradoxically attenuated tumorigenesis some mice strains, including C57BL/6J and B6C3F1, a crossbreed of C3H/HeJ and C57BL/6, whereas promoted in other strains, in particular, BALB/c and CD1 [45,47,95,106,107]. While apparently strain-dependent and not deeply investigated, some authors hypothesized a "feminizing" effect of early PB administration [95,106], also considering the key effects of sex hormones on hepatocarcinogenesis (see Section 2.2.7). Nonetheless, when 4–6 weeks-old mice are initiated with DEN and subsequently submitted to PB exposure (Table 2), the incidence of adenomas and carcinomas increases by 50–90% and 60–100% in a time-dependent and strain-dependent manner compared to animals that only received DEN, indicating marked tumorigenesis promotion [41]. DBA/2, C3H/He, and BALB/c mice showed increased sensibility to PB promotion, while C57BL/6 mice were rather refractory [41,108]. The results were partly attributed to the potential inter-strain differences on (1) PB metabolism, as PB serum levels were increased in DBA/2 and compared to C57BL/6 [41], and (2) PB-induced deregulation of the methylation status of key driver genes, as B6C3F1 is less capable of maintaining methylation balance compared to the C57BL/6 strain [109]. DNA methyltransferase genes (*Dnmt1*, *Dnmt3a*, and *Dnmt3b*) are downregulated in B6C3F1 mice [110] (Figure 1). The enzymes coded by these genes possess CAR response elements (CAREs), reinforcing PB as a CAR agonist. The multiple subsequent genomic events resulting from the deregulation of methylation status may be involved in tumorigenesis in this strain, such as hypomethylated *Hras* and raf upregulation (Figure 1) and alterations other genes involved in cell cycle, apoptosis, angiogenesis, invasion/metastasis [109–111]. The several hepatocarcinogenesis-related susceptibility/resistance loci mapped in these strains may also contribute to the aforementioned differences in response to PB promotion [112] (see Section 2.2.8)

2.2.5. Resistant Hepatocyte Model

One of the most applied models for the study of multistage chemical hepatocarcinogenesis is the Solt-Farber model in rats, which is also known as the "resistant hepatocyte (RH) model" [113,114]. In general, the RH model relies on a chemically induced genotoxic insult as an initiator followed by a regenerative response under a chemically induced selective pressure [115]. While several other chemicals were employed in the 1980s [114,115], the initiation protocol is usually accomplished by a single DEN dose followed by a short-term intragastrical or dietary administration of 2-acetylaminofluorene (2-AAF, PubChem CID: 5897) [113,116]. Under the 2-AAF regimen, rats are subsequently submitted to 70% partial hepatectomy (PH), which was introduced by Higgins and Anderson [117] to induce liver regeneration. 2-AAF administration exerts a mito-inhibitory selective property, thus blocking the proliferation of non-initiated hepatocytes and stimulating the DEN-initiated cells that are "resistant" to 2-AAF toxicity. Under the influence of the PH-induced proliferative stimulus, the selective expansion of these initiated hepatocytes results in preneoplastic foci and hyperplastic nodules, some of which may progress into HCC [113,114]. The model was first established in the susceptible Fisher-344 rat strain and later adapted to other rat strains, including the intermediate susceptible Wistar strain [118]. As the main outcomes of this protocol (Table 2), enzyme-altered preneoplastic lesions featuring an elevated expression of gamma-glutamyltranspeptidase (y-GT) and GST-P, visible primary HCCs and few metastatic tumors are observed in short- or medium-term studies [114,116,119]. About 95–98% of these enzyme-altered foci/nodules are proposed to suffer spontaneous remodeling to normal-appearing hepatocytes, called "remodeling lesions". On the other hand, only a small portion may progress to HCC, denominated as "persistent lesions". These lesions display differences regarding key molecular pathways that could direct their progression. Persistent GST-P-positive lesions have increased proliferative indexes, p53 accumulation, increased anti-apoptotic Bcl-2 staining, and enhanced p65 immunostaining compared to the remodeling ones, which showed increased apoptotic indexes [120] (Figure 2).

Moreover, the stem/progenitor cell origin of HCC has been proposed in this model [121,122]. In rodents, so-called oval cells, small periportal ductular-like progenitor cells that give rise to hepatocyte and bile ductular cell populations, are often observed during the early hepatocarcinogenesis stages in the RH model. The oval cells have been suggested to present natural resistance to mito-inhibitory chemicals and may originate hepatic tumors under the regenerative stimulus. Additionally, HCCs that arise in the RH rat model have shown similar immune–expression of oval cells markers, such as keratin (K)7, K19, and Ov6, indicating its possible progenitor cell derivation [121–123]. Perra et al. [124] have shown the involvement of the Hippo signaling pathway member YAP during the early stages of hepatocarcinogenesis in the RH model. This key transcriptional co-activator was found to be overexpressed at the translational level in both early and late hepatocarcinogenesis stages (Figure 2). In parallel, YAP target genes were also upregulated in preneoplastic foci and in oval cells. Moreover, the experimental disruption of YAP-related transcriptional complexes significantly reduced preneoplastic foci development and oval cell proliferation in rats, indicating the involvement of YAP in liver tumorigenesis. The overexpression of YAP in the early stages was associated with the downregulation of the β-TRCP E3 ligase and miR-375, known to negatively regulate this protein [124] (Figure 2). Of note, enhanced YAP expression was also featured in early human dysplastic nodules and adenomas [124] (Figure 2). Petrelli et al. [125] investigated the involvement of miRNA-gene interactions during the early stages of HR-induced liver carcinogenesis. Noteworthy, 80–85% of the most upregulated/downregulated genes in rat HCC were already altered in early K19-positive preneoplastic nodules. Among the deregulated networks, the activation of the nuclear factor erythroid-related factor 2 (NRF2) pathway and upregulation of the miR-200 family were described in K19-positive nodules. Reinforcing the translational value of the RH model, 78% and 57% of differentially expressed genes and miRNAs in rat HCC have been previously associated with human HCC, respectively. NRF2 pathway upregulation is indeed involved in early Nrf2/Keap1 mutations, which are observed in 71% of early preneoplastic lesions, in 59.3–78.6% of HCCs, and in 50% of lung metastases of HCC-bearing rats (Figure 2). Although the role of NRF2 as a tumor suppressor or oncogene is still controversial, data suggest an oncogenic role of this transcription factor as it may contribute to the clonal expansion of preneoplastic hepatocytes to HCC. Unlike human hepatocarcinogenesis, β-catenin gene mutations do not occur in the early stages of the RH model, and only in 18.5% of HCCs [126] (Figure 2).

2.2.6. Aflatoxin B1

Dietary intervention with low concentrations of aflatoxin B1 (AFB1, PubChem CID: 186907), its metabolites (aflatoxicol), or other aflatoxins (such as G1) has been extensively tested in rodent bioassays for hepatic carcinogenicity [127–130]. Although AFB1 is classified as a group 1 human carcinogen [28], and the consumption of improperly stored aflatoxin-contaminated food is widespread in the world, the identification of human aflatoxin-associated HCC cases is difficult, considering the unclear history of exposure [131]. One of the main molecular alterations caused in humans by AFB1 exposure is the point mutation (G to T) at codon 249 in the TP53 tumor suppressor gene [132]. However, site-specific mutations within the comparable codon in the Tp53 gene are not frequent in AFB1-induced liver (pre)neoplastic lesions in rats [108,133]. In rodents, the early AFB1-related hepatocarcinogenic mechanisms may be associated with increased lipid peroxidation and inflammation, and impaired anti-oxidant response that may contribute to cell injury, DNA damage, and preneoplastic foci growth [134,135]. In AFB1-induced HCC, the transcriptomic analysis revealed that AFB1 accounts for extensive deregulation in the expression of both protein-coding genes and long non-coding RNAs (lncRNAs). Some AFB1-deregulated lncRNAs clusters were associated with modification of apoptosis-, cell cycle-, response to DNA damage stimulus-, and Wnt receptor signaling pathway-related protein-coding genes. Apoptosis is proposed to contribute to AFB1-induced hepatic carcinogenesis since anti-apoptotic (*Bcl2*, *Mapk8*, and *Nfkb1*) and pro-apoptotic genes (*Casp1*, *Il4*, and *Mpo*) were upregulated in the HCC samples [136].

2.2.7. Miscellaneous Chemicals

Many other chemicals, including benzo(a)pyrene (BaP, PubChem CID: 2336), N-methyl-N-nitrosourea (MNU, PubChem CID 114836), and 1,2-dimethylhydrazine (1,2-DMH, PubChem CID: 1322), have been applied in classical bio-assays as initiator chemicals for the induction of enzyme-altered foci and tumors. Considering that these substances are not as efficient in inducing hepatic preneoplastic and/or neoplastic lesions compared to DEN-only protocols, they are usually combined with a chemically induced and/or surgically-induced cell proliferative promoting stimulus [59,137,138]. In a classical colon carcinogenesis bioassay, 1,2-DMH administration in Wistar rats led to increased oxidative stress, impaired anti-oxidant defense, upregulation of pro-apoptotic genes, and the development of few GST-P-positive foci in the liver 24 weeks after the carcinogen regimen [139]. In addition to the role of this hydrazine as an initiator, 1,2-DMH administration was proposed to promote a DEN-initiated bioassay by inducing CYP2E1, enhancing DNA adduct formation in the liver, and increasing the number of GST-P-positive foci [140].

2.2.8. Impact of Genetic Background and Sex

There is a spectrum of paradigms involving mice-specific susceptibility to hepatocarcinogenesis models, not only including the chemically induced bioassays but also the genetically modified ones [112,141] since different mice strains serve as backgrounds for the latter. In this respect, intrinsic genetic factors may contribute to the previously mentioned responses to chemical initiators and promoters. Several quantitative trait loci of susceptibility (Hcs) or resistance (Hcr) have been mapped using recombinant congenic and inbred consomic strains. The greater liver cancer predisposition of the C3H/HeJ compared to C57BL/6J strain is mainly attributed to hepatocarcinogen sensitivity 7 (Hcs7) loci found in chromosome 1 [142]. The $Hcs7^{C3H}$ allele was sufficient to confer susceptible traits to the C57BL/6 strain. Hcs7 may promote hepatocyte growth and proliferation in both normal and preneoplastic hepatocytes, apparently without affecting carcinogen metabolism and subsequent adduct formation [143,144]. Interestingly, Hcs7 encodes transcription factors, regulators of G-protein signaling, a member of the TNF ligand superfamily, and a receptor tyrosine kinase [142]. Other similar studies mapped many sensitivity loci in the C3H/He strain, whereas resistance loci were identified in both C57BL/6 and BALB/c strains, some of them carrying proto-oncogenes (such as c-jun and L-myc) [145–149]. In general, these genetic features may explain the fact that C3H/HeJ mice spontaneously develop HCC in a long-time latency, while incidence is low in crossbred C3B6F1 animals and extremely rare in C57BL/6 males [38]. The crossbred C3B6F1 strain, considered of intermediate susceptibility, is the default mouse strain for the National Toxicology Program. In rats, Hcs and Hcr loci were also identified in backcrosses and intercrosses experiments performed in susceptible F344 rats and resistant Brown Norway (BN) and Copenhagen (Cop) rats [150,151]. Moreover, in DHN strain, which is originated by inbreeding of Donryu colony, the Drh2 cluster located in rat chromosome 4 was closely associated by mapping analysis to suppression of (pre)neoplastic lesions during chemically induced hepatocarcinogenesis, controlling the expansion of GST-P positive foci and the emergence of HCC [152,153]. A general depiction of the main Hcs and Hcr loci in widely-applied mouse strains can be found in Figure 3.

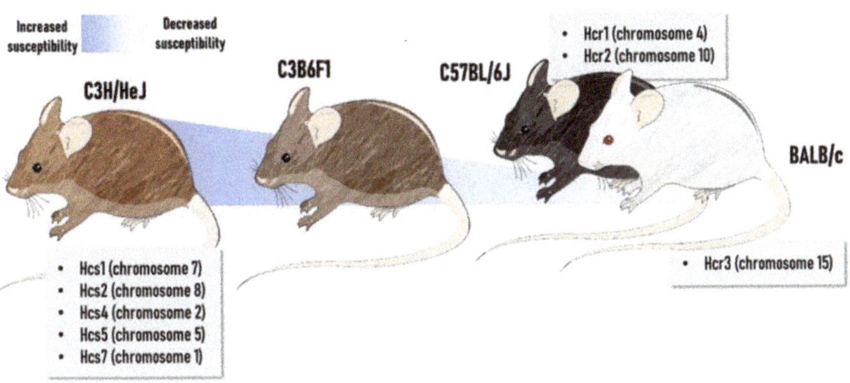

Figure 3. General depiction of some loci potentially involved in hepatocarcinogenesis susceptibility in widely-applied inbred and crossbred mouse strains. Hcs: Hepatocarcinogen susceptibility locus; Hcr: Hepatocarcinogen resistance locus. The figure was composed with the aid of illustrations from the SMART-servier Medical Art available at https://smart.servier.com/ (accessed on 15 January 2021).

One of the main advantages of using chemically induced models of hepatocarcinogenesis is the sex disparity feature reflecting the corresponding human disease. In men, both incidence and mortality rates for HCC are 2.8-fold higher compared to women [1]. In DEN-initiated models in mice, in particular when using CCl_4 or PB as promoters, females develop HCC at a later age and with a lower incidence/multiplicity in comparison to males, in a strain-dependent, dose-dependent, and timepoint-dependent manner [42,55,62]. The roles of sex hormones on hepatocarcinogenesis are not fully understood in both human and animal models. It is reported that 17β-estradiol (E2) exerts an anti-inflammatory effect by inhibiting the nuclear transportation of the p65 subunit of NF-κB in macrophages (RAW 264.7), also reducing NF-κB-related DNA-responsive elements [154]. Heterotopic-engrafted ovariectomized/castrated BALB/c mice treated with E2 featured reduced volume of tumors by suppressing the alternative activation of tumor-associated macrophages into a pro-inflammatory profile in an IL-4-Jak1-Stat6-dependent mechanism [155]. In accordance with these findings, estrogen-related receptor-α (ERR-α) KO mice enhanced DEN-induced hepatocarcinogenesis in a neonatal mice model, increasing the incidence (100 and 25%, respectively) and multiplicity (~7 and ~2 tumor/liver, respectively) of tumors, in comparison to wild-type mice, attesting that ERR-α KO mice are susceptible to HCC initiation and progression. In addition, KO-ERR-α mice display increased nuclear recruitment of p65 subunit, increased level of DNA synthesis, and necrosis occurrence, suggesting a cytokine-driven compensatory proliferation mechanism that promotes hepatocarcinogenesis progression [156]. Thus, E2 is suggested to be one of the mechanisms responsible for the sex disparities observed in epidemiological and in vivo experimental model data, attenuating the HCC progression. Although the genetic basis of female resistance for hepatocarcinogenesis is not fully unveiled, the introgression of Hcs4 from BN rats in F344 background revealed that this locus of chromosome 16 may display resistance genes regulated by sex hormones. The gonadectomy of congenic F344.BN-Hcs4 rats during the establishment of resistant hepatocyte increased the development of (pre)neoplastic lesions in females while decreasing in males. In keeping with these findings, the administration of testosterone to gonadectomized F344.BN-Hcs4 females resulted in enhanced (pre)neoplastic lesion burden, similarly to parental F344 males, whereas the administration of E2 to gonadectomized F344.BN-Hcs4 males decreased (pre)neoplastic lesion emergence, relatable to parental BN females. These effects were accompanied by functional receptor modulation, in special ERR-α, indicating the potential modulation of sex hormone-sensitive gene (s) in this chromosome [151].

It is also suggested that androgens might be responsible for the sex disparities observed in humans and in vivo experimental models. The androgens exert their bioactive function by interacting with androgen receptors (AR), which then act as a transcription factor and induce the expression of key molecules associated with hepatocarcinogenesis [157]. However, it is still uncertain whether androgens and/or AR were responsible for inducing the HCC progression. Accordingly, it is observed that nuclear AR is overexpressed in ~33% of HCC samples when compared to noncancerous liver tissues (~2-fold), correlating to the poorer overall survival of patients and prognostic [158]. Additionally, it is observed that both male and female KO-AR (total or liver-specific) mice submitted to a DEN-induced hepatocarcinogenesis model featured similar serum testosterone levels, in addition to a longer latency period, with reduced incidence and size of tumors, when compared to male, female and littermates wild-type mice. This data suggest that AR rather than testosterone promotes HCC progression by modulating the oxidative-apoptotic axis [159]. Therefore, the sexual disparities observed in epidemiological and reflected in vivo experimental model is mainly related to the mechanisms of the E2 and AR by modulating the inflammatory-oxidant axis that turns the hepatic milieu susceptible to HCC emergence.

2.3. *Diet-Induced Rodent Models*

2.3.1. NAFLD-Associated HCC Models

In the last decade, a variety of suitable preclinical models mimicking NAFLD/NASH-driven HCC have been developed. As reviewed by Febbraio et al. [160], although none of the available models fully reproduce the broad range of complex events of NAFLD/NASH pathogenesis, presenting discrepancies in the presence/absence of obesity, insulin resistance, inflammation/ER stress, and NASH, most mechanistic data on NAFLD-associated hepatocarcinogenesis are derived from these mouse models (Figure 4A). In general, these bioassays are classified into (1) diet-induced, (2) chemically induced, and (3) genetically modified models, and (4) "hybrid" models combining these 3 interventions (Table 3). There is a great diversity of dietary ad libitum intervention models available in the literature, mostly displaying high sugar and/or fat contents. In general, diet-only interventions require a long period of latency, also presenting a highly variable tumor incidence, multiplicity, and size (Table 3). Despite this disadvantage, neoplastic lesions arise as part of the natural disease progression and do not require induction by a chemical carcinogen. In these models, the C57BL/6J strain is widely chosen because of its predisposition to developing insulin resistance and obesity [161]. More recently, Asgharpour et al. [162] showed that B6/129 mice, which are derived from a C57BL/6J and 129S1/SvImJ background, are more insulin-resistant, NASH-prone, and HCC-prone compared to their parental strains. In NASH-driven HCC in B6/129 mice, the transcriptomic analysis revealed the activation of both metabolic and oncogenic pathways, including nitrogen and amino acid metabolism, oxidative stress signaling, inflammation, cell adhesion, and ECM remodeling. Interestingly, tumors featured the upregulation of the proto-oncogene Mertk, which is a tyrosine kinase-coding gene involved in proliferation and invasion, and the downregulation of *Ctnnbip1*, a negative regulator of the β-catenin pathway. The comparison between the transcriptomic signatures of human HCC and NASH-driven HCC in B6/129 mice revealed close similarity to S1/2 subclasses of human HCC, which are characterized by WNT, MYC, and AKT pathway activation [162,163]. As demonstrated by Dowman et al. [164], some of these tumors showed nuclear accumulation of β-catenin protein, indicating Wnt pathway activation in mice as well. In the C57BL/6J strain, NASH-driven HCC featured miRNA deregulation, including the upregulation of miR-155, -193b, -27a, -31, -99b, -484, -574-3p, -125a-5p, and -182, and the downregulation of miR-20a, -200c, -93, -340-5p, and -720. Some of these miRNAs were proposed to have oncogenic or tumor suppressor activities, similar to the corresponding human disease [165] (Figure 4A).

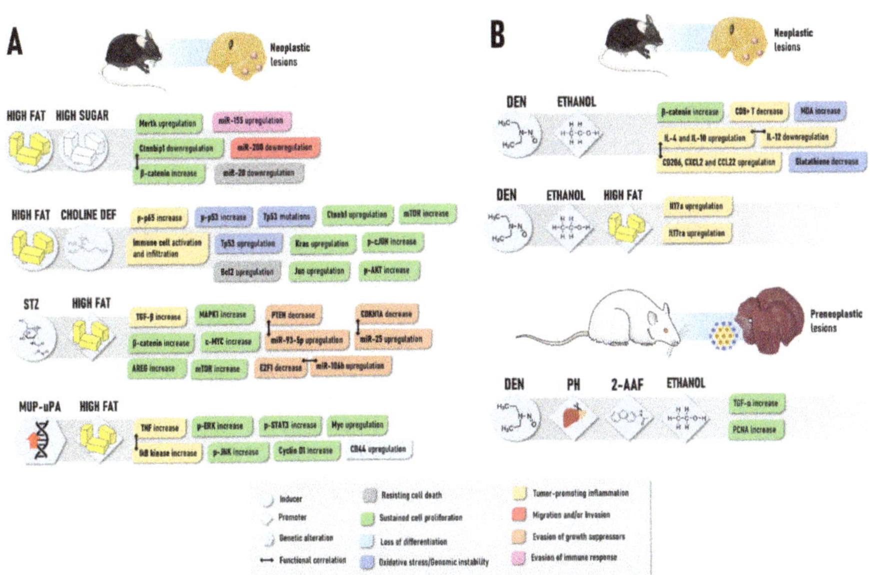

Figure 4. General depiction of the main molecular alterations and functional hallmarks involved in the development of neoplastic (adenomas and carcinomas) lesions in NASH-driven (**A**) or ALD-driven models (**B**) in mice and rats. NASH: Non-alcoholic steatohepatitis; ALD: Alcoholic liver disease. The figure was composed with the aid of illustrations from the SMART-servier Medical Art available at https://smart.servier.com/ (accessd on 15 January 2021).

Table 3. Summary of some of the NASH-induced hepatocarcinogenesis protocols mice.

Model	Procedure	Animal (Species, Strain, Age)	Timepoints and Incidence of Lesions	References
HF/HS diet	High-fat and high sugar diet (~35% hydrogenated coconut oil and soybean oil, ~19% carbohydrate, *w/w*) for 12, 24 or 48 weeks	Juvenile (4 weeks) C57BL/6J mice (male)	Neoplastic lesions: 0% at weeks 12 and 24 and 20% at week 48	[165]
	High-fat diet (~21% fat and 0.1% cholesterol, *w/w*) and high sugar solution (23.1/18.9 g/L of fructose/glucose) for 56 weeks	Juvenile to adult (6–8 weeks) B6/129 mice (male)	Adenomas: 25% at week 56 Well-differentiated Carcinomas: 100% at week 56 Poorly-differentiated Carcinomas: 37.5% at week 56	[162]
	High-fat (23% *w/w*, of which 23% saturated, 34% trans, 31% monounsaturated (cis), 12% polyunsaturated) and high sugar solution (23.1/18.9 g/L of fructose/glucose) for 24 or 48 weeks	Juvenile to adult (6–8 weeks) B6/129 mice (male)	Neoplastic lesions: 0 at week 24 and 40% at week 48	[164]
	High-fat and high sugar diet (21.1% fat, 41% sucrose, and 1.25% cholesterol, *w/w*) and high sugar solution (23.1/18.9 g/L of fructose/glucose) for 12 or 24 weeks	Adult (9 weeks) C57BL/6J mice (male)	Neoplastic lesions: 0% at week 12 and 30% at week 24	[166]
HF/HS diet and CCl_4	High-fat and high sugar diet (21.1% fat, 41% sucrose, and 1.25% cholesterol, *w/w*) and high sugar solution for 12 or 24 weeks CCl_4: Multiple i.p. 0.2 mL/kg b.w. for 24 or 36 weeks (1×/week)		Neoplastic lesions: 0% at week 12 and 100% at week 24	
HF diet	High-fat (~24% fat, lard and soybean oil, *w/w*) for 48 weeks	Juvenile (4–5 weeks) C57BL/6J mice (male and female)	Carcinomas: 2.5% at week 48	[167]
CDHF diet	High-fat (~24% fat, lard and soybean oil, *w/w*) and choline-deficient diet for 48 weeks		Carcinomas: 25% at week 48	
CDAHF diet	High-fat (~35% fat, *w/w*), choline-deficient, L-amino acid-defined, 0.1% methionine diet for 12, 24, 36, 48 or 60 weeks	Juvenile (5 weeks) C57BL/6J mice (male)	Adenomas: 0% at week 12 and 24; 67% at week 36; 100% at weeks 48 and 60 Carcinomas: 0% at week 12 and 24; 17% at week 36; 9% at week 48; 26% at week 60	[168]

Table 3. *Cont.*

Model	Procedure	Animal (Species, Strain, Age)	Timepoints and Incidence of Lesions	References
CDA diet	Choline-deficient L-amino acid-defined diet for 24 or 36 weeks	Juvenile to adult (6–8 weeks) C57BL/6 mice	Carcinomas: 30–40% at weeks 24 and 36	[169]
CDA diet and CCl_4	Choline-deficient L-amino acid-defined diet for 24 or 36 weeks CCl_4: Multiple i.p. 0.2 mL/kg b.w. for 24 or 36 weeks (1×/week)		Carcinomas: 40% at weeks 24; 100% at week 36	
STAM	STZ: Single s.c. 200 µg (day 2) High fat (32%) diet for 20 weeks	Infant (2 days) C57BL/6J mice (male)	Carcinomas: 100% at weeks 14 to 20	[170]
Pten null	-	*Pten* null mice (male and female)	Adenomas: 47% at week 44 and 100% at weeks 74–78; Carcinomas: 66% at weeks 74–78	[171]
MUP-uPA transgenic	High-fat (59% kcal from fat) for 32 or 40 weeks	MUP-uPA mice	Adenomas: 50% at week 32 and 71.4% at week 40; Carcinomas: 16.6% at week 32 and 50% at week 40	[172]

b.wt.: body weight; i.p.: intraperitoneal; s.c.: subcutaneous. HS: high sugar; HF: high fat; CCl_4: carbon tetrachloride; DEN: diethylnitrosamine; CDHF: choline-deficient high fat; CDA: choline-deficient L-amino acid-defined; CDAHF: choline-deficient L-amino acid-defined high fat; STAM: Stelic Animal Model; STZ: streptozotocin; uPA: urokinase plasminogen activator.

Another commonly applied diet-induced bioassay for NAFLD/NASH-driven HCC modeling is the ad libitum feeding with the toxic choline-deficient high-fat (CDHF) diet [167]. The CD diet is known to exacerbate HF-induced NASH, as inadequate choline uptake impairs hepatic lipoprotein secretion and promotes oxidative damage caused by mitochondrial dysfunction and ER stress. Wolf et al. [167] demonstrated that the CDHF diet increased the incidence of HF-induced HCC by 10-fold (Table 3), unraveling an interaction between inflammatory cells (natural killer and CD8+ T lymphocytes) and hepatocytes that lead to liver damage, canonical NF-kB signaling activation hence promoting NASH-to-HCC transition (Figure 4A). Although CDHF-induced HCCs showed heterogeneous patterns of chromosomal aberrations, copy number changes revealed similarity with cryptogenic HCC in humans. These murine tumors also demonstrated deregulated protein and/or mRNA expression of many oncogenes, such as p-AKT, p-cJUN, p65, *Bcl2*, *Ctnnb1*, *Kras*, and *Tp53*, that also presented missense activating mutations. Interventions with L-amino acid-defined diets -in combination with the CDHF diet also fuel the HF-driven HCC burden by enhancing the NASH background [168].

In toxin-based approaches, the so-called Stelic Animal Model (STAM) of NASH-hepatocarcinogenesis is widely established [170,173,174] (Table 3). Low-dose administration of streptozotocin (STZ) in the first days of life of a mouse leads to oxidative injury in pancreatic islets and profound changes in hepatic transcriptomic profile [175,176]. This alkylating agent established diabetic conditions, usually absent in dietary interventions, which promote rapid lipogenesis, fatty acid oxidation, hepatocellular injury, and fibrosis [170,173]. In combination with the HF diet, STZ-administered mice display a higher and faster burden of tumors compared to HF diet-only interventions (Table 3), since mice display at least 4 detectable HCCs, and an average tumor growth rate of 150% from 16 to 20 weeks of age [170]. De Conti et al. [174] further characterized profound deregulation in miRNA-target networks in this model, including the upregulation of many miRNAs and the activation of major oncogenic pathways, including TGF-β, Wnt/β-catenin, ERK1/2, mTOR, and EGF signaling. In particular, E2F1, PTEN, and CDKN1A were directly targeted by the upregulation of miR-106b, miR-93-5p, and miR-25 in NASH-cirrhosis and full-fledged HCCs stages (Figure 4A). Interestingly, some of these upregulated miRNAs were also featured in human HCC, and progressive increase of their expression levels from the NAFLD/NASH (weeks 6–12) to HCC (week 20) stages, eliciting their importance during disease progression. Among the other toxins applied in association with dietary interventions (Table 3), CCl_4 multiple administrations, in similar protocols as described in Table 2, are also chosen to increase not only the inflammatory/fibrotic context of NASH but also the neoplastic lesion burden [85,169].

Concerning NASH-related genetically modified models, the hepatocyte-specific *Pten* deficiency results in a fast induction of steatohepatitis, as hepatocytes acquire adipogenic-like features [171,177]. Given that *Pten* is also a tumor suppressor gene highly implicated in hepatocyte homeostasis, *Pten* null mice also have a high burden of hepatocellular neoplastic lesions (Table 3) [171]. More recently, Nakagawa et al. [172] developed a model combining HF diet feeding and MUP-urokinase plasminogen activator (uPA) transgenic mice. These animals have high levels of uPA, which induces transient ER stress and liver damage, that is also implicated in human NASH, leading to the development of indistinguishable NASH-related morphological and molecular hallmarks. Other authors also noticed a high burden of hepatocellular adenomas and carcinomas (Table 3), evidencing that uPA-induced ER-stress and HF have synergistic roles on both NASH development and HCC progression. These events were highly dependent on TNF production by inflammatory liver macrophages and TNF receptor 1 (TNFR1)-IkB kinase b (IKKb) signaling in hepatocytes.

In general, diet-induced protocols (Section 2.3) usually rely on the administration of "real-world" methodological approaches (fat, sugar and alcohol intake, as the corresponding human habits) as their main advantage. Moreover, key transcriptomic resemblances to human HCC are also observed [161,162]. Nonetheless, one should consider the long latency time to HCC emergence and highly variable incidence as the main disadvantages of these bioassays.

2.3.2. ALD-Associated HCC Models

According to IARC, there is sufficient evidence in both humans and experimental animal models to substantiate the carcinogenicity of ethanol since this toxin is classified in group 1 [28]. Nonetheless, a very low frequency of neoplastic alterations is observed in long-term intervention in rats [178]. In mice, a statistical trend was observed in 2-year-long ethanol intervention in drinking water regarding the incidence of neoplastic lesions, mainly adenomas [179]. Alcohol is usually applied as a promoting or co-carcinogenic agent in chemically induced HCC models in drinking water or as a part of liquid diets (Table 4). Nevertheless, the experimental use of ethanol as a promoter of DEN-initiated models displays controversial results in rats. The cessation of long-term alcohol administration after DEN initiation seemed to enhance GST-P+ foci development. However, intermittent alcohol intake showed to decrease in the number of these preneoplastic lesions. As ethanol exerts suppressing effects on liver regeneration, the cessation of long-term alcohol insult may reactivate hepatocyte proliferation, thereby promoting preneoplastic liver development [180,181]. As these effects were not evaluated in neoplastic lesions and the exact mechanisms were not fully evaluated, a model-dependent effect should not be discarded. Conversely, using the same model, ethanol significantly increased the Ki-67 positivity in GST-P+ foci and incidence/multiplicity of HCC in Cx32 dominant-negative transgenic rats but not in wild-type counterparts. In addition, increased nuclear-phosphorylated Erk1/2 and reduced Erk1/2-inhibitor Dusp1 protein and mRNA were only observed in Cx32 transgenic rats, suggesting enhancing effects of ethanol on DEN-induced hepatocarcinogenesis via Cx32 dysfunction, which is commonly observed in human chronic liver disease [182].

Table 4. Summary of the ALD-induced hepatocarcinogenesis protocols in rodents.

Model	Procedure	Animal (Species, Strain, Age)	Timepoints and Incidence of Lesions	References
Ethanol	Liquid low (1%, *w/w*) or high (3%, *w/w*) ethanol diet for ~100–110 weeks	Juvenile to adult (6–7 weeks) Sprague-Dawley rats (male and female)	Neoplastic lesions: 2% at weeks ~100–110 (for both doses and sexes)	[178]
	Ethanol in drinking water 2.5% or 5% (*v/v*) for 104 weeks	Juvenile (4 weeks) B6C3F1 mice (male)	Neoplastic lesions: 34% (2.5%) and 52% (5%) at week 108 Adenomas: 25.5% (2.5%) and 39.6% (5%) at week 108	[179]
MeIQx and Ethanol	MeIQx: 200 mg/kg diet for 8 weeks Ethanol in drinking water 10 or 20% (*v/v*) for 16 weeks	Juvenile (3 weeks) F344/DuCrj rats (male)	Adenomas: ~80% (10%) or ~100% (20%) at week 27 Carcinomas: ~20% (10%) or ~50% (20%) at week 27	[183]
Resistant hepatocyte and Ethanol	DEN: Single i.p. 200 mg/kg b.w.,at week 6 2-AAF: 200 mg/kg in diet for 3 weeks PH: at week 9 Ethanol in drinking water 5% (*v/v*) for 5 or 15 weeks	Juvenile (4 weeks) Wistar rats (male)	Preneoplastic foci: 63%–100% at week, 18 44%–100% at week 28 * Adenomas: 75% at week 18, 94% at week 28 * Carcinomas: 0% at week 18, 0% at week 28 *	[184]

Table 4. *Cont.*

Model	Procedure	Animal (Species, Strain, Age)	Timepoints and Incidence of Lesions	References
DEN and Ethanol	DEN: Single i.p. 200 mg/kg b.w., at week 9 Ethanol in drinking water 5% (*v/v*) for 16 weeks	Adult (9 weeks) WT rats	Adenomas: 8% at week 25 Carcinomas: 0% at week 25	[182]
		Adult (9 weeks) Cx32 dominant-negative transgenic rats	Adenomas: 25% at week 25 Carcinomas: 25% at week 25	
	DEN: Single i.p.10 mg/kg b.w., at week 2 Lieber-Decarli diet (4.9% of ethanol, *v/v*) for 16 weeks	Infant (2 weeks) C57BL/6 mice (male and female)	Eosinophilic foci: 53% at week 23 Adenomas: 60% at week 23 * Carcinomas: 13% at week 23 *	[185]
	DEN: Single i.p. 1 mg/kg b.w. at weeks 3–4 Ethanol in drinking water 5% (*v/v*) for 3 days, followed by 10% (*v/v*) for 3 days and 10/20% (*v/v*) (alternate days) for 8 weeks, during weeks 16 to 24 or 40 to 48	Juvenile (3–4 weeks) B6C3 mice (male)	Neoplastic lesions: 97.5% at week 48 *	[186]
DEN, Ethanol and HF	DEN: Single i.p. 25 mg/kg b.w. at weeks 2 Liquid ethanol diet (gradually increased from 1% to 2% and 3% (*v/v*, throughout 3 weeks), and maintained at 3.5% (*v/v*)) for 18 or 24 weeks.	Infant (2 weeks) C57BL/6 mice (male)	Carcinomas: ~20% at week 18; ~70% at week 24	[187]

* not statistically different from DEN-only control mice. 2-AAF: 2-Acetylaminefluorene; ALD: alcoholic liver disease; b.wt.: body weight; Cx32: connexin 32; i.p.: intraperitoneal. DEN: diethylnitrosamine; MeIQx: 2-amino-3, 8-dimethylimidazo 4,5-f]quinoxaline; PH: patial hepatectomy.

Ethanol also elevates the abundance of preneoplastic and neoplastic lesions induced by an RH model in Sprague Dawley rats [184] and the 2-amino-3, 8-dimethylimidazo [4,5-f] quinoxaline (MeIQx) carcinogen, a relevant heterocyclic amine found in cooked meat [183,188] (Table 4). In the RH model, a 5% ethanol intervention for 15 weeks enhanced the size and area occupied by GST-P+ preneoplastic foci and the multiplicity of neoplastic lesions while not significantly altering their incidence (Table 4). Furthermore, ethanol increased the proportion and the multiplicity of preneoplastic foci with the double expression of GST-P and transforming-growth factor-alpha (TGF-α) markers, indicating that TGF-α may be a pathway for the promoting activity of ethanol towards hepatocarcinogenesis [184] (Figure 4B). In addition, high doses (10% and 20%) of ethanol dose-dependently increased the incidence and/or the multiplicity of hepatocellular adenoma/carcinoma induced by MeIQx in rats [183]. A low dose intervention increased the number of small MeIQx-induced GST-P+ foci by enhancing proliferating cell nuclear antigen (PCNA) immunohistochemical staining and the levels of 8-hydroxydeoxyguanosine, a marker of oxidative DNA damage [188]. Similar effects on the enhancement of GST-P+ foci emergence were also observed when ethanol was co-administered with MeIQx or N-nitrosomorpholine [189,190].

In mice, the promoting effects are similar to the rat bioassays. In general, alcohol is proposed to increase the multiplicity and/or size of both preneoplastic and neoplastic lesions in DEN-induced models while not having pronounced effects on the incidence of lesions compared to DEN (pair-fed or not)-only counterparts (Table 4). Brandon-Warner et al. [186] reported a slight but significant increase in the number and area occupied by preneoplastic foci mediated by a 10/20% 8-week-long ethanol intervention after multiple DEN administrations in juvenile B6C3F1 mice. Promoting effects on the size of macroscopically-identified neoplastic observed in a 5% ethanol 10-week-long intervention in adult C57BL/6 mice submitted to multiple DEN injections [191]. Likewise, Mercer et al. [185] described an increase in the mean number of lesions, and, specifically, in the number of eosinophilic cell foci and adenomas after a 5% ethanol 16-week-long intervention in a neonatal mouse model. In this case, a significant ethanol-induced enhancing effect on incidence was observed only concerning eosinophilic foci phenotype (Table 4). Mechanistically, the alcohol-promoting effects in DEN-initiated mouse models are related to immune system disturbances, oxidative stress, and sustained cell proliferation hallmarks (Table 4). Ethanol intervention promoted the mRNA expression of epithelial-mesenchymal transition (EMT) biomarkers, such as E-cadherin, Snail, MMP-9, and also favored M2 polarization of tumor-associated macrophages by upregulating IL-4, IL-10, CD206, CXCL2, and CCL22, and downregulating IL-12 mRNA (Figure 4B). Antitumoral CD8+ T cells were decreased in the liver of ethanol-fed mice as well [191]. Ethanol exacerbated DEN-induced oxidative stress in the liver by enhancing malondialdehyde while diminishing glutathione levels [186]. Of note, ethanol-mediated increase in hepatocyte proliferation was associated with a cytoplasmic β-catenin staining pattern in alcohol-associated HCC cases in humans [192], and the ethanol-mediated activation of the β-catenin axis was also implicated in the increase of stemness and metastasis of HCC cells [193]. Likewise, increased expression of Wnt7a, β-catenin, phosphorylated GSK3β, and several targets of Wnt/β-catenin pathway, such as glutamine synthetase, cyclin D1, Wnt1 inducible signaling pathways protein (WISP1), and matrix metalloproteinase-7, were detected in the liver of male Sprague-Dawley rats fed only alcohol for ~5 months via total enteral nutrition [185]. Although the described molecular events are directly correlated with liver tumorigenesis, most investigations did not perform molecular analysis in neoplastic lesions but only in whole liver samples. In a recent investigation that considered the epidemiological evidence of the synergistic effects of alcohol with other common HCC risk factors, Ma et al. [187] showed that the combination of HF and ethanol accelerates by 2-fold or 8-fold the DEN-induced HCC incidence in mice compared to HF-fed only counterparts (Table 4). Further experiments in IL-17 KO mice showed that the expression of many classical tumorigenic genes in HCC, as well as inflammatory, fibrogenic, and lipogenic genes in non-tumoral tissue, is IL-17-dependent. Ethanol-induced IL-17 signaling in steatotic hepatocytes, mediated by Th17 cells and macrophages, promoted (1) lipogenesis via activation of the caspase-2-SP1-SREBP1/2-DHCR7 pathway and (2) hepatocellular damage by preventing TNFR1 exocytosis [187] (Figure 4B).

2.4. Genetically Engineered Mouse Models

2.4.1. Hepatitis Virus Transgenic Mice

First established in the 1990s, HBV and HCV transgenic models enabled a deeper insight into the pathogenetic mechanisms involved in viral-induced chronic liver injury and malignant transformation. Using transgenic technology, the transgenic phenotype is intentionally obtained by the insertion of exogenous viral DNA. To investigate the oncogenic mechanisms involved in HBV infection, transgenic mice expressing different components of the viral particle were developed (Table 5). Kim et al. [194] first introduced transgenic mice containing the entire of HBx protein-coding gene and its transcriptional enhancer via micro-injection into single-cell embryos of the CD1 mice strain. Further studies were also performed in C57BL/6xDBA and C57BL/6xCBA hybrid backgrounds [195,196]. The HBx protein acts as a transcriptional transactivator, promoting the expression of many oncogenes, including FOS, JUN, and MYC, also inducing HBV transcription and replication [197,198]. Most HBx-transgenic mice feature macro and micro fatty changes at 6–11 months of age. Although inflammation was not frequently observed from 6–15 months of age, most mice developed preneoplastic liver dysplasia and adenomas from 6–11 months [196]. Proteomic analysis of these preneoplastic stages revealed that 22 proteins were differentially expressed in dysplasia or adenomas, 5 upregulated and 17 downregulated. Deregulated proteins indicated that alterations in glycolysis and lipogenesis might be critical during the early (preneoplastic) stages of HBx-induced hepatocarcinogenesis, considering the upregulation of fatty acid-binding protein 2 (FABP2) and the downregulation of cytoplasmic malate dehydrogenase (MDH) and mitochondrial 3-ketoacyl-CoA thiolase (HADHA). These alterations were validated in human HCC samples. In addition to metabolism-related proteins, the authors found that raf kinase inhibitory protein (RKIP), a negative regulator of Raf-1, was downregulated at all hepatocarcinogenesis stages [196].

Table 5. Summary of some of the genetically-engineered models established in mice.

Model	Genetic Modification	Timepoints and Incidence of Lesions	References
HBV-transgenic	HBx gene	- Carcinomas: 0% at weeks 16 and 24, 50% at 44–52 weeks, 75% at 60–72 weeks	[195]
		- Adenomas: 8.3% at week 24, 57% at week 44, 12% at week 60 - Carcinomas: 8.3% at week 24, 14% at week 44, 76% at week 60	[196]
	HBx, HBsAg, and pre-S gene	- Carcinomas: 100% at week 80	[199]
		- Preneoplastic foci: 25% at weeks 24–28; ~71% at weeks 36–48; ~83% at weeks 52–80; ~57% at weeks 92–136 - Adenomas: 0% at weeks 24–28; 37.5% at weeks 36–48; ~46% at weeks 52–80; ~75% at weeks 92–136 - Carcinomas: 0% at weeks 24–28; 12.5% at weeks 36–48; ~33% at weeks 52–80; ~25% at weeks 92–136	[200]
	pre-S/S gene (rtA181T/sW172mutation)	- Carcinomas: 8.3% at week 72 (HBsAg low), 23.1% at week 72 (HBsAg high)	[201]
HCV-transgenic	Core gene	- Carcinomas: 25.9–30.8% at week 64	[202]
c-myc transgenic	*cnsfo-myc* overexpression	- Preneoplastic foci: 50–83% at weeks 48–56; 80–100% at weeks 72–80 - Adenomas: 40–66% at weeks 48–56; 30–100% at weeks 72–80 - Carcinomas: 0–37% at weeks 48–56; 10–65% at weeks 72–80	[203]
		- Adenomas: 0% at week 24; 20% at week 32; 33% at week 40; ~50 to 90% at weeks 48–64 - Carcinomas: 0% at week 32; 8% at week 40; ~30 to 60% to weeks 48–64	[204]
		- Carcinomas: 0% at week 36; ~10% at week 40; 25 to 50% at week 48; 50% at week 48	[205]
c-myc/TGF-α transgenic	Double *c-myc*/TGF-α overexpression	- Carcinomas: ~25% at week 12; 50–75% at week 24; ~100% at weeks 32–36	
		- Preneoplastic foci: 40–100% at weeks 32–40 - Adenomas: 40–100% at weeks 32–40 - Carcinomas: 30–100% at weeks 32–40	[203]
E2F-1 transgenic	E2F-1 overexpression	- Preneoplastic foci: 91% at weeks 32–40; 100% at weeks 40–48 - Adenomas: 73% at weeks 32–40; 100% at weeks 40–48 - Carcinomas: 33% at weeks 32–40; 0% at weeks 40–48	[206]

Table 5. *Cont.*

Model	Genetic Modification	Timepoints and Incidence of Lesions	References
c-myc/E2F-1 transgenic	Double *c-myc*/E2F-1 overexpression	- Carcinomas: ~25% at week 24; 100% at week 32	[205]
Apc knockout	*Apc* deletion	- Carcinomas: 67% at week 32	[207]
β-catenin/H-ras mutant	Double *Catnb/Hras* overexpression	- Carcinomas: 100% at week 24	[208]
cMyc + shp53 mice	*c-myc* overexpression, p53 downregulation	- Adenomas + Carcinomas: 38% at week 15	[209]

HBV: hepatitis B virus; HCV: hepatitis C virus; TGF-α: transforming-growth factor-α.

Another transgenic strain called Tg (Alb-1 HBV) Bri44 was designed to contain coding regions for HBx, HBsAg, and pre-S proteins (Table 5). These mice feature a stepwise liver disease with prolonged liver cell injury, death, pronounced inflammation, and elevated compensatory hepatocyte proliferation, which is triggered by the increased production and retention of the HBV large envelope polypeptide. The necro-inflammatory context resembling human disease leads to the progressive development of preneoplastic foci, adenomas, and carcinomas [199,200,210]. Barone et al. [211] revealed the early molecular events in 3 -week-old Tg (Alb-1 HBV) Bri44 mice by transcriptomic analysis. It was found that 25 genes are upregulated, including those involved in NF-κB signal transduction (*Vcam1* and *Cxcr4*), regulation of transcription (*Hmgb2, Nfatc1, Nupr1*, and *Atf3*), cell cycle and proliferation (*Cdkn2d* and *Slfn2*), and negative regulation of apoptosis (NuprI), and 20 downregulated, such as anti-proliferative (*Ghr* and *Erbb3*) and pro-apoptotic (*Bnip3*) genes. As a long latency time is required for Tg (Alb-1 HBV) Bri44 mice to develop neoplastic lesions (Table 5), protocols using DEN or AFB1 may synergistically act with the genetic modification to accelerate neoplastic lesion development, presenting an incidence of 75–90% within 15 months [212]. Lai et al. [201] developed a transgenic mouse model featuring a mutated HBV pre-S/S gene and its promoter in C57BL/6 mice, since rtA181T/sW172* mutation confers resistance to antiviral therapies, exerting an oncogenic potential [213]. Two transgenic strains were developed, both featuring an sW172* mutation but one expressing high and the other low intrahepatic levels of HBV surface antigen (HBsAg). Although the incidence of HCC was low in both strains after 18 months of age (Table 5), enhanced ER stress-related and proliferation-related proteins were increased in the non-neoplastic tissue. The investigation of miRNA-mRNA networks further revealed that transgenic mice also presented microRNA-873-mediated reduced expression of tumor suppressor CUB and Sushi multiple domains 3 (CSMD3) protein.

Concerning HCV, mice expressing the full core gene presented progressive morphological and molecular changes that ultimately resulted in the development of HCC. At 9 and 12 months of age, the mice only showed pronounced steatosis without inflammatory or neoplastic lesions. At 16 months, the major neoplastic changes were eosinophilic adenomas with fatty changes, while between 16 and 19 months of age, some nodules were adenomas or well-differentiated HCCs [202] (Table 5). Proteomic analysis stepwise of this model revealed that at 12 months, proteins related to respiration, electron-transfer system, apoptosis, fatty acid metabolism defense against oxidative stress were upregulated. At 16 months, most differentially expressed proteins were downregulated, including those involved in anti-oxidant defense, β-oxidation, and apoptosis [214]. In another model of mice expressing HCV core-E1-E2 proteins treated with DEN, the genetic modification increased the size of neoplastic alterations, increasing cell proliferation and decreasing apoptosis [215]. Although HBV and HCV transgenic models are usually time-consuming (long latency time for HCC emergence), their main advantages comprise the investigation of virus-related oncogenic mechanisms and the screening of therapeutic options.

2.4.2. Other Gene Expression Systems

A myriad of genetic manipulations to assess the effects of the activation of oncogenes and/or disabling of tumor suppressor genes have been developed in mice in the past decades. As the stepwise hepatocarcinogenesis process involves the acquisition and accumulation of genomic alterations, these models enable the investigation of potential therapeutic targets in preclinical settings. The methodological approach consists of the use of an albumin promoter, the induction of specific genes with molecules, as well as the recently developed hydrodynamic tail vein injection (HTVI) using the Sleeping Beauty (SB) transposase system or CRISPR/Cas9 genome editing tool (Table 6). Developed by Sandgren et al. [216], transgenic mice overexpressing the oncogene MYC, directed by the albumin enhancer/promoter or alpha1-antitrypsin promoter technologies, displayed mild to severe hepatic dysplasia in young mice, while hepatic neoplasia emergence required a long latency time, up to 16 months (Table 5). MYC overexpression leading to tumor development is also achieved by the integration of the Woodchuck Hepatitis Virus (WHV) in the mouse genome [217]. In this model, the overexpression of MYC along with IGF-2 during the neonatal stage, driving a strong proliferative stimulus, is proposed to drive hepatocellular transformation [218]. Frequently found in the corresponding human disease, activating β-catenin gene mutation is featured in 50–55% of WHV or promoter-activated MYC -driven liver carcinogenesis [219].

Table 6. Summary of some of the HTV protocols for hepatocarcinogenesis in mice.

Genes	Plasmids	Strain	Timepoints and/or Lesions	References
NRasV12 and *myr-AKT*	7.5 μg of myr-AKT1; 7.5 μg N-RasV12; SB transposase (25:1)	C57BL/6	HCC formation and progression after 2 to 4 weeks post-injection	[220]
c-Myc and *β-catenin*	10 μg pT3-EF1a-MYC; 10 μg pT3-N90-CTNNB1; 2.5 μg SB13-Luc transposase-encoding vector	C57BL/6	Poorly to moderately differentiated HCC with solid/trabecular pattern, and immunoexpression of CK19 and nuclear β-catenin	[221]
β-catenin and *tert* or *pten*	10 μg pT3-N90-CTNNB1; 10 μg pT3-EF1a-Tert or 10 μg pX330-Pten; 2.5 μg SB13-Luc transposase-encoding vector	C57BL/6	Well to moderately differentiated HCC with trabecular pattern, abundant clear cells, and immuno-expression of glutamine synthetase	[221]
c-Myc and *axin1*	10 μg pT3-EF1a-MYC; 10 μg pX330-Axin1; 2.5 μg SB13-Luc transposase-encoding vector	C57BL/6	Well to moderately differentiated HCCs with trabecular pattern	[221]
c-Myc and *MCL1*	10 μg pT3-EF1α-c-MYC-shLuc; 5 μg pT3-EF1α-Mcl1; SB transposase (25:1)	C57BL/6 FVB/N Balb/C	Liver tumor formation after 5 to 8 weeks post-injection	[222,223]
c-met and *axin1*	20 μg pT3-EF1α-c-Met; 40 μg pX330-Axin1.1; 0.8 μg pCVM/SB	FVB/N	HCC burden at 9 to 12 weeks post-injection showing membranous immunoexpression of E-Cadherin and absence of glutamine synthetase	[224]

Table 6. *Cont.*

Genes	Plasmids	Strain	Timepoints and/or Lesions	References
c-Myc and *myr-AKT*	16–36 μg of mixed plasmids: pT3-EF1a-myrAKT-HA; pT3-EF1α-c-MYC; SB13 transposase-expression plasmid	C57BL/6J	Well to moderately differentiated HCC at 8 to 10 weeks post-injection showing trabecular or nest-like patterns	[225]
myr-AKT and/or *Hras*	16–43 μg of mixed plasmids: pT3-EF1a-myrAKT-HA; cDNA fragments of FLAG-HRASV12; SB13 transposase-expression plasmid	C57BL/6J	*Akt* or *Hras*: multiple HCC associated with lipid accumulation after 20–28 weeks post-injection *Akt* and *Hras:* HCC after 8 weeks post-injection with a higher proliferation rate	[226]
c-met and *β-catenin*	5 μg pT3-EF5a-hMet-V5; 5 μg pT3-EF5α-S33Y-β-catenin-Myc or 5 μg pT3-EF5α-S45Y-β-catenin-Myc; SB transposase (25:1)	FVB/N	Well-differentiated HCC by 6 to 9.5 weeks post-injection	[227–229]
myr-AKT and *c-met*	pT3-EF1α -HA-myr-AKT1; pT3-EF1α-V5-c-Met; SB transposase (25:1)	FVB/N	Lethal burden of HCC within 6 to 8 weeks post-injection showing admixture of clear, lipid-rich and lipid-poor, basophilic cells	[229]
myr-AKT and *β-catenin*	pT3-EF5-AKT; pT3-EF1α-ΔN90-β-catenin.	FVB/N C57BL/6	Progression to HCC only in vivo passage of steatotic tumor cells from hepatocellular adenomas	[228]

Double transgenic mice bearing both albumin enhancer/promoter c-myc and metallothionein 1 promoter TGF-α were developed by Murakami et al. [230] and Santoni-Rugiu et al. [203]. The co-expression of both genes in the mouse liver promoted neoplastic lesion emergence compared to MYC overexpression alone (Table 5). Increased levels of TGF-α mRNA and protein during the early stages of liver carcinogenesis in MYC/TGF-α transgenic mice are proposed to have a key role in the clonal expansion and malignant conversion of the preneoplastic cell population [230]. HCC in c-myc/TGF-α transgenic mice had extensive genomic instability (loss heterozygosity) while displaying a low rate of β-catenin mutation and subsequent nuclear accumulation (12.5%) [205]. As MYC/TGF-α transgenic mice showed increased expression of oncogenic E2F1 and 2 and induction of their target genes [203], E2F-1 and double MYC/E2F-1 transgenic mice were developed [203,205]. Compared to MYC or E2F-1 transgenic mice, double transgenic animals also showed potential cooperation between MYC and E2F-1 oncogenes (Table 5), featuring a high frequency of β-catenin mutational activation and nuclear accumulation in both adenomas and carcinomas [205]. Based on comparative analysis of the transcriptome of these models with the corresponding human disease, Lee et al. [76] proposed that MYC, E2F-1, and double MYC/E2F-1 transgenic mice had similar global gene expression to a group of human HCCs with better survival, whereas MYC/TGF-α transgenic mice reflected a poorer survival HCC group. Regarding the frequently targeted Wnt/β-catenin pathway, a mutant mouse strain displaying adenomatous polyposis coli (*Apc*) deletion by the injection of adenovirus encoding Cre recombinase led to the development of HCC albeit presenting low incidence [207] (Table 5). On the other hand, simultaneous mutations in the *Catnb* (β-catenin) and *Hras* genes by using the same technology cooperatively accelerated HCC development (Table 5), as β-catenin may promote the clonal expansion of *Hras*-induced dysplastic (preneoplastic) cells [208]. More recently, using HT coupled with the SB transposon system, Chung et al. [209] developed a mouse strain transfected with transposons expressing MYC and a short hairpin RNA downregulating p53. Interestingly, tumor incidence and multiplicity were accelerated by the establishment of a CCl_4-induced fibrosis context. Employing the same system, Tao et al. [227] modeled the concomitant HMET overexpression or activation and CTNNB1 mutations found in 9–12.5% of human HCCs by developing hMet/β-catenin point mutant mice. HT method to deliver a CRISPR plasmid DNA expressing Cas9 models targeting both tumor suppressor genes *Pten* and *Tp53* has also been recently employed [231,232], underscoring the potential of this tool for the development of novel hepatocarcinogenesis models.

2.5. Humanized Mouse Models

HCC is an inflammation-driven cancer, and recognition of the involvement of components of the innate and adaptive immune system in the development and progression of HCC has accelerated research into new therapies capable of targeting and/or modulating the immune system [233]. Promising results of combining current HCC treatment modalities, such as locoregional treatment and anti-angiogenic therapy combined with immunotherapy, have resulted in recently approved new treatment strategies for HCC [234,235]. Importantly, preclinical translational research with the ultimate goal of demonstrating therapeutic efficacy with an acceptable safety profile for human disease requires essential components that characterize human (patho)physiology. In HCC, this relates to neoplastic hepatocytes surrounded by a tumor micro-environment (TME) composed of (suppressive) immune cells. Currently used HCC models in immunocompetent mice do not accurately represent the human immune system since significant immunobiological differences exist between mice and humans, including dissimilarities in T cell signaling, immune cell receptor expression, and antigen presentation [236,237]. Xenograft models require immunocompromised mice to avoid rejection of an implanted human cell line or patient-derived material and thus do not provide a solution for immune-oncology studies [16]. More recently, mice harboring a humanized immune system (HIS) have been developed and now opened the field of research towards the use of preclinical models based on patient-derived HCC tissue in the context of an effective human immune system, essential in the evaluation of immune-oncology drug efficacy and safety [16,238]. In addition to the humanization of immunocompetent mice via transgenesis, immunocompromised mice can acquire a humanized immune system through engraftment with human peripheral blood mononuclear cells (PBMCs) or CD34+ hematopoietic stem cells (HPSC) [239].

In the PBMC-humanized mouse model, PBMCs from healthy donors are intravenously, intraperitoneally, or intrasplenically injected within several days after implantation of a human cell line or patient-derived xenograft, CDX and PDX, respectively, in an immunodeficient mouse [240,241]. Despite comprising several types of immune cells of both the lymphoid and myeloid lineage at transplantation, PBMC engraftments give rise to an almost exclusively T cell-oriented humanized immune system, since appropriate signals for the survival and expansion of B, NK, and myeloid cells are lacking [240,242]. The engrafted lymphocytes are functionally mature, and HIS mice can be utilized almost immediately after PBMC injection for therapy evaluation. However, in addition to its restriction to the evaluation of T cell-based immunity, the use of this model is limited to short-term experiments because injection of PBMCs elicits xenogeneic graft-versus-host disease (GvHD) a few weeks after engraftment [239]. Delay of GvHD development and improvement of the overall immune functionality can be obtained through genetic enhancements of the recipient mice, including replacement of murine major histocompatibility complex (MHC) by human leukocyte antigen (HLA) expression [242]. A new mouse strain lacking murine MHC molecules (NSG-(KbDb)null(IA)null) has been created in which human PBMCs can be engrafted without the development of acute xenogeneic GvHD [243]. The model has effectively been used to evaluate immune checkpoint inhibitors in human cancer xenograft models and is ideally suited to evaluate anti-cancer (immuno)therapy [244]. Despite being relatively straightforward, fast, and cost-effective compared to other HIS models, the PBMC-humanized mouse model has thus far only been applied as the HCC CDX model [245,246]. As an alternative to PBMCs engraftments, CD34+ HSCs derived from multiple potential sources, including granulocyte colony-stimulating factor (G-CSF)-mobilized peripheral blood, adult bone marrow, fetal liver, and umbilical cord blood can be engrafted in immunocompromised mice [239,241].

The CD34+-humanized mouse model has the advantage of displaying a more complete representation of the human immune system. However, due to the lack of human cytokines and growth factors, the developed T, B, NK, and myeloid cells all exhibit functional impairments [247]. Transgenic immunocompromised mouse strains expressing human cytokines and growth factors could further enhance the engraftment of CD34+ HPSC and support the development of functional human immune cells. In this respect, transgenic expression of human IL-3 and granulocyte-macrophage colony-stimulating factor (GM-CSF) in NOD/Shi-scid IL2rγnull (NOG) mice is beneficial for the development of myeloid cells [240]. Importantly, since human T cells derived from engrafted HSCs undergo a positive and negative selection on murine MHC molecules during development in the thymus, tolerance towards the murine host is established [248]. However, as murine thymic epithelial cells do not express HLA, the resultant T cells are not able to recognize antigens in an HLA-restricted manner. Consequently, these HIS mice elicit an inappropriate T cell response against human HCC xenografts [249]. To overcome this issue, transgenic mice expressing HLA molecules matched to the PDX donor should be used [250]. To this end, Serra-Hassoun et al. [251] created a new lymphoid mouse strain which, in addition to the replacement of murine MHC by HLA, expresses human signal regulatory protein alpha (hSIRPα) on murine phagocytes to enable human HSC engraftment [251]. In contrast to the PBMC-humanized mouse model, which can be utilized almost immediately after PBMC engraftment, it takes up to 10 to 12 weeks to develop a robust humanized immune system following HPSC engraftment. Moreover, recipient mice need to be irradiated to ensure engraftment of human HPSC [239].

The success of HPSC engraftment is dependent on the age, sex, and strain of the mice, the route of engraftment, and the source of the CD34+ HPSC [239,247]. Despite having the advantage of enabling stable and long-term humanization and representing a substantial portion of the human immune system, notable limitations, such as incomplete immune cell development and the time required to establish the humanized immune system, have limited the utilization of this HIS model concerning HCC [242]. Hitherto, only Zhao et al. [19] effectively developed and used a PDX humanized HCC model for investigation of the human-specific TME and immunotherapeutic treatment strategies. In this model, 1 to 3 days old NOD-Prkdc-scid IL2rgnull (NSG) pups were irradiated and intrahepatically injected with human fetal liver-derived CD45+ HSCs. The created HIS mice were subcutaneously transplanted with an HLA-matched PDX 8 to 10 weeks after HPSC engraftment. Effective human immune responses and the therapeutic efficacy of immune checkpoint inhibitors were demonstrated [19]. Since the HCC tumor interacts with both infiltrating immune cells and liver-resident cells, orthotopic implantation of a PDX in the hepatic micro-environment of CD34+-humanized mice would represent an even more attractive platform for preclinical evaluation of immunotherapeutic and other HCC treatment strategies but remains to be developed.

A final HIS model that has not yet found its way to the field of HCC research is the bone marrow/liver/thymus (BLT) model. In this model, fetal liver and/or bone marrow-derived CD34+ HSC engraftment is preceded by transplantation of a small fragment of human fetal thymic and liver tissue under the kidney capsule of immunocompromised mice [242]. As it provides human thymic tissue, the BLT model enables improved human T cell development. However, since the positive selection of T cells occurs solely on human MHC molecules and T cells with an affinity for murine MHC are not eliminated, the incidence of xenogeneic GvHD is higher in this model compared to the CD34+- model [247]. This problem could be overcome by using a transgenic MHC-deficient mouse strain. The technical difficulty of creating the mice and the necessity of human fetal tissue are 2 major factors that currently impede the utilization of the BLT model over the PBMC or CD34+-humanized mouse model [242].

3. In Vitro Models of HCC

3.1. Primary Hepatocytes

Because of their high resemblance to the in vivo phenotype, primary human hepatocytes (PHH) (Table 7) are considered the gold standard for in vitro studies on biotransformation, toxicity and drug-induced liver injury [252–255]. The liver is a major target for chemical-induced injury caused by carcinogens [256]. The application potential of primary hepatocytes from rodent or human origin in HCC studies focuses on the initiation and promotion stages of cancer [254,256,257]. In this respect, PHH have been used for genotoxicity assays, albeit to a lesser extent compared to their rodent counterparts [254,256,257]. More recently, PHH have been used to elucidate the molecular pathways or genomic effects in response to carcinogens [246,258–262]. In addition, PHH have been addressed to evaluate the molecular and cellular events of HCC initiation through in vitro transformation based on lentiviral-transduction of oncogenic Harvey-RAS, simian virus 40 (SV40) small T antigen, and SV40 large T antigen [263]. The major drawbacks of PHH are the high donor-to-donor variability [253] and their progressive dedifferentiation, which makes them less suitable for long-term culture [264]. This dedifferentiation process is already initiated during the isolation process [265]. To overcome dedifferentiation, sandwich cultures or spheroid models of PHH can be used [261,266–269]. Although spheroid models of PHH have not been used in genotoxicity studies, sandwich cultures of PHH have been applied to elucidate the epigenetic effects of AFB1 in HCC initiation [262] and to assess the modulation of AFB1-mediated genotoxicity by chemopreventive chemicals [270].

Table 7. Advantages, disadvantages and applications of commonly used liver-based in vitro models in liver cancer research.

In Vivo Model	Advantages	Disadvantages	Applications	References
PHH monolayer	- Similar to in vivo phenotype - High biotransformation capacity	- High donor-to-donor variability - Progressive dedifferentiation - Fail to represent complex in vivo environment	- Biotransformation studies - Toxicity studies - Drug-induced liver injury studies - Studies related to initiation and promotion of HCC - Liver disease studies	[252–257,264,266,271]
PHH sandwich cultures	- Prolonged viability - Retained morphology	- Altered protein expression over time - High donor-to-donor variability	- Studies of HCC initiation by carcinogens - Drug-induced liver injury studies	[252,261,268,272,273]
Liver cell lines monolayer	- Easy to use - Stable phenotype - Reproducible - Fit for high-throughput and high-content analyses - Allow genetic manipulation - Low cost	- Fail to represent intertumor and intratumor diversity - Less differentiated than PHH - Reduced or absent biotransformation capacity - Fail to represent complex in vivo environment	- Drug-screening - Safety testing - Genotoxicity studies - HCC biology studies - Studies assessing molecular and (epi)genetic modifications in HCC - Overexpression/silencing studies	[253,254,256,274–280]

Table 7. *Cont.*

In Vivo Model	Advantages	Disadvantages	Applications	References
Co-culture models	- Closer resemblance to in vivo cell heterogeneity and tumor-specific micro-environment - More pathologically relevant model by allowing cell-cell interactions	- Lack of standard protocols - High interlaboratory variability	- Studying influences of non-parenchymal cells and micro-environments on HCC	[253,280–285]
Stem cell-derived model	- High culture stability - Expandable - Reproducible - Metabolism that resembles PHH	- Lack of standard protocols for isolation of primary CSC	- HCC therapy studies - Studies related to initiation, promotion and drug resistance of HCC	[286–288]
Spheroid models	- Retained morphology and phenotypic functions - Higher biotransformation capacity compared 2D culture - Display oxygen and nutrient gradients - Closer resemblance to in vivo tumor-specific micro-environment - Closer resemblance to in vivo cell heterogeneity when combining co-culture techniques with spheroid models	- High donor-to-donor variability of PHH - More time-consuming and expensive than 2D culture - Lacks uniformity depending on spheroid-forming method - Low throughput depending on spheroid-forming method - Long-term culture is difficult	- Liver function studies - (Geno)toxicity studies - Liver disease studies - Drug delivery and efficacy studies - Studies investigating role of stem cells - Tumor-growth studies - Angiogenesis studies - Immunotherapy studies	[226,253,254,267,268,289–296]
Organoid model	- Mimic functionality and architecture of native tissue - Fit for high-throughput analyses - Expandable - Cryopreservation is possible - Allow genetic manipulation - Retention of tumor heterogeneity - Limited starting material is required - Based on healthy or tumorigenic material	- High cost - Time-consuming - Average success rate	- Regenerative medicine - Personalized drug discovery - Toxicity studies - Gene therapy studies - HBV-related carcinogenesis - Model liver cancer initiation - Molecular and cellular characterization of HCC - Study of inter- and intratumor diversity in HCC	[220,280,284,297–303]
Precision-cut-liver slices	- Capture complex in vivo micro-environment - Retained polarized morphology - Low cost - Automation possible - Closely resemble in vivo gene expression - Based on healthy or tumorigenic material	- Less suitable for high-throughput analyses - Labor-intensive preparation and incubation - Variable culture conditions between studies jeopardizing reproducibility - Reduced albumin and cytochrome P450 expression over time - Technically difficult - Limited availability	- Immunological studies - Toxicity studies - Genotoxicity assessment - Drug-screening - Liver disease studies	[266,288,304–316]

3.2. *Hepatic Cell Lines*

Cell lines (Tables 8 and 9) are popular models for drug screening studies [256], drug safety testing [176], and studies related to liver disease [253] due to their ease of use, phenotypic stability and reproducibility. Human liver cell lines are typically immortalized by genetic engineering or selected based on their tumorigenic phenotype, implying they are derived from human tumors [253,274]. Upon genetic engineering, PHH are most commonly transfected or transduced to overexpress human TERT or viral oncogenes, such as SV40 large T antigen or human papillomavirus16 E6/E7 genes [317]. Immortalized cell lines are often used in parallel with tumor-derived cell lines to assess side effects and therapeutic effects of various compounds [275,318–320] or in studies pinpointing molecular, genetic and epigenetic alterations in HCC [74,276,277,321]. In addition, immortalized cell lines are used to study the influence of tumor micro-environments and overexpression of genes leading to cancer initiation [322,323]. Tumor-derived cell lines are the most widely used cell lines in HCC research [256,324]. Indeed, they are harnessed to assess cancer characteristics, such as cell proliferation, migration, metastasis, evasion of cell death and invasion [278,281,325–329]. Moreover, liver cancer cell lines are frequently used for elucidating molecular mechanisms of HCC during overexpression or silencing gene studies [282,326,327,329]. Liver cancer cell lines can be equally used to provide insight into genetic changes of HCC [277], for the identification of new drug targets [326], and for testing anti-cancer therapies [325,326,330,331]. However, their clinical relevance for the latter application may be questioned [279]. Since each cell line is derived from a single donor, they fail to represent the well-known intertumor and intratumor diversity that hampers the development of HCC therapies [256,279]. This flaw can be circumvented by using several cell lines in parallel [274,324]. Based on their genetic characteristics, a panel of cell lines could be selected to represent different HCC subclasses [274,277,324]. Although more frequently used to elucidate HCC biology and progression, liver cancer cell lines are also employed in genotoxicity testing [254,256,277]. The majority of genotoxicity studies are performed in mouse lymphoma cells, human lymphoblast cells, or Chinese hamster lung cells, which are metabolically incompetent, making their human relevance questionable [254]. In fact, p53-deficient rodent cell lines give rise to more false positives compared to human cell models [332]. Therefore, the use of human liver cell lines in genotoxicity/mutagenicity assays is gaining increasing attention [254]. An emerging tool includes the human hepatoma HepaRG cell line [333]. In contrast, compared to HepG2 cells, which are by far the most widely used cells in liver studies, HepaRG cells functionally express biotransformation enzymes at a level that is comparable to PHH and are therefore a more sensitive cellular system for (geno)toxicity testing [256,333–335]. In addition, HepaRG cells have been shown valuable for understanding the mechanism of action of mutagenic/carcinogenic compounds [336].

Table 8. Overview of commonly used human liver cancer cell lines.

Cell Line	Cancer Type	HBV/HCV	Gender	Age	Race	References
HepG2	Hepatoblastoma	-/-	Male	15	African	[324,337,338]
Huh-7	HCC	-/-	Male	57	Asian	[324,337]
Hep3B	HCC	+/-	Male	8	African-American	[324,337]
HepaRG	HCC	-/+	Female	/	European	[324,333,337]
MHCC97	HCC	+/unknown	Male	39	Asian	[324,337,339]
PLC/PRF/5	HCC	+/-	Male	24	African	[324]
SK-HEP-1	Adenocarcinoma	-/unknown	male	52	European	[324,340]
SNU-475	HCC	+/-	Male	43	Asian	[324]
SNU-423	HCC	+/-	Male	40	Asian	[324]
SNU-449	HCC	+/-	Male	52	Asian	[324]
C3A	Hepatoblastoma	-/-	Male	15	European	[277]
SNU-387	HCC	+/-	Female	41	Asian	[324]

Table 9. Overview of commonly used immortalized human liver cell lines.

Cell Line	Immortalization Method	Gender	Age	Race	References
Fa2N-4	SV40 large T antigen Transfection	Female	12	Unknown	[317,341]
NeHepLxHT	hTERT Retroviral vector	Male	<1 month	Unknown	[337,342]
THLE-2	SV40 large T antigen Retroviral vector	Male	Adult	Unknown	[343]
PH5CH	SV40 large T antigen Lipid mediated gene transfer	Male	58	Unknown	[317,344]

3.3. Co-Cultures

As much as 70–80% of the human liver consists of hepatocytes, while 5–6% comprises non-parenchymal cells such as Kupffer cells, hepatic stellate cells (HSC), and sinusoidal endothelial cells [345,346]. To mimic this in vivo cellular heterogeneity, co-cultures can be set up in which PHH [347,348] or liver cell lines are seeded together with non-parenchymal liver cells, such as HSC [282], endothelial cells [281,349], stem cells [292], or immune cells (Table 7) [283]. In such co-culture settings, transwell chambers can be used, allowing communication between both cell populations [350]. Co-cultures of hepatic cell lines with tumor-specific cells, like cancer-associated fibroblasts or tumor-specific neutrophils [285], provide in vivo-like tumor-specific microenvironments [284,285]. This leads to a more relevant in vitro model that is capable of manifesting the changing cancer characteristics of HCC cell lines under the influence of other cell types [351]. Moreover, co-culture systems provide a valuable model in immune cell therapy studies and research concerning crosstalk between HCC cells and the immune system. In this respect, co-cultures of HepG2 cells with natural killer cells display anti-proliferative effects, and the addition of M1 macrophages reduces HCC viability, invasion, therapy resistance, and migration, whereas M2 macrophages promote tumor invasiveness in HCC co-cultures [283,352,353]. Co-culturing HepG2 cells with endothelial cells yield an in vitro system appropriate for studying angiogenesis [281].

3.4. Stem Cell-Derived Models

Several studies have indicated that HCC develops from cancer stem cells (CSC). CSC are self-renewing and give rise to the different cell lineages in HCCs. Since CSC possess the capacity to form tumors, they are the major drivers of chemotherapeutic resistance, metastasis, and post-treatment tumor recurrence [354]. In this respect, understanding the malignant reprogramming that occurs in these cell types is crucial for the development of effective HCC therapies [355,356]. Induced pluripotent stem cell technology is an ideal tool to model the reprogramming that leads to tumorigenesis and cancer progression (Table 7). Induced pluripotent stem cells are even regarded as a potential therapy for HCC treatment by targeting CSC-related genes. Induced pluripotent CSC can be generated from liver cancer cells to specifically model liver CSC [286]. As such, various liver cancer cell lines have been reprogrammed via retroviral particles, introducing 4 stem cell transcription factor genes, namely KLF4, Sox2, Myc, and Oct4, to better understand the reprogramming process [356]. Together with NANOG and LIN28, these transcription factors have been detected in HCC and are associated with negative clinical outcomes. Besides induced pluripotent CSC, cells with stemness properties are also extracted from primary tumor material or cell lines to be used in studies that evaluate the molecular and epigenetic mechanisms of HCC [357]. Furthermore, stem cells are not only used to model tumorigenesis but also to test patient-specific direct reprogramming therapies and to identify molecular targets for blocking tumor initiation [358–360].

3.5. Spheroid and Organoid Models

While 2D models are valuable tools in liver studies, they lack in vivo-like cell density and a complex microenvironment [253]. 3D models (Table 7), such as spheroid cultures and organoid cultures, have been developed to overcome these flaws. Spheroids can be derived from PHH and hepatic cell lines [361]. Most liver cell lines are less differentiated compared to PHH, which is mainly reflected by a lack of biotransformation capacity [176,253,256]. Throughout the years, 3D spheroid models have been introduced to tackle this shortcoming [291,293]. As such, 3D spheroid cultures of human hepatoma HepG2 cells express more albumin and phase I and II biotransformation enzymes compared to 2D counterparts, making them more applicable for genotoxicity studies [290]. Besides genotoxicity assays, 3D spheroid models of liver cancer cell lines have been used for evaluating anti-cancer agents and drug sensitivity [294,362–365]. In addition, various studies have been set up in which 3D spheroid culture techniques are combined with co-culture approaches [292,295,296,348,365]. In this respect, a combination of primary HCC cells with extracellular matrix (ECM), endothelial cells, and fibroblasts in spheroid cultures creates a model with enhanced tumor-related and neo-angiogenesis markers to study potential HCC therapies [289]. Compared to unicellular 3D spheroid cultures, co-culture spheroid models display a more in vivo-like microenvironment, thereby creating a more pathologically relevant HCC model for drug-screening studies [365,366].

Organoids are self-renewing and self-organizing 3D tissues derived from various stem cell types, such as embryonic stem cells, induced pluripotent stem cells, organ-restricted adult stem cells, and primary tissue (Table 7) [297–299]. The difference with 3D spheroid models is that organoids contain various tissue-specific cells, which are all developed from the stem cell starting material through in vivo-like processes mediated by the provided ECM, namely Matrigel® [367,368]. The latter is an extract of an Engelbreth-Holm-Swarm mouse sarcoma that contains many ECM proteins and some less defined biochemical signaling molecules, including growth factors, necessary for the development of organoids [367–370]. This model can mimic the functionality and architecture of native liver tissue [298]. Liver organoids can be vascularized when hepatic endoderm cells directly differentiated from induced pluripotent stem cells are co-cultured with endothelial and mesenchymal cells [371]. Since the model is based on primary healthy or tumorigenic material, organoids create new possibilities for regenerative medicine, personalized drug discovery, toxicity studies, and gene therapy [297,300]. Liver organoids have been used to model liver cancer initiation [299,372], to study HBV-related hepatocarcinogenesis [301], to investigate drug sensitivity on patient-derived models [302] and to perform molecular and cellular characterization of HCC [299]. Patient-derived liver organoid models permit the generation of tumor biobanks that enable the profiling of genomic diversity as well as drug sensitivity studies [367]. Furthermore, they can not only assist in the assessment of genomic intertumor diversity but equally of intratumor diversity, allowing the study of variations in drug responses within one single tumor [303].

3.6. Precision-Cut Liver Slices

Precision-cut liver slices from rat, mouse, or human origin are ex vivo tools that retain the complex native liver environment containing interactions between all liver cell types and the ECM (Table 7) [309–311]. Liver slices are usually 100–250 μm thick and have a diameter around 5 mm allowing nutrients and oxygen to easily be diffused across all cell layers [312]. Although precision-cut liver slices have been used for various purposes, their use in HCC studies is rather limited. Precision-cut liver slices prepared from primary tumor material have been used as a model to study anti-cancer drugs and oncolytic virotherapy [313,314,373]. It has been suggested that precision-cut liver slices could represent a valuable model to evaluate patient-specific drug responses and therapy resistance in addition to predicting side-effects on adjacent healthy tissue [314,374].

4. Therapeutic Relevance of the HCC Models

The treatments conventionally used in HCC patients, such as tumor resection, chemotherapy, radioembolization, and liver transplantation, are highly dependent on the cancer stage. In advanced-stage HCC, these procedures become unfeasible, requiring the use of systemic therapies [375,376]. However, HCC cells show high resistance to conventional chemotherapy. Sorafenib, a multi-tyrosine kinase inhibitor that reduces tumor growth and angiogenesis, was developed in 2008. Despite this drug having prolonged patient survival in a few months [377], its clinical use has been limited to adverse side effects and refractory drug response, mainly associated with genetic heterogeneity of HCC [378–380]. Recently, novel classes of multi-tyrosine kinase inhibitors have been developed, such as lenvatinib, regorafenib and cabozantinib; however, the prognosis of HCC patients is still poor [381–383]. This scenario urgently drives the search for new therapeutic targets, drugs and therapies, such as immuno and gene therapies [384]. From sorafenib to immunotherapy, in vivo and in vitro models have shown fundamental importance in the pre-clinical phase for anti-HCC therapies.

The use of syngeneic and xenographic models has been widely used in studies of the combination of drugs used in HCC therapy, such as sorafenib, and drugs that enhance its effects by reducing tumor resistance to treatment [272,385–387], in immunotherapeutic studies [273,388–390] and new drug trials [391,392]. In the context of the combination of drugs with known potentialities, chemical induction of HCC by DEN in Fisher and Wistar rats (50 mg/kg, once a week, for 12–14 weeks) was used respectively to evaluate the effects of treatment with sorafenib + ARQ (AKT inhibitor) and sorafenib + fluvastatin (cholesterol-lowering), showing that the combination of drugs increased apoptosis, reducing cell proliferation, angiogenesis and activation of HSC [393,394]. The DEN-induced models have also been applied in mice to test new drugs, such as Romidepsin, a histone deacetylase inhibitor, cell cycle inhibitor and apoptosis inducer [395]. The ability to induce HCC as a late event of CCl_4 induction was used in mice (16% in corn oil, 3x/week/8 weeks) to evaluate the treatment of Sorafenib+ MAPK/ERK pathway inhibitors [396]. On the other hand, studies also associate the carcinogenic action of DEN with the fibro-cirrhotic capacity of CCl_4 in the chemo [397] and immunotherapeutic tests [398]. TAA has also been used in preclinical tests in mice (200 mg/Kg ip or 200 mg/L in drinking water) associated or not with syngeneic and xenographic models, enabling the evaluation of drugs with known or promising potential, chemo or immunotherapy [386,388,399].

Diet is one of the factors influencing the development of HCC, and models that combine STZ and high-calorie diet to mimic HCC resulting from the late stages of NAFLD/NASH are commonly crucial tools in the prevention and treatment of this process. The preclinical use of this model in mice in the evaluation of Liroglutide, a Glucagon-like peptide-1 (GLP-1) receptor used to control glycemia, observed improvement in NASH and suppression of hepatocarcinogenicity [400]. Furthermore, Berberine, known for the treatment of gastroenteritis and for its promising anticancer potential, in this model, reduced tumorigenesis, angiogenesis and inflammation [401]. Genetic engineering enabled the creation of transgenic animals with structural and metabolic alterations totally directed to the specificity of the questions formulated for the object of study. In the preclinical researches of HCC, these animals can be inserted in different models, directing the results to a specific signaling pathway both in the testing of new drugs [386,388,396,397] and in new therapies [388–390,398].

Humanized models allow, for example, artificially created human antibodies to prove their efficiency in humanized mice, increasing the reliability of the clinical trial results. Bi and collaborators constructed a bispecific antibody for GPC3/CD3 and tested the antitumor activity in several cell lines Huh-7, HepG2, Hep3B, SK-Hep-1, and SK-Hep-1-GPC3 and in a xenographic model of subcutaneous inoculation of Huh7 cell line in NOD-SCD mouse. The results not only proved the efficient destruction of CPC3 positive cells but also proved that CPC3 is not present in normal cells and can be an HCC-specific antigen and, therefore, an excellent therapeutic target [390]. Currently, preclinical studies combine numerous tools such as liver cell lines, syngeneic and xenographic models, chemical induction and transgenic animals [386,388,393,396,397,399] in search of treatment alternatives for HCC.

5. Conclusions and Perspectives

Despite recent advances in HCC treatment, only 18% of patients survive more than 5 years after initial diagnosis, a percentage significantly lower compared to other cancer types [402]. The poor prognosis is usually attributed to late diagnosis and lack of response to adjuvant therapies [403]. The inefficiency of anti-neoplastic drugs can be attributed to the high molecular heterogeneity of HCC [404], which increases the need to identify new molecular targets based on signaling pathways activated in hepatocarcinogenesis according to etiologies. In addition, the low translational value of preclinical models could be directly associated with high rates of drug failure in human clinical trials [405]. Besides recapitulating the pathophysiology of liver cancer, the ideal model should be reliable, highly reproducible, technically simple, and at a low cost. The experimental modeling of HCC is particularly challenging due to the molecular heterogeneity and tumor microenvironment with a fibrotic and chronic inflammation background.

Next-generation sequencing has shown a high diversity of genetic and epigenetic alterations in HCCs, allowing the classification in subclasses according to their molecular signatures [13]. Several chemical-induced and/or diet-induced HCC models have been developed to induce all stages of hepatocarcinogenesis in rodents, which usually do not reproduce all molecular alterations observed in human HCC. The use of hybrid models combining classical HCC models and genetically engineered animals has been developed to overcome this critical barrier. More recently, the HTVI methodology has been applied to delivery transposon-based or CRISPR-Cas9 vectors to overexpress or delete/mutate tumor suppressor genes, respectively. This represents an innovative genetic manipulation method to unravel the role of cancer driver genes, specifically in hepatocytes. Furthermore, humanized PDX mice models have been proposed as a promising tool to study the immunological response in human HCC. This model opens new avenues to test novel immunotherapeutic targets and identify mechanisms of immune escape and resistance to immunotherapies. However, several technical limitations still need to be overcome. In vitro liver models have been extensively applied for toxicity studies and drug screening due to their relatively low cost and easy-to-do performance. In the last years, these models have progressively evolved from monolayer monocultures to highly complex 3D co-cultures to recapitulate the tumor micro-environment. In addition, recent advances in genetic manipulation can be easily applied to delete or overexpress target genes, develop new HCC cell lines, and reproduce the molecular heterogeneity of liver cancer in vitro.

In the light of the spectra of in vivo and in vitro models available, and in order to provide a clearer understanding, their main advantages and limitations are summarized in Figure 5 and Table 7, respectively. The choice of a preclinical model must be a thoughtful and clearly defined process, weighting all the summarized aspects to provide relevant, translatable scientific data towards the understanding of hepatocarcinogenesis. In conclusion, the current in vitro-based and in vivo-based HCC models have shown several advantages and disadvantages according to the main application. Despite considerable advances in the HCC modeling, the lack of effective anti-neoplastic therapies urgently needs the establishment of more reliable, translational, and fast-induced HCC preclinical models.

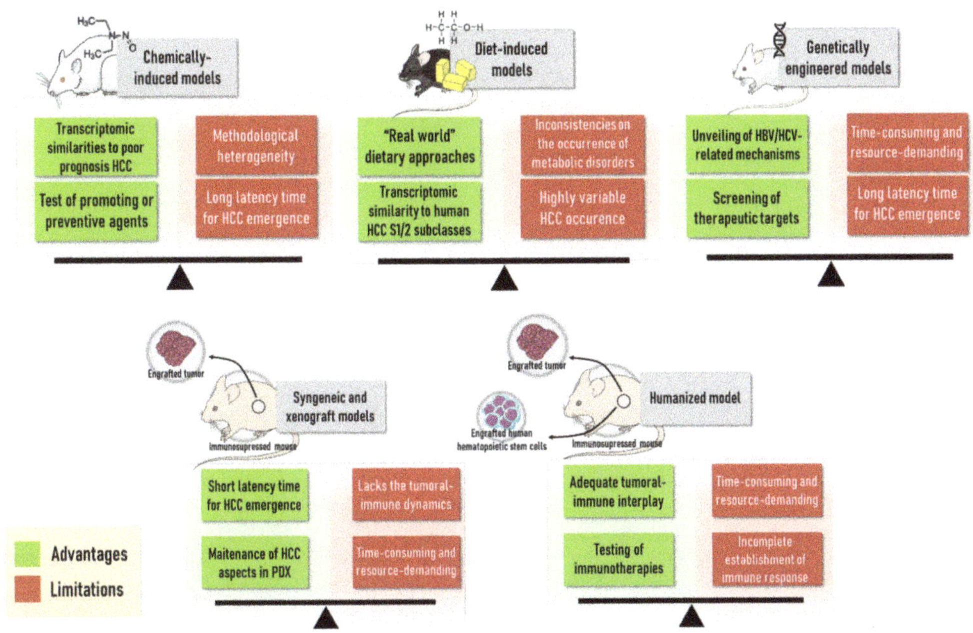

Figure 5. Main advantages and limitations of the main in vivo HCC mouse models available in the literature. The choice of the animal model must weight all the summarized aspects to provide relevant, translatable scientific data towards the understanding of hepatocarcinogenesis. The figure was composed with the aid of illustrations from the SMART-servier Medical Art available at https://smart.servier.com/ (accessed on 15 January 2021).

Author Contributions: Conceptualization, B.C. and M.V.; writting—original draft preparation, G.R.R., G.B.P., C.J.S.C., T.C.d.S., B.V., L.D. (in vivo models), K.L. (in vitro models); writing—review and editing, G.R.R., K.L., B.V., L.D., N.O.S.C., W.A., L.F.B., B.C. and M.V. All authors have read and agreed to the published version of the manuscript.

Funding: This study was financially supported by the Fundação de Amparo à Pesquisa do Estado de São Paulo (FAPESP-FWO Grant 18/10953-9; FAPESP regular grant 16/14420-0 and scholarship 16/12015-0), the Conselho Nacional de Desenvolvimento Científico e Tecnológico (CNPq grant 310557/2019-4), the Coordenação de Aperfeiçoamento de Pessoal de Nível Superior (CAPES), Brasil - Finance code 001, the Fund for Scientific Research-Flanders (FWO Vlaanderen Grants G009514N and G010214N), the University Hospital of the Vrije Universiteit Brussel-Belgium (Willy Gepts Fonds UZ-Brussel) and the Center for Alternatives to Animal Testing at Johns Hopkins University-USA (Grant 2018-13).

Conflicts of Interest: The authors declare no conflict of interest.

References

1. Bray, F.; Ferlay, J.; Soerjomataram, I.; Siegel, R.L.; Torre, L.A.; Jemal, A. Global cancer statistics 2018: GLOBOCAN estimates of incidence and mortality worldwide for 36 cancers in 185 countries. *CA Cancer J. Clin.* **2018**, *68*, 394–424. [CrossRef] [PubMed]
2. Siegel, R.L.; Miller, K.D.; Jemal, A. Cancer statistics, 2020. *CA Cancer J. Clin.* **2020**, *70*, 7–30. [CrossRef]
3. Greten, T.F.; Papendorf, F.; Bleck, J.S.; Kirchhoff, T.; Wohlberedt, T.; Kubicka, S.; Klempnauer, J.; Galanski, M.; Manns, M.P. Survival rate in patients with hepatocellular carcinoma: A retrospective analysis of 389 patients. *Br. J. Cancer* **2005**, *92*, 1862–1868. [CrossRef] [PubMed]
4. Op den Winkel, M.; Nagel, D.; Sappl, J.; op den Winkel, P.; Lamerz, R.; Zech, C.J.; Straub, G.; Nickel, T.; Rentsch, M.; Stieber, P.; et al. Prognosis of Patients with Hepatocellular Carcinoma. Validation and Ranking of Established Staging-Systems in a Large Western HCC-Cohort. *PLoS ONE* **2012**, *7*, e45066. [CrossRef]
5. Yang, J.D.; Kim, W.R.; Coelho, R.; Mettler, T.A.; Benson, J.T.; Sanderson, S.O.; Therneau, T.M.; Kim, B.; Roberts, L.R. Cirrhosis Is Present in Most Patients With Hepatitis B and Hepatocellular Carcinoma. *Clin. Gastroenterol. Hepatol.* **2011**, *9*, 64–70. [CrossRef] [PubMed]

6. Baecker, A.; Liu, X.; La Vecchia, C.; Zhang, Z.-F. Worldwide incidence of hepatocellular carcinoma cases attributable to major risk factors. *Eur. J. Cancer Prev.* **2018**, *27*, 205–212. [CrossRef]
7. Makarova-rusher, O.V.; Altekruse, S.F.; Mcneel, T.S.; Ulahannan, S.; Duffy, A.G.; Graubard, B.I.; Greten, T.F.; Mcglynn, K.A. Population Attributable Fractions of Risk Factors for Hepatocellular Carcinoma in the United States. *Cancer* **2016**, 1757–1765. [CrossRef] [PubMed]
8. Bravi, F.; Bosetti, C.; Tavani, A.; Gallus, S.; La Vecchia, C. Coffee Reduces Risk for Hepatocellular Carcinoma: An Updated Meta-analysis. *Clin. Gastroenterol. Hepatol.* **2013**, *11*, 1413–1421.e1. [CrossRef] [PubMed]
9. Bravi, F.; Tavani, A.; Bosetti, C.; Boffetta, P.; La Vecchia, C. Coffee and the risk of hepatocellular carcinoma and chronic liver disease: A systematic review and meta-analysis of prospective studies. *Eur. J. Cancer Prev.* **2017**, *26*, 368–377. [CrossRef]
10. Marquardt, J.U.; Andersen, J.B.; Thorgeirsson, S.S. Functional and genetic deconstruction of the cellular origin in liver cancer. *Nat. Rev. Cancer* **2015**, *15*, 653–667. [CrossRef] [PubMed]
11. Coleman, W.B. Mechanisms of Human Hepatocarcinogenesis. *Curr. Mol. Med.* **2003**, *3*, 573–588. [CrossRef] [PubMed]
12. Liu, M.; Jiang, L.; Guan, X.-Y. The genetic and epigenetic alterations in human hepatocellular carcinoma: A recent update. *Protein Cell* **2014**, *5*, 673–691. [CrossRef]
13. Llovet, J.M.; Kelley, R.K.; Villanueva, A.; Singal, A.G.; Pikarsky, E.; Roayaie, S.; Lencioni, R.; Koike, K.; Zucman-Rossi, J.; Finn, R.S. Hepatocellular carcinoma. *Nat. Rev. Dis. Prim.* **2021**, *7*, 6. [CrossRef]
14. Ally, A.; Balasundaram, M.; Carlsen, R.; Chuah, E.; Clarke, A.; Dhalla, N.; Holt, R.A.; Jones, S.J.M.; Lee, D.; Ma, Y.; et al. Comprehensive and Integrative Genomic Characterization of Hepatocellular Carcinoma. *Cell* **2017**, *169*, 1327–1341.e23. [CrossRef]
15. Hanahan, D.; Weinberg, R.A. Hallmarks of Cancer: The Next Generation. *Cell* **2011**, *144*, 646–674. [CrossRef]
16. Brown, Z.J.; Heinrich, B.; Greten, T.F. Mouse models of hepatocellular carcinoma: An overview and highlights for immunotherapy research. *Nat. Rev. Gastroenterol. Hepatol.* **2018**, *15*, 536–554. [CrossRef]
17. Shultz, L.D.; Ishikawa, F.; Greiner, D.L. Humanized mice in translational biomedical research. *Nat. Rev. Immunol.* **2007**, *7*, 118–130. [CrossRef] [PubMed]
18. Gu, Q.; Zhang, B.; Sun, H.; Xu, Q.; Tan, Y.; Wang, G.; Luo, Q.; Xu, W.; Yang, S.; Li, J.; et al. Genomic characterization of a large panel of patient-derived hepatocellular carcinoma xenograft tumor models for preclinical development. *Oncotarget* **2015**, *6*, 20160–20176. [CrossRef] [PubMed]
19. Zhao, Y.; Shuen, T.W.H.; Toh, T.B.; Chan, X.Y.; Liu, M.; Tan, S.Y.; Fan, Y.; Yang, H.; Lyer, S.G.; Bonney, G.K.; et al. Development of a new patient-derived xenograft humanised mouse model to study human-specific tumour microenvironment and immunotherapy. *Gut* **2018**, *67*, 1845–1854. [CrossRef]
20. Bosma, G.C.; Custer, R.P.; Bosma, M.J. A severe combined immunodeficiency mutation in the mouse. *Nature* **1983**, *301*, 527–530. [CrossRef] [PubMed]
21. Shultz, L.D.; Schweitzer, P.A.; Christianson, S.W.; Gott, B.; Schweitzer, I.B.; Tennent, B.; McKenna, S.; Mobraaten, L.; Rajan, T.V.; Greiner, D. Multiple defects in innate and adaptive immunologic function in NOD/LtSz-scid mice. *J. Immunol.* **1995**, *154*, 180–191. [PubMed]
22. Sun, F.X.; Tang, Z.Y.; Liu, K.D.; Ye, S.L.; Xue, Q.; Gao, D.M.; Ma, Z.C. Establishment of a metastatic model of human hepatocellular carcinoma in nude mice via orthotopic implantation of histologically intact tissues. *Int. J. Cancer* **1996**, *66*, 239–243. [CrossRef]
23. Genda, T.; Sakamoto, M.; Ichida, T.; Asakura, H.; Kojiro, M.; Narumiya, S.; Hirohashi, S. Cell motility mediated by rho and rho-associated protein kinase plays a critical role in intrahepatic metastasis of human hepatocellular carcinoma. *Hepatology* **1999**, *30*, 1027–1036. [CrossRef] [PubMed]
24. Blumer, T.; Fofana, I.; Matter, M.S.; Wang, X.; Montazeri, H.; Calabrese, D.; Coto-Llerena, M.; Boldanova, T.; Nuciforo, S.; Kancherla, V.; et al. Hepatocellular Carcinoma Xenografts Established From Needle Biopsies Preserve the Characteristics of the Originating Tumors. *Hepatol. Commun.* **2019**, *3*, 971–986. [CrossRef]
25. Shultz, L.D.; Lyons, B.L.; Burzenski, L.M.; Gott, B.; Chen, X.; Chaleff, S.; Kotb, M.; Gillies, S.D.; King, M.; Mangada, J.; et al. Human Lymphoid and Myeloid Cell Development in NOD/LtSz- scid IL2R γ null Mice Engrafted with Mobilized Human Hemopoietic Stem Cells. *J. Immunol.* **2005**, *174*, 6477–6489. [CrossRef]
26. Aruga, A.; Takasaki, K.; Hanyu, F. Establishment and characterization of liver metastatic model of human hepatoma in nude mice. *Int. Hepatol. Commun.* **1993**, *1*, 138–145. [CrossRef]
27. Yan, M.; Li, H.; Zhao, F.; Zhang, L.; Ge, C.; Yao, M.; Li, J. Establishment of NOD/SCID mouse models of human hepatocellular carcinoma via subcutaneous transplantation of histologically intact tumor tissue. *Chinese J. Cancer Res.* **2013**, *25*, 289–298. [CrossRef]
28. International Agency for Research on Cancer IARC monographs on the evaluation of the carcinogenic risk of chemicals to humans: Chemical Agents and Related Occu-pations. *IARC Monogr. Eval. Carcinog. Risks Chem. Man.* **2012**, *10F*, 1–628.
29. Tricker, A.R.; Kubacki, S.J. Review of the occurrence and formation of non-volatile N -nitroso compounds in foods. *Food Addit. Contam.* **1992**, *9*, 39–69. [CrossRef]
30. Lijinsky, W. N-Nitroso compounds in the diet. *Mutat. Res. Toxicol. Environ. Mutagen.* **1999**, *443*, 129–138. [CrossRef]
31. Tsuda, S. DNA Damage Induced by Red Food Dyes Orally Administered to Pregnant and Male Mice. *Toxicol. Sci.* **2001**, *61*, 92–99. [CrossRef]
32. Herrmann, S.S.; Granby, K.; Duedahl-Olesen, L. Formation and mitigation of N-nitrosamines in nitrite preserved cooked sausages. *Food Chem.* **2015**, *174*, 516–526. [CrossRef]

33. Kang, J.S.; Wanibuchi, H.; Morimura, K.; Gonzalez, F.J.; Fukushima, S. Role of CYP2E1 in Diethylnitrosamine-Induced Hepatocarcinogenesis in vivo. *Cancer Res.* **2007**, *67*, 11141–11146. [CrossRef] [PubMed]
34. Gao, J.; Wang, Z.; Wang, G.-J.; Zhang, H.-X.; Gao, N.; Wang, J.; Wang, C.-E.; Chang, Z.; Fang, Y.; Zhang, Y.-F.; et al. Higher CYP2E1 Activity Correlates with Hepatocarcinogenesis Induced by Diethylnitrosamine. *J. Pharmacol. Exp. Ther.* **2018**, *365*, 398–407. [CrossRef] [PubMed]
35. Verna, L. N-Nitrosodiethylamine mechanistic data and risk assessment: Bioactivation, DNA-adduct formation, mutagenicity, and tumor initiation. *Pharmacol. Ther.* **1996**, *71*, 57–81. [CrossRef]
36. Swenberg, J.A.; Dyroff, M.C.; Bedell, M.A.; Popp, J.A.; Huh, N.; Kirstein, U.; Rajewsky, M.F. O4-ethyldeoxythymidine, but not O6-ethyldeoxyguanosine, accumulates in hepatocyte DNA of rats exposed continuously to diethylnitrosamine. *Proc. Natl. Acad. Sci. USA* **1984**, *81*, 1692–1695. [CrossRef]
37. Aleksic, K.; Lackner, C.; Geigl, J.B.; Schwarz, M.; Auer, M.; Ulz, P.; Fischer, M.; Trajanoski, Z.; Otte, M.; Speicher, M.R. Evolution of genomic instability in diethylnitrosamine-induced hepatocarcinogenesis in mice. *Hepatology* **2011**, *53*, 895–904. [CrossRef] [PubMed]
38. Connor, F.; Rayner, T.F.; Aitken, S.J.; Feig, C.; Lukk, M.; Santoyo-Lopez, J.; Odom, D.T. Mutational landscape of a chemically-induced mouse model of liver cancer. *J. Hepatol.* **2018**, *69*, 840–850. [CrossRef]
39. Aparicio-Bautista, D.I.; Pérez-Carreón, J.I.; Gutiérrez-Nájera, N.; Reyes-Grajeda, J.P.; Arellanes-Robledo, J.; Vásquez-Garzón, V.R.; Jiménez-García, M.N.; Villa-Treviño, S. Comparative proteomic analysis of thiol proteins in the liver after oxidative stress induced by diethylnitrosamine. *Biochim. Biophys. Acta-Proteins Proteom.* **2013**, *1834*, 2528–2538. [CrossRef]
40. Ogawa, K. Molecular pathology of early stage chemically induced hepatocarcinogenesis. *Pathol. Int.* **2009**, *59*, 605–622. [CrossRef]
41. Diwan, B.A.; Rice, J.M.; Ohshima, M.; Ward, J.M. Interstrain differences in susceptibility to liver carcinogenesis initiated by N-nitrosodiethylamine and its promotion by phenobarbital in C57BL/6NCr, C3H/HeNCr MTV- and DBA/2NCr mice. *Carcinogenesis* **1986**, *7*, 215–220. [CrossRef] [PubMed]
42. Vesselinovitch, S.D.; Koka, M.; Mihailovich, N.; Rao, K.V.N. Carcinogenicity of diethylnitrosamine in newborn, infant, and adult mice. *J. Cancer Res. Clin. Oncol.* **1984**, *108*, 60–65. [CrossRef]
43. Klaunig, J.E.; Weghorst, T.R.; Weghorst, C.M. Liver tumor promoting ability of corn oil gavage in B6C3F1 male mice. *Cancer Lett.* **1990**, *50*, 215–219. [CrossRef]
44. Kushida, M.; Kamendulis, L.M.; Peat, T.J.; Klaunig, J.E. Dose-Related Induction of Hepatic Preneoplastic Lesions by Diethylnitrosamine in C57BL/6 Mice. *Toxicol. Pathol.* **2011**, *39*, 776–786. [CrossRef] [PubMed]
45. Klaunig, J.E.; Pereira, M.A.; Ruch, R.J.; Weghorst, C.M. Dose-Response Relationship of Diethylnitrosamine-Initiated Tumors in Neonatal Balb/c Mice: Effect of Phenobarbital Promotion. *Toxicol. Pathol.* **1988**, *16*, 381–385. [CrossRef] [PubMed]
46. Weghorst, C.M.; Pereira, M.A.; Klaunig, E. Strain differences in hepatic tumor promotion by phenobarbital in diethylnitrosamine- and dimethylnitrosamine-initiated infant male mice. *Carcinogenesis* **1989**, *10*, 1409–1412. [CrossRef]
47. Goldsworthy, T.L.; Fransson-Steen, R. Quantitation of the Cancer Process in C57BL/6J, B6C3F1 and C3H/HeJ Mice. *Toxicol. Pathol.* **2002**, *30*, 97–105. [CrossRef]
48. Uehara, T.; Ainslie, G.R.; Kutanzi, K.; Pogribny, I.P.; Muskhelishvili, L.; Izawa, T.; Yamate, J.; Kosyk, O.; Shymonyak, S.; Bradford, B.U.; et al. Molecular Mechanisms of Fibrosis-Associated Promotion of Liver Carcinogenesis. *Toxicol. Sci.* **2013**, *132*, 53–63. [CrossRef]
49. Memon, A.; Pyao, Y.; Jung, Y.; Lee, J.I.; Lee, W.K. A Modified Protocol of Diethylnitrosamine Administration in Mice to Model Hepatocellular Carcinoma. *Int. J. Mol. Sci.* **2020**, *21*, 5461. [CrossRef]
50. Pereira, M.A.; Herren-Freund, S.L.; Long, R.E. Dose-response relationship of phenobarbital promotion of diethylnitrosamine initiated tumors in rat liver. *Cancer Lett.* **1986**, *32*, 305–311. [CrossRef]
51. Sugie, S.; Okamoto, K.; Watanabe, T.; Tanaka, T.; Mori, H. Suppressive effect of irsogladine maleate on N-methyl-N-nitro-N-nitrosoguanidine (MNNG)-initiated and glyoxal-promoted gastric carcinogenesis in rats. *Toxicology* **2001**, *166*, 53–61. [CrossRef]
52. Tan, Y.; Yin, P.; Tang, L.; Xing, W.; Huang, Q.; Cao, D.; Zhao, X.; Wang, W.; Lu, X.; Xu, Z.; et al. Metabolomics Study of Stepwise Hepatocarcinogenesis From the Model Rats to Patients: Potential Biomarkers Effective for Small Hepatocellular Carcinoma Diagnosis. *Mol. Cell. Proteom.* **2012**, *11*, M111.010694. [CrossRef]
53. Ding, Y.; Wu, Z.; Wei, Y.; Shu, L.; Peng, Y. Hepatic inflammation-fibrosis-cancer axis in the rat hepatocellular carcinoma induced by diethylnitrosamine. *J. Cancer Res. Clin. Oncol.* **2017**, *143*, 821–834. [CrossRef]
54. Liu, J.; Divoux, A.; Sun, J.; Zhang, J.; Clément, K.; Glickman, J.N.; Sukhova, G.K.; Wolters, P.J.; Du, J.; Gorgun, C.Z.; et al. Genetic deficiency and pharmacological stabilization of mast cells reduce diet-induced obesity and diabetes in mice. *Nat. Med.* **2009**, *15*, 940–945. [CrossRef] [PubMed]
55. Romualdo, G.R.; Prata, G.B.; da Silva, T.C.; Fernandes, A.A.H.; Moreno, F.S.; Cogliati, B.; Barbisan, L.F. Fibrosis-associated hepatocarcinogenesis revisited: Establishing standard medium-term chemically-induced male and female models. *PLoS ONE* **2018**, *13*, e0203879. [CrossRef]
56. Zalatnai, A.; Lapis, K. Decreased hepatocarcinogenic effect of diethylnitrosamine in experimentally induced liver cirrhosis in rat: Delay or inhibition? *Cancer Lett.* **1994**, *79*, 1–7. [CrossRef]
57. Romualdo, G.R.; Grassi, T.F.; Goto, R.L.; Tablas, M.B.; Bidinotto, L.T.; Fernandes, A.A.H.; Cogliati, B.; Barbisan, L.F. An integrative analysis of chemically-induced cirrhosis-associated hepatocarcinogenesis: Histological, biochemical and molecular features. *Toxicol. Lett.* **2017**, *281*, 84–94. [CrossRef]

58. Park, T.J.; Kim, H.S.; Byun, K.H.; Jang, J.J.; Lee, Y.S.; Lim, I.K. Sequential changes in hepatocarcinogenesis induced by diethylnitrosamine plus thioacetamide in Fischer 344 rats: Induction of gankyrin expression in liver fibrosis, pRB degradation in cirrhosis, and methylation ofp16INK4A exon 1 in hepatocellular carcinom. *Mol. Carcinog.* **2001**, *30*, 138–150. [CrossRef] [PubMed]
59. Solt, D.B.; Cayama, E.; Tsuda, H.; Enomoto, K.; Lee, G.; Farber, E. Promotion of liver cancer development by brief exposure to dietary 2-acetylaminofluorene plus partial hepatectomy or carbon tetrachloride. *Cancer Res.* **1983**, *43*, 188–191.
60. Vesselinovitch, S.D.; Hacker, H.J.; Bannasch, P. Histochemical characterization of focal hepatic lesions induced by single diethylnitrosamine treatment in infant mice. *Cancer Res.* **1985**, *45*, 2774–2780. [PubMed]
61. Septer, S.; Edwards, G.; Gunewardena, S.; Wolfe, A.; Li, H.; Daniel, J.; Apte, U. Yes-associated protein is involved in proliferation and differentiation during postnatal liver development. *Am. J. Physiol. Liver Physiol.* **2012**, *302*, G493–G503. [CrossRef] [PubMed]
62. Naugler, W.E.; Sakurai, T.; Kim, S.; Maeda, S.; Kim, K.; Elsharkawy, A.M.; Karin, M. Gender Disparity in Liver Cancer Due to Sex Differences in MyD88-Dependent IL-6 Production. *Science* **2007**, *317*, 121–124. [CrossRef] [PubMed]
63. Watanabe, T.; Tanaka, G.; Hamada, S.; Namiki, C.; Suzuki, T.; Nakajima, M.; Furihata, C. Dose-dependent alterations in gene expression in mouse liver induced by diethylnitrosamine and ethylnitrosourea and determined by quantitative real-time PCR. *Mutat. Res. Toxicol. Environ. Mutagen.* **2009**, *673*, 9–20. [CrossRef] [PubMed]
64. Bannasch, P.; Klimek, F.; Mayer, D. Early bioenergetic changes in hepatocarcinogenesis: Preneoplastic phenotypes mimic responses to insulin and thyroid hormone. *J. Bioenerg. Biomembr.* **1997**, *29*, 303–313. [CrossRef] [PubMed]
65. Su, Q.; Bannasch, P. Relevance of Hepatic Preneoplasia for Human Hepatocarcinogenesis. *Toxicol. Pathol.* **2003**, *31*, 126–133. [CrossRef]
66. Lahm, H.; Gittner, K.; Krebs, O.; Sprague, L.; Deml, E.; Oesterle, D.; Hoeflich, A.; Wanke, R.; Wolf, E. Diethylnitrosamine induces long-lasting re-expression of insulin-like growth factor II during early stages of liver carcinogenesis in mice. *Growth Horm. IGF Res.* **2002**, *12*, 69–79. [CrossRef] [PubMed]
67. Bauer-Hofmann, R.; Klimek, F.; Buchmann, A.; Müller, O.; Bannasch, P.; Schwarz, M. Role of mutations at codon 61 of the c-Ha-ras gene during diethylnitrosamine-induced hepatocarcinogenesis in C3H/He mice. *Mol. Carcinog.* **1992**, *6*, 60–67. [CrossRef]
68. Yamamoto, M.; Tanaka, H.; Xin, B.; Nishikawa, Y.; Yamazaki, K.; Shimizu, K.; Ogawa, K. Role of the Braf V637E mutation in hepatocarcinogenesis induced by treatment with diethylnitrosamine in neonatal B6C3F1 mice. *Mol. Carcinog.* **2017**, *56*, 478–488. [CrossRef]
69. Cast, A.; Valanejad, L.; Wright, M.; Nguyen, P.; Gupta, A.; Zhu, L.; Shin, S.; Timchenko, N. C/EBPα-dependent preneoplastic tumor foci are the origin of hepatocellular carcinoma and aggressive pediatric liver cancer. *Hepatology* **2018**, *67*, 1857–1871. [CrossRef] [PubMed]
70. Chappell, G.; Kutanzi, K.; Uehara, T.; Tryndyak, V.; Hong, H.-H.; Hoenerhoff, M.; Beland, F.A.; Rusyn, I.; Pogribny, I.P. Genetic and epigenetic changes in fibrosis-associated hepatocarcinogenesis in mice. *Int. J. Cancer* **2014**, *134*, 2778–2788. [CrossRef]
71. Marrone, A.K.; Shpyleva, S.; Chappell, G.; Tryndyak, V.; Uehara, T.; Beland, F.A.; Rusyn, I.; Pogribny, I.P. Differentially Expressed MicroRNAs Provide Mechanistic Insight into Fibrosis-Associated Liver Carcinogenesis in Mice. *Mol. Carcinog.* **2016**, *55*, 808–817. [CrossRef]
72. He, Q.; Wang, F.; Honda, T.; Lindquist, D.M.; Dillman, J.R.; Timchenko, N.A.; Redington, A.N. Intravenous miR-144 inhibits tumor growth in diethylnitrosamine-induced hepatocellular carcinoma in mice. *Tumor Biol.* **2017**, *39*, 1–8. [CrossRef] [PubMed]
73. Dow, M.; Pyke, R.M.; Tsui, B.Y.; Alexandrov, L.B.; Nakagawa, H.; Taniguchi, K.; Seki, E.; Harismendy, O.; Shalapour, S.; Karin, M.; et al. Integrative genomic analysis of mouse and human hepatocellular carcinoma. *Proc. Natl. Acad. Sci. USA* **2018**, *115*, E9879–E9888. [CrossRef]
74. Chen, X.; Yamamoto, M.; Fujii, K.; Nagahama, Y.; Ooshio, T.; Xin, B.; Okada, Y.; Furukawa, H.; Nishikawa, Y. Differential reactivation of fetal/neonatal genes in mouse liver tumors induced in cirrhotic and non-cirrhotic conditions. *Cancer Sci.* **2015**, *106*, 972–981. [CrossRef]
75. Yim, S.Y.; Shim, J.-J.; Shin, J.-H.; Jeong, Y.S.; Kang, S.-H.; Kim, S.-B.; Eun, Y.G.; Lee, D.J.; Conner, E.A.; Factor, V.M.; et al. Integrated Genomic Comparison of Mouse Models Reveals Their Clinical Resemblance to Human Liver Cancer. *Mol. Cancer Res.* **2018**, *16*, 1713–1723. [CrossRef] [PubMed]
76. Lee, J.-S.; Chu, I.-S.; Mikaelyan, A.; Calvisi, D.F.; Heo, J.; Reddy, J.K.; Thorgeirsson, S.S. Application of comparative functional genomics to identify best-fit mouse models to study human cancer. *Nat. Genet.* **2004**, *36*, 1306–1311. [CrossRef]
77. Liu, Y.-F.; Zha, B.-S.; Zhang, H.-L.; Zhu, X.-J.; Li, Y.-H.; Zhu, J.; Guan, X.-H.; Feng, Z.-Q.; Zhang, J.-P. Characteristic gene expression profiles in the progression from liver cirrhosis to carcinoma induced by diethylnitrosamine in a rat model. *J. Exp. Clin. Cancer Res.* **2009**, *28*, 107. [CrossRef]
78. Moreno, F.S.; Rizzi, M.B.S.L.; Dagli, M.L.Z.; Penteado, M.V.C. Inhibitory effects of β-carotene on preneoplastic lesions induced in Wistar rats by the resistant hepatocyte model. *Carcinogenesis* **1991**, *12*, 1817–1822. [CrossRef]
79. Romualdo, G.R.; Goto, R.L.; Henrique Fernandes, A.A.; Cogliati, B.; Barbisan, L.F. Dietary zinc deficiency predisposes mice to the development of preneoplastic lesions in chemically-induced hepatocarcinogenesis. *Food Chem. Toxicol.* **2016**, *96*, 280–289. [CrossRef] [PubMed]
80. Romualdo, G.R.; Prata, G.B.; da Silva, T.C.; Evangelista, A.F.; Reis, R.M.; Vinken, M.; Moreno, F.S.; Cogliati, B.; Barbisan, L.F. The combination of coffee compounds attenuates early fibrosis-associated hepatocarcinogenesis in mice: Involvement of miRNA profile modulation. *J. Nutr. Biochem.* **2020**, *85*, 108479. [CrossRef]

81. Miranda, M.L.P.; Furtado, K.S.; de Oliveira Andrade, F.; Heidor, R.; da Cruz, R.S.; Nogueira, M.S.; de Castro, I.A.; Purgatto, E.; Barbisan, L.F.; Moreno, F.S. β-ionone inhibits nonalcoholic fatty liver disease and its association with hepatocarcinogenesis in male Wistar rats. *Chem. Biol. Interact.* **2019**, *308*, 377–384. [CrossRef]
82. De Conti, A.; Tryndyak, V.; Heidor, R.; Jimenez, L.; Moreno, F.S.; Beland, F.A.; Rusyn, I.; Pogribny, I.P. Butyrate-containing structured lipids inhibit RAC1 and epithelial-to-mesenchymal transition markers: A chemopreventive mechanism against hepatocarcinogenesis. *J. Nutr. Biochem.* **2020**, *86*, 108496. [CrossRef]
83. Sarmiento-Machado, L.M.; Romualdo, G.R.; Zapaterini, J.R.; Tablas, M.B.; Fernandes, A.A.H.; Moreno, F.S.; Barbisan, L.F. Protective Effects of Dietary Capsaicin on the Initiation Step of a Two-Stage Hepatocarcinogenesis Rat Model. *Nutr. Cancer* **2021**, *73*, 817–828. [CrossRef] [PubMed]
84. Weber, L.W.D.; Boll, M.; Stampfl, A. Hepatotoxicity and Mechanism of Action of Haloalkanes: Carbon Tetrachloride as a Toxicological Model. *Crit. Rev. Toxicol.* **2003**, *33*, 105–136. [CrossRef]
85. Tsuchida, T.; Friedman, S.L. Mechanisms of hepatic stellate. *Nat. Publ. Gr.* **2017**. [CrossRef]
86. Cho, K.-J.; Jang, J.-J. Effects of carbon tetrachloride, ethanol and acetaldehyde on diethylnitrosamine-induced hepatocarcinogenesis in rats. *Cancer Lett.* **1993**, *70*, 33–39. [CrossRef]
87. Hajovsky, H.; Hu, G.; Koen, Y.; Sarma, D.; Cui, W.; Moore, D.S.; Staudinger, J.L.; Hanzlik, R.P. Metabolism and Toxicity of Thioacetamide and Thioacetamide S -Oxide in Rat Hepatocytes. *Chem. Res. Toxicol.* **2012**, *25*, 1955–1963. [CrossRef] [PubMed]
88. Tatematsu, M.; Tsuda, H.; Shirai, T.; Masui, T.; Ito, N. Placental Glutathione S-Transferase (GST-P) as a New Marker for Hepatocarcinogenesis: In Vivo Short-Term Screening for Hepatocarcinogens. *Toxicol. Pathol.* **1987**, *15*, 60–68. [CrossRef]
89. Ito, S.; Tateno, C.; Tuda, M.; Yoshitake, A. Immunohistochemical Demonstration of the Gap Junctional Protein Connexin 32 and Proliferating Cell Nuclear Antigen in Glutathione S-Transferase Placental Form-Negative Lesions of Rat Liver Induced by Diethylnitrosamine and Clofibrate. *Toxicol. Pathol.* **1996**, *24*, 690–695. [CrossRef]
90. Kimura, M.; Fujii, Y.; Yamamoto, R.; Yafune, A.; Hayashi, S.; Suzuki, K.; Shibutani, M. Involvement of multiple cell cycle aberrations in early preneoplastic liver cell lesions by tumor promotion with thioacetamide in a two-stage rat hepatocarcinogenesis model. *Exp. Toxicol. Pathol.* **2013**, *65*, 979–988. [CrossRef] [PubMed]
91. Tsuchiya, T.; Wang, L.; Yafune, A.; Kimura, M.; Ohishi, T.; Suzuki, K.; Mitsumori, K.; Shibutani, M. Disruptive cell cycle regulation involving epigenetic downregulation of Cdkn2a (p16Ink4a) in early-stage liver tumor-promotion facilitating liver cell regeneration in rats. *Toxicology* **2012**, *299*, 146–154. [CrossRef] [PubMed]
92. Mizukami, S.; Yafune, A.; Watanabe, Y.; Nakajima, K.; Jin, M.; Yoshida, T.; Shibutani, M. Identification of epigenetically downregulated Tmem70 and Ube2e2 in rat liver after 28-day treatment with hepatocarcinogenic thioacetamide showing gene product downregulation in hepatocellular preneoplastic and neoplastic lesions produced by tumor promoti. *Toxicol. Lett.* **2017**, *266*, 13–22. [CrossRef]
93. Uehara, T.; Hirode, M.; Ono, A.; Kiyosawa, N.; Omura, K.; Shimizu, T.; Mizukawa, Y.; Miyagishima, T.; Nagao, T.; Urushidani, T. A toxicogenomics approach for early assessment of potential non-genotoxic hepatocarcinogenicity of chemicals in rats. *Toxicology* **2008**, *250*, 15–26. [CrossRef] [PubMed]
94. Omura, K.; Uehara, T.; Morikawa, Y.; Hayashi, H.; Mitsumori, K.; Minami, K.; Kanki, M.; Yamada, H.; Ono, A.; Urushidani, T. Detection of initiating potential of non-genotoxic carcinogens in a two-stage hepatocarcinogenesis study in rats. *J. Toxicol. Sci.* **2014**, *39*, 785–794. [CrossRef]
95. Diwan, B.A.; Rice, J.M.; Ward, J.M.; Ohshima, M.; Lynch, P.H. Inhibition by phenobarbital and lack of effect of amobarbital on the development of liver tumors induced by N-nitrosodiethylamine in juvenile B6C3F1 mice. *Cancer Lett.* **1984**, *23*, 223–234. [CrossRef]
96. Greaves, P.; Irisarri, E.; Monro, A.M. Hepatic foci of cellular and enzymatic alteration and nodules in rats treated with clofibrate or diethylnitrosamine followed by phenobarbital: Their rate of onset and their reversibility. *J. Natl. Cancer Inst.* **1986**, *76*, 475–484.
97. Jang, J.-J.; Henneman, J.R.; Kurata, Y.; Uno, H.; Ward, J.M. Alterations in populations of GST-p-immunoreactive single hepatocytes and hepatocellular foci after a single injection of N-nitrosodiethlyamine with or without phenobarbital promotion in male F344NCr rats. *Cancer Lett.* **1993**, *71*, 89–95. [CrossRef]
98. Aydinlik, H.; Nguyen, T.D.; Moennikes, O.; Buchmann, A.; Schwarz, M. Selective pressure during tumor promotion by phenobarbital leads to clonal outgrowth of β-catenin-mutated mouse liver tumors. *Oncogene* **2001**, *20*, 7812–7816. [CrossRef]
99. Lee, G. Review Article: Paradoxical Effects of Phenobarbital on Mouse Hepatocarcinogenesis. *Toxicol. Pathol.* **2000**, *28*, 215–225. [CrossRef]
100. Braeuning, A.; Schwarz, M. Is the question of phenobarbital as potential liver cancer risk factor for humans really resolved? *Arch. Toxicol.* **2016**, *90*, 1525–1526. [CrossRef] [PubMed]
101. Aiub, C.A.F.; Gadermaier, G.; Oliveira, I.; Felzenszwalb, I.; Ferreira, F.; Pinto, L.F.R.; Eckl, P. N-Nitrosodiethylamine genotoxicity in primary rat hepatocytes: Effects of cytochrome P450 induction by phenobarbital. *Toxicol. Lett.* **2011**, *206*, 139–143. [CrossRef] [PubMed]
102. Yamamoto, Y.; Moore, R.; Goldsworthy, T.L.; Negishi, M.; Maronpot, R.R. The Orphan Nuclear Receptor Constitutive Active/Androstane Receptor Is Essential for Liver Tumor Promotion by Phenobarbital in Mice. *Cancer Res.* **2004**, *64*, 7197–7200. [CrossRef] [PubMed]

103. Braeuning, A.; Heubach, Y.; Knorpp, T.; Kowalik, M.A.; Templin, M.; Columbano, A.; Schwarz, M. Gender-Specific Interplay of Signaling through β-Catenin and CAR in the Regulation of Xenobiotic-Induced Hepatocyte Proliferation. *Toxicol. Sci.* **2011**, *123*, 113–122. [CrossRef] [PubMed]
104. Moennikes, O.; Buchmann, A.; Romualdi, A.; Ott, T.; Werringloer, J.; Willecke, K.; Schwarz, M. Lack of phenobarbital-mediated promotion of hepatocarcinogenesis in connexin32-null mice. *Cancer Res.* **2000**, *60*, 5087–5091. [PubMed]
105. Marx-Stoelting, P.; Mahr, J.; Knorpp, T.; Schreiber, S.; Templin, M.F.; Ott, T.; Buchmann, A.; Schwarz, M. Tumor Promotion in Liver of Mice with a Conditional Cx26 Knockout. *Toxicol. Sci.* **2008**, *103*, 260–267. [CrossRef]
106. Pereira, M.A.; Klaunig, J.E.; Herren-Freund, S.L.; Ruch, R.J. Effect of Phenobarbital on the Development of Liver Tumors in Juvenile and Adult Mice. *JNCI J. Natl. Cancer Inst.* **1986**, *77*, 449–452. [CrossRef] [PubMed]
107. Tamano, S.; Merlino, G.T.; Ward, J.M. Rapid development of hepatic tumors in transforming growth factor α transgenic mice associated with increased cell proliferation in precancerous hepatocellular lesions initiated by N -nitrosodiethylamine and promoted by phenobarbital. *Carcinogenesis* **1994**, *15*, 1791–1798. [CrossRef]
108. Lee, C.-C.; Liu, J.-Y.; Lin, J.-K.; Chu, J.-S.; Shew, J.-Y. p53 point mutation enhanced by hepatic regeneration in aflatoxin B1-induced rat liver tumors and preneoplastic lesions. *Cancer Lett.* **1998**, *125*, 1–7. [CrossRef]
109. Ray, J.S.; Harbison, M.L.; McClain, R.M.; Goodman, J.I. Alterations in the methylation status and expression of theraf oncogene in phenobarbital-induced and spontaneous B6C3F1 mouse liver tumors. *Mol. Carcinog.* **1994**, *9*, 155–166. [CrossRef]
110. Phillips, J.M.; Burgoon, L.D.; Goodman, J.I. Phenobarbital Elicits Unique, Early Changes in the Expression of Hepatic Genes that Affect Critical Pathways in Tumor-Prone B6C3F1 Mice. *Toxicol. Sci.* **2009**, *109*, 193–205. [CrossRef] [PubMed]
111. Vorce, R.L.; Goodman, J.I. Hypomethylation of ras oncogenes in chemically induced and spontaneous b6c3f1 mouse liver tumors. *J. Toxicol. Environ. Health* **1991**, *34*, 367–384. [CrossRef] [PubMed]
112. Maronpot, R.R. Biological Basis of Differential Susceptibility to Hepatocarcinogenesis among Mouse Strains. *J. Toxicol. Pathol.* **2009**, *22*, 11–33. [CrossRef]
113. Solt, D.; Farber, E. New principle for the analysis of chemical carcinogenesis. *Nature* **1976**, *263*, 701–703. [CrossRef]
114. Solt, D.B.; Medline, A.; Farber, E. Rapid emergence of carcinogen-induced hyperplastic lesions in a new model for the sequential analysis of liver carcinogenesis. *Am. J. Pathol.* **1977**, *88*, 595–618. [PubMed]
115. Tsuda, H.; Lee, G.; Farber, E. Induction of resistant hepatocytes as a new principle for a possible short-term in vivo test for carcinogens. *Cancer Res.* **1980**, *40*, 1157–1164.
116. Semple-Roberts, E.; Hayes, M.A.; Armstrong, D.; Becker, R.A.; Racz, W.J.; Farser, E. Alternative methods of selecting rat hepatocellular noduli resistant to 2-acetylaminofluorene. *Int. J. Cancer* **1987**, *40*, 643–645. [CrossRef]
117. Higgins, G.F.; Anderson, R.M.; Higgins, G.M.; Anderson, R.M. Experimental pathology of liver: Restoration of the liver of the white rat following partial surgical removal. *Arch. Pathol.* **1931**, *12*, 186–202.
118. Naves, M.M.V.; Silveira, E.R.; Dagli, M.L.Z.; Moreno, F.S. Effects of β-carotene and vitamin A on oval cell proliferation and connexin 43 expression during hepatic differentiation in the rat11This work was supported by grants from Fundação de Amparo à Pesquisa do Estado de São Paulo (FAPESP—process no. 1996/7566–. *J. Nutr. Biochem.* **2001**, *12*, 685–692. [CrossRef]
119. Potter, V.R. Use of two sequential applications of initiators in the production of hepatomas in the rat: An examination of the Solt-Farber protocol. *Cancer Res.* **1984**, *44*, 2733–2736.
120. Mazzantini, R.P.; de Conti, A.; Moreno, F.S. Persistent and remodeling hepatic preneoplastic lesions present differences in cell proliferation and apoptosis, as well as in p53, Bcl-2 and NF-κB pathways. *J. Cell. Biochem.* **2008**, *103*, 538–546. [CrossRef]
121. Faris, R.A.; Monfils, B.A.; Dunsford, H.A.; Hixson, D.C. Antigenic relationship between oval cells and a subpopulation of hepatic foci, nodules, and carcinomas induced by the "resistant hepatocyte" model system. *Cancer Res.* **1991**, *51*, 1308–1317.
122. Sell, S.; Dunsford, H.A. Evidence for the stem cell origin of hepatocellular carcinoma and cholangiocarcinoma. *Am. J. Pathol.* **1989**, *134*, 1347–1363.
123. Andersen, J.B.; Loi, R.; Perra, A.; Factor, V.M.; Ledda-Columbano, G.M.; Columbano, A.; Thorgeirsson, S.S. Progenitor-derived hepatocellular carcinoma model in the rat. *Hepatology* **2010**, *51*, 1401–1409. [CrossRef]
124. Perra, A.; Kowalik, M.A.; Ghiso, E.; Ledda-Columbano, G.M.; Di Tommaso, L.; Angioni, M.M.; Raschioni, C.; Testore, E.; Roncalli, M.; Giordano, S.; et al. YAP activation is an early event and a potential therapeutic target in liver cancer development. *J. Hepatol.* **2014**, *61*, 1088–1096. [CrossRef] [PubMed]
125. Petrelli, A.; Perra, A.; Cora, D.; Sulas, P.; Menegon, S.; Manca, C.; Migliore, C.; Kowalik, M.A.; Ledda-Columbano, G.M.; Giordano, S.; et al. MicroRNA/gene profiling unveils early molecular changes and nuclear factor erythroid related factor 2 (NRF2) activation in a rat model recapitulating human hepatocellular carcinoma (HCC). *Hepatology* **2014**, *59*, 228–241. [CrossRef] [PubMed]
126. Zavattari, P.; Perra, A.; Menegon, S.; Kowalik, M.A.; Petrelli, A.; Angioni, M.M.; Follenzi, A.; Quagliata, L.; Ledda-Columbano, G.M.; Terracciano, L.; et al. Nrf2, but not β-catenin, mutation represents an early event in rat hepatocarcinogenesis. *Hepatology* **2015**, *62*, 851–862. [CrossRef]
127. Newberne, P.M.; Wogan, G.N. Sequential morphologic changes in aflatoxin B carcinogenesis in the rat. *Cancer Res.* **1968**, *28*, 770–781. [PubMed]
128. Butler, W.H.; Greenblatt, M.; Lijinsky, W. Carcinogenesis in rats by aflatoxins B1, G1, and B2. *Cancer Res.* **1969**, *29*, 2206–2211.
129. Wogan, G.N.; Paglialunga, S.; Newberne, P.M. Carcinogenic effects of low dietary levels of aflatoxin B1 in rats. *Food Cosmet. Toxicol.* **1974**, *12*, 681–685. [CrossRef]

130. Nixon, J.E.; Hendricks, J.D.; Pawloswki, N.E.; Loveland, P.M.; Sinnhuber, R.O. Carcinogenicity of Aflatoxlcol in Fischer 344 Rats2, 3, 4. *JNCI J. Natl. Cancer Inst.* **1981**, *66*, 1159–1163. [CrossRef]
131. Williams, J.H.; Phillips, T.D.; Jolly, P.E.; Stiles, J.K.; Jolly, C.M.; Aggarwal, D. Human aflatoxicosis in developing countries: A review of toxicology, exposure, potential health consequences, and interventions. *Am. J. Clin. Nutr.* **2004**, *80*, 1106–1122. [CrossRef]
132. Aguilar, F.; Hussain, S.P.; Cerutti, P. Aflatoxin B1 induces the transversion of G->T in codon 249 of the p53 tumor suppressor gene in human hepatocytes. *Proc. Natl. Acad. Sci. USA* **1993**, *90*, 8586–8590. [CrossRef] [PubMed]
133. Hulla, J.E.; Chen, Z.Y.; Eaton, D.L. Aflatoxin B1-induced rat hepatic hyperplastic nodules do not exhibit a site-specific mutation within the p53 gene. *Cancer Res.* **1993**, *53*, 9–11. [PubMed]
134. Shen, H.M.; Shi, C.Y.; Lee, H.P.; Ong, C.N. Aflatoxin B1-Induced Lipid Peroxidation in Rat Liver. *Toxicol. Appl. Pharmacol.* **1994**, *127*, 145–150. [CrossRef] [PubMed]
135. Singh, K.B.; Maurya, B.K.; Trigun, S.K. Activation of oxidative stress and inflammatory factors could account for histopathological progression of aflatoxin-B1 induced hepatocarcinogenesis in rat. *Mol. Cell. Biochem.* **2015**, *401*, 185–196. [CrossRef]
136. Shi, J.; He, J.; Lin, J.; Sun, X.; Sun, F.; Ou, C.; Jiang, C. Distinct response of the hepatic transcriptome to Aflatoxin B1 induced hepatocellular carcinogenesis and resistance in rats. *Sci. Rep.* **2016**, *6*, 31898. [CrossRef] [PubMed]
137. Columbano, A.; Rajalakshmi, S.; Sarma, D.S. Requirement of cell proliferation for the initiation of liver carcinogenesis as assayed by three different procedures. *Cancer Res.* **1981**, *41*, 2079–2083.
138. Sakai, H.; Tsukamoto, T.; Yamamoto, M.; Shirai, N.; Lidaka, T.; Yanai, T.; Masegi, T.; Tatematsu, M. Differential Effects of Partial Hepatectomy and Carbon Tetrachloride Administration on Induction of Liver Cell Foci in a Model for Detection of Initiation Activity. *Japanese J. Cancer Res.* **2001**, *92*, 1018–1025. [CrossRef]
139. Ramos Caetano, B.F.; Baptista Tablas, M.; Ribeiro Romualdo, G.; Marchesan Rodrigues, M.A.; Barbisan, L.F. Early molecular events associated with liver and colon sub-acute responses to 1,2-dimethylhydrazine: Potential implications on preneoplastic and neoplastic lesion development. *Toxicol. Lett.* **2020**, *329*, 67–79. [CrossRef]
140. Punvittayagul, C.; Chariyakornkul, A.; Chewonarin, T.; Jarukamjorn, K.; Wongpoomchai, R. Augmentation of diethylnitrosamine–induced early stages of rat hepatocarcinogenesis by 1,2-dimethylhydrazine. *Drug Chem. Toxicol.* **2019**, *42*, 641–648. [CrossRef]
141. Pascale, R.M.; Simile, M.M.; Peitta, G.; Seddaiu, M.A.; Feo, F.; Calvisi, D.F. Experimental Models to Define the Genetic Predisposition to Liver Cancer. *Cancers* **2019**, *11*, 1450. [CrossRef] [PubMed]
142. Bilger, A.; Bennett, L.M.; Carabeo, R.A.; Chiaverotti, T.A.; Dvorak, C.; Liss, K.M.; Schadewald, S.A.; Pitot, H.C.; Drinkwater, N.R. A Potent Modifier of Liver Cancer Risk on Distal Mouse Chromosome 1 This article is dedicated to the memory of our late colleague, Kristin M. Liss. *Genetics* **2004**, *167*, 859–866. [CrossRef]
143. Drinkwater, N.R. Genetic Control of Hepatocarcinogenesis In C3H Mice. *Drug Metab. Rev.* **1994**, *26*, 201–208. [CrossRef]
144. Hanigan, M.H.; Kemp, C.J.; Ginsler, J.J.; Drinkwater, N.R. Rapid growth of preneoplastic lesions in hepatocarcinogen-Sensitive C3H/HeJ male mice relative to C57BL/6J male mice. *Carcinogenesis* **1988**, *9*, 885–890. [CrossRef] [PubMed]
145. Gariboldi, M.; Manenti, G.; Canzian, F.; Falvella, F.S.; Pierotti, M.A.; Della Porta, G.; Binelli, G.; Dragani, T.A. Chromosome mapping of murine susceptibility loci to liver carcinogenesis. *Cancer Res.* **1993**, *53*, 209–211.
146. Lee, G.H.; Bennett, L.M.; Carabeo, R.A.; Drinkwater, N.R. Identification of hepatocarcinogen-resistance genes in DBA/2 mice. *Genetics* **1995**, *139*, 387–395. [CrossRef]
147. Dragani, T.A.; Manenti, G.; Gariboldi, M.; De Gregorio, L.; Pierotti, M.A. Genetics of liver tumor susceptibility in mice. *Toxicol. Lett.* **1995**, *82–83*, 613–619. [CrossRef]
148. Manenti, G.; Binelli, G.; Gariboldi, M.; Canzian, F.; De Gregorio, L.; Falvella, F.S.; Dragani, T.A.; Pierotti, M.A. Multiple Loci Affect Genetic Predisposition to Hepatocarcinogenesis in Mice. *Genomics* **1994**, *23*, 118–124. [CrossRef] [PubMed]
149. Manenti, G.; Galvan, A.; Falvella, F.S.; Pascale, R.M.; Spada, E.; Milani, S.; Neira, A.G.; Feo, F.; Dragani, T.A. Genetic control of resistance to hepatocarcinogenesis by the mouse Hpcr3 locus. *Hepatology* **2008**, *48*, 617–623. [CrossRef]
150. De Miglio, M.R.; Pascale, R.M.; Simile, M.M.; Muroni, M.R.; Virdis, P.; Kwong, K.M.T.; Wong, L.K.L.; Bosinco, G.M.; Pulina, F.R.; Calvisi, D.F.; et al. Polygenic control of hepatocarcinogenesis in Copenhagen × F344 rats. *Int. J. Cancer* **2004**, *111*, 9–16. [CrossRef]
151. De Miglio, M.R.; Virdis, P.; Calvisi, D.F.; Frau, M.; Muroni, M.R.; Simile, M.M.; Daino, L.; Careddu, G.M.; Sanna-Passino, E.; Pascale, R.M.; et al. Mapping a Sex Hormone–Sensitive Gene Determining Female Resistance to Liver Carcinogenesis in a Congenic F344.BN- Hcs4 Rat. *Cancer Res.* **2006**, *66*, 10384–10390. [CrossRef] [PubMed]
152. Yan, Y. Resistance of DRH strain rats to chemical carcinogenesis of liver: Genetic analysis of later progression stage. *Carcinogenesis* **2002**, *23*, 189–196. [CrossRef] [PubMed]
153. Higashi, K.; Denda, A.; Higashi, T.; Hiai, H. Genetic resistance to chemical hepatocarcinogenesis in the DRH rat strain. *Comp. Med.* **2004**, *54*, 373–377.
154. Ghisletti, S.; Meda, C.; Maggi, A.; Vegeto, E. 17β-Estradiol Inhibits Inflammatory Gene Expression by Controlling NF-κB Intracellular Localization. *Mol. Cell. Biol.* **2005**, *25*, 2957–2968. [CrossRef] [PubMed]
155. Yang, W.; Lu, Y.; Xu, Y.; Xu, L.; Zheng, W.; Wu, Y.; Li, L.; Shen, P. Estrogen represses hepatocellular carcinoma (HCC) Growth via Inhibiting Alternative Activation of Tumor-associated Macrophages (TAMs). *J. Biol. Chem.* **2012**, *287*, 40140–40149. [CrossRef]
156. Hong, E.-J.; Levasseur, M.-P.; Dufour, C.R.; Perry, M.-C.; Giguere, V. Loss of estrogen-related receptor promotes hepatocarcinogenesis development via metabolic and inflammatory disturbances. *Proc. Natl. Acad. Sci. USA* **2013**, *110*, 17975–17980. [CrossRef]

157. Kalra, M.; Mayes, J.; Assefa, S.; Kaul, A.K.; Kaul, R. Role of sex steroid receptors in pathobiology of hepatocellular carcinoma. *World J. Gastroenterol.* **2008**, *14*, 5945. [CrossRef]
158. Zhang, H.; Li, X.-X.; Yang, Y.; Zhang, Y.; Wang, H.-Y.; Zheng, X.F.S. Significance and mechanism of androgen receptor overexpression and androgen receptor/mechanistic target of rapamycin cross-talk in hepatocellular carcinoma. *Hepatology* **2018**, *67*, 2271–2286. [CrossRef]
159. Ma, W.; Hsu, C.; Wu, M.; Wu, C.; Wu, C.; Lai, J.; Jou, Y.; Chen, C.; Yeh, S.; Chang, C. Androgen Receptor Is a New Potential Therapeutic Target for the Treatment of Hepatocellular Carcinoma. *Gastroenterology* **2008**, *135*, 947–955.e5. [CrossRef]
160. Febbraio, M.A.; Reibe, S.; Shalapour, S.; Ooi, G.J.; Watt, M.J.; Karin, M. Preclinical Models for Studying NASH-Driven HCC: How Useful Are They? *Cell Metab.* **2019**, *29*, 18–26. [CrossRef]
161. Almind, K.; Kahn, C.R. Genetic Determinants of Energy Expenditure and Insulin Resistance in Diet-Induced Obesity in Mice. *Diabetes* **2004**, *53*, 3274–3285. [CrossRef]
162. Asgharpour, A.; Cazanave, S.C.; Pacana, T.; Seneshaw, M.; Vincent, R.; Banini, B.A.; Kumar, D.P.; Daita, K.; Min, H.; Mirshahi, F.; et al. A diet-induced animal model of non-alcoholic fatty liver disease and hepatocellular cancer. *J. Hepatol.* **2016**, *65*, 579–588. [CrossRef] [PubMed]
163. Hoshida, Y.; Nijman, S.M.B.; Kobayashi, M.; Chan, J.A.; Brunet, J.-P.; Chiang, D.Y.; Villanueva, A.; Newell, P.; Ikeda, K.; Hashimoto, M.; et al. Integrative Transcriptome Analysis Reveals Common Molecular Subclasses of Human Hepatocellular Carcinoma. *Cancer Res.* **2009**, *69*, 7385–7392. [CrossRef]
164. Dowman, J.K.; Hopkins, L.J.; Reynolds, G.M.; Nikolaou, N.; Armstrong, M.J.; Shaw, J.C.; Houlihan, D.D.; Lalor, P.F.; Tomlinson, J.W.; Hübscher, S.G.; et al. Development of Hepatocellular Carcinoma in a Murine Model of Nonalcoholic Steatohepatitis Induced by Use of a High-Fat/Fructose Diet and Sedentary Lifestyle. *Am. J. Pathol.* **2014**, *184*, 1550–1561. [CrossRef] [PubMed]
165. Tessitore, A.; Cicciarelli, G.; Del Vecchio, F.; Gaggiano, A.; Verzella, D.; Fischietti, M.; Mastroiaco, V.; Vetuschi, A.; Sferra, R.; Barnabei, R.; et al. MicroRNA expression analysis in high fat diet-induced NAFLD-NASH-HCC progression: Study on C57BL/6J mice. *BMC Cancer* **2016**, *16*, 3. [CrossRef]
166. Tsuchida, T.; Lee, Y.A.; Fujiwara, N.; Ybanez, M.; Allen, B.; Martins, S.; Fiel, M.I.; Goossens, N.; Chou, H.; Hoshida, Y.; et al. A simple diet- and chemical-induced murine NASH model with rapid progression of steatohepatitis, fibrosis and liver cancer. *J. Hepatol.* **2018**, *69*, 385–395. [CrossRef]
167. Wolf, M.J.; Adili, A.; Piotrowitz, K.; Abdullah, Z.; Boege, Y.; Stemmer, K.; Ringelhan, M.; Simonavicius, N.; Egger, M.; Wohlleber, D.; et al. Metabolic Activation of Intrahepatic CD8+ T Cells and NKT Cells Causes Nonalcoholic Steatohepatitis and Liver Cancer via Cross-Talk with Hepatocytes. *Cancer Cell* **2014**, *26*, 549–564. [CrossRef] [PubMed]
168. Ikawa-Yoshida, A.; Matsuo, S.; Kato, A.; Ohmori, Y.; Higashida, A.; Kaneko, E.; Matsumoto, M. Hepatocellular carcinoma in a mouse model fed a choline-deficient, L-amino acid-defined, high-fat diet. *Int. J. Exp. Pathol.* **2017**, *98*, 221–233. [CrossRef] [PubMed]
169. De Minicis, S.; Agostinelli, L.; Rychlicki, C.; Sorice, G.P.; Saccomanno, S.; Candelaresi, C.; Giaccari, A.; Trozzi, L.; Pierantonelli, I.; Mingarelli, E.; et al. HCC Development Is Associated to Peripheral Insulin Resistance in a Mouse Model of NASH. *PLoS ONE* **2014**, *9*, e97136. [CrossRef]
170. Takakura, K.; Koido, S.; Fujii, M.; Hashiguchi, T.; Shibazaki, Y.; Yoneyama, H.; Katagi, H.; Kajihara, M.; Misawa, T.; Homma, S.; et al. Characterization of non-alcoholic steatohepatitis-derived hepatocellular carcinoma as a human stratification model in mice. *Anticancer Res.* **2014**, *34*, 4849–4855.
171. Horie, Y.; Suzuki, A.; Kataoka, E.; Sasaki, T.; Hamada, K.; Sasaki, J.; Mizuno, K.; Hasegawa, G.; Kishimoto, H.; Iizuka, M.; et al. Hepatocyte-specific Pten deficiency results in steatohepatitis and hepatocellular carcinomas. *J. Clin. Investig.* **2004**, *113*, 1774–1783. [CrossRef] [PubMed]
172. Nakagawa, H.; Umemura, A.; Taniguchi, K.; Font-Burgada, J.; Dhar, D.; Ogata, H.; Zhong, Z.; Valasek, M.A.; Seki, E.; Hidalgo, J.; et al. ER Stress Cooperates with Hypernutrition to Trigger TNF-Dependent Spontaneous HCC Development. *Cancer Cell* **2014**, *26*, 331–343. [CrossRef]
173. Fujii, M.; Shibazaki, Y.; Wakamatsu, K.; Honda, Y.; Kawauchi, Y.; Suzuki, K.; Arumugam, S.; Watanabe, K.; Ichida, T.; Asakura, H.; et al. A murine model for non-alcoholic steatohepatitis showing evidence of association between diabetes and hepatocellular carcinoma. *Med. Mol. Morphol.* **2013**, *46*, 141–152. [CrossRef]
174. De Conti, A.; Ortega, J.F.; Tryndyak, V.; Dreval, K.; Moreno, F.S.; Rusyn, I.; Beland, F.A.; Pogribny, I.P. MicroRNA deregulation in nonalcoholic steatohepatitis-associated liver carcinogenesis. *Oncotarget* **2017**, *8*, 88517–88528. [CrossRef] [PubMed]
175. Bolzán, A.D.; Bianchi, M.S. Genotoxicity of Streptozotocin. *Mutat. Res. Mutat. Res.* **2002**, *512*, 121–134. [CrossRef]
176. Guo, S.; Mao, X.; Yan, Y.; Zhang, Y.; Ming, L. Changes of liver transcriptome profiles following oxidative stress in streptozotocin-induced diabetes in mice. *PeerJ* **2020**, *8*, e8983. [CrossRef] [PubMed]
177. Muir, K.; Hazim, A.; He, Y.; Peyressatre, M.; Kim, D.-Y.; Song, X.; Beretta, L. Proteomic and Lipidomic Signatures of Lipid Metabolism in NASH-Associated Hepatocellular Carcinoma. *Cancer Res.* **2013**, *73*, 4722–4731. [CrossRef]
178. Holmberg, B.; Ekström, T. The effects of long-term oral administration of ethanol on Sprague-Dawley rats — A condensed report. *Toxicology* **1995**, *96*, 133–145. [CrossRef]
179. Beland, F.A.; Benson, R.W.; Mellick, P.W.; Kovatch, R.M.; Roberts, D.W.; Fang, J.-L.; Doerge, D.R. Effect of ethanol on the tumorigenicity of urethane (ethyl carbamate) in B6C3F1 mice. *Food Chem. Toxicol.* **2005**, *43*, 1–19. [CrossRef] [PubMed]

180. Tanaka, T.; Hirota, Y.; Kuriyama, M.; Nishiguchi, S.; Otani, S. Cessation of Long-term Alcohol Administration and Two-day Cycling of Exposure Respectively Promote and Inhibit Hepatocarcinogenesis in Rats. *Asian Pac. J. Cancer Prev.* **2000**, *1*, 325–328.
181. Wanibuchi, H.; Zhang, Y.; Kinoshita, A.; Wei, M.; Kang, J.S.; Fukushima, S. Effects of cessation of alcohol exposure on rat hepatocarcinogenesis. *Asian Pac. J. Cancer Prev.* **2006**, *7*, 122–126. [PubMed]
182. Kato, H.; Naiki-Ito, A.; Naiki, T.; Suzuki, S.; Yamashita, Y.; Sato, S.; Sagawa, H.; Kato, A.; Kuno, T.; Takahashi, S. Connexin 32 dysfunction promotes ethanol-related hepatocarcinogenesis via activation of Dusp1-Erk axis. *Oncotarget* **2016**, *7*, 2009–2021. [CrossRef]
183. Kushida, M.; Wanibuchi, H.; Morimura, K.; Kinoshita, A.; Kang, J.S.; Puatanachokchai, R.; Wei, M.; Funae, Y.; Fukushima, S. Dose-dependence of promotion of 2-amino-3,8-dimethylimidazo [4,5-f] quinoxaline-induced rat hepatocarcinogenesis by ethanol: Evidence for a threshold. *Cancer Sci.* **2005**, *96*, 747–757. [CrossRef]
184. Pires, P.W.; Furtado, K.S.; Justullin, L.A.; Rodrigues, M.A.M.; Felisbino, S.L.; Barbisan, L.F. Chronic ethanol intake promotes double gluthatione S-transferase/transforming growth factor-α-positive hepatocellular lesions in male Wistar rats. *Cancer Sci.* **2008**, *99*, 221–228. [CrossRef] [PubMed]
185. Mercer, K.E.; Hennings, L.; Sharma, N.; Lai, K.; Cleves, M.A.; Wynne, R.A.; Badger, T.M.; Ronis, M.J.J. Alcohol Consumption Promotes Diethylnitrosamine-Induced Hepatocarcinogenesis in Male Mice through Activation of the Wnt/β-Catenin Signaling Pathway. *Cancer Prev. Res.* **2014**, *7*, 675–685. [CrossRef] [PubMed]
186. Brandon-Warner, E.; Walling, T.L.; Schrum, L.W.; McKillop, I.H. Chronic Ethanol Feeding Accelerates Hepatocellular Carcinoma Progression in a Sex-Dependent Manner in a Mouse Model of Hepatocarcinogenesis. *Alcohol. Clin. Exp. Res.* **2012**, *36*, 641–653. [CrossRef]
187. Ma, H.-Y.; Yamamoto, G.; Xu, J.; Liu, X.; Karin, D.; Kim, J.Y.; Alexandrov, L.B.; Koyama, Y.; Nishio, T.; Benner, C.; et al. IL-17 signaling in steatotic hepatocytes and macrophages promotes hepatocellular carcinoma in alcohol-related liver disease. *J. Hepatol.* **2020**, *72*, 946–959. [CrossRef] [PubMed]
188. Karim, M.R.; Wanibuchi, H.; Wei, M.; Morimura, K.; Salim, E.I.; Fukushima, S. Enhancing risk of ethanol on MeIQx-induced rat hepatocarcinogenesis is accompanied with increased levels of cellular proliferation and oxidative stress. *Cancer Lett.* **2003**, *192*, 37–47. [CrossRef]
189. Wanibuchi, H.; Wei, M.; Karim, M.R.; Morimura, K.; Doi, K.; Kinoshita, A.; Fukushima, S. Existence of No Hepatocarcinogenic Effect Levels of 2-amino-3,8-dimethylimidazo [4,5-f] quinoxaline with or without Coadministration with Ethanol. *Toxicol. Pathol.* **2006**, *34*, 232–236. [CrossRef]
190. Tatsuta, M.; Iishi, H.; Baba, M.; Yano, H.; Iseki, K.; Uehara, H.; Nakaizumi, A. Enhancement by ethyl alcohol of experimental hepatocarcinogenesis induced by N-nitrosomorpholine. *Int. J. Cancer* **1997**, *71*, 1045–1048. [CrossRef]
191. Yan, G.; Wang, X.; Sun, C.; Zheng, X.; Wei, H.; Tian, Z.; Sun, R. Chronic Alcohol Consumption Promotes Diethylnitrosamine-Induced Hepatocarcinogenesis via Immune Disturbances. *Sci. Rep.* **2017**, *7*, 2567. [CrossRef]
192. Edamoto, Y.; Hara, A.; Biernat, W.; Terracciano, L.; Cathomas, G.; Riehle, H.-M.; Matsuda, M.; Fujii, H.; Scoazec, J.-Y.; Ohgaki, H. Alterations of RB1, p53 and Wnt pathways in hepatocellular carcinomas associated with hepatitis C, hepatitis B and alcoholic liver cirrhosis. *Int. J. Cancer* **2003**, *106*, 334–341. [CrossRef]
193. Chen, D.; Yan, Y.; Wang, X.; Li, S.; Liu, Y.; Yu, D.; He, Y.; Deng, R.; Liu, Y.; Xu, M.; et al. Chronic alcohol exposure promotes HCC stemness and metastasis through β-catenin/miR-22-3p/TET2 axis. *Aging* **2021**, *13*, 14433–14455. [CrossRef] [PubMed]
194. Kim, C.-M.; Koike, K.; Saito, I.; Miyamura, T.; Jay, G. HBx gene of hepatitis B virus induces liver cancer in transgenic mice. *Nature* **1991**, *351*, 317–320. [CrossRef]
195. Yu, D.-Y.; Moon, H.-B.; Son, J.-K.; Jeong, S.; Yu, S.-L.; Yoon, H.; Han, Y.-M.; Lee, C.-S.; Park, J.-S.; Lee, C.-H.; et al. Incidence of hepatocellular carcinoma in transgenic mice expressing the hepatitis B virus X-protein. *J. Hepatol.* **1999**, *31*, 123–132. [CrossRef]
196. Kim, S.-Y.; Lee, P.Y.; Shin, H.-J.; Kim, D.H.; Kang, S.; Moon, H.-B.; Kang, S.W.; Kim, J.-M.; Park, S.G.; Park, B.C.; et al. Proteomic analysis of liver tissue from HBx -transgenic mice at early stages of hepatocarcinogenesis. *Proteomics* **2009**, *9*, 5056–5066. [CrossRef]
197. Benn, J.; Su, F.; Doria, M.; Schneider, R.J. Hepatitis B virus HBx protein induces transcription factor AP-1 by activation of extracellular signal-regulated and c-Jun N-terminal mitogen-activated protein kinases. *J. Virol.* **1996**, *70*, 4978–4985. [CrossRef]
198. Tang, H.; Delgermaa, L.; Huang, F.; Oishi, N.; Liu, L.; He, F.; Zhao, L.; Murakami, S. The Transcriptional Transactivation Function of HBx Protein Is Important for Its Augmentation Role in Hepatitis B Virus Replication. *J. Virol.* **2005**, *79*, 5548–5556. [CrossRef] [PubMed]
199. Dunsford, H.A.; Sell, S.; Chisari, F. V Hepatocarcinogenesis due to chronic liver cell injury in hepatitis B virus transgenic mice. *Cancer Res.* **1990**, *50*, 3400–3407.
200. Toshkov, I.; Chisari, F.V.; Bannasch, P. Hepatic preneoplasia in hepatitis B virus transgenic mice. *Hepatology* **1994**, *20*, 1162–1172. [CrossRef] [PubMed]
201. Lai, M.-W.; Liang, K.-H.; Lin, W.-R.; Huang, Y.-H.; Huang, S.-F.; Chen, T.-C.; Yeh, C.-T. Hepatocarcinogenesis in transgenic mice carrying hepatitis B virus pre-S/S gene with the sW172* mutation. *Oncogenesis* **2016**, *5*, e273. [CrossRef]
202. Moriya, K.; Fujie, H.; Shintani, Y.; Yotsuyanagi, H.; Tsutsumi, T.; Ishibashi, K.; Matsuuras, Y.; Kimura, S.; Miyamura, T.; Koike, K. The core protein of hepatitis C virus induces hepatocellular carcinoma in transgenic mice. *Nat. Med.* **1998**, *4*, 1065–1067. [CrossRef] [PubMed]
203. Santoni-Rugiu, E.; Nagy, P.; Jensen, M.R.; Factor, V.M.; Thorgeirsson, S.S. Evolution of neoplastic development in the liver of transgenic mice co-expressing c-myc and transforming growth factor-alpha. *Am. J. Pathol.* **1996**, *149*, 407–428. [PubMed]

204. Thorgeirsson, S.S.; Santoni-Rugiu, E. Transgenic mouse models in carcinogenesis: Interaction of c- myc with transforming growth factor α and hepatocyte growth factor in hepatocarcinogenesis. *Br. J. Clin. Pharmacol.* **1996**, *42*, 43–52. [CrossRef] [PubMed]
205. Calvisi, D.F.; Factor, V.M.; Ladu, S.; Conner, E.A.; Thorgeirsson, S.S. Disruption of β-catenin pathway or genomic instability define two distinct categories of liver cancer in transgenic mice. *Gastroenterology* **2004**, *126*, 1374–1386. [CrossRef]
206. Conner, E.A.; Lemmer, E.R.; Omori, M.; Wirth, P.J.; Factor, V.M.; Thorgeirsson, S.S. Dual functions of E2F-1 in a transgenic mouse model of liver carcinogenesis. *Oncogene* **2000**, *19*, 5054–5062. [CrossRef]
207. Colnot, S.; Decaens, T.; Niwa-Kawakita, M.; Godard, C.; Hamard, G.; Kahn, A.; Giovannini, M.; Perret, C. Liver-targeted disruption of Apc in mice activates -catenin signaling and leads to hepatocellular carcinomas. *Proc. Natl. Acad. Sci. USA* **2004**, *101*, 17216–17221. [CrossRef]
208. Harada, N.; Oshima, H.; Katoh, M.; Tamai, Y.; Oshima, M.; Taketo, M.M. Hepatocarcinogenesis in Mice with β-Catenin and Ha-Ras Gene Mutations. *Cancer Res.* **2004**, *64*, 48–54. [CrossRef]
209. Chung, S.I.; Moon, H.; Kim, D.Y.; Cho, K.J.; Ju, H.; Kim, D.Y.; Ahn, S.H.; Han, K.; Ro, S.W. Development of a transgenic mouse model of hepatocellular carcinoma with a liver fibrosis background. *BMC Gastroenterol.* **2016**, 1–9. [CrossRef]
210. Chisari, F.V.; Filippi, P.; Buras, J.; McLachlan, A.; Popper, H.; Pinkert, C.A.; Palmiter, R.D.; Brinster, R.L. Structural and pathological effects of synthesis of hepatitis B virus large envelope polypeptide in transgenic mice. *Proc. Natl. Acad. Sci. USA* **1987**, *84*, 6909–6913. [CrossRef] [PubMed]
211. Barone, M.; Spano, D.; D'Apolito, M.; Centra, M.; Lasalandra, C.; Capasso, M.; Di Leo, A.; Volinia, S.; Arcelli, D.; Rosso, N.; et al. Gene Expression Analysis in HBV Transgenic Mouse Liver: A Model to Study Early Events Related to Hepatocarcinogenesis. *Mol. Med.* **2006**, *12*, 115–123. [CrossRef]
212. Sell, S.; Hunt, J.M.; Dunsford, H.A.; Chisari, F. V Synergy between hepatitis B virus expression and chemical hepatocarcinogens in transgenic mice. *Cancer Res.* **1991**, *51*, 1278–1285.
213. Lai, M.; Huang, S.; Hsu, C.-W.; Chang, M.-H.; Liaw, Y.-F.; Yeh, C.-T. Identification of nonsense mutations in hepatitis B virus S gene in patients with hepatocellular carcinoma developed after lamivudine therapy. *Antivir. Ther.* **2009**, *14*, 249–261. [PubMed]
214. Ichibangase, T.; Moriya, K.; Koike, K.; Imai, K. A Proteomics Method Revealing Disease-Related Proteins in Livers of Hepatitis-Infected Mouse Model. *J. Proteome Res.* **2007**, *6*, 2841–2849. [CrossRef]
215. Kamegaya, Y.; Hiasa, Y.; Zukerberg, L.; Fowler, N.; Blackard, J.T.; Lin, W.; Choe, W.H.; Schmidt, E.V.; Chung, R.T. Hepatitis C virus acts as a tumor accelerator by blocking apoptosis in a mouse model of hepatocarcinogenesis. *Hepatology* **2005**, *41*, 660–667. [CrossRef]
216. Sandgren, E.P.; Quaife, C.J.; Pinkert, C.A.; Palmiter, R.D.; Brinster, R.L. Oncogene-induced liver neoplasia in transgenic mice. *Oncogene* **1989**, *4*, 715–724. [PubMed]
217. Etiemble, J.; Degott, C.; Renard, C.A.; Fourel, G.; Shamoon, B.; Vitvitski-Trépo, L.; Hsu, T.Y.; Tiollais, P.; Babinet, C.; Buendia, M.A. Liver-specific expression and high oncogenic efficiency of a c-myc transgene activated by woodchuck hepatitis virus insertion. *Oncogene* **1994**, *9*, 727–737.
218. Liu, P.; Terradillos, O.; Renard, C.A.; Feldmann, G.; Buendia, M.A.; Bernuau, D. Hepatocarcinogenesis in woodchuck hepatitis virus/c-myc mice: Sustained cell proliferation and biphasic activation of insulin-like growth factor II. *Hepatology* **1997**, *25*, 874–883. [CrossRef] [PubMed]
219. Coste, A.d.L.; Romagnolo, B.; Billuart, P.; Renard, C.-A.; Buendia, M.-A.; Soubrane, O.; Fabre, M.; Chelly, J.; Beldjord, C.; Kahn, A.; et al. Somatic mutations of the -catenin gene are frequent in mouse and human hepatocellular carcinomas. *Proc. Natl. Acad. Sci. USA* **1998**, *95*, 8847–8851. [CrossRef]
220. Molina-Sánchez, P.; Ruiz de Galarreta, M.; Yao, M.A.; Lindblad, K.E.; Bresnahan, E.; Bitterman, E.; Martin, T.C.; Rubenstein, T.; Nie, K.; Golas, J.; et al. Cooperation Between Distinct Cancer Driver Genes Underlies Intertumor Heterogeneity in Hepatocellular Carcinoma. *Gastroenterology* **2020**, *159*, 2203–2220.e14. [CrossRef]
221. Méndez-Lucas, A.; Lin, W.; Driscoll, P.C.; Legrave, N.; Novellasdemunt, L.; Xie, C.; Charles, M.; Wilson, Z.; Jones, N.P.; Rayport, S.; et al. Identifying strategies to target the metabolic flexibility of tumours. *Nat. Metab.* **2020**, *2*, 335–350. [CrossRef]
222. Méndez-Lucas, A.; Li, X.; Hu, J.; Che, L.; Song, X.; Jia, J.; Wang, J.; Xie, C.; Driscoll, P.C.; Tschaharganeh, D.F.; et al. Glucose Catabolism in Liver Tumors Induced by c-MYC Can Be Sustained by Various PKM1/PKM2 Ratios and Pyruvate Kinase Activities. *Cancer Res.* **2017**, *77*, 4355–4364. [CrossRef] [PubMed]
223. Qiao, Y.; Wang, J.; Karagoz, E.; Liang, B.; Song, X.; Shang, R.; Evert, K.; Xu, M.; Che, L.; Evert, M.; et al. Axis inhibition protein 1 (Axin1) Deletion–Induced Hepatocarcinogenesis Requires Intact β-Catenin but Not Notch Cascade in Mice. *Hepatology* **2019**, *70*, 2003–2017. [CrossRef]
224. Yamamoto, M.; Xin, B.; Watanabe, K.; Ooshio, T.; Fujii, K.; Chen, X.; Okada, Y.; Abe, H.; Taguchi, Y.; Miyokawa, N.; et al. Oncogenic Determination of a Broad Spectrum of Phenotypes of Hepatocyte-Derived Mouse Liver Tumors. *Am. J. Pathol.* **2017**, *187*, 2711–2725. [CrossRef]
225. Xin, B.; Yamamoto, M.; Fujii, K.; Ooshio, T.; Chen, X.; Okada, Y.; Watanabe, K.; Miyokawa, N.; Furukawa, H.; Nishikawa, Y. Critical role of Myc activation in mouse hepatocarcinogenesis induced by the activation of AKT and RAS pathways. *Oncogene* **2017**, *36*, 5087–5097. [CrossRef]
226. Zhan, N.; Michael, A.A.; Wu, K.; Zeng, G.; Bell, A.; Tao, J.; Monga, S.P. The Effect of Selective c-MET Inhibitor on Hepatocellular Carcinoma in the MET-Active, β-Catenin-Mutated Mouse Model. *Gene Expr.* **2018**, *18*, 135–147. [CrossRef] [PubMed]

227. Tao, J.; Xu, E.; Zhao, Y.; Singh, S.; Li, X.; Couchy, G.; Chen, X.; Zucman-Rossi, J.; Chikina, M.; Monga, S.P.S. Modeling a human hepatocellular carcinoma subset in mice through coexpression of met and point-mutant β-catenin. *Hepatology* **2016**, *64*, 1587–1605. [CrossRef]
228. Stauffer, J.K.; Scarzello, A.J.; Andersen, J.B.; De Kluyver, R.L.; Back, T.C.; Weiss, J.M.; Thorgeirsson, S.S.; Wiltrout, R.H. Coactivation of AKT and β-Catenin in Mice Rapidly Induces Formation of Lipogenic Liver Tumors. *Cancer Res.* **2011**, *71*, 2718–2727. [CrossRef] [PubMed]
229. Hu, J.; Che, L.; Li, L.; Pilo, M.G.; Cigliano, A.; Ribback, S.; Li, X.; Latte, G.; Mela, M.; Evert, M.; et al. Co-activation of AKT and c-Met triggers rapid hepatocellular carcinoma development via the mTORC1/FASN pathway in mice. *Sci. Rep.* **2016**, *6*, 1–12. [CrossRef]
230. Murakami, H.; Sanderson, N.D.; Nagy, P.; Marino, P.A.; Merlino, G.; Thorgeirsson, S.S. Transgenic mouse model for synergistic effects of nuclear oncogenes and growth factors in tumorigenesis: Interaction of c-myc and transforming growth factor alpha in hepatic oncogenesis. *Cancer Res.* **1993**, *53*, 1719–1723. [PubMed]
231. Xue, W.; Chen, S.; Yin, H.; Tammela, T.; Papagiannakopoulos, T.; Joshi, N.S.; Cai, W.; Yang, G.; Bronson, R.; Crowley, D.G.; et al. CRISPR-mediated direct mutation of cancer genes in the mouse liver. *Nature* **2014**, *514*, 380–384. [CrossRef]
232. Liu, Y.; Qi, X.; Zeng, Z.; Wang, L.; Wang, J.; Zhang, T.; Xu, Q.; Shen, C.; Zhou, G.; Yang, S.; et al. CRISPR/Cas9-mediated p53 and Pten dual mutation accelerates hepatocarcinogenesis in adult hepatitis B virus transgenic mice. *Sci. Rep.* **2017**, *7*, 1–11. [CrossRef]
233. Sachdeva, M. Immunology of hepatocellular carcinoma. *World J. Hepatol.* **2015**, *7*, 2080. [CrossRef]
234. Johnston, M.P.; Khakoo, S.I. Immunotherapy for hepatocellular carcinoma: Current and future. *World J. Gastroenterol.* **2019**, *25*, 2977–2989. [CrossRef]
235. Kole, C.; Charalampakis, N.; Tsakatikas, S.; Vailas, M.; Moris, D.; Gkotsis, E.; Kykalos, S.; Karamouzis, M.V.; Schizas, D. Immunotherapy for Hepatocellular Carcinoma: A 2021 Update. *Cancers* **2020**, *12*, 2859. [CrossRef] [PubMed]
236. Mestas, J.; Hughes, C.C.W. Of Mice and Not Men: Differences between Mouse and Human Immunology. *J. Immunol.* **2004**, *172*, 2731–2738. [CrossRef] [PubMed]
237. Platzer, B.; Stout, M.; Fiebiger, E. Antigen Cross-Presentation of Immune Complexes. *Front. Immunol.* **2014**, *5*, 1–10. [CrossRef]
238. Macek Jilkova, Z.; Kurma, K.; Decaens, T. Animal Models of Hepatocellular Carcinoma: The Role of Immune System and Tumor Microenvironment. *Cancers* **2019**, *11*, 1487. [CrossRef] [PubMed]
239. Li, E.; Lin, L.; Chen, C.-W.; Ou, D.-L. Mouse Models for Immunotherapy in Hepatocellular Carcinoma. *Cancers* **2019**, *11*, 1800. [CrossRef]
240. Verma, B.; Wesa, A. Establishment of Humanized Mice from Peripheral Blood Mononuclear Cells or Cord Blood CD34+ Hematopoietic Stem Cells for Immune-Oncology Studies Evaluating New Therapeutic Agents. *Curr. Protoc. Pharmacol.* **2020**, *89*, 1–19. [CrossRef]
241. Pearson, T.; Greiner, D.L.; Shultz, L.D. Creation of "Humanized" Mice to Study Human Immunity. *Curr. Protoc. Immunol.* **2008**, *81*, 1–21. [CrossRef]
242. Morillon, Y.M.; Sabzevari, A.; Schlom, J.; Greiner, J.W. The Development of Next-generation PBMC Humanized Mice for Preclinical Investigation of Cancer Immunotherapeutic Agents. *Anticancer Res.* **2020**, *40*, 5329–5341. [CrossRef]
243. Brehm, M.A.; Kenney, L.L.; Wiles, M.V.; Low, B.E.; Tisch, R.M.; Burzenski, L.; Mueller, C.; Greiner, D.L.; Shultz, L.D. Lack of acute xenogeneic graft- versus -host disease, but retention of T-cell function following engraftment of human peripheral blood mononuclear cells in NSG mice deficient in MHC class I and II expression. *FASEB J.* **2019**, *33*, 3137–3151. [CrossRef]
244. Yao, L.-C.; Cheng, M.; Shultz, L.D.; Keck, J.G. Abstract 5619: PBMC humanized NSG-(K b D b) null (IA) null mouse model to evaluate immune-oncology drug efficacy. In Proceedings of the AACR Annual Meeting 2020, American Association for Cancer Research, Philadelphia, PA, USA, 27–28 April 2020.
245. Su, S.; Zhou, H.; Xue, M.; Liu, J.-Y.; Ding, L.; Cao, M.; Zhou, Z.-X.; Hu, H.-M.; Wang, L.-X. Anti-tumor Efficacy of a Hepatocellular Carcinoma Vaccine Based on Dendritic Cells Combined with Tumor-derived Autophagosomes in Murine Models. *Asian Pac. J. Cancer Prev.* **2013**, *14*, 3109–3116. [CrossRef]
246. Zhang, Y.; Zhang, H.; Wei, M.; Mou, T.; Shi, T.; Ma, Y.; Cai, X.; Li, Y.; Dong, J.; Wei, J. Recombinant Adenovirus Expressing a Soluble Fusion Protein PD-1/CD137L Subverts the Suppression of CD8+ T Cells in HCC. *Mol. Ther.* **2019**, *27*, 1906–1918. [CrossRef]
247. De La Rochere, P.; Guil-Luna, S.; Decaudin, D.; Azar, G.; Sidhu, S.S.; Piaggio, E. Humanized Mice for the Study of Immuno-Oncology. *Trends Immunol.* **2018**, *39*, 748–763. [CrossRef] [PubMed]
248. Zumwalde, N.A.; Gumperz, J.E. Modeling Human Antitumor Responses In Vivo Using Umbilical Cord Blood-Engrafted Mice. *Front. Immunol.* **2018**, *9*, 1–7. [CrossRef]
249. Shultz, L.D.; Saito, Y.; Najima, Y.; Tanaka, S.; Ochi, T.; Tomizawa, M.; Doi, T.; Sone, A.; Suzuki, N.; Fujiwara, H.; et al. Generation of functional human T-cell subsets with HLA-restricted immune responses in HLA class I expressing NOD/SCID/IL2r null humanized mice. *Proc. Natl. Acad. Sci. USA* **2010**, *107*, 13022–13027. [CrossRef] [PubMed]
250. Walsh, N.C.; Kenney, L.L.; Jangalwe, S.; Aryee, K.-E.; Greiner, D.L.; Brehm, M.A.; Shultz, L.D. Humanized Mouse Models of Clinical Disease. *Annu. Rev. Pathol. Mech. Dis.* **2017**, *12*, 187–215. [CrossRef]
251. Serra-Hassoun, M.; Bourgine, M.; Boniotto, M.; Berges, J.; Langa, F.; Michel, M.-L.; Freitas, A.A.; Garcia, S. Human Hematopoietic Reconstitution and HLA-Restricted Responses in Nonpermissive Alymphoid Mice. *J. Immunol.* **2014**, *193*, 1504–1511. [CrossRef] [PubMed]

252. Wilkening, S.; Stahl, F.; Bader, A. Comparison of primary human hepatocytes and hepatoma cell line hepg2 with regard to their biotransformation properties. *Drug. Metab. DIspos.* **2003**, *31*, 1035–1042. [CrossRef] [PubMed]
253. Collins, S.D.; Yuen, G.; Tu, T.; Budzinska, M.A.; Spring, K.; Bryant, K.; Shackel, N.A. In Vitro Models of the Liver: Disease Modeling, Drug Discovery and Clinical Applications. In *Tijdschrift voor Geneeskunde*; Tirnitz-Parker, J.E.E., Ed.; Codon Publications: Brisbane, Australia, 2019; Volume 38, pp. 505–507. ISBN 9780994438188.
254. Guo, X.; Seo, J.-E.; Li, X.; Mei, N. Genetic toxicity assessment using liver cell models: Past, present, and future. *J. Toxicol. Environ. Health Part B* **2020**, *23*, 27–50. [CrossRef] [PubMed]
255. Vilas-Boas, V.; Cooreman, A.; Gijbels, E.; Van Campenhout, R.; Gustafson, E.; Ballet, S.; Annaert, P.; Cogliati, B.; Vinken, M. Primary hepatocytes and their cultures for the testing of drug-induced liver injury. *Adv. Pharmacol.* **2019**, *85*, 1–30. [CrossRef]
256. Arellanes-Robledo, J.; Hernández, C.; Camacho, J.; Pérez-Carreón, J.I. In Vitro Models of HCC. In *Liver Pathophysiology*; Muriel, P., Ed.; Elsevier: Amsterdam, The Netherlands, 2017; pp. 563–579. ISBN 9780128043219.
257. Brambilla, G.; Martelli, A. Human hepatocytes in genotoxicity assays. *Pharmacol. Res.* **1990**, *22*, 381–392. [CrossRef]
258. Aniagu, S.O.; Williams, T.D.; Chipman, J.K. Changes in gene expression and assessment of DNA methylation in primary human hepatocytes and HepG2 cells exposed to the environmental contaminants—Hexabromocyclododecane and 17-β oestradiol. *Toxicology* **2009**, *256*, 143–151. [CrossRef]
259. Ayed-Boussema, I.; Pascussi, J.-M.; Maurel, P.; Bacha, H.; Hassen, W. Effect of Aflatoxin B1 on Nuclear Receptors PXR, CAR, and AhR and Their Target Cytochromes P450 mRNA Expression in Primary Cultures of Human Hepatocytes. *Int. J. Toxicol.* **2012**, *31*, 86–93. [CrossRef]
260. Ayed-Boussema, I.; Pascussi, J.M.; Rjiba, K.; Maurel, P.; Bacha, H.; Hassen, W. The mycotoxin, patulin, increases the expression of PXR and AhR and their target cytochrome P450s in primary cultured human hepatocytes. *Drug Chem. Toxicol.* **2012**, *35*, 241–250. [CrossRef] [PubMed]
261. Ayed-Boussema, I.; Pascussi, J.M.; Zaied, C.; Maurel, P.; Bacha, H.; Hassen, W. Ochratoxin A induces CYP3A4, 2B6, 3A5, 2C9, 1A1, and CYP1A2 gene expression in primary cultured human hepatocytes: A possible activation of nuclear receptors. *Drug Chem. Toxicol.* **2012**, *35*, 71–80. [CrossRef]
262. Rieswijk, L.; Claessen, S.M.H.; Bekers, O.; van Herwijnen, M.; Theunissen, D.H.J.; Jennen, D.G.J.; de Kok, T.M.C.M.; Kleinjans, J.C.S.; van Breda, S.G.J. Aflatoxin B1 induces persistent epigenomic effects in primary human hepatocytes associated with hepatocellular carcinoma. *Toxicology* **2016**, *350–352*, 31–39. [CrossRef] [PubMed]
263. Pez, F.; Gifu, P.; Degli-Esposti, D.; Fares, N.; Lopez, A.; Lefrançois, L.; Michelet, M.; Rivoire, M.; Bancel, B.; Sylla, B.S.; et al. In vitro transformation of primary human hepatocytes: Epigenetic changes and stemness properties. *Exp. Cell Res.* **2019**, *384*, 111643. [CrossRef] [PubMed]
264. Kiamehr, M.; Heiskanen, L.; Laufer, T.; Düsterloh, A.; Kahraman, M.; Käkelä, R.; Laaksonen, R.; Aalto-Setälä, K. Dedifferentiation of Primary Hepatocytes is Accompanied with Reorganization of Lipid Metabolism Indicated by Altered Molecular Lipid and miRNA Profiles. *Int. J. Mol. Sci.* **2019**, *20*, 2910. [CrossRef] [PubMed]
265. Elaut, G.; Henkens, T.; Papeleu, P.; Snykers, S.; Vinken, M.; Vanhaecke, T.; Rogiers, V. Molecular Mechanisms Underlying the Dedifferentiation Process of Isolated Hepatocytes and Their Cultures. *Curr. Drug Metab.* **2006**, *7*, 629–660. [CrossRef] [PubMed]
266. Gijbels, E.; Vanhaecke, T.; Vinken, M. Establishment of Sandwich Cultures of Primary Human Hepatocytes. In *Experimental Cholestasis Research*; Methods in Molecular Biology; Springer: New York, NY, USA, 2019; Volume 1981, pp. 99–115. ISBN 978-1-4939-9419-9.
267. Bell, C.C.; Hendriks, D.F.G.; Moro, S.M.L.; Ellis, E.; Walsh, J.; Renblom, A.; Fredriksson Puigvert, L.; Dankers, A.C.A.; Jacobs, F.; Snoeys, J.; et al. Characterization of primary human hepatocyte spheroids as a model system for drug-induced liver injury, liver function and disease. *Sci. Rep.* **2016**, *6*, 25187. [CrossRef]
268. Bell, C.C.; Lauschke, V.M.; Vorrink, S.U.; Palmgren, H.; Duffin, R.; Andersson, T.B.; Ingelman-Sundberg, M. Transcriptional, Functional, and Mechanistic Comparisons of Stem Cell–Derived Hepatocytes, HepaRG Cells, and Three-Dimensional Human Hepatocyte Spheroids as Predictive In Vitro Systems for Drug-Induced Liver Injury. *Drug Metab. Dispos.* **2017**, *45*, 419–429. [CrossRef] [PubMed]
269. Bell, C.C.; Dankers, A.C.A.; Lauschke, V.M.; Sison-Young, R.; Jenkins, R.; Rowe, C.; Goldring, C.E.; Park, K.; Regan, S.L.; Walker, T.; et al. Comparison of Hepatic 2D Sandwich Cultures and 3D Spheroids for Long-term Toxicity Applications: A Multicenter Study. *Toxicol. Sci.* **2018**, *162*, 655–666. [CrossRef]
270. Gross-Steinmeyer, K.; Stapleton, P.L.; Tracy, J.H.; Bammler, T.K.; Strom, S.C.; Buhler, D.R.; Eaton, D.L. Modulation of Aflatoxin B1–Mediated Genotoxicity in Primary Cultures of Human Hepatocytes by Diindolylmethane, Curcumin, and Xanthohumols. *Toxicol. Sci.* **2009**, *112*, 303–310. [CrossRef]
271. Green, C.J.; Parry, S.A.; Gunn, P.J.; Ceresa, C.D.L.; Rosqvist, F.; Piché, M.-E.; Hodson, L. Studying non-alcoholic fatty liver disease: The ins and outs of in vivo, ex vivo and in vitro human models. *Horm. Mol. Biol. Clin. Investig.* **2018**, *41*, 1–22. [CrossRef]
272. Saraswati, S.; Alhaider, A.; Abdelgadir, A.M.; Tanwer, P.; Korashy, H.M. Phloretin attenuates STAT-3 activity and overcomes sorafenib resistance targeting SHP-1–mediated inhibition of STAT3 and Akt/VEGFR2 pathway in hepatocellular carcinoma. *Cell Commun. Signal.* **2019**, *17*, 127. [CrossRef]
273. Li, D.; Li, N.; Zhang, Y.-F.; Fu, H.; Feng, M.; Schneider, D.; Su, L.; Wu, X.; Zhou, J.; Mackay, S.; et al. Persistent Polyfunctional Chimeric Antigen Receptor T Cells That Target Glypican 3 Eliminate Orthotopic Hepatocellular Carcinomas in Mice. *Gastroenterology* **2020**, *158*, 2250–2265.e20. [CrossRef] [PubMed]

274. Zeilinger, K.; Freyer, N.; Damm, G.; Seehofer, D.; Knöspel, F. Cell sources for in vitro human liver cell culture models. *Exp. Biol. Med.* **2016**, *241*, 1684–1698. [CrossRef]
275. Czauderna, C.; Palestino-Dominguez, M.; Castven, D.; Becker, D.; Zanon-Rodriguez, L.; Hajduk, J.; Mahn, F.L.; Herr, M.; Strand, D.; Strand, S.; et al. Ginkgo biloba induces different gene expression signatures and oncogenic pathways in malignant and non-malignant cells of the liver. *PLoS ONE* **2018**, *13*, e0209067. [CrossRef]
276. Chen, Z.; Zhuang, W.; Wang, Z.; Xiao, W.; Don, W.; Li, X.; Chen, X. MicroRNA-450b-3p inhibits cell growth by targeting phosphoglycerate kinase 1 in hepatocellular carcinoma. *J. Cell. Biochem.* **2019**, *120*, 18805–18815. [CrossRef]
277. Hirschfield, H.; Bian, C.B.; Higashi, T.; Nakagawa, S.; Zeleke, T.Z.; Nair, V.D.; Fuchs, B.C.; Hoshida, Y. In vitro modeling of hepatocellular carcinoma molecular subtypes for anti-cancer drug assessment. *Exp. Mol. Med.* **2018**, *50*, e419. [CrossRef] [PubMed]
278. Ran, L.-K.; Chen, Y.; Zhang, Z.-Z.; Tao, N.-N.; Ren, J.-H.; Zhou, L.; Tang, H.; Chen, X.; Chen, K.; Li, W.-Y.; et al. SIRT6 Overexpression Potentiates Apoptosis Evasion in Hepatocellular Carcinoma via BCL2-Associated X Protein–Dependent Apoptotic Pathway. *Clin. Cancer Res.* **2016**, *22*, 3372–3382. [CrossRef]
279. Gillet, J.-P.; Varma, S.; Gottesman, M.M. The Clinical Relevance of Cancer Cell Lines. *JNCI J. Natl. Cancer Inst.* **2013**, *105*, 452–458. [CrossRef] [PubMed]
280. Zanoni, M.; Cortesi, M.; Zamagni, A.; Arienti, C.; Pignatta, S.; Tesei, A. Modeling neoplastic disease with spheroids and organoids. *J. Hematol. Oncol.* **2020**, *13*, 97. [CrossRef]
281. Feng, P.-C.; Ke, X.-F.; Kuang, H.-L.; Pan, L.-L.; Ye, Q.; Wu, J.-B. BMP2 secretion from hepatocellular carcinoma cell HepG2 enhances angiogenesis and tumor growth in endothelial cells via activation of the MAPK/p38 signaling pathway. *Stem Cell Res. Ther.* **2019**, *10*, 237. [CrossRef]
282. Zhang, R.; Gao, X.; Zuo, J.; Hu, B.; Yang, J.; Zhao, J.; Chen, J. STMN1 upregulation mediates hepatocellular carcinoma and hepatic stellate cell crosstalk to aggravate cancer by triggering the MET pathway. *Cancer Sci.* **2020**, *111*, 406–417. [CrossRef] [PubMed]
283. Hosseinzadeh, F.; Ai, J.; Ebrahimi-Barough, S.; Seyhoun, I.; Hajifathali, A.; Muhammadnejad, S.; Hosseinzadeh, F.; Shadnoush, M.; Dabiri Oskouei, F.; Verdi, J. Natural Killer Cell Expansion with Autologous Feeder Layer and Anti-CD3 Antibody for Immune Cell Therapy of Hepatocellular Carcinoma. *Asian Pac. J. Cancer Prev.* **2019**, *20*, 3797–3803. [CrossRef]
284. Sun, L.; Wang, Y.; Wang, L.; Yao, B.; Chen, T.; Li, Q.; Liu, Z.; Liu, R.; Niu, Y.; Song, T.; et al. Resolvin D1 prevents epithelial-mesenchymal transition and reduces the stemness features of hepatocellular carcinoma by inhibiting paracrine of cancer-associated fibroblast-derived COMP. *J. Exp. Clin. Cancer Res.* **2019**, *38*, 170. [CrossRef] [PubMed]
285. Zhou, S.; Yin, D.; Hu, Z.; Luo, C.; Zhou, Z.; Xin, H.; Yang, X.; Shi, Y.; Wang, Z.; Huang, X.; et al. A Positive Feedback Loop Between Cancer Stem-Like Cells and Tumor-Associated Neutrophils Controls Hepatocellular Carcinoma Progression. *Hepatology* **2019**, *70*, 1214–1230. [CrossRef] [PubMed]
286. Wuputra, K.; Lin, C.-S.; Tsai, M.-H.; Ku, C.-C.; Lin, W.-H.; Yang, Y.-H.; Kuo, K.-K.; Yokoyama, K.K. Cancer cell reprogramming to identify the genes competent for generating liver cancer stem cells. *Inflamm. Regen.* **2017**, *37*, 15. [CrossRef]
287. Marquardt, J.U.; Factor, V.M.; Thorgeirsson, S.S. Epigenetic regulation of cancer stem cells in liver cancer: Current concepts and clinical implications. *J. Hepatol.* **2010**, *53*, 568–577. [CrossRef]
288. Gehart, H.; Clevers, H. Stem Cell-Derived Liver Cells. In *The Liver*; Wiley: Hoboken, NJ, USA, 2020; pp. 1015–1021.
289. Wang, Y.; Takeishi, K.; Li, Z.; Cervantes-Alvarez, E.; Collin de l'Hortet, A.; Guzman-Lepe, J.; Cui, X.; Zhu, J. Microenvironment of a tumor-organoid system enhances hepatocellular carcinoma malignancy-related hallmarks. *Organogenesis* **2017**, *13*, 83–94. [CrossRef] [PubMed]
290. Štampar, M.; Tomc, J.; Filipič, M.; Žegura, B. Development of in vitro 3D cell model from hepatocellular carcinoma (HepG2) cell line and its application for genotoxicity testing. *Arch. Toxicol.* **2019**, *93*, 3321–3333. [CrossRef] [PubMed]
291. Mišík, M.; Nersesyan, A.; Ropek, N.; Huber, W.W.; Haslinger, E.; Knasmueller, S. Use of human derived liver cells for the detection of genotoxins in comet assays. *Mutat. Res. Toxicol. Environ. Mutagen.* **2019**, *845*, 402995. [CrossRef]
292. Liu, C.; Liu, Y.; Xu, X.; Guo, X.; Sun, G.; Ma, X. Mesenchymal stem cells enhance the metastasis of 3D-cultured hepatocellular carcinoma cells. *BMC Cancer* **2016**, *16*, 566. [CrossRef]
293. Shah, U.-K.; de Oliveira Mallia, J.; Singh, N.; Chapman, K.E.; Doak, S.H.; Jenkins, G.J.S. A three-dimensional in vitro HepG2 cells liver spheroid model for genotoxicity studies. *Mutat. Res. Toxicol. Environ. Mutagen.* **2018**, *825*, 51–58. [CrossRef]
294. Eilenberger, C.; Rothbauer, M.; Ehmoser, E.-K.; Ertl, P.; Küpcü, S. Effect of Spheroidal Age on Sorafenib Diffusivity and Toxicity in a 3D HepG2 Spheroid Model. *Sci. Rep.* **2019**, *9*, 4863. [CrossRef]
295. Chiew, G.G.Y.; Wei, N.; Sultania, S.; Lim, S.; Luo, K.Q. Bioengineered three-dimensional co-culture of cancer cells and endothelial cells: A model system for dual analysis of tumor growth and angiogenesis. *Biotechnol. Bioeng.* **2017**, *114*, 1865–1877. [CrossRef]
296. Yang, T.; Zhang, W.; Wang, L.; Xiao, C.; Wang, L.; Gong, Y.; Huang, D.; Guo, B.; Li, Q.; Xiang, Y.; et al. Co-culture of dendritic cells and cytokine-induced killer cells effectively suppresses liver cancer stem cell growth by inhibiting pathways in the immune system. *BMC Cancer* **2018**, *18*, 1–10. [CrossRef]
297. Clevers, H. Modeling Development and Disease with Organoids. *Cell* **2016**, *165*, 1586–1597. [CrossRef] [PubMed]
298. Fatehullah, A.; Tan, S.H.; Barker, N. Organoids as an in vitro model of human development and disease. *Nat. Cell Biol.* **2016**, *18*, 246–254. [CrossRef]

299. Van Tienderen, G.S.; Groot Koerkamp, B.; IJzermans, J.N.M.; van der Laan, L.J.W.; Verstegen an Tienderen, M.M.A. Recreating Tumour Complexity in a Dish: Organoid Models to Study Liver Cancer Cells and their Extracellular Environment. *Cancers* **2019**, *11*, 1706. [CrossRef] [PubMed]
300. Tharehalli, U.; Svinarenko, M.; Lechel, A. Remodelling and Improvements in Organoid Technology to Study Liver Carcinogenesis in a Dish. *Stem Cells Int.* **2019**, *2019*, 1–8. [CrossRef]
301. Torresi, J.; Tran, B.M.; Christiansen, D.; Earnest-Silveira, L.; Schwab, R.H.M.; Vincan, E. HBV-related hepatocarcinogenesis: The role of signalling pathways and innovative ex vivo research models. *BMC Cancer* **2019**, *19*, 707. [CrossRef]
302. Nuciforo, S.; Fofana, I.; Matter, M.S.; Blumer, T.; Calabrese, D.; Boldanova, T.; Piscuoglio, S.; Wieland, S.; Ringnalda, F.; Schwank, G.; et al. Organoid Models of Human Liver Cancers Derived from Tumor Needle Biopsies. *Cell Rep.* **2018**, *24*, 1363–1376. [CrossRef] [PubMed]
303. Li, L.; Knutsdottir, H.; Hui, K.; Weiss, M.J.; He, J.; Philosophe, B.; Cameron, A.M.; Wolfgang, C.L.; Pawlik, T.M.; Ghiaur, G.; et al. Human primary liver cancer organoids reveal intratumor and interpatient drug response heterogeneity. *JCI Insight* **2019**, *4*, 1–16. [CrossRef]
304. Renwick, A.B.; Watts, P.S.; Edwards, R.J.; Barton, P.T.; Guyonnet, I.; Price, R.J.; Tredger, J.M.; Pelkonen, O.; Boobis, A.R.; Lake, B. Differential maintenance of cytochrome P450 enzymes in cultured precision-cut human liver slices. *Drug Metab. Dispos.* **2020**, *28*, 1202–1209.
305. Wu, X.; Roberto, J.B.; Knupp, A.; Kenerson, H.L.; Truong, C.D.; Yuen, S.Y.; Brempelis, K.J.; Tuefferd, M.; Chen, A.; Horton, H.; et al. Precision-cut human liver slice cultures as an immunological platform. *J. Immunol. Methods* **2018**, *455*, 71–79. [CrossRef] [PubMed]
306. Janssen, A.W.F.; Betzel, B.; Stoopen, G.; Berends, F.J.; Janssen, I.M.; Peijnenburg, A.A.; Kersten, S. The impact of PPARα activation on whole genome gene expression in human precision cut liver slices. *BMC Genom.* **2015**, *16*, 760. [CrossRef]
307. Plazar, J.; Filipič, M.; Groothuis, G.M.M. Antigenotoxic effect of Xanthohumol in rat liver slices. *Toxicol. Vitr.* **2008**, *22*, 318–327. [CrossRef]
308. Boess, F.; Kamber, M.; Romer, S.; Gasser, R.; Muller, D.; Albertini, S.; Suter, L. Gene expression in two hepatic cell lines, cultured primary hepatocytes, and liver slices compared to the in vivo liver gene expression in rats: Possible implications for toxicogenomics use of in vitro systems. *Toxicol. Sci.* **2003**, *73*, 386–402. [CrossRef] [PubMed]
309. Van Grunsven, L.A. 3D in vitro models of liver fibrosis. *Adv. Drug Deliv. Rev.* **2017**, *121*, 133–146. [CrossRef]
310. Prins, G.; Luangmonkong, T.; Oosterhuis, D.; Mutsaers, H.; Dekker, F.; Olinga, P. A Pathophysiological Model of Non-Alcoholic Fatty Liver Disease Using Precision-Cut Liver Slices. *Nutrients* **2019**, *11*, 507. [CrossRef]
311. Gore, E.; Bigaeva, E.; Oldenburger, A.; Kim, Y.O.; Rippmann, J.F.; Schuppan, D.; Boersema, M.; Olinga, P. PI3K inhibition reduces murine and human liver fibrogenesis in precision-cut liver slices. *Biochem. Pharmacol.* **2019**, *169*, 113633. [CrossRef] [PubMed]
312. Crespo Yanguas, S.; Cogliati, B.; Willebrords, J.; Maes, M.; Colle, I.; van den Bossche, B.; de Oliveira, C.P.M.S.; Andraus, W.; Alves, V.A.; Leclercq, I.; et al. Experimental models of liver fibrosis. *Arch. Toxicol.* **2016**, *90*, 1025–1048. [CrossRef]
313. Zimmermann, M.; Armeanu, S.; Smirnow, I.; Kupka, S.; Wagner, S.; Wehrmann, M.; Rots, M.G.; Groothuis, G.M.M.; Weiss, T.S.; Königsrainer, A.; et al. Human precision-cut liver tumor slices as a tumor patient-individual predictive test system for oncolytic measles vaccine viruses. *Int. J. Oncol.* **2009**, *34*, 1247–1256. [CrossRef] [PubMed]
314. Palma, E.; Doornebal, E.J.; Chokshi, S. Precision-cut liver slices: A versatile tool to advance liver research. *Hepatol. Int.* **2019**, *13*, 51–57. [CrossRef]
315. Granitzny, A.; Knebel, J.; Schaudien, D.; Braun, A.; Steinberg, P.; Dasenbrock, C.; Hansen, T. Maintenance of high quality rat precision cut liver slices during culture to study hepatotoxic responses: Acetaminophen as a model compound. *Toxicol. Vitr.* **2017**, *42*, 200–213. [CrossRef]
316. Paish, H.L.; Reed, L.H.; Brown, H.; Bryan, M.C.; Govaere, O.; Leslie, J.; Barksby, B.S.; Garcia Macia, M.; Watson, A.; Xu, X.; et al. A Bioreactor Technology for Modeling Fibrosis in Human and Rodent Precision-Cut Liver Slices. *Hepatology* **2019**, *70*, 1377–1391. [CrossRef] [PubMed]
317. Ramboer, E.; Vanhaecke, T.; Rogiers, V.; Vinken, M. Immortalized Human Hepatic Cell Lines for In Vitro Testing and Research Purposes. In *Protocols in In Vitro Hepatocyte Research*; Vinken, M., Rogiers, V., Eds.; Springer: New York, NY, USA, 2015; pp. 53–76. ISBN 978-1-4939-2074-7.
318. Zheng, Y.-H.; Yin, L.-H.; Grahn, T.H.M.; Ye, A.-F.; Zhao, Y.-R.; Zhang, Q.-Y. Anticancer Effects of Baicalein on Hepatocellular Carcinoma Cells. *Phyther. Res.* **2014**, *28*, 1342–1348. [CrossRef]
319. Zhou, T.; Ye, L.; Bai, Y.; Sun, A.; Cox, B.; Liu, D.; Li, Y.; Liotta, D.; Snyder, J.P.; Fu, H.; et al. Autophagy and Apoptosis in Hepatocellular Carcinoma Induced by EF25-(GSH)2: A Novel Curcumin Analog. *PLoS ONE* **2014**, *9*, e107876. [CrossRef]
320. Tripathi, K.P.; Granata, I.; Guarracino, M.R. A computational integrative approach based on alternative splicing analysis to compare immortalized and primary cancer cells. *Int. J. Biochem. Cell Biol.* **2017**, *91*, 116–123. [CrossRef]
321. Du, X.; Wu, L.; Ur Rahman, M.S.; Teng, X.; Teng, L.; Ye, J.; Cao, J. Promoter Hypomethylation Is Responsible for Upregulated Expression of HAI-1 in Hepatocellular Carcinoma. *Dis. Markers* **2019**, *2019*, 1–12. [CrossRef]
322. He, M.; Qin, H.; Poon, T.C.W.; Sze, S.-C.; Ding, X.; Co, N.N.; Ngai, S.-M.; Chan, T.-F.; Wong, N. Hepatocellular carcinoma-derived exosomes promote motility of immortalized hepatocyte through transfer of oncogenic proteins and RNAs. *Carcinogenesis* **2015**, *36*, 1008–1018. [CrossRef]

323. Kongkavitoon, P.; Butta, P.; Sanpavat, A.; Bhattarakosol, P.; Tangtanatakul, P.; Wongprom, B.; Tangkijvanich, P.; Hirankarn, N.; Palaga, T. Regulation of periostin expression by Notch signaling in hepatocytes and liver cancer cell lines. *Biochem. Biophys. Res. Commun.* **2018**, *506*, 739–745. [CrossRef]
324. Caruso, S.; Calatayud, A.-L.; Pilet, J.; La Bella, T.; Rekik, S.; Imbeaud, S.; Letouzé, E.; Meunier, L.; Bayard, Q.; Rohr-Udilova, N.; et al. Analysis of Liver Cancer Cell Lines Identifies Agents With Likely Efficacy Against Hepatocellular Carcinoma and Markers of Response. *Gastroenterology* **2019**, *157*, 760–776. [CrossRef] [PubMed]
325. Ogasawara, S.; Mihara, Y.; Kondo, R.; Kusano, H.; Akiba, J.; Yano, H. Antiproliferative Effect of Lenvatinib on Human Liver Cancer Cell Lines In Vitro and In Vivo. *Anticancer Res.* **2019**, *39*, 5973–5982. [CrossRef] [PubMed]
326. Wang, C.; Wang, H.; Lieftink, C.; du Chatinier, A.; Gao, D.; Jin, G.; Jin, H.; Beijersbergen, R.L.; Qin, W.; Bernards, R. CDK12 inhibition mediates DNA damage and is synergistic with sorafenib treatment in hepatocellular carcinoma. *Gut* **2020**, *69*, 727–736. [CrossRef]
327. Wu, X.-S.; Bao, T.-H.; Ke, Y.; Sun, D.-Y.; Shi, Z.-T.; Tang, H.-R.; Wang, L. Hint1 suppresses migration and invasion of hepatocellular carcinoma cells in vitro by modulating girdin activity. *Tumor Biol.* **2016**, *37*, 14711–14719. [CrossRef] [PubMed]
328. Lu, P.; Ding, Q.; Li, X.; Ji, X.; Li, L.; Fan, Y.; Xia, Y.; Tian, D.; Liu, M. SWELL1 promotes cell growth and metastasis of hepatocellular carcinoma in vitro and in vivo. *EBioMedicine* **2019**, *48*, 100–116. [CrossRef]
329. Zhuang, H.; Cao, G.; Kou, C.; Liu, T. CCL2/CCR2 axis induces hepatocellular carcinoma invasion and epithelial-mesenchymal transition inï¿½vitro through activation of the Hedgehog pathway. *Oncol. Rep.* **2017**, *39*, 21–30. [CrossRef] [PubMed]
330. Xing, W.; Chen, D.-T.; Pan, J.-H.; Chen, Y.-H.; Yan, Y.; Li, Q.; Xue, R.-F.; Yuan, Y.-F.; Zeng, W.-A. Lidocaine Induces Apoptosis and Suppresses Tumor Growth in Human Hepatocellular Carcinoma Cells In Vitro and in a Xenograft Model In Vivo. *Anesthesiology* **2017**, *126*, 868–881. [CrossRef]
331. Sanaei, M.; Kavoosi, F.; Roustazadeh, A.; Shahsavani, H. In vitro effect of the Histone deacetylase inhibitor valproic acid on viability and apoptosis of the PLC/PRF5 human hepatocellular carcinoma cell line. *Asian Pac. J. Cancer Prev.* **2018**, *19*, 2507–2510. [CrossRef]
332. Fowler, P.; Smith, K.; Young, J.; Jeffrey, L.; Kirkland, D.; Pfuhler, S.; Carmichael, P. Reduction of misleading ("false") positive results in mammalian cell genotoxicity assays. I. Choice of cell type. *Mutat. Res. Toxicol. Environ. Mutagen.* **2012**, *742*, 11–25. [CrossRef] [PubMed]
333. Guillouzo, A.; Corlu, A.; Aninat, C.; Glaise, D.; Morel, F.; Guguen-Guillouzo, C. The human hepatoma HepaRG cells: A highly differentiated model for studies of liver metabolism and toxicity of xenobiotics. *Chem. Biol. Interact.* **2007**, *168*, 66–73. [CrossRef]
334. Yokoyama, Y.; Sasaki, Y.; Terasaki, N.; Kawataki, T.; Takekawa, K.; Iwase, Y.; Shimizu, T.; Sanoh, S.; Ohta, S. Comparison of Drug Metabolism and Its Related Hepatotoxic Effects in HepaRG, Cryopreserved Human Hepatocytes, and HepG2 Cell Cultures. *Biol. Pharm. Bull.* **2018**, *41*, 722–732. [CrossRef] [PubMed]
335. Seo, J.-E.; Tryndyak, V.; Wu, Q.; Dreval, K.; Pogribny, I.; Bryant, M.; Zhou, T.; Robison, T.W.; Mei, N.; Guo, X. Quantitative comparison of in vitro genotoxicity between metabolically competent HepaRG cells and HepG2 cells using the high-throughput high-content CometChip assay. *Arch. Toxicol.* **2019**, *93*, 1433–1448. [CrossRef]
336. Doktorova, T.Y.; Yildirimman, R.; Vinken, M.; Vilardell, M.; Vanhaecke, T.; Gmuender, H.; Bort, R.; Brolen, G.; Holmgren, G.; Li, R.; et al. Transcriptomic responses generated by hepatocarcinogens in a battery of liver-based in vitro models. *Carcinogenesis* **2013**, *34*, 1393–1402. [CrossRef] [PubMed]
337. Bairoch, A. The Cellosaurus, a Cell-Line Knowledge Resource. *J. Biomol. Tech.* **2018**, *29*, 25–38. [CrossRef] [PubMed]
338. López-Terrada, D.; Cheung, S.W.; Finegold, M.J.; Knowles, B.B. Hep G2 is a hepatoblastoma-derived cell line. *Hum. Pathol.* **2009**, *40*, 1512–1515. [CrossRef]
339. Tian, J.; Tang, Z.; Xue, Q. [Expressions of the metastasis-associated factors of a new human hepatocellular carcinoma cell line with highly metastatic potential]. *Zhonghua Yi Xue Za Zhi* **1999**, *79*, 470–472. [CrossRef] [PubMed]
340. Heffelfinger, S.C.; Hawkins, H.H.; Barrish, J.; Taylor, L.; Darlington, G.J. SK HEP-1: A human cell line of endothelial origin. *Vitr. Cell. Dev. Biol.-Anim.* **1992**, *28*, 136–142. [CrossRef]
341. Mills, J.B.; Rose, K.A.; Sadagopan, N.; Sahi, J.; de Morais, S.M.F. Induction of Drug Metabolism Enzymes and MDR1 Using a Novel Human Hepatocyte Cell Line. *J. Pharmacol. Exp. Ther.* **2004**, *309*, 303–309. [CrossRef]
342. Reid, Y.; Gaddipati, J.P.; Yadav, D.; Kantor, J. Establishment of a human neonatal hepatocyte cell line. *Vitr. Cell. Dev. Biol.-Anim.* **2009**, *45*, 535–542. [CrossRef]
343. Pfeifer, A.M.A.; Cole, K.E.; Smoot, D.T.; Weston, A.; Groopman, J.D.; Shields, P.G.; Vignaud, J.M.; Juillerat, M.; Lipsky, M.M.; Trump, B.F. Simian virus 40 large tumor antigen-immortalized normal human liver epithelial cells express hepatocyte characteristics and metabolize chemical carcinogens. *Proc. Natl. Acad. Sci. USA* **1993**, *90*, 5123–5127. [CrossRef] [PubMed]
344. Noguchi, M.; Hirohashi, S. Cell lines from non-neoplastic liver and hepatocellular carcinoma tissue from a single patient. *Vitr. Cell. Dev. Biol.-Anim.* **1996**, *32*, 135–137. [CrossRef]
345. Bogdanos, D.P.; Gao, B.; Gershwin, M.E. Liver Immunology. *Compr Physiol.* **2013**, *2*, 567–598. [CrossRef]
346. Ding, C.; Li, Y.; Guo, F.; Jiang, Y.; Ying, W.; Li, D.; Yang, D.; Xia, X.; Liu, W.; Zhao, Y.; et al. A Cell-type-resolved Liver Proteome. *Mol. Cell. Proteom.* **2016**, *15*, 3190–3202. [CrossRef]
347. Doumba, P.P.; Nikolopoulou, M.; Gomatos, I.P.; Konstadoulakis, M.M.; Koskinas, J. Co-culture of primary human tumor hepatocytes from patients with hepatocellular carcinoma with autologous peripheral blood mononuclear cells: Study of their in vitro immunological interactions. *BMC Gastroenterol.* **2013**, *13*, 17. [CrossRef]

348. Baze, A.; Parmentier, C.; Hendriks, D.F.G.; Hurrell, T.; Heyd, B.; Bachellier, P.; Schuster, C.; Ingelman-Sundberg, M.; Richert, L. Three-Dimensional Spheroid Primary Human Hepatocytes in Monoculture and Coculture with Nonparenchymal Cells. *Tissue Eng. Part C Methods* **2018**, *24*, 534–545. [CrossRef]
349. Tomizawa, M.; Shinozaki, F.; Motoyoshi, Y.; Sugiyama, T.; Yamamoto, S.; Ishige, N. Cell death in a co-culture of hepatocellular carcinoma cells and human umbilical vascular endothelial cells in a medium lacking glucose and arginine. *Oncol. Lett.* **2017**, *13*, 258–262. [CrossRef] [PubMed]
350. Lin, J.; Cao, S.; Wang, Y.; Hu, Y.; Liu, H.; Li, J.; Chen, J.; Li, P.; Liu, J.; Wang, Q.; et al. Long non-coding RNA UBE2CP3 enhances HCC cell secretion of VEGFA and promotes angiogenesis by activating ERK1/2/HIF-1α/VEGFA signalling in hepatocellular carcinoma. *J. Exp. Clin. Cancer Res.* **2018**, *37*, 113. [CrossRef] [PubMed]
351. Serhal, R.; Saliba, N.; Hilal, G.; Moussa, M.; Hassan, G.; El Atat, O.; Alaaeddine, N. Effect of adipose-derived mesenchymal stem cells on hepatocellular carcinoma: In vitro inhibition of carcinogenesis. *World J. Gastroenterol.* **2019**, *25*, 567–583. [CrossRef]
352. Yin, Z.; Ma, T.; Lin, Y.; Lu, X.; Zhang, C.; Chen, S.; Jian, Z. IL-6/STAT3 pathway intermediates M1/M2 macrophage polarization during the development of hepatocellular carcinoma. *J. Cell. Biochem.* **2018**, *119*, 9419–9432. [CrossRef] [PubMed]
353. Yeung, O.W.H.; Lo, C.-M.; Ling, C.-C.; Qi, X.; Geng, W.; Li, C.-X.; Ng, K.T.P.; Forbes, S.J.; Guan, X.-Y.; Poon, R.T.P.; et al. Alternatively activated (M2) macrophages promote tumour growth and invasiveness in hepatocellular carcinoma. *J. Hepatol.* **2015**, *62*, 607–616. [CrossRef]
354. Schulte, L.-A.; López-Gil, J.C.; Sainz, B.; Hermann, P.C. The Cancer Stem Cell in Hepatocellular Carcinoma. *Cancers* **2020**, *12*, 684. [CrossRef]
355. Zheng, Y.-W. Cellular reprogramming and hepatocellular carcinoma development. *World J. Gastroenterol.* **2013**, *19*, 8850. [CrossRef]
356. Kim, H.J.; Jeong, J.; Park, S.; Jin, Y.-W.; Lee, S.-S.; Lee, S.B.; Choi, D. Establishment of Hepatocellular Cancer Induced Pluripotent Stem Cells Using a Reprogramming Technique. *Gut Liver* **2017**, *11*, 261–269. [CrossRef]
357. Gao, X.; Sheng, Y.; Yang, J.; Wang, C.; Zhang, R.; Zhu, Y.; Zhang, Z.; Zhang, K.; Yan, S.; Sun, H.; et al. Osteopontin alters DNA methylation through up-regulating DNMT1 and sensitizes CD133+/CD44+ cancer stem cells to 5 azacytidine in hepatocellular carcinoma. *J. Exp. Clin. Cancer Res.* **2018**, *37*, 179. [CrossRef] [PubMed]
358. Moriguchi, H.; Chung, R.T.; Sato, C. An Identification of Novel Therapy for Human Hepatocellular Carcinoma by Using Human Induced Pluripotent Stem Cells. *Hepatology* **2010**, *51*, 1089–1090. [CrossRef] [PubMed]
359. Moriguchi, H.; Madson, J. The reprogramming therapy for a patient with advanced hepatocellular carcinoma by using human-induced pluripotent stem (iPS) cells technology. *Case Reports* **2013**, *2013*, bcr2013008950. [CrossRef]
360. Liu, M.; Yan, Q.; Sun, Y.; Nam, Y.; Hu, L.; Loong, J.H.C.; Ouyang, Q.; Zhang, Y.; Li, H.-L.; Kong, F.-E.; et al. A hepatocyte differentiation model reveals two subtypes of liver cancer with different oncofetal properties and therapeutic targets. *Proc. Natl. Acad. Sci. USA* **2020**, *117*, 6103–6113. [CrossRef] [PubMed]
361. Basu, A.; Dydowiczová, A.; Trosko, J.E.; Bláha, L.; Babica, P. Ready to go 3D? A semi-automated protocol for microwell spheroid arrays to increase scalability and throughput of 3D cell culture testing. *Toxicol. Mech. Methods* **2020**, *30*, 590–604. [CrossRef]
362. Kim, S.M.; Han, J.M.; Le, T.T.; Sohng, J.K.; Jung, H.J. Anticancer and Antiangiogenic Activities of Novel α-Mangostin Glycosides in Human Hepatocellular Carcinoma Cells via Downregulation of c-Met and HIF-1α. *Int. J. Mol. Sci.* **2020**, *21*, 4043. [CrossRef] [PubMed]
363. Ceballos, M.P.; Angel, A.; Delprato, C.B.; Livore, V.I.; Ferretti, A.C.; Lucci, A.; Comanzo, C.G.; de Lujan Alvarez, M.; Quiroga, A.D.; Mottino, A.D.; et al. Sirtuin 1 and 2 inhibitors enhance the inhibitory effect of sorafenib in hepatocellular carcinoma cells. *Eur. J. Pharmacol.* **2021**, *892*. [CrossRef]
364. Varan, G.; Akkın, S.; Demirtürk, N.; Benito, J.M.; Bilensoy, E. Erlotinib entrapped in cholesterol-depleting cyclodextrin nanoparticles shows improved antitumoral efficacy in 3D spheroid tumors of the lung and the liver. *J. Drug Target.* **2021**, *29*, 439–453. [CrossRef]
365. Song, Y.; Lee, S.-Y.; Kim, S.; Choi, I.; Kim, S.-H.; Shum, D.; Heo, J.; Kim, A.-R.; Kim, K.M.; Seo, H.R. Inhibitors of Na+/K+ ATPase exhibit antitumor effects on multicellular tumor spheroids of hepatocellular carcinoma. *Sci. Rep.* **2020**, *10*, 5318. [CrossRef] [PubMed]
366. Song, Y.; Kim, J.-S.; Kim, S.-H.; Park, Y.K.; Yu, E.; Kim, K.-H.; Seo, E.-J.; Oh, H.-B.; Lee, H.C.; Kim, K.M.; et al. Patient-derived multicellular tumor spheroids towards optimized treatment for patients with hepatocellular carcinoma. *J. Exp. Clin. Cancer Res.* **2018**, *37*, 109. [CrossRef]
367. Ranga, A.; Gjorevski, N.; Lutolf, M.P. Drug discovery through stem cell-based organoid models. *Adv. Drug Deliv. Rev.* **2014**, *69–70*, 19–28. [CrossRef] [PubMed]
368. Prior, N.; Inacio, P.; Huch, M. Liver organoids: From basic research to therapeutic applications. *Gut* **2019**, *68*, 2228–2237. [CrossRef]
369. Yin, X.; Mead, B.E.; Safaee, H.; Langer, R.; Karp, J.M.; Levy, O. Engineering Stem Cell Organoids. *Cell Stem Cell* **2016**, *18*, 25–38. [CrossRef]
370. Jee, J.H.; Lee, D.H.; Ko, J.; Hahn, S.; Jeong, S.Y.; Kim, H.K.; Park, E.; Choi, S.Y.; Jeong, S.; Lee, J.W.; et al. Development of Collagen-Based 3D Matrix for Gastrointestinal Tract-Derived Organoid Culture. *Stem Cells Int.* **2019**, *2019*, 1–15. [CrossRef] [PubMed]
371. Takebe, T.; Sekine, K.; Enomura, M.; Koike, H.; Kimura, M.; Ogaeri, T.; Zhang, R.-R.; Ueno, Y.; Zheng, Y.-W.; Koike, N.; et al. Vascularized and functional human liver from an iPSC-derived organ bud transplant. *Nature* **2013**, *499*, 481–484. [CrossRef] [PubMed]

372. Sun, L.; Wang, Y.; Cen, J.; Ma, X.; Cui, L.; Qiu, Z.; Zhang, Z.; Li, H.; Yang, R.-Z.; Wang, C.; et al. Modelling liver cancer initiation with organoids derived from directly reprogrammed human hepatocytes. *Nat. Cell Biol.* **2019**, *21*, 1015–1026. [CrossRef]
373. Kern, M.A.; Haugg, A.M.; Eiteneuer, E.; Konze, E.; Drebber, U.; Dienes, H.P.; Breuhahn, K.; Schirmacher, P.; Kasper, H.U. Ex vivo analysis of antineoplastic agents in precision-cut tissue slices of human origin: Effects of cyclooxygenase-2 inhibition in hepatocellular carcinoma. *Liver Int.* **2006**, *26*, 604–612. [CrossRef]
374. Vaira, V.; Fedele, G.; Pyne, S.; Fasoli, E.; Zadra, G.; Bailey, D.; Snyder, E.; Faversani, A.; Coggi, G.; Flavin, R.; et al. Preclinical model of organotypic culture for pharmacodynamic profiling of human tumors. *Proc. Natl. Acad. Sci. USA* **2010**, *107*, 8352–8356. [CrossRef]
375. Forner, A.; Reig, M.; Bruix, J. Hepatocellular carcinoma. *Lancet* **2018**, *391*, 1301–1314. [CrossRef]
376. Marquardt, J.U.; Thorgeirsson, S.S. SnapShot: Hepatocellular Carcinoma. *Cancer Cell* **2014**, *25*, 550.e1. [CrossRef]
377. Llovet, J.M.; Ricci, S.; Mazzaferro, V.; Hilgard, P.; Gane, E.; Blanc, J.-F.; de Oliveira, A.C.; Santoro, A.; Raoul, J.-L.; Forner, A.; et al. Sorafenib in Advanced Hepatocellular Carcinoma. *N. Engl. J. Med.* **2008**, *359*, 378–390. [CrossRef] [PubMed]
378. Lencioni, R.; Llovet, J.M.; Han, G.; Tak, W.Y.; Yang, J.; Guglielmi, A.; Paik, S.W.; Reig, M.; Kim, D.Y.; Chau, G.-Y.; et al. Sorafenib or placebo plus TACE with doxorubicin-eluting beads for intermediate stage HCC: The SPACE trial. *J. Hepatol.* **2016**, *64*, 1090–1098. [CrossRef]
379. Kudo, M.; Han, G.; Finn, R.S.; Poon, R.T.P.; Blanc, J.-F.; Yan, L.; Yang, J.; Lu, L.; Tak, W.-Y.; Yu, X.; et al. Brivanib as adjuvant therapy to transarterial chemoembolization in patients with hepatocellular carcinoma: A randomized phase III trial. *Hepatology* **2014**, *60*, 1697–1707. [CrossRef] [PubMed]
380. Labeur, T.A.; Ten Cate, D.W.G.; Bart Takkenberg, R.; Azahaf, H.; van Oijen, M.G.H.; van Delden, O.M.; de Man, R.A.; van Vugt, J.L.A.; IJzermans, J.N.M.; Eskens, F.A.L.M.; et al. Are we SHARP enough? The importance of adequate patient selection in sorafenib treatment for hepatocellular carcinoma. *Acta Oncol.* **2018**, *57*, 1467–1474. [CrossRef]
381. Kudo, M.; Finn, R.S.; Qin, S.; Han, K.-H.; Ikeda, K.; Piscaglia, F.; Baron, A.; Park, J.-W.; Han, G.; Jassem, J.; et al. Lenvatinib versus sorafenib in first-line treatment of patients with unresectable hepatocellular carcinoma: A randomised phase 3 non-inferiority trial. *Lancet* **2018**, *391*, 1163–1173. [CrossRef]
382. Bruix, J.; Qin, S.; Merle, P.; Granito, A.; Huang, Y.-H.; Bodoky, G.; Pracht, M.; Yokosuka, O.; Rosmorduc, O.; Breder, V.; et al. Regorafenib for patients with hepatocellular carcinoma who progressed on sorafenib treatment (RESORCE): A randomised, double-blind, placebo-controlled, phase 3 trial. *Lancet* **2017**, *389*, 56–66. [CrossRef]
383. Abou-Alfa, G.K.; Meyer, T.; Cheng, A.-L.; El-Khoueiry, A.B.; Rimassa, L.; Ryoo, B.-Y.; Cicin, I.; Merle, P.; Chen, Y.; Park, J.-W.; et al. Cabozantinib in Patients with Advanced and Progressing Hepatocellular Carcinoma. *N. Engl. J. Med.* **2018**, *379*, 54–63. [CrossRef] [PubMed]
384. Pardoll, D.M. The blockade of immune checkpoints in cancer immunotherapy. *Nat. Rev. Cancer* **2012**, *12*, 252–264. [CrossRef]
385. Groenendijk, F.H.; Mellema, W.W.; van der Burg, E.; Schut, E.; Hauptmann, M.; Horlings, H.M.; Willems, S.M.; van den Heuvel, M.M.; Jonkers, J.; Smit, E.F.; et al. Sorafenib synergizes with metformin in NSCLC through AMPK pathway activation. *Int. J. Cancer* **2015**, *136*, 1434–1444. [CrossRef]
386. Huang, D.Q.; Muthiah, M.D.; Zhou, L.; Jumat, H.; Tan, W.X.; Lee, G.H.; Lim, S.G.; Kow, A.; Bonney, G.; Shridhar, I.; et al. Predicting HCC Response to Multikinase Inhibitors With In Vivo Cirrhotic Mouse Model for Personalized Therapy. *Cell. Mol. Gastroenterol. Hepatol.* **2021**, *11*, 1313–1325. [CrossRef]
387. Shen, J.; Cai, W.; Ma, Y.; Xu, R.; Huo, Z.; Song, L.; Qiu, X.; Zhang, Y.; Li, A.; Cao, W.; et al. hGC33-Modified and Sorafenib-Loaded Nanoparticles have a Synergistic Anti-Hepatoma Effect by Inhibiting Wnt Signaling Pathway. *Nanoscale Res. Lett.* **2020**, *15*, 220. [CrossRef] [PubMed]
388. Rodríguez, M.M.; Onorato, A.; Cantero, M.J.; Domínguez, L.; Bayo, J.; Fiore, E.; García, M.; Atorrasagasti, C.; Canbay, A.; Malvicini, M.; et al. 4-methylumbelliferone-mediated polarization of M1 macrophages correlate with decreased hepatocellular carcinoma aggressiveness in mice. *Sci. Rep.* **2021**, *11*, 6310. [CrossRef] [PubMed]
389. Zhou, Z.-F.; Peng, F.; Li, J.-Y.; Ye, Y.-B. Intratumoral IL-12 Gene Therapy Inhibits Tumor Growth In A HCC-Hu-PBL-NOD/SCID Murine Model. *Onco. Targets. Ther.* **2019**, *12*, 7773–7784. [CrossRef]
390. Bi, Y.; Jiang, H.; Wang, P.; Song, B.; Wang, H.; Kong, X.; Li, Z. Treatment of hepatocellular carcinoma with a GPC3-targeted bispecific T cell engager. *Oncotarget* **2017**, *8*, 52866–52876. [CrossRef] [PubMed]
391. Nishina, S.; Yamauchi, A.; Kawaguchi, T.; Kaku, K.; Goto, M.; Sasaki, K.; Hara, Y.; Tomiyama, Y.; Kuribayashi, F.; Torimura, T.; et al. Dipeptidyl Peptidase 4 Inhibitors Reduce Hepatocellular Carcinoma by Activating Lymphocyte Chemotaxis in Mice. *Cell. Mol. Gastroenterol. Hepatol.* **2019**, *7*, 115–134. [CrossRef] [PubMed]
392. Zhou, Y.; Fu, C.; Kong, Y.; Pan, D.; Wang, Y.; Huang, S.; Li, Z.; Ning, Z.; Lu, X.; Shan, S.; et al. Antitumor and immunomodulatory effects of a novel multitarget inhibitor, CS2164, in mouse hepatocellular carcinoma models. *Anticancer. Drugs* **2019**, *30*, 909–916. [CrossRef] [PubMed]
393. Jilkova, Z.M.; Kuyucu, A.Z.; Kurma, K.; Ahmad Pour, S.T.; Roth, G.S.; Abbadessa, G.; Yu, Y.; Schwartz, B.; Sturm, N.; Marche, P.N.; et al. Combination of AKT inhibitor ARQ 092 and sorafenib potentiates inhibition of tumor progression in cirrhotic rat model of hepatocellular carcinoma. *Oncotarget* **2018**, *9*, 11145–11158. [CrossRef]
394. Cheng, Y.; Luo, R.; Zheng, H.; Wang, B.; Liu, Y.; Liu, D.; Chen, J.; Xu, W.; Li, A.; Zhu, Y. Synergistic anti-tumor efficacy of sorafenib and fluvastatin in hepatocellular carcinoma. *Oncotarget* **2017**, *8*, 23265–23276. [CrossRef]

395. Afaloniati, H.; Angelopoulou, K.; Giakoustidis, A.; Hardas, A.; Pseftogas, A.; Makedou, K.; Gargavanis, A.; Goulopoulos, T.; Iliadis, S.; Papadopoulos, V.; et al. HDAC1/2 Inhibitor Romidepsin Suppresses DEN-Induced Hepatocellular Carcinogenesis in Mice. *Onco. Targets. Ther.* **2020**, *13*, 5575–5588. [CrossRef]
396. Sung, Y.-C.; Liu, Y.-C.; Chao, P.-H.; Chang, C.-C.; Jin, P.-R.; Lin, T.-T.; Lin, J.-A.; Cheng, H.-T.; Wang, J.; Lai, C.P.; et al. Combined delivery of sorafenib and a MEK inhibitor using CXCR4-targeted nanoparticles reduces hepatic fibrosis and prevents tumor development. *Theranostics* **2018**, *8*, 894–905. [CrossRef]
397. Wang, Q.; Bin, C.; Xue, Q.; Gao, Q.; Huang, A.; Wang, K.; Tang, N. GSTZ1 sensitizes hepatocellular carcinoma cells to sorafenib-induced ferroptosis via inhibition of NRF2/GPX4 axis. *Cell Death Dis.* **2021**, *12*, 426. [CrossRef]
398. Chung, A.S.; Mettlen, M.; Ganguly, D.; Lu, T.; Wang, T.; Brekken, R.A.; Hsiehchen, D.; Zhu, H. Immune Checkpoint Inhibition is Safe and Effective for Liver Cancer Prevention in a Mouse Model of Hepatocellular Carcinoma. *Cancer Prev. Res.* **2020**, *13*, 911–922. [CrossRef] [PubMed]
399. Reszegi, A.; Horváth, Z.; Fehér, H.; Wichmann, B.; Tátrai, P.; Kovalszky, I.; Baghy, K. Protective Role of Decorin in Primary Hepatocellular Carcinoma. *Front. Oncol.* **2020**, *10*, 1–15. [CrossRef] [PubMed]
400. Kojima, M.; Takahashi, H.; Kuwashiro, T.; Tanaka, K.; Mori, H.; Ozaki, I.; Kitajima, Y.; Matsuda, Y.; Ashida, K.; Eguchi, Y.; et al. Glucagon-Like Peptide-1 Receptor Agonist Prevented the Progression of Hepatocellular Carcinoma in a Mouse Model of Nonalcoholic Steatohepatitis. *Int. J. Mol. Sci.* **2020**, *21*, 5722. [CrossRef] [PubMed]
401. Luo, Y.; Tian, G.; Zhuang, Z.; Chen, J.; You, N.; Zhuo, L.; Liang, B.; Song, Y.; Zang, S.; Liu, J.; et al. Berberine prevents non-alcoholic steatohepatitis-derived hepatocellular carcinoma by inhibiting inflammation and angiogenesis in mice. *Am. J. Transl. Res.* **2019**, *11*, 2668–2682.
402. Chow, A.K.-M.; Yau, S.W.-L.; Ng, L. Novel molecular targets in hepatocellular carcinoma. *World J. Clin. Oncol.* **2020**, *11*, 589–605. [CrossRef]
403. Chen, Z.; Xie, H.; Hu, M.; Huang, T.; Hu, Y.; Sang, N.; Zhao, Y. Recent progress in treatment of hepatocellular carcinoma. *Am. J. Cancer Res.* **2020**, *10*, 2993–3036. [CrossRef]
404. Xue, R.; Li, R.; Guo, H.; Guo, L.; Su, Z.; Ni, X.; Qi, L.; Zhang, T.; Li, Q.; Zhang, Z.; et al. Variable Intra-Tumor Genomic Heterogeneity of Multiple Lesions in Patients With Hepatocellular Carcinoma. *Gastroenterology* **2016**, *150*, 998–1008. [CrossRef]
405. Chan, S.L. Drug Development for Hepatocellular Carcinoma: Knowing the Past Helps to Understand the Future. *Oncologist* **2014**, *19*, 1115–1117. [CrossRef]

cancers

MDPI

Review

Valproic Acid and Breast Cancer: State of the Art in 2021

Anna Wawruszak [1,*], Marta Halasa [1], Estera Okon [1], Wirginia Kukula-Koch [2] and Andrzej Stepulak [1]

1 Department of Biochemistry and Molecular Biology, Medical University of Lublin, 20-093 Lublin, Poland; marta.halasa@umlub.pl (M.H.); estera.okon@umlub.pl (E.O.); andrzej.stepulak@umlub.pl (A.S.)
2 Department of Pharmacognosy, Medical University of Lublin, 20-093 Lublin, Poland; wirginia.kukulakoch@umlub.pl
* Correspondence: anna.wawruszak@umlub.pl; Tel.: +48-81448-6350

Simple Summary: Breast cancer (BC) is the most common cancer diagnosed among women worldwide. Despite numerous studies, the pathogenesis of BC is still poorly understood, and effective therapy of this disease remains a challenge for medicine. This article provides the current state of knowledge of the impact of valproic acid (VPA) on different histological subtypes of BC, used in monotherapy or in combination with other active agents in experimental studies in vitro and in vivo. The comprehensive review highlights the progress that has been made on this topic recently.

Abstract: Valproic acid (2-propylpentanoic acid, VPA) is a short-chain fatty acid, a member of the group of histone deacetylase inhibitors (HDIs). VPA has been successfully used in the treatment of epilepsy, bipolar disorders, and schizophrenia for over 50 years. Numerous in vitro and in vivo pre-clinical studies suggest that this well-known anticonvulsant drug significantly inhibits cancer cell proliferation by modulating multiple signaling pathways. Breast cancer (BC) is the most common malignancy affecting women worldwide. Despite significant progress in the treatment of BC, serious adverse effects, high toxicity to normal cells, and the occurrence of multi-drug resistance (MDR) still limit the effective therapy of BC patients. Thus, new agents which improve the effectiveness of currently used methods, decrease the emergence of MDR, and increase disease-free survival are highly needed. This review focuses on in vitro and in vivo experimental data on VPA, applied individually or in combination with other anti-cancer agents, in the treatment of different histological subtypes of BC.

Keywords: breast cancer; valproic acid (VPA); histone deacetylase inhibitor (HDI); histone acetylation; histone deacetylases (HDACs); epigenetics; targeted therapy

Citation: Wawruszak, A.; Halasa, M.; Okon, E.; Kukula-Koch, W.; Stepulak, A. Valproic Acid and Breast Cancer: State of the Art in 2021. *Cancers* **2021**, *13*, 3409. https://doi.org/10.3390/cancers13143409

Academic Editors: Felisbina Queiroga and Bruno Cogliati

Received: 26 April 2021
Accepted: 5 July 2021
Published: 7 July 2021

Publisher's Note: MDPI stays neutral with regard to jurisdictional claims in published maps and institutional affiliations.

Copyright: © 2021 by the authors. Licensee MDPI, Basel, Switzerland. This article is an open access article distributed under the terms and conditions of the Creative Commons Attribution (CC BY) license (https://creativecommons.org/licenses/by/4.0/).

1. Introduction

Regardless of socioeconomic status and level of development of societies, cancer is one of the most common causes of morbidity and mortality worldwide [1,2]. According to the GLOBOCAN, one in six women and one in five men were diagnosed with cancer in 2018 [1]. Unfortunately, the rates of cancer incidence and mortality are still rising. It is estimated that 13 million people will die from cancer in 2030 [3].

Breast cancer (BC) is the most common cancer diagnosed among women in all regions in the world except in the eastern areas of Africa where cervical cancer occurs most often. Over the last twenty years, there has been an approximate 30% increase in the incidence rate of this disease [4]. Moreover, BC is the leading cause of neoplasms death in over 100 countries all over the world. In 2018, over 2 million new BC cases were diagnosed and nearly 630,000 deaths from BC have been reported worldwide [1,4].

Understanding the biological landscape of BC and its phenotypic heterogeneity is a key element in developing novel targeted therapies [5]. The integration of nucleic acid and peptide sequencing based on mass spectrometry and advanced biomolecular analysis allowing to define the post-translational modifications, provide a better understanding of the pathophysiology of BC, and help to develop new more effective therapeutic strategies in the treatment of this disease [6]. Unfortunately, despite numerous studies, the pathogenesis of BC is still unknown, and effective therapy of this disease is one of the most important challenges of medicine.

Both genetic and epigenetic modifications are responsible for the progression of BC. Unlike irreversible genetic alterations, epigenetic modifications can be reversible. This suggests that epigenetic changes are favored in therapeutic applications. DNA methyltransferases and histone deacetylases are the main targets for epigenetic therapy. Several inhibitors of DNA methyltransferases and histone deacetylases have been approved by the US Food and Drug Administration (FDA) as anti-cancer drugs [7,8]. Reversible histone acetylation, catalyzed by histone acetyltransferases (HATs) and histone deacetylases (HDACs), plays an important role in epigenetic regulation of gene expression. An imbalance between HAT and HDAC expression leads to the development of numerous cancers [8,9]. In most cancer cell lines, a reduction in histone acetylation levels was observed due to overexpression of HDACs activity [10]. Histone deacetylase inhibitors (HDIs) are promising new generation cytostatics that increase histone acetylation. HDIs modulate the structure of chromatin, leading to changes in the expression of genes involved in numerous signaling pathways, including induction of apoptosis, cell cycle arrest, and inhibition of angiogenesis [11]. However, the mechanism of antitumor activity and the specificity of HDIs have not been fully understood.

In our review article, we described the current state of knowledge of the use of valproic acid (VPA), short-chain fatty acid, representative of the HDIs which has been successfully used in the treatment of epilepsy, bipolar disorders, and schizophrenia for over 50 years, individually or in combination with other active agents, in the treatment of BC, with particular emphasis on the progress that have been done in this topic recently.

2. Molecular Subtypes of Breast Cancers and Limitations in the Therapy of Patients Harboring These Subtypes

BC is a complex group of diseases with specific pathological features and clinical implications. Extensive evidence suggests that BCs with varied biological and histopathological characteristics develop differently, resulting in miscellaneous responses to the treatment, and therefore various therapeutic strategies should be used [12–14].

Classic immunohistochemistry markers such as expression of estrogen (ER), progesterone (PG), and human epidermal growth factor (HER2) receptors; and clinicopathological factors, like tumor grade, size, nodal involvement, are conventionally used to select therapy and to predict disease progression. The widespread use of high-throughput techniques for gene expression analysis has shown that the response of cancer cells to treatment is not due to prognostic factors of anatomical origin, but to internal molecular characteristics of BCs [12,15].

Five molecular subtypes of BC, including luminal A, luminal B, HER2-overexpressed, triple-negative and normal-like, were identified (Figure 1) [12,15].

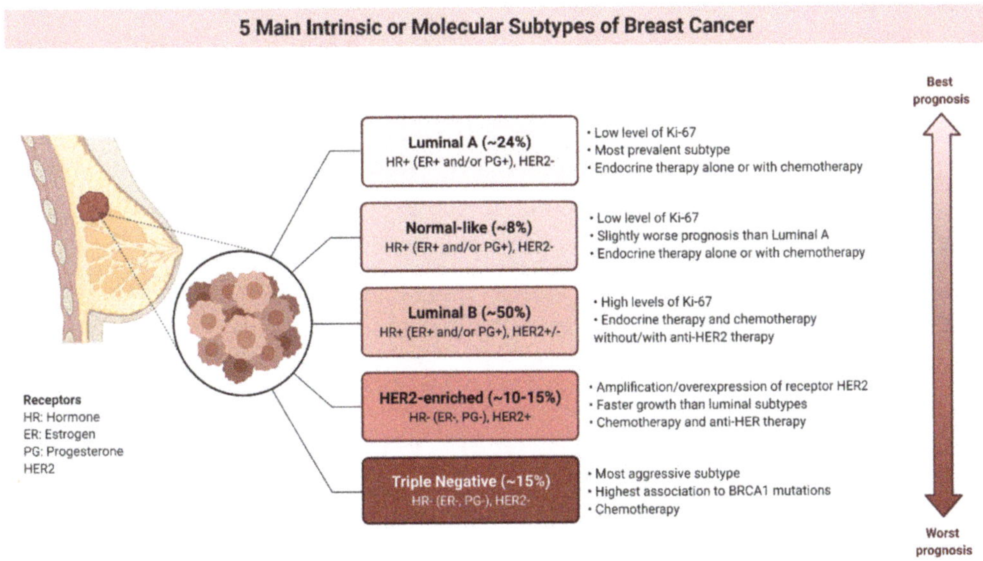

Figure 1. 5 main intrinsic or molecular subtypes of breast cancer (BC) (ER-estrogen receptor, PG-progesterone receptor, HER2-human epidermal growth factor receptor, Ki67-proliferation index marker) [12,13,16,17].

The hormone receptors (ER and PG) positive subtypes of BC are the most common types of breast carcinoma, among these luminal A and luminal B forms are more prevalent [18]. Luminal A BCs have a higher level of expression estrogen-related genes and lower expression of proliferative markers (Ki67 < 20%) compared to luminal B type (Ki67 $\geq$ 20%). Moreover, luminal B cancers are usually characterized by higher histological grade than luminal A tumors. At the molecular level, luminal A subtypes are associated with somatic mutations in *GATA3* (GATA binding protein 3), *PIK3CA* (phosphatidylinositol-4,5-bisphosphate 3-kinase catalytic subunit alpha) and *MAP3K1* (mitogen-activated protein kinase kinase 1) genes, and often exhibit *cyclin D1* overexpression. Luminal B tumors show frequent mutations in the *TP53* and *PIK3CA* genes as well as dysregulations in the retinoblastoma and MAPK (mitogen-activated protein kinase) signaling pathways [13]. Luminal cancers respond well to hormone-related therapies [12,19,20].

Unlike the luminal type, HER2-overexpressed and triple-negative breast cancer (TNBC) subtypes are characterized by a lack of expression of ER and PG receptors, as well as high aggressiveness [18]. Although HER2-overexpressed BCs carry a poorer clinical prognosis compared to luminal subtypes, they have a much better response and sensitivity to anthracycline [21] and taxane-based neoadjuvant chemotherapy [22,23]. Even though the therapy of patients expressing HER2 receptors has been revolutionized by the introduction of anti-HER2 monoclonal antibodies (e.g., trastuzumab, bevacizumab, lapatinib) [24–26], recurrence and development of metastasis are serious clinical issues. In addition, not all patients with HER2-overexpression respond to therapy with trastuzumab. C-X-C chemokine receptor type 4 (CXCR4) up-regulation and phosphatase and tensin homolog (PTEN) loss are associated with resistance to treatment with trastuzumab. Therefore, new therapies are being sought for the treatment of cancers resistant to anti-HER2 monoclonal antibodies [12,27].

Approximately 15% of patients suffering from BC are diagnosed with its most severe form-TNBC. TNBC characterizes lack or low expression of hormone and HER2 receptors as well as a high level of basal markers, such as keratin or epidermal growth factor receptor (EGFR). There is also increased activation of the *WNT* signaling pathway and frequent mutations in the *TP53* and *BRCA1* genes [13]. Therefore, standard hormone therapies and targeted therapy directed against HER2 are excluded. TNBC characterizes a very aggressive clinical course, and a higher risk of local and systemic relapse [12,28–30]. TNBC has the worst prognosis of all the BC subtypes and is treated with systemic chemotherapy to which it responds better than other subtypes. Unfortunately, the use of traditional cytostatics (cisplatin, paclitaxel) is limited by numerous side effects (bone marrow damage, severe renal failure, peripheral neuropathies), as well as the occurrence of resistance to therapies [31–34]. Due to the lack of recognized molecular targets for therapy, TNBC is an object of interest for clinical trials with novel treatment approaches [13].

Normal-like type of BC accounts 7.8% of all cancer cases and characterizes similar immunohistochemistry status to the luminal A subtype (HR+ (ER+ and/or PG+), HER, low Ki-67) and normal breast tissue profiling. The tumor necrosis factor alpha (TNFα) pathway activity increased gradually from luminal A, luminal B, normal-like, HER2-enriched and TNBC subtypes [35]. Still, while normal-like BC has relatively good prognosis, its outlook is slightly worse than luminal A cancers' prognosis [12]. Similar to the luminal A subtype, normal-like signature was found significantly less expressed in metastatic tumors than in primary tumors. It has been demonstrated that both normal-like and luminal A signatures show a negative correlation between time to tumor recurrence (TTR) and the magnitude of gene/signature expression changes between primary and metastatic disease [36]. Interestingly, normal-like cancer is less sensitive to paclitaxel- and doxorubicin-containing preoperative chemotherapy than the TNBC and HER2+ subtypes [37].

The presence or absence of receptors characteristic of BC allows to use of specific targeted therapies and the personalized treatment of BC patients. Targeted agents acting at the epigenetic level are currently being investigated in the treatment of different hematological malignancies and solid tumors.

3. Histone Deacetylase Inhibitors (HDIs)

The abnormal histone acetylation profile leads to numerous cellular disorders, including tumor initiation and progression [38]. It has been shown that histone acetylation disturbances are an important factor in the progression of BC. Studies linked with abnormal acetylation level of histones in BC focus on molecular mechanisms of BC development, identification of novel biomarkers for prediction aggressiveness of the tumor, and therapeutic potential [39].

Histone acetylation modifying enzymes control the transcription process by changing the status of histone acetylation as well as other transcription factors occurring mainly in the promoter region (Figure 2) [38]. Equilibrium in the activity of the opposing enzymes: HATs and HDACs is necessary to maintain epigenetic regulation of gene expression [40]. HATs catalyze the reversible acetylation reaction at the ε-amino group of lysine residues. Neutralization of the positive charge of lysine residues due to histone acetylation is correlated with chromatin relaxation and increased transcriptional activity of genes. Unlike HATs, HDACs remove acetyl groups leading to condensation of chromatin and silencing the transcriptional activity of genes (Figure 2) [8,41].

Based on the yeast protein homology and functional criteria, HDACs were divided into four classes: zinc- (I, II and IV) and NAD-dependent (III). HDACs 1, 2, 3 and 8 belong to the I class. The II class is divided into two subclasses IIa (HDAC4, 5, 7, 9) and IIb (HDAC6, 10). Class III due to homology to silent information regulator 2 (SIR2) of *Saccharomyces cerevisiae* is called sirtuins and includes SIRT1-SIRT7. Class IV contains only one member HDAC11 (Table 1) [8,41].

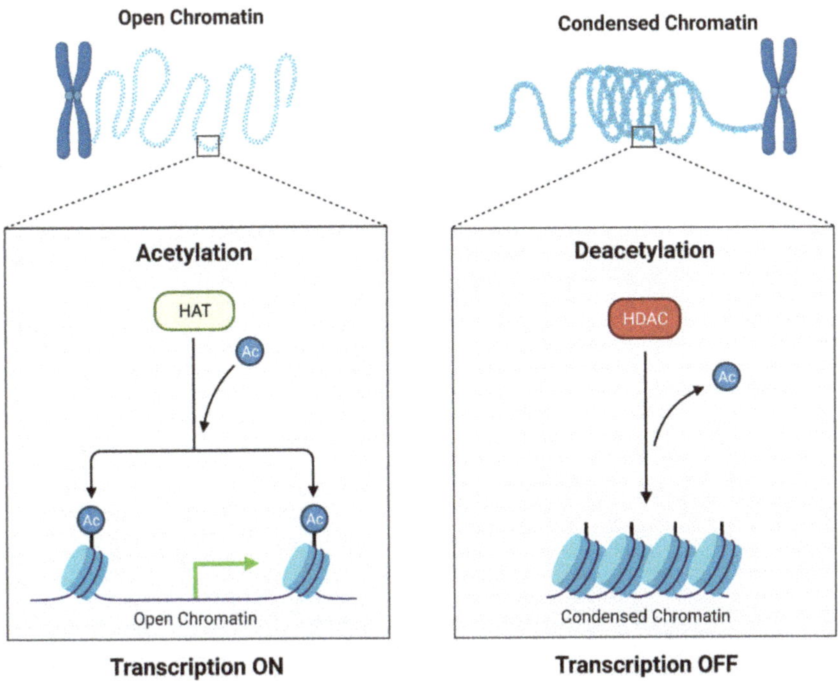

Figure 2. Histone acetylation modifying enzymes (HAT histone acetyltransferase, HDAD-histone deacetylase) control the transcription process by changing the status of histone acetylation and conformation of chromatin [38].

Table 1. Classification of histone deacetylases (HDACs) [8,41].

Class of HDAC	HDAC	Zinc/Nicotinamide (NAD)-Dependent
I	HDAC1, 2, 3, 8	zinc-dependent
II	IIa: HDAC4, 5, 7, 9 IIb: HDAC6, 10	zinc-dependent
III	SIRT1-SIRT7	NAD-dependent
IV	HDAC 11	zinc-dependent

Due to the fact that HDACs exert a significant effect on chromatin remodeling, their inhibitors (HDIs) have become an interesting field of study. HDIs are divided into four classes: hydroxamic acids (i.e., trichostatin A (TSA), vorinostat (SAHA), belinostat (PXD-101), panobinostat (LBH-589), resminostat (4SC-201)); short chain fatty acids (i.e. sodium butyrate (NaB), phenylbutyrate (PBA), valproic acid (VPA)); cyclic peptides (i.e. apicidin (CAS183506-66-3), romidepsin (FK228)); benzamides (i.e. mocetinostat (MGCD103), entinostat (MS-275), domatinostat (4SC-202)) (Table 2) [10].

So far, four HDIs have been approved by the US Food and Drug Administration (FDA) for the treatment of certain types of cancer: SAHA-for the treatment of cutaneous manifestations of cutaneous T-cell lymphoma (CTCL) in patients with the progressive, persistent, or relapsing disease [42]; LBH-589-in polytherapy with bortezomib and dexamethasone for therapy of patients with relapsed and/or refractory multiple myeloma [43]; FK228 and PXD-101-for the treatment of peripheral T-cell lymphomas (PTCLs), a rare disease belonging to non-Hodgkin lymphomas [44,45]. Therefore, HDACs modulators may also be used as potential drugs in the BC treatment [39]. HDIs via inhibition of HDACs activity, increase the acetylation level of both histone and non-histone proteins [46,47] maintaining a global cellular acetylation profile which enables the activation of genes responsible for inhibiting the progression of BC. Results from pre-clinical and clinical studies have shown that HDIs can induce different anti-cancer mechanisms in many types of BC [8,40,41]. Since VPA, as a psychoneurological drug, crossing the blood-brain barrier (BBC), it could also effectively eliminate metastatic BC cells in the brain of patients (Figure 3) [48].

Table 2. Classes of histone deacetylase inhibitors (HDIs) (CTCL-cutaneous T-cell lymphoma, FDA-Food and Drug Administration, PTCL-peripheral T-cell lymphoma) [10].

Class of HDIs	HDI	Abrreviation	FDA Approval for Cancer Treatment
Hydroxamic acids	Tichostatin A Vorinostat Belinostat Panobinostat Resminostat	TSA SAHA PXD-101 LBH-589 4SC-201	Approved for CTCL treatment Approved for PTCLs treatment Approved for multiple myeloma treatment
Short chain fatty acids	Sodium butyrate Phenylbutyrate Valproic acid	NaB PBA VPA	
Cyclic peptides	Apicidin Romidepsin	CAS183506-66-3 FK228	Approved for PTCLs treatment
Benzamides	Mocetinostat Entinostat Domatinostat	MGCD103 MS-275 4SC-202	

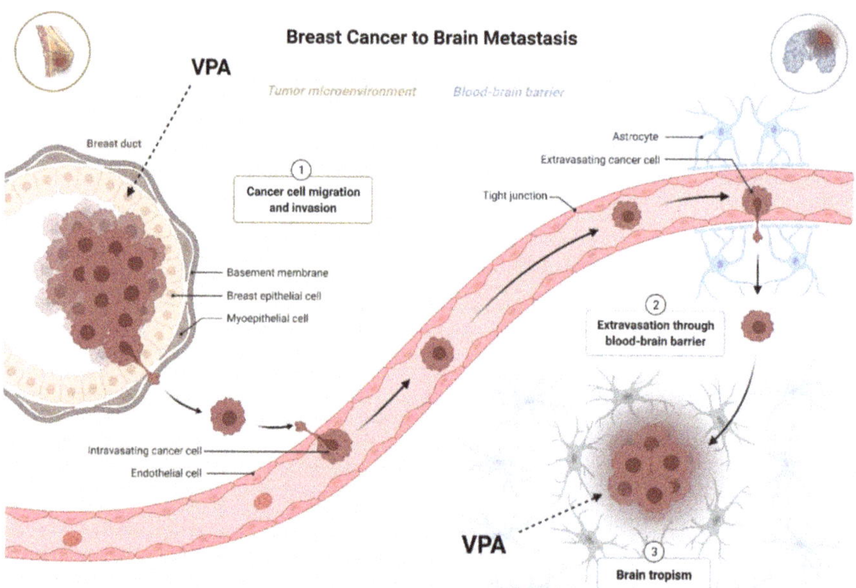

Figure 3. The potential effects of valproic acid (VPA) on breast cancer cells and metastatic breast cancer (BC) cells in the brain.

HDIs are able to inhibit proliferation and induce the differentiation and apoptosis of tumor cells resistant to different cytostatic drugs by regulating the expression several genes. It was already demonstrated in 2004 that administration of TSA to BC cells resistant to tamoxifen caused an increase in estrogen receptor expression, which in turn allowed for re-sensitization of these cells to the administered drug [49]. In addition, it has been shown that new synthetic HDI-FA17 overcome multidrug resistance (MDR) in BC cells of the MCF-7/MTX insensitive to methotrexate cell line [50]. In general, HDIs can induce tumor growth inhibition and apoptosis of tumor cells. Interestingly, in opposite to standard cytostatic agents, HDIs show significantly lower toxicity to normal cells [8,40,41].

4. Valproic Acid and Breast Cancer

Valproic acid (2-propylpentanoic acid, VPA) belongs to the group of short-chain fatty acids. VPA causes acetylation of the N-terminal tails on histones H3 and H4, and inhibits the activity of HDAC I and II, probably by binding to the catalytic center, and in consequence, blocking access to the substrate [8,51]. VPA has been approved by the FDA for the treatment of epilepsy and other convulsive diseases and has been used successfully in the therapy of these diseases for over five decades [52]. It has been demonstrated that VPA shows anticancer activity (Figure 4) in a diversity of human cancers [53–55], including breast carcinoma [56–59].

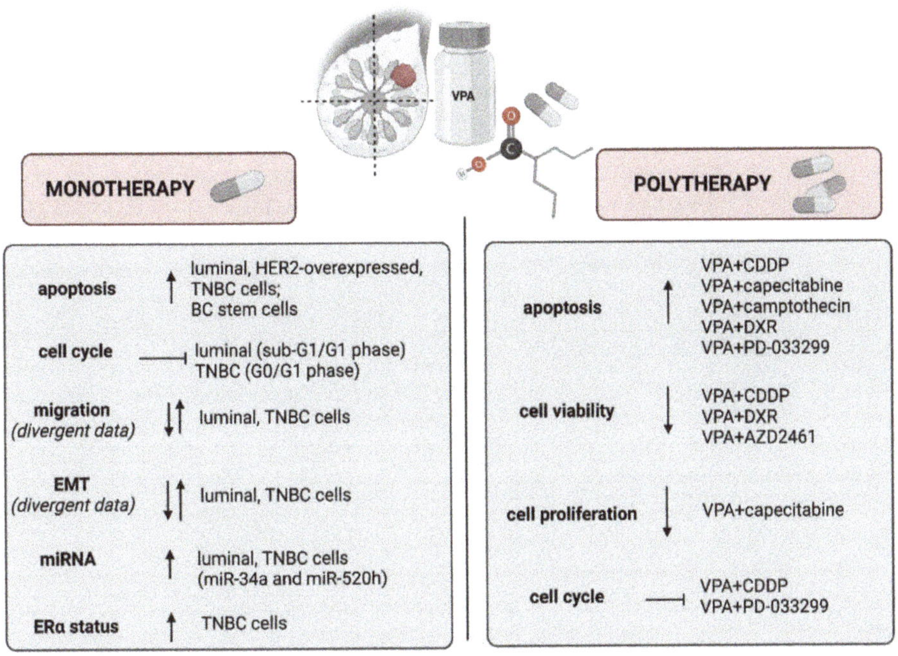

Figure 4. The functional mechanism of valproic acid in breast cancer treatment (BC-breast cancer, CDDP-cisplatin, DXR-doxorubicin, ER-estrogen receptor, EMT-epithelial-mesenchymal transition, HER2-human epidermal growth factor receptor 2, TNBC-triple-negative breast cancer, VPA-valproic acid) (↑—increase, ↓—decrease, ⊣—stop).

4.1. VPA Induces Apoptosis and Inhibits Cell Cycle

VPA decreases cell viability through arresting of the cell cycle in G1 or sub-G1 phases, induction of p21 protein expression and apoptosis by upregulation of Bak, downregulation of Bcl-2 expression, increasing Bax/Bcl-2 ratio, and, as a consequence, decreasing telomerase activity in estrogen-positive MCF7 BC cells. Telomerase is a ribonucleoprotein reverse transcriptase involved in the elongation of the telomeres and is responsible for the phenomenon of resistance to apoptosis in cancer cells [60,61]. VPA reduced proliferation not only MCF-7 BC cells but also MCF-7 BC stem cells in a time (24, 48, 72 h) and dose (0.6–20 mM) dependent manner. Cancer stem cells (CSCs) are a subpopulation of cells that reinitiate carcinogenesis, induce resistance to chemotherapy, are prone to develop metastases, and lead to disease relapse due to acquired resistance to apoptosis. Epigenetic alterations play a pivotal role in the regulation of stemness and also have been implicated in the development of drug resistance. It has been detected that MCF-7 stem cells were much more resistant to VPA than MCF-7 cells. Moreover, VPA increased levels of M30 protein (cytokeratin 18 neoepitope), caspase 3 and 7 activations, annexin-V-FITC positivity, suggesting apoptosis induction in BC stem cells. The late stage of apoptosis (secondary necrosis) was also evidenced by nuclear pyknosis with propidium iodide staining [61].

Similar to receptor-positive BCs, VPA induces cell cycle inhibition and apoptosis in BC cells with HER2 overexpression. HER2-overexpressed BC cells are more sensitive to VPA than HER2-negative ones. It has been demonstrated that the anti-proliferative mechanism of VPA in BC cells is related to their HER2-expression status. Therefore, VPA may synergize with drugs used in the therapy of HER2-overexpressed BC, like anti-HER2 monoclonal antibodies (e.g., trastuzumab, bevacizumab, lapatinib) or anthracycline and taxane-based neoadjuvant chemotherapy to inhibit HER2-overexpressing BC cell proliferation more effectively. The antiproliferative effect of VPA results from Hsp90 dysfunction which is involved in hyperacetylation of Hsp70 (non-histone protein acetylation). Hyperacetylation of Hsp70 directly affects the HER2 receptor protein, which is the client of the Hsp90 protein. The loss of Hsp90 function leads to the degradation of Hsp90 client proteins and the process of apoptosis. The alteration of the level of cyclin-dependent kinase inhibitor p21/WAF1, cleaved caspase-3, acetylated heat shock protein (Hsp) 70, acetylated Hsp90, and acetylated α-tubulin by VPA was determined in SKBR3 HER2-overexpressing BC cells. It has been observed that VPA upregulates expression and induces targeting of p21 WAF1, cleaved caspase-3, upregulates Hsp 70 acetylation, inhibits differentiation, and exhibits antiproliferative activity in BC cells in a dose- and time-dependent manner [8,62]. It has been demonstrated that VPA also remarkably inhibits the growth and triggers apoptotic cell death through G0/G1 arrest in MDA-MB-231 TNBC cells (Table 3) [57].

Table 3. Mechanism of action of valproic acid (VPA) in in vitro breast cancer (BC) pre-clinical setting (BC-breast cancer, EMT-epithelial-mesenchymal transition, ER-estrogen receptor, HER-2-human epidermal growth factor receptor 2).

Cellular Process	Sub-Type of BC	Cell Line	Mechanism of Action	References
Apoptosis	Luminal	MCF7	↑apoptosis (↑p21, ↑Bak, ↑Bax/Bcl-2 ratio, ↓Bcl-2 proteins expression, ↓telomerase activity)	[60]
		MCF7 tem cells	↑apoptosis (↑M30 protein expression, ↑caspase 3 and 7 activation, ↑nuclear pycnosis)	[61]
	HER-2-overexpressed	SKBR3	↑apoptosis (↑cleaves caspase 3, ↑Hsp70 acetyaltion)	[62]
	TNBC	MDA-MB-231	↑apoptosis	[57]
Cell cycle	Luminal	MCF7	cell cycle arrest in sub-G1 phase	[60]
		ZR-75-1	cell cycle arrest in G1 phase	[60]
	HER-2-overexpressed	SKBR3	↑p21WAF1 protein expression	[62]
	TNBC	MDA-MB-231	cell cycle arrest in G0/G1 phase	[57]
Migration	Luminal	MCF7	↑migration	[63]
		MCF7 T47D	↓migration	[64]
	TNBC	MDA-MB-231	↓migration (↑*nm23H1* gene expression)	[65]
		MDA-MB-231 MDA-MB-468	↓migration	[64]
		MDA-MB-231	↑migration	[57,63]

Table 3. *Cont.*

Cellular Process	Sub-Type of BC	Cell Line	Mechanism of Action	References
EMT	Luminal	MCF7	↑EMT (↑*Snail*, ↑*Zeb-2* genes expression)	[63]
		MCF7 T47D	↓EMT (↑*E-cadherin* gene and protein expresssion)	[64]
	TNBC	MDA-MB-231	↑EMT (↑*Snail*, ↑*Zeb-2* genes expression)	[63]
			↑EMT (↑*Snail*, ↓*E-cadherin*, ↓*GKS3β* genes expression)	[57]
		MDA-MB-468	↑EMT (↑*N-cadherin* gene and protein expression)	[64]
miRNA	Luminal	MCF7	↑miR-34a, ↑miR-520h expression	[56]
	TNBC	MDA-MB-231	↑miR-34a, ↑miR-520h expression	[56]
ER receptor status	TNBC	MDA-MB-231	↑*ERα*, ↑*FoxA1* genes and proteins expression	[66]
Metabolic pathways	Luminal	MCF7	↑furfural expression, alteration in alanine, taurine and hypotaurine metabolism	[67]

↑—increase, ↓—decrease.

4.2. VPA Regulates Migration and Epithelial-Mesenchymal Transition (EMT)

Sodium valproate (VPA-derivative) at concentrations of 0.8–3.2 mM inhibits migration of MDA-MB-231 TNBC cells in a dose-dependent manner by upregulation of nm23H1 gene expression [65]. Nm23 (non-metastatic clone 23), also known as ndpk (nucleoside diphosphate kinase), is a metastasis suppressor gene locating on codon 21.3 of the long arm of chromosome 17. The Nm23H1 protein acts as an upstream regulator that modulates downstream metastasis-related genes, which results in tumor metastasis inhibition. Overexpression of nm23H1 gene decreases proliferation, invasion, and metastasis of cancer cells, probably mediated by nm23h1 regulation by the HDACs. It has been reported, that silencing of nm23H1 resulted in an increased in rac (Rac family small GTPase 1) gene expression and, in consequence, in the invasive ability of TNBC cells. Overexpression of nm23H1 can be a promising prognostic indicator linked with longer overall survival of patients harboring various types of cancers, including BC. However, the mechanism by which Nm23H1 participates in tumor metastasis is not fully understood [65,68,69].

Besides inhibition of cell migration, VPA affects the epithelial to mesenchymal transition (EMT). EMT is an important process of transdifferentiation in solid cancers progression and the development of metastasis. During EMT polarized, immotile epithelial cells are transformed into migratory mesenchymal-like cells prone to migration, metastasis formation, drug resistance, and BC stemness features development [70–73]. Numerous signaling pathways are involved in the EMT process, including: cadherin [74], notch [75], transforming growth factor-β (TGF-β) [76], matrix metalloproteinases [77], urokinase plasminogen activator [78] and WNT/beta-catenin [79,80] pathways [70]. However, understanding of the crosstalk of multisignaling pathways as well as assemblies of key transcription factors involved in the EMT process remains incomplete [10,70]. It has been demonstrated that VPA in concentration 1 mM does not affect cancer cell proliferation, whereas significantly increases the migration and induces EMT-like properties of MCF7 luminal and MDA-MB-231 TNBC cells via upregulation of Snail and Zeb1 transcription factors expression. Moreover, knockdown of Snail and Zeb1 attenuate VPA induced cell migration and EMT process. VPA increases the Snail protein stability through suppression of its phosphorylation at serine 11 (Ser 11). VPA also increases the transcription and promoter activity of Zeb1 via HDAC2-dependent manner. HDAC2 overexpression blocks VPA-induced Zeb1 expression [63]. In line with these findings, another research group confirmed that VPA induces cell migration and EMT process in TNBC cells through a significant increase of Snail expression and downregulation of E-cadherin and GKS3β levels. Interestingly, the levels of β-catenin and AKT were reduced after VPA treatment, suggesting that AKT/GSK3β/β-catenin signaling pathway does not mediate EMT activation [57]. Increased migration and EMT are usually associated with the worst prognosis tumors; thus, the use of VPA in monotherapy of metastatic BC may be limited. The EMT is a process characteristic of solid tumors, therefore the low therapeutic effectiveness of HDIs in this type of neoplasms may be associated with the activation of EMT process.

Changes in the expression of cadherins, so-called cadherin switches, are used very often to monitor the EMT process in development and tumor progression, in particular migration and invasion potential. It has been demonstrated that VPA inhibits the proliferation and migration in a time- and dose-dependent fashion, regardless of the BC cell type. However, BC cells with the more mesenchymal phenotype (MDA-MB-468) were found to overexpress N-cadherin, whereas BC lines with an epithelial phenotype (T47D, MCF7) responded to HDI treatment through a significant increase in E-cadherin expression. Therefore, the authors conclude that HDI induction or reversal of EMT is not a universal mechanism, yet inhibition of cell migration is, and thus EMT should not be considered as the only measurement for tumor aggressiveness [64]. Taking into account the very divergent results regarding the influence of VPA on the EMT process this phenomenon should be thoroughly re-examined using the entire panel of BC cell lines and in vivo models (Table 3).

4.3. VPA Affects microRNAs (miRNAs) Expression

MicroRNAs (miRNAs) are small RNAs that suppress gene expression through their interaction with 3'untranslated region in specific target mRNAs. Non-coding RNAs (ncRNAs), including miRNAs exert critical function in the regulation of cellular processes that are involved in the EMT, as a result, some miRNAs impact cancer stemness and drug resistance. Therefore, understanding the relationship between EMT and miRNAs is beneficial to both basic research and clinical treatment [81–84]. The impact of VPA on miR-34a [85], miR-520h [81], and their target gene HDAC1 expression, as well as their involvement in the induction of apoptosis in MCF-7 and MDA-MB-231 BC cell lines, were evaluated. miRNA-34a is a well-described EMT-inhibiting miRNA [85]. In the beginning, possible target genes of miR-34a and miR-520h as well as their role in apoptosis regulation were investigated using bioinformatics analyses. Potential targets of hsa-miR-34a-5p and hsa-miR-520h-5p were in silico evaluated using predictive databases, then, a functional enrichment analysis was performed with the resulting target genes, to determine genes involved in apoptosis pathway and expressed in BC tissue. Furthermore, the interactions of the potential target genes with each other as well as with hsa-miR-34a-5p and hsa-miR-520h-5p were evaluated using STRING where 56 potential interactions between the above-mentioned miRNAs and apoptosis genes were described. It has been demonstrated that VPA increases the expression of miR-34a and miR-520h and decreases HDAC1 expression in MCF-7 cells. In turn, VPA decreased the expression of these miRNAs and increased the HDAC1 expression in MDA-MB-231 BC cells. Similarly like in MDA-MB-231 BC cells, in cancer tissue the expression of miR-34a and miR-520h significantly decreased, while the expression of HDAC1 increased after VPA treatment in vivo. This raises the possibility that VPA differently regulates the expression of the same genes, depending on BC cancer type or their molecular profiling (Table 3) [56].

Changes in the miRNA level were observed after VPA treatment also in pancreatic [86], colon [87] or thymic carcinomas [88], as well as acute myeloid leukemia [89]. VPA treatment induced expression of ErbB family members-targeting microRNAs (miR-133a, miR-133b, miR-125a, miR-125b, and miR-205) in pancreatic cancer cells without altering mRNAs levels of EGFR, ErbB2, and ErbB3 [86]. A role for these miRNAs has been demonstrated in BC. miR-133 targets YES1 proto-oncogene and inhibits the growth of TNBC cells [90]. MiR-205 suppresses the malignant behaviors of BC cells by targeting CLDN11 via modulation of the EMT [91]. It has been also demonstrated that the expression of the tumor-suppressing microRNA-125b decreased in samples of BC expressing HER2 and ER, and in TNBC [92]. In turn, treating colon cancer cells with VPA reduces the levels of precursor-miR17-92a and mature miR-92a, as well as c-Myc [87]. The upregulation of miR-92a-3p was detected in tamoxifen-resistant BC cells, suggesting that lower level of miR-92a-3p could effectively improve the therapy with this drug [93]. It was also found that VPA treatment of TC1889 cells (thymic carcinoma) led to miR-145-5p up-regulation and concomitant down-regulation of miR-145-5p target genes and exhibited antitumor effects, including cell cycle arrest and the reduction of cell viability, migration capability and, colony-forming ability [88]. miR-145-5p suppresses BC progression by inhibition of (sex-determining region Y)-box 2 (SOX2) [94], a transcription factor that is essential for maintaining self-renewal and pluripotency in BC cells [95]. VPA treatment downregulated expression of CHEK1, RAD51 as well as TYMS genes, which were identified as putative targets of miR-15a and miR-16 in acute myeloid leukemia [89]. It was demonstrated that miR-15a/miR-16 induces mitochondrial-dependent apoptosis in BC cells by suppressing oncogene BMI1 [96]. The expression and role of individual mRNAs depend on the type of tumor; thus the potential effect of VPA on individual miRNAs requires further research.

4.4. VPA and Estrogen Receptor Status

It has been demonstrated that VPA affects not only cell migration, proliferation, and cancer cell survival, but also the expression and activation of hormone receptors in BC cells observed in the pre-clinical and clinical studies [97]. Therefore, VPA can be valuable drugs in BC therapy where ER receptor is silenced by epigenetic modifications. It has been reported that ER receptor can be indirectly activated by sub-therapeutic doses of VPA. Sub-therapeutic concentrations of VPA (100 μM) can mimic estrogen and induce growth in an estrogen-depleted medium. This effect was abolished by adding an estrogen receptor antagonist. Nonetheless, therapeutic doses of VPA act via mechanisms unrelated to the stimulation of estrogen [98]. Interestingly, MDA-MB-231 ER-negative BC cells re-activate ER receptor expression and function after VPA treatment. It has been reported that VPA induced mRNA and protein expression of ER-alpha, while did not modify the level of ER-beta. Consequently, VPA increased expression of the ER-related transcription factor FoxA1, induced inhibitory effect of tamoxifen on cell growth, and caused estradiol-induced up-regulation of estrogen-regulated genes (e.g., pS2, progesterone receptor). Summarizing, VPA inducing ER-alpha and FoxA1, conferred to MDA-MB 231 cells an estrogen-sensitive phenotype, restoring their sensitivity to anti-estrogen therapy (Table 3) [66].

4.5. VPA Affects Metabolic Pathways

Metabolomics is a post-genomic research area comprising different analytical methods for small molecules analysis [99]. Metabolomics is a promising strategy to explain the pathogenesis of cancer and identify new targets for cancer diagnosis and therapy [67]. The effect of VPA on metabolites and metabolic pathways in BC cells was determined. Metabolomic analysis based on UPLC-MS/MS allowed for the identification of 3137 differential metabolites of VPA in MCF-7 cells and 2472 metabolites in MDA-MB-231 BC cells after VPA treatment. VPA particularly affected the beta-alanine, taurine, and hypotaurine metabolism pathways. Expression of furfural was up-regulated after VPA treatment in both BC cell lines. All these findings can contribute to the identification of new targets for BC treatment (Table 3) [67].

5. "Valproic Acid et al." and Breast Cancer

Despite the significant progress in the development of novel therapeutic options, standard chemotherapy of cancers still does not bring satisfactory results [58,100,101]. Chemotherapy with the use of standard cytostatics or their derivatives is limited due to many adverse effects, high toxicity to normal cells, or the occurrence of chemotolerance [102]. Therefore, combinations of established anticancer chemotherapeutics and new active agents with different mechanisms of action are being tested to increase the effectiveness of the therapy and improve clinical outcomes of oncological patients [58,103]. New active compounds approved for the treatment of BC, which effectively and selectively eliminate BC cells, and additionally enhance anticancer properties of currently used chemotherapeutics without destroying healthy tissue, are of great importance [11,59]. It has been demonstrated that new active agents not only reduce the effective doses of chemotherapeutic drugs but also sensitize cancer cells to standard cytostatics, increase the combined therapeutic activity of both active compounds and limit cytotoxic effect in relation to human normal cells through lowering the therapeutic doses of standard cytostatics by partially replacing them with new less toxic active agents. Additionally, combined therapy with new active agents can diminish multidrug-resistance (MDR) of currently used chemotherapeutic regimens [103]. In this context, many natural [103–106] and synthetic chemical compounds, including HDIs [11,58,59], have been identified and have become an interesting class of active agents for combined anti-cancer therapy.

Cisplatin (CDDP) or cis-diamminedichloroplatinum (II) is a well-known chemotherapeutic drug widely used in the therapy of numerous human cancers including lung, head and neck, bladder, ovarian, testicular, or TNBC. The mechanism of action of CDDP is linked to its ability to crosslink with the purine bases on the DNA, interfering with DNA repair mechanisms, causing DNA damage, and subsequently inducing apoptosis pathways in cancer cells. However, due to drug resistance and numerous adverse effects such as kidney insufficiency, gastrointestinal disorders, allergic reactions, hemorrhage, decreased immunity to infections, and hearing loss, the use of CDDP is limited [107]. The combination of VPA and CDDP resulted in the induction of apoptosis and cell cycle arrest in G1 phase in receptor-positive as well as TNBC cell lines in comparison to CDDP-monotherapy. CDDP with VPA applied together at a fixed-ratio of 1:1 exerted additive or additive with the tendency towards synergy interactions in the viability of MCF7 and T47D cells, respectively. In contrast, antagonistic (sub-additive) interaction was observed for the combination of CDDP with VPA in MDA-MB-231 TNBC cell line [59]. Interestingly, combined treatment of VPA and CDDP in MDA-MB-231 BC cells with the altered (increased or decreased) activity of Notch1 receptor yielded additive interaction. Therefore, VPA might be considered as potential therapeutic agents in combination therapy with CDDP against TNBC with altered Notch1 activity (Table 4) [58].

Table 4. Mechanism of action of valproic acid (VPA) and other anti-cancer drugs in combination in in vitro and in vivo breast cancer (BC) pre-clinical setting (CDDP-cisplatin, CDK-cyclin-dependent kinase, DXR-doxorubicin, 5-FU-5-fluorouracil, HER2-human epidermal growth factor receptor 2, N/A-not analyzed, PARP-poly (ADP-ribose) polymerase 1, TNBC-triple-negative breast cancer).

Drug-Drug Combination	BC-Subtype	Cell Line	Mechanism of Action	Type of Pramacological Interaction	References
VPA+CDDP	Luminal	MCF7	↑apoptosis, ↓cell viability, cell cycle arrest	additivity	[59]
		T47D		additivity with tendency towards synergism	[59]
	TNBC	MDA-MB-231	↑apoptosis, ↓cell viability, cell cycle arrest	antagonism	[59]
		MDA-MB-231 with decreased and increased Notch1 activity	↓cell viability	additivity	[58]
VPA+5-FU	TNBC	MDA-MB-468	sensitization of BC cells insensitive to 5-FU	N/A	[108]
VPA+capecitabine	Luminal, HER2-overexpressed, TNBC	MCF7, SKBR3, MDA-MB-231, MDA-MB-468	↓proliferation, ↑apoptosis, ↑thymidine phosphorylase gene and protein expression	synergism, additivity	[109]

Table 4. *Cont.*

Drug-Drug Combination	BC-Subtype	Cell Line	Mechanism of Action	Type of Pramacological Interaction	References
VPA+camptothecin	Luminal	MCF7	↑apoptosis (↓BcL-xl protein espression)	synergism (more than additivity)	[110]
VPA+DXR	TNBC	Hs578T	↓viability, ↑cytotoxicity, ↑apoptosis, ↑Cx43 protein expression	N/A	[111]
VPA+AZD2461 (PARP1 inhibitor)	Luminal	MCF7	↓viability	mild antagonism	[112]
VPA+PD-033299 (CDK inhibitor)	Luminal, HER2-overexpressed, TNBC	Panel of BC cells (MCF7, T47D, BT474, MDA-MB-361, SKBR3, HCC1143, HCC38, HCC1806, BT483, BT549, MDA-MB-435, MDA-MB-453,) and 3D cultures from pleural effusion of patients	↑apoptosis, cell cycle arrest, overexpression of CDKN1C gene	synergism	[113]

↑—increase, ↓—decrease.

5-fluorouracil (5-FU) is one of the oldest chemotherapeutic drugs routinely used in single or combined modalities with other chemotherapeutic agents in the therapy of a variety of solid tumors, including BC. Mechanism of action of 5-FU has been attributed to the production of cytotoxic metabolites that are incorporated into DNA and RNA, and inhibiting thymidylate synthase, finally leading to apoptosis and cell cycle arrest in cancer cells [114]. Resistance to 5-FU is a serious clinical problem in cancer therapy, and overcoming it is a challenge for chemotherapy. HDIs can overcome resistance to various anti-cancer drugs in vitro. It has been reported that VPA increases the sensitivity of MDA-MB-468 TNBC cells to 5-FU in 5-FU sensitive and 5-FU resistant BC cells (Table 4) [108].

Capecitabine is an oral prodrug of FU, which is approved for the treatment of metastatic BC in different settings [115]. The combined therapy of VPA and capecitabine resulted in synergistic or additive antiproliferative and pro-apoptotic effects in BC cells in vitro and in vivo. It has been demonstrated that low anticonvulsant dosage of VPA induces the time- and dose-dependent up-regulation of thymidine phosphorylase (TP) gene and its protein expression in BC cells, however, TP level remains unchanged in the non-tumorigenic MCF-10A cells. TP is a key enzyme requires for its conversion of FU to 5-FU in tumors. HDAC3 was the main isoform whose inhibition was involved in the modulation of TP activity. Thereby, the combination of VPA and capecitabine could be regarded as an innovative anti-cancer strategy for the therapy of BC (Table 4) [8,109].

Camptothecin is a naturally occurring alkaloid derived from the Camptotheca acuminate. Camptothecin forms a stable ternary complex, prevents normal DNA re-ligation, and causes the complex to collide with the replication fork, leading to a DNA double-strand break [116,117]. It has also been demonstrated that VPA and camptothecin applied together induce caspase-dependent apoptosis through modulation of anti- and pro-apoptotic gene expression and loss of the mitochondrial membrane potential in MCF7 BC cells, whereas neither compound alone could efficiently induce apoptosis. It has been demonstrated that Bcl-xL expression was induced in MCF-7 BC cells treated with camptothecin alone, in contrast to cells treated with camptothecin and VPA together. Induction of apoptosis was completely suppressed by the ectopic of Bcl-xL overexpression in MCF-7 cells. Camptothecin and Bcl-xL inactivation with using siRNA or BH3 mimetic caused efficient induction of apoptosis in these cells. The cytotoxic effect of camptothecin in combination with VPA was more than additive in MCF-7 BC cells, therefore simultaneous administration of VPA and camptothecin can be a useful strategy for therapy of BC (Table 4) [110].

Doxorubicin (DXR) is a member of the anthracycline family and is currently the most effective chemotherapeutic drug used in the treatment of early and advanced breast cancer. Unfortunately, it has been demonstrated that DXR can induce drug resistance which limits the effectiveness of the agent in single-drug treatment regimes. However, the exact mechanisms of drug resistance are still poorly understood [118,119]. Sodium valproate significantly enhanced the cytotoxicity of DXR and stimulated apoptosis induced by DXR in vitro. Exposure to sodium valproate and DXR in combination resulted in significantly increased early and late cell apoptosis rate and lowered cell viability compared with doxorubicin treatment alone. Moreover, western blotting analysis demonstrated that sodium valproate increased connexin 43 (Cx43) protein expression in Hs578T BC cells [120]. Cx43 is a prominent gap junction protein within the normal human breast tissue. Cx43 plays a tumor-suppressive role in primary tumors (Table 4) [111].

Poly (ADP-ribose) polymerase 1 (PARP1) and cyclin-dependent kinase (CDK) inhibitors. PARP1 inhibitors are newly developed anticancer active agents which target cells with defects in the homologous recombination (HR) pathway. Newly developed PARP1 inhibitor AZD2461 and VPA can effectively reduce the growth of MCF-7 BC cells with no fundamental DNA repair defect. VPA and AZD2461 applied together decreased cell viability of MCF-7 cells, where IC_{50} values for VPA and AZD2461 were 4.89 mM and 42.83 μM, respectively after 48 h of treatment with active agents. Also, the trypan blue exclusion assay results showed a time- and dose-dependent decrease in cell viability in luminal BC cells after treatment with both compounds. Unfortunately, combination analysis showed a mild antagonism between VPA and AZD2461 while γ-H2AX levels were found not to be significantly increased in MCF-7 cells treated with VPA and AZD2461 together compared to each compound alone [112]. However, it has been reported that HDAC (VPA) and cyclin-dependent kinase (CDK) inhibitors (PD-033299) show synergistic interaction in BC cells and 3D cultures from pleural effusions of patients (Table 4) [113].

6. Valproic Acid Derivatives and Drug Carriers

VPA derivatives are promising antiproliferative agents targeting the HDAC8. Unfortunately, most of these compounds are poorly soluble in water. Therefore, G4 PAMAM, four generations of polyamidoamine dendrimers, were used to improve VPA derivatives' water solubility. It has been demonstrated that G4 PAMAM dendrimers are capable of transporting weakly water-soluble aryl-VPA derivate compounds to increase their cytotoxicity against BC cell lines. VPA/CF-G4 PAMAM dendrimer complex shows anti-proliferative activity against MCF-7 and 3T3-L1, as well as MDA-MB-231 BC cells in the micro- and millimolar concentrations, respectively. Molecular docking and molecular dynamics simulations as well as HPLC-UV/VIS, MALDI-TOF, 1H NMR, and atomic force microscopy were used to evaluate the affinity of VPA, and its derivatives on G4 PAMAM, and then to establish the formation of the drug-G4 PAMAM complexes. HPLC UV/VIS experiments demonstrated an increase in the drug water solubility which was proportional to the G4 PAMAM amount [121]. Thus, chemical modification of VPA derivatives together with carrier development could provide a new treatment concept in the future.

7. Clinical Trials

Several HDIs are now being tested in BC patients in different clinical trials ranging from early phase I to randomized phase III either as single agents or in combination with approved cytostatic agents [122–127]. Despite the proven activity and high effectiveness of the use of some HDIs in hematologic malignancies [42–45], the single-agent activity seems more limited in solid tumors. Somehow, results from certain clinical trials are promising, especially those that employed VPA in combination with other chemotherapeutic drugs. Combination of VPA and epirubicin or FEC100 (5-fluorouracil, epirubicin, and cyclophosphamide), an approved regimen for BC patients, were determined [128,129]. The I phase of the study enrolled 44 patients with different solid tumors to determine the safety, toxicity, and maximum-tolerated dose of a sequence-specific combination of VPA and epirubicin. Patients were treated with increasing doses of VPA (days 1–3) followed by epirubicin (day 3) in 3-week cycles. The maximum-tolerated and recommended doses for II phase were determined (VPA 140 mg/kg/d for 48 h followed by epirubicin 100 mg/m^2). Interestingly, sustained plasma concentrations of VPA exceeding those required for in vitro synergy were achieved with acceptable toxicity. Moreover, anticancer activity of VPA and epirubicin was observed in patients with anthracycline-resistant tumors [128]. In the II phase dose expansion enrolled 15 patients with locally advanced (IIIC) or metastatic (IV) BC (14 evaluable for response). Patients in the dose-expansion group were treated with a 120 mg/kg/day VPA loading dose followed by 60 mg/kg given every 12 h for 5 doses followed by FEC100. At dose-expansion, 9 of 14 (64%) evaluable patients had an objective response. In the trial, it has been demonstrated that a combination of VPA and FEC100 has an acceptable toxicity profile and antitumor efficacy [129]. In addition to VPA, its derivatives have been tested in clinical trials. The biological and clinical efficacy as well as the safety of magnesium valproate and hydralazine (methyltransferase inhibitor) in combination with doxorubicin cyclophosphamide in the neoadjuvant treatment of locally advanced BC, were determined. BC patients were treated with 182 mg or 83 mg of hydralazine for rapid- or slow-acetylators, respectively; and 30 mg/kg of magnesium valproate, starting from the day-7 until chemotherapy ended, the latter consisting of four cycles of doxorubicin 60 mg/m^2 and cyclophosphamide 600 mg/m^2 every 21 days. Needle biopsies were taken from primary tumors at diagnosis and day 8 of treatment with valproate and hydralazine. Regarding the safety of cytotoxic chemotherapy-associated magnesium valproate and hydralazine, this treatment was well-tolerated. Drowsiness in the majority of patients was the toxicity that could be attributed to the experimental therapy with valproate, however, it was not interfering with patient functioning in daily living. Interestingly, magnesium valproate and hydralazine in combination up- and down-regulated 89 and 1091 genes at least 3-fold, respectively. The results of this study demonstrate that therapy with valproate and hydralazine is safe, that it achieves the molecular changes expected from the use of HDI and demethylating agents, and appear to increase the efficacy of conventional cytostatic drugs [130]. M.D. Anderson Cancer Center is currently recruiting participants for the phase I clinical trial to determine the side effects and the maximum tolerated doses (MTDs) and dose-limiting toxicities (DLTs) of bevacizumab and temsirolimus alone or in combination with VPA or cetuximab in patients with advanced or metastatic malignancy, including BC (III and IV stages). In the clinical trial patients receive temsirolimus intravenously (i.v.) over 30–60 min on days 1, 8, 15, and 22, bevacizumab i.v. over 30–90 min on days 1 and 15, and VPA orally (p.o.) daily on days 1–7 and 15–21. The purpose of the study is the preliminary assessment of the anti-tumor efficacy of each combination, assessment of the pharmacokinetics, pharmacodynamic markers of target inhibition, and correlates of response [131].

8. Discussion

BC is one of the leading causes of cancer-related death among women worldwide. A significant challenge in treating BC is the limited array of therapeutic options and the rapid development of resistance against currently used agents, especially in TNBC, the most aggressive subtype of BC [97]. The idea of treating BC patients with active agents able to re-establish expression of tumor suppressor genes silenced by epigenetic mechanisms is currently being tested [132]. Up-regulated HDAC activity is associated with a closed chromatin assembly and subsequent gene repression, forming a characteristic feature of malignantly transformed cells [133,134]. Histone acetylation prevents chromatin condensation as well as makes promoters and other control elements of chromatin more accessible to different transcription factors, and thus seems to be the most important mechanism in HDIs anticancer action [135]. HDIs as the epigenetic modifiers have pleiotropic effects on many biological processes such as cancer cell growth arrest, proliferation, differentiation, angiogenesis, invasion, metastasis or immunogenicity [135]. VPA is a clinically available HDI that notably increases apoptosis, induces cell cycle arrest, and abolishes drug resistance in BC cells in vitro as well as decreases tumor growth and metastases in animal models [135]. However, divergent data on the effects of VPA on different signaling pathways, including the EMT process or miRNA pathways limit the use of this compound as a single agent in the therapy of BC, although VPA is currently still prescribed worldwide as a well-tolerated, relatively safe and effective anticonvulsant and mood stabilizer. Even though other HDIs have demonstrated more promising antitumor effects, VPA was investigated for anti-cancer activity based on its low toxicity profile and availability. It has been shown excellent tolerability of VPA within the serum range of 50–100 μg/mL based on experience from its use as an antiepileptic agent [136]. VPA was rapidly absorbed after oral administration, with peak serum levels occurring approximately 1–4 h after a single oral dose [135]. Due to the fact, that VPA has been used in clinical practice in nontoxic therapeutic concentrations in many seizure types and syndromes, and remains a mainstay for treatment of epilepsy of all age groups except for infants, as well as mania in bipolar disorders, migraine prophylaxys, neuropathic pain and schizophrenia, for more than four decades, its pharmacokinetic profile, side effects and toxicity are thus well documented [135]. Moreover, therapy with VPA is widespread, relatively cheap, and available. Since VPA, as a psychoneurological drug, crossing the blood-brain barrier, it could also effectively eliminate metastatic BC cells in the brain of patients. There is currently no effective therapy for treating metastatic TNBC in the brain. The diagnosis of an oncological disease may cause a serious psychological crisis, and the use of VPA in the treatment of BC could simultaneously reduce the symptoms of neuropsychiatric diseases. VPA is generally well tolerated by patients. However, neurological side effects such as dizziness, sedation, and tremor as well as mild gastrointestinal toxicities usually occur early during treatment. Fatal hepatotoxicity is very rare and mainly occurs in children aged less than 2 years who are treated with multiple drugs [137]. VPA is also a known human teratogen and its prescription during pregnancy (especially in the first trimester) may cause multiple birth defects that are overall designated as fetal valproate syndrome. The major congenital abnormalities are neural tube defects, facial dysmorphism, growth retardation, delay in postnatal cognitive development, and autism [135]. Despite the side effects of VPA, especially in pregnant women, clinical trials are still ongoing for its potential application in the treatment of several types of cancers, including solid and non-solid tumors [138]. While VPA administered alone demonstrated an anti-cancer effect in the pre-clinical setting, little improvement was observed with VPA in monotherapy in the clinical setting. These findings suggest that VPA needs to be combined with hormonal therapy agents or traditional chemotherapy agents in the BC setting in the future. The promising pre-clinical data suggest that VPA can be repurposed as an adjunctive agent in combination with many cytotoxic, hormonal, and immunotherapeutic agents for the treatment of BC [97]. The effect of VPA in monotherapy on induction of apoptosis, inhibition of the cell cycle, EMT or miRNA pathways do not differ significantly between the histological subtypes of BC. However, it has been demonstrated that HER2-overexpressed BC cells are more sensitive to VPA than HER2-negative. It is known that the anti-proliferative mechanism of VPA in BC cells is related to their HER2-expression status. Therefore, VPA may synergize with drugs used in the therapy of HER2-overexpressed BC, like anti-HER2 monoclonal antibodies or anthracycline and taxane-based neoadjuvant chemotherapy. Interestingly, a different therapeutic effect was observed in the VPA-combined therapy depending on the type of BC. It has been demonstrated that the anticancer effect of VPA in combination with other active agents, is highly cell-type specific. Additivity or additivity with a tendency towards synergism was demonstrated between VPA and CDDP in luminal BC cells, while antagonism was evidenced in TNBC cells between the same drug combination. Interestingly, changes in Notch1 activity in MDA-MB-231 TNBC cells caused additive interaction. Therefore, the therapy with VPA and CDDP can be a promising regimen in patients with the most aggressive type of BC-TNBC with increased Notch1 activity. Synergistic type of pharmacological interaction was also demonstrated between VPA and capecitabine or PD-033299 (CDK inhibitor) in luminal, HER2-overexpressed, and TNBC cells as well as with camptothecine in luminal BC cells. In contrast, mild antagonism was evidenced in luminal BC cells treated with VPA in combination with AZD2461 (PARP1 inhibitor).

However, further investigations are warranted to evaluate the efficacy and to provide optimal treatment.

9. Conclusions

Despite significant progress in the therapy of BC patients, serious side effects, as well as high toxicity of standard chemotherapeutics to normal cells limit the effectiveness of the therapy. Moreover, the existence of de novo drug resistance (tumor does not respond to treatment since the beginning of therapy) or acquired resistance (response to the drug disappears over time) support the failure of standard therapies and do not bring satisfactory results [139–141]. Thus, a new generation of cytostatics effective in the treatment of BC is being sought, the use of which will not only reduce the doses of standard chemotherapeutics but also eliminate the phenomenon of drug resistance. Clinical trials investigating new targeted drugs as well as therapeutic combinations with their use have led to significant advances in BC therapy [7]. Epigenetic regulation of histone and non-histone proteins may be a novel approach and hold significant progress for the successful treatment of BC. HDIs are a promising class of anti-neoplastic agents that induce differentiation and apoptosis in many types of cancer cells, including breast carcinoma cells [142]. VPA is a clinically available HDI that notably inhibits migration, increases apoptosis, cell cycle arrest, and abolishes drug resistance [108] in BC cells. Unfortunately, the IC_{50} doses of VPA are relatively high compared to other HDIs in in vitro studies [59]. Moreover, divergent data on the effects of VPA on different signaling pathways, including the EMT process or miRNA pathways limit the use of this compound in monotherapy of BC. Combined therapy with the use of VPA and standard cytostatic drugs to reduce the doses of VPA and limit the adverse effects caused by standard chemotherapeutics seems to be a promising strategy in the future. Unfortunately, data on the VPA activity in combination with other anti-cancer drugs from the in vivo models is still not sufficient. Therefore, results obtained from in vitro studies should be thoroughly validated in in vivo models. Results from in vivo experiments might offer a rationale for clinical studies of a new combined therapy, to improve the clinical outcome of patients with BC.

Author Contributions: Conceptualization, A.W. and A.S.; methodology, A.W., M.H., E.O., W.K.-K.; software, A.W.; resources, A.W., M.H., E.O., W.K.-K.; writing—original draft preparation, A.W., M.H., E.O., W.K.-K.; writing—review and editing, A.S.; visualization, A.W.; supervision, A.S.; project administration, A.W.; funding acquisition, A.W., M.H., A.S. All authors have read and agreed to the published version of the manuscript.

Funding: This research was funded by Medical University of Lublin, grant number DS440/2020 and DS440/2021-2022; The Iwanowska Programme, The Polish National Agency for Academic Ex-change, grant numbers PPN/IWA/2018/1/00005 and PPN/IWA/2019/1/00160.

Acknowledgments: The authors thank Agnieszka Styczynska for the editorial assistance and proof readings. All figures were created with www.biorender.com (accessed on 25 April 2021).

Conflicts of Interest: The authors declare no conflict of interest.

References

1. Bray, F.; Ferlay, J.; Soerjomataram, I.; Siegel, R.L.; Torre, L.A.; Jemal, A. Global cancer statistics 2018: GLOBOCAN estimates of incidence and mortality worldwide for 36 cancers in 185 countries. *CA Cancer J. Clin.* **2018**, *68*, 394–424. [CrossRef] [PubMed]
2. Momenimovahed, Z.; Salehiniya, H. Epidemiological characteristics of and risk factors for breast cancer in the world. *Breast Cancer Targets Ther.* **2019**, *11*, 151–164. [CrossRef] [PubMed]
3. The Lancet GLOBOCAN 2018: Counting the toll of cancer. *Lancet* **2018**, *392*, 985. [CrossRef]
4. Ferlay, J.; Colombet, M.; Soerjomataram, I.; Mathers, C.; Parkin, D.M.; Piñeros, M.; Znaor, A.; Bray, F. Estimating the global cancer incidence and mortality in 2018: GLOBOCAN sources and methods. *Int. J. Cancer* **2019**, *144*, 1941–1953. [CrossRef] [PubMed]
5. Janni, W. Targeted Therapy of Breast Cancer. *Oncol. Res. Treat.* **2016**, *39*, 100–101. [CrossRef]
6. Nagini, S. Breast Cancer: Current Molecular Therapeutic Targets and New Players. *Anticancer Agents Med. Chem.* **2017**, *17*, 152–163. [CrossRef]
7. Cai, F.F.; Kohler, C.; Zhang, B.; Wang, M.H.; Chen, W.J.; Zhong, X.Y. Epigenetic therapy for breast cancer. *Int. J. Mol. Sci.* **2011**, *12*, 4465–4476. [CrossRef]

8. Damaskos, C.; Valsami, S.; Kontos, M.; Spartalis, E.; Kalampokas, T.; Kalampokas, E.; Athanasiou, A.; Moris, D.; Daskalopoulou, A.; Davakis, S.; et al. Histone deacetylase inhibitors: An attractive therapeutic strategy against breast cancer. *Anticancer Res.* **2017**, *37*, 35–46. [CrossRef]
9. Gediya, P.; Parikh, P.K.; Vyas, V.K.; Ghate, M.D. Histone deacetylase 2: A potential therapeutic target for cancer and neurodegenerative disorders. *Eur. J. Med. Chem.* **2021**, *216*. [CrossRef] [PubMed]
10. Wawruszak, A.; Kalafut, J.; Okon, E.; Czapinski, J.; Halasa, M.; Przybyszewska, A.; Miziak, P.; Okla, K.; Rivero-Muller, A.; Stepulak, A. Histone deacetylase inhibitors and phenotypical transformation of cancer cells. *Cancers* **2019**, *11*, 148. [CrossRef]
11. Gumbarewicz, E.; Luszczki, J.J.; Wawruszak, A.; Dmoszynska-Graniczka, M.; Grabarska, A.J.; Jarzab, A.M.; Polberg, K.; Stepulak, A. Isobolographic analysis demonstrates additive effect of cisplatin and HDIs combined treatment augmenting their anti-cancer activity in lung cancer cell lines. *Am. J. Cancer Res.* **2016**, *6*, 2831–2845.
12. Dai, X.; Li, T.; Bai, Z.; Yang, Y.; Liu, X.; Zhan, J.; Shi, B. Breast cancer intrinsic subtype classification, clinical use and future trends. *Am. J. Cancer Res.* **2015**, *5*, 2929–2943. [PubMed]
13. Szymiczek, A.; Lone, A.; Akbari, M.R. Molecular intrinsic versus clinical subtyping in breast cancer: A comprehensive review. *Clin. Genet.* **2021**, *99*, 613–637. [CrossRef]
14. McDonald, E.S.; Clark, A.S.; Tchou, J.; Zhang, P.; Freedman, G.M. Clinical diagnosis and management of breast cancer. *J. Nucl. Med.* **2016**, *57*, 9S–16S. [CrossRef] [PubMed]
15. Bonacho, T.; Rodrigues, F.; Liberal, J. Immunohistochemistry for diagnosis and prognosis of breast cancer: A review. *Biotech. Histochem.* **2020**, *95*, 71–91. [CrossRef]
16. Cheang, M.C.U.; Chia, S.K.; Voduc, D.; Gao, D.; Leung, S.; Snider, J.; Watson, M.; Davies, S.; Bernard, P.S.; Parker, J.S.; et al. Ki67 index, HER2 status, and prognosis of patients with luminal B breast cancer. *J. Natl. Cancer Inst.* **2009**, *101*, 736–750. [CrossRef]
17. Smid, M.; Wang, Y.; Zhang, Y.; Sieuwerts, A.M.; Yu, J.; Klijn, J.G.M.; Foekens, J.A.; Martens, J.W.M. Subtypes of breast cancer show preferential site of relapse. *Cancer Res.* **2008**, *68*, 3108–3114. [CrossRef] [PubMed]
18. Gong, Y.; Liu, Y.R.; Ji, P.; Hu, X.; Shao, Z.M. Impact of molecular subtypes on metastatic breast cancer patients: A SEER population-based study. *Sci. Rep.* **2017**, *7*. [CrossRef]
19. Perou, C.M.; Sørile, T.; Eisen, M.B.; Van De Rijn, M.; Jeffrey, S.S.; Ress, C.A.; Pollack, J.R.; Ross, D.T.; Johnsen, H.; Akslen, L.A.; et al. Molecular portraits of human breast tumours. *Nature* **2000**, *406*, 747–752. [CrossRef] [PubMed]
20. Sørlie, T.; Perou, C.M.; Tibshirani, R.; Aas, T.; Geisler, S.; Johnsen, H.; Hastie, T.; Eisen, M.B.; Van De Rijn, M.; Jeffrey, S.S.; et al. Gene expression patterns of breast carcinomas distinguish tumor subclasses with clinical implications. *Proc. Natl. Acad. Sci. USA* **2001**, *98*, 10869–10874. [CrossRef]
21. Van Ramshorst, M.S.; van der Voort, A.; van Werkhoven, E.D.; Mandjes, I.A.; Kemper, I.; Dezentjé, V.O.; Oving, I.M.; Honkoop, A.H.; Tick, L.W.; van de Wouw, A.J.; et al. Neoadjuvant chemotherapy with or without anthracyclines in the presence of dual HER2 blockade for HER2-positive breast cancer (TRAIN-2): A multicentre, open-label, randomised, phase 3 trial. *Lancet Oncol.* **2018**, *19*, 1630–1640. [CrossRef]
22. Harbeck, N.; Gluz, O. Neoadjuvant therapy for triple negative and HER2-positive early breast cancer. *Breast* **2017**, *34*, S99–S103. [CrossRef]
23. Wuerstlein, R.; Harbeck, N. Neoadjuvant Therapy for HER2-positive Breast Cancer. *Rev. Recent Clin. Trials* **2017**, *12*, 81–92. [CrossRef] [PubMed]
24. Maximiano, S.; Magalhães, P.; Guerreiro, M.P.; Morgado, M. Trastuzumab in the Treatment of Breast Cancer. *BioDrugs* **2016**, *30*, 75–86. [CrossRef]
25. Cameron, D.; Piccart-Gebhart, M.J.; Gelber, R.D.; Procter, M.; Goldhirsch, A.; de Azambuja, E.; Castro, G.; Untch, M.; Smith, I.; Gianni, L.; et al. 11 years' follow-up of trastuzumab after adjuvant chemotherapy in HER2-positive early breast cancer: Final analysis of the HERceptin Adjuvant (HERA) trial. *Lancet* **2017**, *389*, 1195–1205. [CrossRef]
26. Drooger, J.C.; van Tinteren, H.; de Groot, S.M.; ten Tije, A.J.; de Graaf, H.; Portielje, J.E.A.; Jager, A.; Honkoop, A.; Linn, S.C.; Kroep, J.R.; et al. A randomized phase 2 study exploring the role of bevacizumab and a chemotherapy-free approach in HER2-positive metastatic breast cancer: The HAT study (BOOG 2008–2003), a Dutch Breast Cancer Research Group trial. *Cancer* **2016**, *122*, 2961–2970. [CrossRef] [PubMed]
27. Nagata, Y.; Lan, K.H.; Zhou, X.; Tan, M.; Esteva, F.J.; Sahin, A.A.; Klos, K.S.; Li, P.; Monia, B.P.; Nguyen, N.T.; et al. PTEN activation contributes to tumor inhibition by trastuzumab, and loss of PTEN predicts trastuzumab resistance in patients. *Cancer Cell* **2004**, *6*, 117–127. [CrossRef]
28. Marotti, J.D.; de Abreu, F.B.; Wells, W.A.; Tsongalis, G.J. Triple-Negative Breast Cancer: Next-Generation Sequencing for Target Identification. *Am. J. Pathol.* **2017**, *187*, 2133–2138. [CrossRef]
29. Jitariu, A.A.; Cîmpean, A.M.; Ribatti, D.; Raica, M. Triple negative breast cancer: The kiss of death. *Oncotarget* **2017**, *8*, 46652–46662. [CrossRef] [PubMed]
30. Jhan, J.R.; Andrechek, E.R. Triple-negative breast cancer and the potential for targeted therapy. *Pharmacogenomics* **2017**, *18*, 1595–1609. [CrossRef] [PubMed]
31. Chun, K.H.; Park, J.H.; Fan, S. Predicting and overcoming chemotherapeutic resistance in breast cancer. In *Advances in Experimental Medicine and Biology*; Springer LLC: New York, NY, USA, 2017; Volume 1026, pp. 59–104.
32. Reilly, D.O.; al Sendi, M.; Kelly, C.M. Overview of recent advances in metastatic triple negative breast cancer. *World J. Clin. Oncol.* **2021**, *12*, 164–182. [CrossRef]

33. Mezi, S.; Botticelli, A.; Pomati, G.; Cerbelli, B.; Scagnoli, S.; Amirhassankhani, S.; D'amati, G.; Marchetti, P. Standard of care and promising new agents for the treatment of mesenchymal triple-negative breast cancer. *Cancers* **2021**, *13*, 1080. [CrossRef] [PubMed]
34. Loibl, S.; Poortmans, P.; Morrow, M.; Denkert, C.; Curigliano, G. Breast cancer. *Lancet* **2021**. [CrossRef]
35. Liu, D.; Vadgama, J.; Wu, Y. Basal-like breast cancer with low TGFβ and high TNFα pathway activity is rich in activated memory CD4 T cells and has a good prognosis. *Int. J. Biol. Sci.* **2021**, *17*, 670–682. [CrossRef]
36. Cejalvo, J.M.; De Dueñas, E.M.; Galván, P.; García-Recio, S.; Gasión, O.B.; Paré, L.; Antolín, S.; Martinello, R.; Blancas, I.; Adamo, B.; et al. Intrinsic subtypes and gene expression profiles in primary and metastatic breast cancer. *Cancer Res.* **2017**, *77*, 2213–2221. [CrossRef] [PubMed]
37. Liu, Z.; Zhang, X.S.; Zhang, S. Breast tumor subgroups reveal diverse clinical prognostic power. *Sci. Rep.* **2014**, *4*, 1–9. [CrossRef] [PubMed]
38. Riaz, S.K.; Saeed, M.; Malik, M.F.A. Clinical and therapeutic implications of histone acetylation in breast cancer. *West Indian Med. J.* **2016**, *65*, 337–344.
39. Guo, P.; Chen, W.; Li, H.; Li, M.; Li, L. The Histone Acetylation Modifications of Breast Cancer and their Therapeutic Implications. *Pathol. Oncol. Res.* **2018**, *24*, 807–813. [CrossRef] [PubMed]
40. Ediriweera, M.K.; Tennekoon, K.H.; Samarakoon, S.R. Emerging role of histone deacetylase inhibitors as anti-breast-cancer agents. *Drug Discov. Today* **2019**, *24*, 685–702. [CrossRef]
41. Garmpis, N.; Damaskos, C.; Garmpi, A.; Kalampokas, E.; Kalampokas, T.; Spartalis, E.; Daskalopoulou, A.; Valsami, S.; Kontos, M.; Nonni, A.; et al. Histone deacetylases as new therapeutic targets in triple-negative breast cancer: Progress and promises. *Cancer Genom. Proteom.* **2017**, *14*, 299–313.
42. Mann, B.S.; Johnson, J.R.; Cohen, M.H.; Justice, R.; Pazdur, R. FDA Approval Summary: Vorinostat for Treatment of Advanced Primary Cutaneous T-Cell Lymphoma. *Oncologist* **2007**, *12*, 1247–1252. [CrossRef] [PubMed]
43. Eleutherakis-Papaiakovou, E.; Kanellias, N.; Kastritis, E.; Gavriatopoulou, M.; Terpos, E.; Dimopoulos, M.A. Efficacy of Panobinostat for the Treatment of Multiple Myeloma. *J. Oncol.* **2020**. [CrossRef]
44. Iyer, S.P.; Foss, F.F. Romidepsin for the Treatment of Peripheral T-Cell Lymphoma. *Oncologist* **2015**, *20*, 1084–1091. [CrossRef]
45. Lee, H.Z.; Kwitkowski, V.E.; Del Valle, P.L.; Ricci, M.S.; Saber, H.; Habtemariam, B.A.; Bullock, J.; Bloomquist, E.; Shen, Y.L.; Chen, X.H.; et al. FDA approval: Belinostat for the treatment of patients with relapsed or refractory peripheral T-cell lymphoma. *Clin. Cancer Res.* **2015**, *21*, 2666–2670. [CrossRef] [PubMed]
46. Glozak, M.A.; Sengupta, N.; Zhang, X.; Seto, E. Acetylation and deacetylation of non-histone proteins. *Gene* **2005**, *363*, 15–23. [CrossRef] [PubMed]
47. Spange, S.; Wagner, T.; Heinzel, T.; Krämer, O.H. Acetylation of non-histone proteins modulates cellular signalling at multiple levels. *Int. J. Biochem. Cell Biol.* **2009**, *41*, 185–198. [CrossRef] [PubMed]
48. Wang, Z.; Leng, Y.; Tsai, L.K.; Leeds, P.; Chuang, D.M. Valproic acid attenuates blood-brain barrier disruption in a rat model of transient focal cerebral ischemia: The roles of HDAC and MMP-9 inhibition. *J. Cereb. Blood Flow Metab.* **2011**, *31*, 52–57. [CrossRef] [PubMed]
49. Jang, E.R.; Lim, S.J.; Lee, E.S.; Jeong, G.; Kim, T.Y.; Bang, Y.J.; Lee, J.S. The histone deacetylase inhibitor trichostatin a sensitizes estrogen receptor α-negative breast cancer cells to tamoxifen. *Oncogene* **2004**, *23*, 1724–1736. [CrossRef]
50. Chen, S.Y.; Zheng, X.W.; Cai, J.X.; Zhang, W.P.; You, H.S.; Xing, J.F.; Dong, Y.L. Histone deacetylase inhibitor reverses multidrug resistance by attenuating the nucleophosmin level through PI3K/Akt pathway in breast cancer. *Int. J. Oncol.* **2016**, *49*, 294–304. [CrossRef]
51. Lipska, K.; Gumieniczek, A.; Filip, A.A. Anticonvulsant valproic acid and other short-chain fatty acids as novel anticancer therapeutics: Possibilities and challenges. *Acta Pharm.* **2020**, *70*, 291–301. [CrossRef]
52. Romoli, M.; Mazzocchetti, P.; D'Alonzo, R.; Siliquini, S.; Rinaldi, V.E.; Verrotti, A.; Calabresi, P.; Costa, C. Valproic Acid and Epilepsy: From Molecular Mechanisms to Clinical Evidences. *Curr. Neuropharmacol.* **2018**, *17*, 926–946. [CrossRef] [PubMed]
53. Sanaei, M.; Kavoosi, F. Effect of Valproic Acid on the Class I Histone Deacetylase 1, 2 and 3, Tumor Suppressor Genes p21WAF1/CIP1 and p53, and Intrinsic Mitochondrial Apoptotic Pathway, Pro- (Bax, Bak, and Bim) and anti- (Bcl-2, Bcl-xL, and Mcl-1) Apoptotic Genes Expression, Ce. *Asian Pacific J. Cancer Prev.* **2021**, *22*, 89–95. [CrossRef] [PubMed]
54. Pang, B.; Zhang, J.; Zhang, X.; Yuan, J.; Shi, Y.; Qiao, L. Inhibition of lipogenesis and induction of apoptosis by valproic acid in prostate cancer cells via the C/EBPα/SREBP-1 pathway. *Acta Biochim. Biophys. Sin.* **2021**, *53*, 354–364. [CrossRef]
55. Jahani, M.; Khanahmad, H.; Nikpour, P. Evaluation of the Effects of Valproic Acid Treatment on Cell Survival and Epithelial-Mesenchymal Transition-Related Features of Human Gastric Cancer Cells. *J. Gastrointest. Cancer* **2020**. [CrossRef]
56. Injinari, N.; Amini-Farsani, Z.; Yadollahi-Farsani, M.; Teimori, H. Apoptotic effects of valproic acid on miR-34a, miR-520h and HDAC1 gene in breast cancer. *Life Sci.* **2021**, *269*, 119027. [CrossRef]
57. Ozman, Z.; Ozbek Iptec, B.; Sahin, E.; Guney Eskiler, G.; Deveci Ozkan, A.; Kaleli, S. Regulation of valproic acid induced EMT by AKT/GSK3β/β-catenin signaling pathway in triple negative breast cancer. *Mol. Biol. Rep.* **2021**, *48*, 1335–1343. [CrossRef] [PubMed]
58. Wawruszak, A.; Luszczki, J.J.; Kalafut, J.; Okla, K.; Halasa, M.; Rivero-Muller, A.; Stepulak, A. Additive pharmacological interaction between cisplatin (CDDP) and histone deacetylase inhibitors (HDIs) in MDA-MB-231 triple negative breast cancer (TNBC) cells with altered notch1 activity—an isobolographic analysis. *Int. J. Mol. Sci.* **2019**, *20*, 3663. [CrossRef]

59. Wawruszak, A.; Luszczki, J.J.; Grabarska, A.; Gumbarewicz, E.; Dmoszynska-Graniczka, M.; Polberg, K.; Stepulak, A. Assessment of interactions between cisplatin and two histone deacetylase inhibitors in MCF7, T47D and MDA-MB-231 human breast cancer cell lines—An isobolographic analysis. *PLoS ONE* **2015**, *10*, e0143013. [CrossRef]
60. Fortunati, N.; Bertino, S.; Costantino, L.; Bosco, O.; Vercellinatto, I.; Catalano, M.G.; Boccuzzi, G. Valproic acid is a selective antiproliferative agent in estrogen-sensitive breast cancer cells. *Cancer Lett.* **2008**, *259*, 156–164. [CrossRef]
61. Aztopal, N.; Erkisa, M.; Erturk, E.; Ulukaya, E.; Tokullugil, A.H.; Ari, F. Valproic acid, a histone deacetylase inhibitor, induces apoptosis in breast cancer stem cells. *Chem. Biol. Interact.* **2018**, *280*, 51–58. [CrossRef]
62. Mawatari, T.; Ninomiya, I.; Inokuchi, M.; Harada, S.; Hayashi, H.; Oyama, K.; Makino, I.; Nakagawara, H.; Miyashita, T.; Tajima, H.; et al. Valproic acid inhibits proliferation of HER2-expressing breast cancer cells by inducing cell cycle arrest and apoptosis through Hsp70 acetylation. *Int. J. Oncol.* **2015**, *47*, 2073–2081. [CrossRef] [PubMed]
63. Zhang, S.; Tang, Z.; Qing, B.; Tang, R.; Duan, Q.; Ding, S.; Deng, D. Valproic acid promotes the epithelial-to-mesenchymal transition of breast cancer cells through stabilization of Snail and transcriptional upregulation of Zeb1. *Eur. J. Pharmacol.* **2019**, *865*. [CrossRef]
64. Wawruszak, A.; Gumbarewicz, E.; Okon, E.; Jeleniewicz, W.; Czapinski, J.; Halasa, M.; Okla, K.; Smok-Kalwat, J.; Bocian, A.; Rivero-Muller, A.; et al. Histone deacetylase inhibitors reinforce the phenotypical markers of breast epithelial or mesenchymal cancer cells but inhibit their migratory properties. *Cancer Manag. Res.* **2019**, *11*, 8345–8358. [CrossRef]
65. Li, G.F.; Qian, T.L.; Li, G.S.; Yang, C.X.; Qin, M.; Huang, J.; Sun, M.; Han, Y.Q. Sodium valproate inhibits MDA-MB-231 breast cancer cell migration by upregulating NM23H1 expression. *Genet. Mol. Res.* **2012**, *11*, 77–86. [CrossRef]
66. Fortunati, N.; Bertino, S.; Costantino, L.; De Bortoli, M.; Compagnone, A.; Bandino, A.; Catalano, M.G.; Boccuzzi, G. Valproic acid restores ERα and antiestrogen sensitivity to ERα-negative breast cancer cells. *Mol. Cell. Endocrinol.* **2010**, *314*, 17–22. [CrossRef]
67. Zhou, X.; Li, Z.; Wang, X.; Jiang, G.; Shan, C.; Liu, S. Metabolomics reveals the effect of valproic acid on MCF-7 and MDA-MB-231 cells. *Xenobiotica* **2020**, *50*, 252–260. [CrossRef] [PubMed]
68. Han, W.; Zhang, C.; Cao, F.-Y.; Cao, F.; Jiang, L.; Ding, H.Z. Prognostic and clinicopathological value of NM23 expression in patients with breast cancer: A systematic review and meta-analysis. *Curr. Probl. Cancer* **2017**, *41*, 80–93. [CrossRef]
69. Mátyási, B.; Farkas, Z.; Kopper, L.; Sebestyén, A.; Boissan, M.; Mehta, A.; Takács-Vellai, K. The Function of NM23-H1/NME1 and Its Homologs in Major Processes Linked to Metastasis. *Pathol. Oncol. Res.* **2020**, *26*, 49–61. [CrossRef]
70. Das, V.; Bhattacharya, S.; Chikkaputtaiah, C.; Hazra, S.; Pal, M. The basics of epithelial–mesenchymal transition (EMT): A study from a structure, dynamics, and functional perspective. *J. Cell. Physiol.* **2019**, *234*, 14535–14555. [CrossRef] [PubMed]
71. Chaffer, C.L.; San Juan, B.P.; Lim, E.; Weinberg, R.A. EMT, cell plasticity and metastasis. *Cancer Metastasis Rev.* **2016**, *35*, 645–654. [CrossRef] [PubMed]
72. Kotiyal, S.; Bhattacharya, S. Breast cancer stem cells, EMT and therapeutic targets. *Biochem. Biophys. Res. Commun.* **2014**, *453*, 112–116. [CrossRef]
73. Song, K.; Farzaneh, M. Signaling pathways governing breast cancer stem cells behavior. *Stem Cell Res. Ther.* **2021**, *12*, 245. [CrossRef]
74. Kaszak, I.; Witkowska-Piłaszewicz, O.; Niewiadomska, Z.; Dworecka-Kaszak, B.; Toka, F.N.; Jurka, P. Role of cadherins in cancer—A review. *Int. J. Mol. Sci.* **2020**, *21*, 7624. [CrossRef] [PubMed]
75. Kar, R.; Jha, N.K.; Jha, S.K.; Sharma, A.; Dholpuria, S.; Asthana, N.; Chaurasiya, K.; Singh, V.K.; Burgee, S.; Nand, P. A "NOTCH" deeper into the epithelial-to-mesenchymal transition (EMT) program in breast cancer. *Genes* **2019**, *10*, 961. [CrossRef] [PubMed]
76. Nawshad, A.; Lagamba, D.; Hay, E.D. Transforming growth factor β (TGFβ) signalling in palatal growth, apoptosis and epithelial mesenchymal transformation (EMT). *Arch. Oral Biol.* **2004**, *49*, 675–689. [CrossRef] [PubMed]
77. Santibanez, J.F.; Obradović, H.; Kukolj, T.; Krstić, J. Transforming growth factor-β, matrix metalloproteinases, and urokinase-type plasminogen activator interaction in the cancer epithelial to mesenchymal transition. *Dev. Dyn.* **2018**, *247*, 382–395. [CrossRef] [PubMed]
78. Jaiswal, R.K.; Varshney, A.K.; Yadava, P.K. Diversity and functional evolution of the plasminogen activator system. *Biomed. Pharmacother.* **2018**, *98*, 886–898. [CrossRef]
79. Makena, M.R.; Gatla, H.; Verlekar, D.; Sukhavasi, S.; Pandey, M.K.; Pramanik, K.C. Wnt/β-catenin signaling: The culprit in pancreatic carcinogenesis and therapeutic resistance. *Int. J. Mol. Sci.* **2019**, *20*, 4242. [CrossRef] [PubMed]
80. Lei, Y.; Chen, L.; Zhang, G.; Shan, A.; Ye, C.; Liang, B.; Sun, J.; Liao, X.; Zhu, C.; Chen, Y.; et al. MicroRNAs target the Wnt/β-catenin signaling pathway to regulate epithelial-mesenchymal transition in cancer. *Oncol. Rep.* **2020**, *44*, 1299–1313. [CrossRef]
81. Su, C.M.; Wang, M.Y.; Hong, C.C.; Chen, H.A.; Su, Y.H.; Wu, C.H.; Huang, M.T.; Chang, Y.W.; Jiang, S.S.; Sung, S.Y.; et al. MiR-520h is crucial for DAPK2 regulation and breast cancer progression. *Oncogene* **2016**, *35*, 1134–1142. [CrossRef]
82. Hussen, B.M.; Shoorei, H.; Mohaqiq, M.; Dinger, M.E.; Hidayat, H.J.; Taheri, M.; Ghafouri-Fard, S. The Impact of Non-coding RNAs in the Epithelial to Mesenchymal Transition. *Front. Mol. Biosci.* **2021**, *8*. [CrossRef]
83. Pan, G.; Liu, Y.; Shang, L.; Zhou, F.; Yang, S. EMT-associated microRNAs and their roles in cancer stemness and drug resistance. *Cancer Commun.* **2021**, *41*, 199–217. [CrossRef] [PubMed]
84. Dong, B.; Li, S.; Zhu, S.; Yi, M.; Luo, S.; Wu, K. MiRNA-mediated EMT and CSCs in cancer chemoresistance. *Exp. Hematol. Oncol.* **2021**, *10*. [CrossRef] [PubMed]

85. Lim, D.; Cho, J.G.; Yun, E.; Lee, A.; Ryu, H.Y.; Lee, Y.J.; Yoon, S.; Chang, W.; Lee, M.S.; Kwon, B.S.; et al. MicroRNA 34A–AXL axis regulates vasculogenic mimicry formation in breast cancer cells. *Genes* **2021**, *12*, 9. [CrossRef] [PubMed]
86. Lin, T.; Ren, Q.; Zuo, W.; Jia, R.; Xie, L.; Lin, R.; Zhao, H.; Chen, J.; Lei, Y.; Wang, P.; et al. Valproic acid exhibits anti-tumor activity selectively against EGFR/ErbB2/ErbB3-coexpressing pancreatic cancer via induction of ErbB family members-targeting microRNAs. *J. Exp. Clin. Cancer Res.* **2019**, *38*. [CrossRef]
87. Hu, S.; Liu, L.; Chang, E.B.; Wang, J.Y.; Raufman, J.P. Butyrate inhibits pro-proliferative miR-92a by diminishing c-Myc-induced miR-17-92a cluster transcription in human colon cancer cells. *Mol. Cancer* **2015**, *14*. [CrossRef]
88. Bellissimo, T.; Ganci, F.; Gallo, E.; Sacconi, A.; Tito, C.; De Angelis, L.; Pulito, C.; Masciarelli, S.; Diso, D.; Anile, M.; et al. Thymic Epithelial Tumors phenotype relies on miR-145-5p epigenetic regulation. *Mol. Cancer* **2017**, *16*. [CrossRef]
89. Rücker, F.G.; Lang, K.M.; Fütterer, M.; Komarica, V.; Schmid, M.; Döhner, H.; Schlenk, R.F.; Döhner, K.; Knudsen, S.; Bullinger, L. Molecular dissection of valproic acid effects in acute myeloid leukemia identifies predictive networks. *Epigenetics* **2016**, *11*, 517–525. [CrossRef] [PubMed]
90. Zhang, G.; Wang, J.; Zheng, R.; Song, B.; Huang, L.; Liu, Y.; Hao, Y.; Bai, X. MiR-133 Targets YES1 and Inhibits the Growth of Triple-Negative Breast Cancer Cells. *Technol. Cancer Res. Treat.* **2020**, *19*. [CrossRef]
91. Shen, Y.; Xu, Y.; Huang, L.; Chi, Y.; Meng, L. MiR-205 suppressed the malignant behaviors of breast cancer cells by targeting CLDN11 via modulation of the epithelial-to-mesenchymal transition. *Aging* **2021**, *13*. [CrossRef]
92. Chernyi, V.S.; Tarasova, P.V.; Kozlov, V.V.; Saik, O.V.; Kushlinskii, N.E.; Gulyaeva, L.F. Search of MicroRNAs Regulating the Receptor Status of Breast Cancer In Silico and Experimental Confirmation of Their Expression in Tumors. *Bull. Exp. Biol. Med.* **2017**, *163*, 655–659. [CrossRef] [PubMed]
93. Cun, J.; Yang, Q. Bioinformatics-based interaction analysis of miR-92a-3p and key genes in tamoxifen-resistant breast cancer cells. *Biomed. Pharmacother.* **2018**, *107*, 117–128. [CrossRef] [PubMed]
94. Tang, W.; Zhang, X.; Tan, W.; Gao, J.; Pan, L.; Ye, X.; Chen, L.; Zheng, W. miR-145-5p Suppresses Breast Cancer Progression by Inhibiting SOX_2. *J. Surg. Res.* **2019**, *236*, 278–287. [CrossRef]
95. Chaudhary, S.; Islam, Z.; Mishra, V.; Rawat, S.; Ashraf, G.M.; Kolatkar, P.R. Sox_2: A Regulatory Factor in Tumorigenesis and Metastasis. *Curr. Protein Pept. Sci.* **2019**, *20*, 495–504. [CrossRef] [PubMed]
96. Patel, N.; Garikapati, K.R.; Ramaiah, M.J.; Polavarapu, K.K.; Bhadra, U.; Bhadra, M.P. miR-15a/miR-16 induces mitochondrial dependent apoptosis in breast cancer cells by suppressing oncogene BMI1. *Life Sci.* **2016**, *164*, 60–70. [CrossRef]
97. Heers, H.; Stanislaw, J.; Harrelson, J.; Lee, M.W. Valproic acid as an adjunctive therapeutic agent for the treatment of breast cancer. *Eur. J. Pharmacol.* **2018**, *835*, 61–74. [CrossRef] [PubMed]
98. Olsen, C.M.; Meussen-Elholm, E.T.M.; Røste, L.S.; Taubøll, E. Antiepileptic drugs inhibit cell growth in the human breast cancer cell line MCF7. *Mol. Cell. Endocrinol.* **2004**, *213*, 173–179. [CrossRef]
99. Tayanloo-Beik, A.; Sarvari, M.; Payab, M.; Gilany, K.; Alavi-Moghadam, S.; Gholami, M.; Goodarzi, P.; Larijani, B.; Arjmand, B. OMICS insights into cancer histology; Metabolomics and proteomics approach. *Clin. Biochem.* **2020**, *84*, 13–20. [CrossRef]
100. Sarin, N.; Engel, F.; Rothweiler, F.; Cinatl, J.; Michaelis, M.; Frötschl, R.; Fröhlich, H.; Kalayda, G. Key Players of Cisplatin Resistance: Towards a Systems Pharmacology Approach. *Int. J. Mol. Sci.* **2018**, *19*, 767. [CrossRef] [PubMed]
101. Al-malky, H.S.; Al Harthi, S.E.; Osman, A.M.M. Major obstacles to doxorubicin therapy: Cardiotoxicity and drug resistance. *J. Oncol. Pharm. Pract.* **2020**, *26*, 434–444. [CrossRef]
102. Alfarouk, K.O.; Stock, C.M.; Taylor, S.; Walsh, M.; Muddathir, A.K.; Verduzco, D.; Bashir, A.H.H.; Mohammed, O.Y.; Elhassan, G.O.; Harguindey, S.; et al. Resistance to cancer chemotherapy: Failure in drug response from ADME to P-gp. *Cancer Cell Int.* **2015**, *15*, 71. [CrossRef] [PubMed]
103. Okon, E.; Luszczki, J.J.; Kukula-Koch, W.; Halasa, M.; Jarzab, A.; Khurelbat, D.; Stepulak, A.; Wawruszak, A. Synergistic or Additive Pharmacological Interactions between Magnoflorine and Cisplatin in Human Cancer Cells of Different Histological Origin. *Int. J. Mol. Sci.* **2020**, *21*, 2848. [CrossRef] [PubMed]
104. Kukula-Koch, W.; Grabarska, A.; Łuszczki, J.; Czernicka, L.; Nowosadzka, E.; Gumbarewicz, E.; Jarząb, A.; Audo, G.; Upadhyay, S.; Głowniak, K.; et al. Superior anticancer activity is demonstrated by total extract of Curcuma longa L. as opposed to individual curcuminoids separated by centrifugal partition chromatography. *Phyther. Res.* **2018**, *32*, 933–942. [CrossRef] [PubMed]
105. Wróblewska-łuczka, P.; Grabarska, A.; Łuszczki, J.J.; Florek-łuszczki, M.; Plewa, Z. Synergy, additivity, and antagonism between cisplatin and selected coumarins in human melanoma cells. *Int. J. Mol. Sci.* **2021**, *22*, 537. [CrossRef] [PubMed]
106. Jarzab, A.; Luszczki, J.; Guz, M.; Skalicka-Wozniak, K.; Halasa, M.; Smok-Kalwat, J.; Polberg, K.; Stepulak, A. Combination of osthole and cisplatin against rhabdomyosarcoma TE671 cells yielded additive pharmacologic interaction by means of isobolographic analysis. *Anticancer Res.* **2018**, *38*, 205–210. [PubMed]
107. Dasari, S.; Bernard Tchounwou, P. Cisplatin in cancer therapy: Molecular mechanisms of action. *Eur. J. Pharmacol.* **2014**, *740*, 364–378. [CrossRef]
108. Minegaki, T.; Suzuki, A.; Mori, M.; Tsuji, S.; Yamamoto, S.; Watanabe, A.; Tsuzuki, T.; Tsunoda, T.; Yamamoto, A.; Tsujimoto, M.; et al. Histone deacetylase inhibitors sensitize 5-fluorouracil-resistant MDA-MB-468 breast cancer cells to 5-fluorouracil. *Oncol. Lett.* **2018**, *16*, 6202–6208. [CrossRef]
109. Terranova-Barberio, M.; Roca, M.S.; Zotti, A.I.; Leone, A.; Bruzzese, F.; Vitagliano, C.; Scogliamiglio, G.; Russo, D.; D'Angelo, G.; Franco, R.; et al. Valproic acid potentiates the anticancer activity of capecitabine in vitro and in vivo in breast cancer models via induction of thymidine phosphorylase expression. *Oncotarget* **2016**, *7*, 7715–7731. [CrossRef]

110. Arakawa, Y.; Saito, S.; Yamada, H.; Aiba, K. Simultaneous treatment with camptothecin and valproic acid suppresses induction of Bcl-XL and promotes apoptosis of MCF-7 breast cancer cells. *Apoptosis* **2009**, *14*, 1076–1085. [CrossRef]
111. Phillips, S.L.; Williams, C.B.; Zambrano, J.N.; Williams, C.J.; Yeh, E.S. Connexin 43 in the development and progression of breast cancer: What's the connection? *Int. J. Oncol.* **2017**, *51*, 1005–1013. [CrossRef]
112. Sargazi, S.; Kooshkaki, O.; Zavar Reza, J.; Saravani, R.; Zarei Jaliani, H.; Mirinejad, S.; Meshkini, F. Mild antagonistic effect of Valproic acid in combination with AZD2461 in MCF-7 breast cancer cells. *Med. J. Islamic Repub. Iran* **2019**, *33*, 29. [CrossRef]
113. Soldi, R.; Cohen, A.L.; Cheng, L.; Sun, Y.; Moos, P.J.; Bild, A.H. A genomic approach to predict synergistic combinations for breast cancer treatment. *Pharm. J.* **2013**, *13*, 94–104. [CrossRef]
114. Ponce-Cusi, R.; Calaf, G.M. Apoptotic activity of 5-fluorouracil in breast cancer cells transformed by low doses of ionizing α-particle radiation. *Int. J. Oncol.* **2016**, *48*, 774–782. [CrossRef] [PubMed]
115. Varshavsky-Yanovsky, A.N.; Goldstein, L.J. Role of Capecitabine in Early Breast Cancer. *J. Clin. Oncol.* **2020**, *38*, 179–182. [CrossRef]
116. Zunino, F.; Pratesi, G. Camptothecins in clinical development. *Expert Opin. Investig. Drugs* **2004**, *13*, 269–284. [CrossRef]
117. Ulukan, H.; Swaan, P.W. Camptothecins: A review of their chemotherapeutic potential. *Drugs* **2002**, *62*, 2039–2057. [CrossRef]
118. Lovitt, C.J.; Shelper, T.B.; Avery, V.M. Doxorubicin resistance in breast cancer cells is mediated by extracellular matrix proteins. *BMC Cancer* **2018**, *18*. [CrossRef] [PubMed]
119. Christowitz, C.; Davis, T.; Isaacs, A.; Van Niekerk, G.; Hattingh, S.; Engelbrecht, A.M. Mechanisms of doxorubicin-induced drug resistance and drug resistant tumour growth in a murine breast tumour model. *BMC Cancer* **2019**, *19*, 757. [CrossRef] [PubMed]
120. Tong, X.H.; Zheng, C.; Jiang, G.J.; Dong, S.Y. Sodium valproate enhances doxorubicin cytotoxicity in breast cancer cells in vitro. *Nan Fang Yi Ke Da Xue Xue Bao* **2015**, *35*, 62–65. Available online: https://pubmed.ncbi.nlm.nih.gov/25613611/ (accessed on 21 April 2021).
121. Muñoz, A.M.; Fragoso-Vázquez, M.J.; Martel, B.P.; Chávez-Blanco, A.; Dueñas-González, A.; García-Sánchez, J.R.; Bello, M.; Romero-Castro, A.; Correa-Basurto, J. Targeting Breast Cancer Cells with G4 PAMAM Dendrimers and Valproic Acid Derivative Complexes. *Anticancer Agents Med. Chem.* **2020**, *20*, 1857–1872. [CrossRef]
122. Ribociclib&Belinostat in Patients W Metastatic Triple Neg Breast Cancer & Recurrent Ovarian Cancer W Response Prediction by Genomics. Available online: https://clinicaltrials.gov/ct2/show/NCT04315233?term=Histone+deacetylase+inhibitor&cond=Breast+Cancer&draw=3&rank=5 (accessed on 17 June 2021).
123. Pembrolizumab and Tamoxifen with or without Vorinostat for the Treatment of Estrogen Receptor Positive Breast Cancer. Available online: https://clinicaltrials.gov/ct2/show/NCT04190056?term=Histone+deacetylase+inhibitor&cond=Breast+Cancer&draw=3 (accessed on 17 June 2021).
124. A Pilot Study of the Combination of Entinostat With Capecitabine in High Risk Breast Cancer After Neo-adjuvant Therapy. Available online: https://clinicaltrials.gov/ct2/show/NCT03473639?term=Histone+deacetylase+inhibitor&cond=Breast+Cancer&draw=3&rank=36 (accessed on 17 June 2021).
125. Talazoparib in Combination With Belinostat for Metastatic Breast Cancer, Metastatic Castration Resistant Prostate Cancer, and Metastatic Ovarian Cancer. Available online: https://clinicaltrials.gov/ct2/show/NCT04703920?term=Histone+deacetylase+inhibitor&cond=Breast+Cancer&draw=3&rank=32 (accessed on 17 June 2021).
126. BN-Brachyury, Entinostat, Adotrastuzumab Emtansine and M7824 in Advanced Stage Breast Cancer (BrEAsT). Available online: https://clinicaltrials.gov/ct2/show/NCT04296942?term=Histone+deacetylase+inhibitor&cond=Breast+Cancer&draw=3&rank=17 (accessed on 17 June 2021).
127. Olaparib in Combination With Vorinostat in Patients With Relapsed/Refractory and/or Metastatic Breast Cancer. Available online: https://clinicaltrials.gov/ct2/show/NCT03742245?term=Histone+deacetylase+inhibitor&cond=Breast+Cancer&draw=3&rank=11 (accessed on 17 June 2021).
128. Münster, P.; Marchion, D.; Bicaku, E.; Schmitt, M.; Ji, H.L.; DeConti, R.; Simon, G.; Fishman, M.; Minton, S.; Garrett, C.; et al. Phase I trial of histone deacetylase inhibition by valproic acid followed by the topoisomerase II inhibitor epirubicin in advanced solid tumors: A clinical and translational study. *J. Clin. Oncol.* **2007**, *25*, 1979–1985. [CrossRef]
129. Munster, P.; Marchion, D.; Bicaku, E.; Lacevic, M.; Kim, J.; Centeno, B.; Daud, A.; Neuger, A.; Minton, S.; Sullivan, D. Clinical and biological effects of valproic acid as a histone deacetylase inhibitor on tumor and surrogate tissues: Phase i/ii trial of valproic acid and epirubicin/FEC. *Clin. Cancer Res.* **2009**, *15*, 2488–2496. [CrossRef]
130. Arce, C.; Pérez, C.; González-Fierro, A.; de la Cruz-Hernńdez, E.; Revilla-Vázquez, A.; Chávez-Blanco, A.; Trejo-Becerril, C.; Pérez-Cárdenas, E.; Taja-Chayeb, L.; Bargallo, E.; et al. A proof-of-principle study of epigenetics therapy added to neoadjuvant doxorubicin cyclophosphamide for locally advanced breast cancer. *PLoS ONE* **2006**, *1*, e98. [CrossRef] [PubMed]
131. Bevacizumab and Temsirolimus Alone or in Combination With Valproic Acid or Cetuximab in Treating Patients With Advanced or Metastatic Malignancy or Other Benign Disease. Available online: https://clinicaltrials.gov/ct2/show/study/NCT01552434?term=valproic+acid+and+breast+cancer&draw=2&rank=7 (accessed on 17 June 2021).
132. Fardi, M.; Solali, S.; Farshdousti Hagh, M. Epigenetic mechanisms as a new approach in cancer treatment: An updated review. *Genes Dis.* **2018**, *5*, 304–311. [CrossRef] [PubMed]
133. Souza, C.; Chatterji, B. HDAC Inhibitors as Novel Anti-Cancer Therapeutics. *Recent Pat. Anticancer Drug Discov.* **2015**, *10*, 145–162. [CrossRef]

134. Diederich, M.; Chateauvieux, S.; Morceau, F.; Dicato, M. Molecular and therapeutic potential and toxicity of valproic acid. *J. Biomed. Biotechnol.* **2010**, *2010*. [CrossRef]
135. Činčárová, L.; Zdráhal, Z.; Fajkus, J. New perspectives of valproic acid in clinical practice. *Expert Opin. Investig. Drugs* **2013**, *22*, 1535–1547. [CrossRef]
136. Goyal, J.; Rodriguez, R. Evidence from clinical trials for the use of valproic acid in solid tumors: Focus on prostate cancer. *Clin. Investig.* **2013**, *3*, 467–478. [CrossRef]
137. Duenas-Gonzalez, A.; Candelaria, M.; Perez-Plascencia, C.; Perez-Cardenas, E.; de la Cruz-Hernandez, E.; Herrera, L.A. Valproic acid as epigenetic cancer drug: Preclinical, clinical and transcriptional effects on solid tumors. *Cancer Treat. Rev.* **2008**, *34*, 206–222. [CrossRef]
138. Ponzano, A.; Tiboni, G.M. Teratology of valproic acid: An updated review of the possible mediating mechanisms. *Minerva Ginecol.* **2018**, *70*, 303–322.
139. Meisel, J.L.; Venur, V.A.; Gnant, M.; Carey, L. Evolution of Targeted Therapy in Breast Cancer: Where Precision Medicine Began. *Am. Soc. Clin. Oncol. Educ. B* **2018**, *38*, 78–86. [CrossRef]
140. Dhritlahre, R.K.; Saneja, A. Recent advances in HER2-targeted delivery for cancer therapy. *Drug Discov. Today* **2020**. [CrossRef]
141. Mitsogianni, M.; Trontzas, I.P.; Gomatou, G.; Ioannou, S.; Syrigos, N.K.; Kotteas, E.A. The changing treatment of metastatic her2-positive breast cancer. *Oncol. Lett.* **2021**, *21*. [CrossRef] [PubMed]
142. Catalano, M.G.; Fortunati, N.; Pugliese, M.; Poli, R.; Bosco, O.; Mastrocola, R.; Aragno, M.; Boccuzzi, G. Valproic acid, a histone deacetylase inhibitor, enhances sensitivity to doxorubicin in anaplastic thyroid cancer cells. *J. Endocrinol.* **2006**, *191*, 465–472. [CrossRef] [PubMed]

MDPI
St. Alban-Anlage 66
4052 Basel
Switzerland
www.mdpi.com

Cancers Editorial Office
E-mail: cancers@mdpi.com
www.mdpi.com/journal/cancers

Disclaimer/Publisher's Note: The statements, opinions and data contained in all publications are solely those of the individual author(s) and contributor(s) and not of MDPI and/or the editor(s). MDPI and/or the editor(s) disclaim responsibility for any injury to people or property resulting from any ideas, methods, instructions or products referred to in the content.

www.ingramcontent.com/pod-product-compliance
Lightning Source LLC
Chambersburg PA
CBHW040804250926
51850CB00007BA/1081
9783039286171